G. M. Henkin · J. Leiterer

Theory of Functions on Complex Manifolds

Mathematische Lehrbücher und Monographien
Herausgegeben von der Akademie der Wissenschaften der DDR
Institut für Mathematik

II. Abteilung
Mathematische Monographien
Band 60

Theory of Functions on Complex Manifolds

von G. M. Henkin und J. Leiterer

Theory of Functions on Complex Manifolds

by G. M. Henkin and J. Leiterer

Akademie-Verlag Berlin
1984

Verfasser:

Dr. sc. Gennadi Markovič Henkin
Zentrales Ökonomisch-Mathematisches Institut
der Akademie der Wissenschaften der UdSSR, Moskau

Dr. sc. Jürgen Leiterer
Institut für Mathematik
der Akademie der Wissenschaften der DDR, Berlin

ISSN 0076-5430

Erschienen im Akademie-Verlag, DDR-1086 Berlin, Leipziger Str. 3—4

Lizenznummer: 202 · 100/402/83
Printed in the German Democratic Republic
Gesamtherstellung: VEB Druckerei „Thomas Müntzer“,
5820 Bad Langensalza
Lektor: Dipl.-Math. Gesine Reiher
Einband- und Schutzumschlaggestaltung: Dietmar Kunz
LSV 1035
Bestellnummer: 763 048 0 (6672)
04200

Preface

Till the early fifties the theory of functions of several complex variables was mainly developed by constructive methods of analysis. We emphasize the work of A. WEIL in 1935 and of K. OKA in the period from 1936 till 1951. WEIL generalized the Cauchy integral formula to polynomial polyhedra in $\mathbb{C}^n$ and obtained an analogue of the Runge approximation theorem for such polyhedra. Equipped with the Weil formula OKA solved the so-called fundamental problems (Cousin problem, Levi problem, et al.).

In the fifties H. CARTAN, J. P. SERRE and H. GRAUERT discovered that by means of the theory of sheaves introduced in 1945 by J. LERAY the constructive methods of analysis in the theory of Oka can be reduced to a minimum and, moreover, that the theory of Oka admits far-reaching generalizations. In the sixties L. HÖRMANDER, J. J. KOHN and C. B. MORREY deduced the main results of OKA with the help of methods from the theory of partial differential equations and obtained, in addition, estimates in certain weighted L_2-metrics for solutions of the Cauchy-Riemann equations.

During the fifties and sixties it seemed that the method of integral representations, which works so successfully in the case of one variable, is not suitable to the case of several variables, because it is troublesome and gives only very special results.

However, in the seventies integral representations turned out to be the natural method for solving several problems related to Oka's theory, which are connected with the boundary behaviour of holomorphic functions. The basic tool is an integral representation formula for holomorphic functions discovered in 1955 by J. LERAY, which contains the Weil formula as a special case. Certain developments of this formula made it possible to solve several of such problems that are not easily obtained with other methods. Moreover, it turned out that by means of these formulas one can build up a large part of the theory of functions of several complex variables in a new and more constructive way.

It is the aim of this book to present such a new introduction to the theory of functions of several complex variables, where the main results will be obtained in a strengthened form — uniform estimates for solutions of the Cauchy-Riemann equations, uniform estimates for extensions of holomorphic functions from submanifolds, uniform approximation of holomorphic functions that are continuous on the boundary, et al.

It has been assumed that the reader has a certain knowledge of the theory of functions of one complex variable and the calculus of differential forms (Stokes' formula).

Chapter 1 starts with facts concerning holomorphic functions, plurisubharmonic functions, domains of holomorphy and pseudoconvex domains. Then we deduce from Stokes' formula the Martinelli-Bochner formula and the Leray formula as well as their generalizations to differential forms in $\mathbb{C}^n$ (Koppelman formula and Koppelman-Leray formula).

In Chapter 2 first the Cauchy-Riemann equations are solved by means of integral formulas in pseudoconvex open sets in $\mathbb{C}^n$. Then we prove this result on Stein manifolds, where an inductive procedure with respect to the levels of a strictly plurisubharmonic exhausting function will be used. For strictly pseudoconvex open sets with C^2-boundary, solutions of the Cauchy-Riemann equations with 1/2-Hölder estimates are obtained. The identity of domains of holomorphy (Stein manifolds) and pseudoconvex open sets in $\mathbb{C}^n$ (complex manifolds with strictly plurisubharmonic exhausting function) is proved, that is, the Levi problem is solved. Further, uniform approximation theorems are proved.

Chapter 3 is devoted to strictly pseudoconvex open sets in $\mathbb{C}^n$ with not necessarily smooth boundary. By means of integral formulas the Cauchy-Riemann equations are solved with uniform estimates in such sets. A uniform approximation theorem is proved for functions which are continuous on a strictly pseudoconvex compact set and holomorphic in the inner points. Further, for strictly pseudoconvex open sets D with not necessarily smooth boundary, an integral formula is constructed which gives bounded holomorphic extensions to D for bounded holomorphic functions defined on the intersection of D with a complex plane. In Chapter 4 this result will be generalized to the case of an intersection with an arbitrary closed complex submanifold in some neighbourhood of $\bar{D}$.

Chapters 1—3 are self-contained. Here we do not use without proof any result from the theory of functions of several complex variables. Only in Chapter 4 we use without proof some special results from the theory of coherent analytic sheaves, for the proof of which we can refer to several books devoted to this subject.

In Chapter 4 our principal aim is to extend the integral formulas introduced in the preceding sections to Stein manifolds. Moreover, in this chapter the Weil formula for analytic polyhedra as well as its generalization to differential forms and a more general class of polyhedra in Stein manifolds is proved. Some applications of these formulas are given. In the Notes at the end some further applications are outlined.

In our opinion Chapters 1 and 2 can be used as an elementary introduction to the theory of functions of several complex variables. Chapters 3 and 4 contain more special and more difficult results obtained only recently by means of complicated estimations, and references to the theory of coherent analytic sheaves. They can be used as an introduction to one of the actual fields of research in complex analysis.

There is also another way to develop the theory of functions of several complex variables by means of integral formulas. This way was outlined in 1961 by E. Bishop and is based on the concept of special analytic polyhedra. In distinction to the approach presented in this book, the way of Bishop is suitable not only for smooth complex manifolds but also for analytic spaces with singularities. However, this way seems to be more complicated and, above all, does not give uniform estimates, whereas in our opinion the latter is the main advantage of the method of integral formulas.

Finally, we point out that in our opinion there are also further interesting possibilities for applying the method of integral formulas, for example to the theory of CR-functions and to problems of complex analysis and integral geometry on projective manifolds connected with the theory of R. Penrose.

We thank Dr. B. Jöricke (Berlin) who helped improve parts of the manuscript. We thank also Dr. R. Höppner and G. Reiher from the Akademie-Verlag Berlin for support and cooperation. We are greatly indebted to Prof. H. Boas (New York) for proof reading and removing a lot of mistakes (including the worst English ones).

Berlin and Moscow, July 1981 G. M. Henkin and J. Leiterer

Contents

1. Elementary properties of functions of several complex variables

Summary. In Section 1.1 we define holomorphic functions of several complex variables and prove simple properties of these functions. In Section 1.2, by means of a simple extension of the Cauchy-Green formula to several variables, we solve the inhomogeneous Cauchy-Riemann equations for some special cases. As a consequence we obtain a theorem of Hartogs which gives examples of open sets in $\mathbb{C}^n$ $(n \geqq 2)$ where all holomorphic functions can be continued holomorphically to a larger open set. Open sets for which this is not possible are called domains of holomorphy. These are investigated in Section 1.3. Section 1.4 is devoted to continuous plurisubharmonic and strictly plurisubharmonic C^2-functions. In Section 1.5, by means of these functions, pseudoconvex and strictly pseudoconvex open sets are introduced. We prove that every domain of holomorphy is pseudoconvex, but the converse (Levi's problem) is left to Section 2.7. Sections 1.6—1.12 are devoted to integral representation formulas for functions as well as for differential forms n $\mathbb{C}^n$. These formulas form the basic tool for the methods developed in the present book.

1.1. Holomorphic functions

We assume that the reader knows a certain amount of the theory of functions of one complex variable. Nevertheless we begin with a proof of the Cauchy-Green formula for one complex variable, because this proof is the model for the proofs of the integral representation formulas for several complex variables which form the basis of this book.

Notation. Let $\mathbb{C}^1$ be the complex plane. By x_1, x_2 we denote the real coordinates in $\mathbb{C}^1$ such that $\mathbb{C}^1 \ni z = x_1 + \mathrm{i}x_2$. For every complex-valued continuous function f in an open set in $\mathbb{C}^1$ we define (in the sense of distributions)

$$\frac{\partial f}{\partial z} := \frac{1}{2}\left(\frac{\partial f}{\partial x_1} + \frac{1}{\mathrm{i}}\frac{\partial f}{\partial x_2}\right), \qquad \frac{\partial f}{\partial \bar{z}} := \frac{1}{2}\left(\frac{\partial f}{\partial x_1} - \frac{1}{\mathrm{i}}\frac{\partial f}{\partial x_2}\right),$$

$$\partial f := \frac{\partial f}{\partial z}\,\mathrm{d}z \quad \text{and} \quad \bar{\partial} f := \frac{\partial f}{\partial \bar{z}}\,\mathrm{d}\bar{z}\,,$$

where $\bar{z} := x_1 - \mathrm{i}x_2$. Then the differential $\mathrm{d}f$ of f can be expressed as $\mathrm{d}f = \partial f + \bar{\partial} f$. We also write $\mathrm{d}_z f$, $\partial_z f$ and $\bar{\partial}_z f$ instead of $\mathrm{d}f$, ∂f and $\bar{\partial} f$ to make clear that the differentiation is with respect to z if f depends on other variables, too.

Recall that a C^1-function f is holomorphic if and only if $\bar{\partial} f = 0$ (the Cauchy-Riemann equations).

1.1.1. Theorem (Cauchy-Green formula). *Let $D \subset\subset \mathbb{C}^1$ be an open set with C^1-oundary ∂D, and let f be a complex-valued continuous function on $\bar{D}$ such that $\bar{\partial} f$ is* *also continuous on* $\bar{D}$[1]). *Then*

$$f(z) = \frac{1}{2\pi i} \int_{\partial D} \frac{f(\zeta)\, d\zeta}{\zeta - z} - \frac{1}{2\pi i} \int_{D} \frac{\bar{\partial} f(\zeta) \wedge d\zeta}{\zeta - z}, \qquad z \in D. \tag{1.1.1}$$

Proof. Fix $z \in D$. Then, for $\zeta \in \bar{D} \setminus z$, $d_\zeta[(f(\zeta)\, d\zeta)/(\zeta - z)] = \bar{\partial} f(\zeta) \wedge d\zeta/(\zeta - z)$ and, therefore, by Stokes' formula, for sufficiently small $\varepsilon > 0$

$$\frac{1}{2\pi i} \int_{|\zeta - z| = \varepsilon} \frac{f(\zeta)\, d\zeta}{\zeta - z} = \frac{1}{2\pi i} \int_{\zeta \in \partial D} \frac{f(\zeta)\, d\zeta}{\zeta - z} - \frac{1}{2\pi i} \int_{\zeta \in D, |\zeta - z| > \varepsilon} \frac{\bar{\partial} f(\zeta) \wedge d\zeta}{\zeta - z}.$$

It is clear that the right-hand side of this relation tends to the right-hand side of (1.1.1) when $\varepsilon \to 0$. Consider the left-hand side. Since $d\bar{\zeta} \wedge d\zeta/2i$ is the Lebesgue measure on $\mathbb{C}^1$, Stokes' formula gives

$$\int_{|\zeta - z| = \varepsilon} \frac{d\zeta}{\zeta - z} = \frac{1}{\varepsilon^2} \int_{|\zeta - z| = \varepsilon} (\bar{\zeta} - \bar{z})\, d\zeta = \frac{1}{\varepsilon^2} \int_{|\zeta - z| < \varepsilon} d\bar{\zeta} \wedge d\zeta = 2\pi i.$$

Further

$$\left| \int_{|\zeta - z| = \varepsilon} \frac{f(\zeta) - f(z)}{\zeta - z}\, d\zeta \right| \leqq 2\pi \max_{|\zeta - z| = \varepsilon} |f(\zeta) - f(z)| \to 0 \quad \text{for} \quad \varepsilon \to 0.$$

Hence

$$\lim_{\varepsilon \to 0} \int_{|\zeta - z| = \varepsilon} \frac{f(\zeta)\, d\zeta}{\zeta - z} = 2\pi i f(z). \blacksquare$$

Since the first integral on the right-hand side of the Cauchy-Green formula (1.1.1) depends holomorphically on $z \in D$, and since $\bar{\partial} f = 0$ for holomorphic functions, we obtain:

1.1.2. Corollary. *Let f be a continuous complex-valued function in an open set $D \subseteq \mathbb{C}^1$. Then f is holomorphic in D if and only if $\bar{\partial} f = 0$ in D.*

1.1.3. Theorem. *Let $D \subset\subset \mathbb{C}^1$ be a bounded open set, and let f be a bounded continuous complex-valued function in D. Then the continuous function*

$$u(z) := -\frac{1}{2\pi i} \int_{D} \frac{f(\zeta)\, d\bar{\zeta} \wedge d\zeta}{\zeta - z}, \qquad z \in D, \tag{1.1.1'}$$

is a solution of $\frac{\partial u}{\partial \bar{z}} = f$ *in* D.

Proof. First we consider the case that f is continuously differentiable in D. Fix $\xi \in D$. It is sufficient to prove that $\partial u/\partial \bar{z} = f$ in some neighbourhood of ξ. Choose a C^∞-function χ on $\mathbb{C}^1$ such that $\chi = 1$ in a neighbourhood $V_\xi \subset\subset D$ of ξ and $\chi = 0$ in some neighbourhood of $\mathbb{C}^1 \setminus D$. Then $u = u_1 + u_2$, where

$$u_1(z) := -\frac{1}{2\pi i} \int_{D} \frac{\chi(\zeta) f(\zeta)\, d\bar{\zeta} \wedge d\zeta}{\zeta - z}, \qquad u_2(z) := -\frac{1}{2\pi i} \int_{D} \frac{(1 - \chi(\zeta)) f(\zeta)\, d\bar{\zeta} \wedge d\zeta}{\zeta - z}.$$

[1]) This means, $\partial f/\partial \bar{z}$ which is defined in the sense of distributions in D is continuous in D and admits a continuous continuation to $\bar{D}$.

Since $1 - \chi = 0$ in V_ξ, u_2 is holomorphic in V_ξ, that is, $\bar\partial u_2 = 0$ in V_ξ. Therefore it remains to prove that $\partial u_1/\partial\bar z = f$ in V_ξ. Since $\chi = 0$ in a neighbourhood of $\mathbb{C}^1 \setminus D$, the function χf can be continued by zero to $\mathbb{C}^1$, and we obtain

$$u_1(z) = -\frac{1}{2\pi i}\int\limits_{\mathbb{C}^1}\frac{\chi(\zeta)f(\zeta)\,d\bar\zeta\wedge d\zeta}{\zeta - z} = -\frac{1}{2\pi i}\int\limits_{\mathbb{C}^1}\frac{\chi(\zeta+z)f(\zeta+z)\,d\bar\zeta\wedge d\zeta}{\zeta}.$$

Taking into account that $\partial/\partial\bar z[\chi(\zeta+z)f(\zeta+z)]\,d\bar\zeta = \bar\partial_\zeta[\chi(\zeta+z)f(\zeta+z)]$, we conclude that

$$\frac{\partial u_1(z)}{\partial\bar z} = -\frac{1}{2\pi i}\int\limits_{\mathbb{C}^1}\frac{\bar\partial_\zeta[\chi(\zeta+z)f(\zeta+z)]\wedge d\zeta}{\zeta} = -\frac{1}{2\pi i}\int\limits_{\mathbb{C}^1}\frac{\bar\partial_\zeta[\chi(\zeta)f(\zeta)]\wedge \partial\zeta}{\zeta - z}.$$

Application of Theorem 1.1.1 with f replaced by χf and D equal to a disc containing the support of χ now gives that $\partial u_1/\partial\bar z = f$ in V_ξ.

Now we consider the case that f is an arbitrary continuous and bounded function on D. Choose a sequence f_k of continuously differentiable functions which converges uniformly on D to f. Then the sequence

$$u_k(z) := -\frac{1}{2\pi i}\int\limits_D\frac{f_k(\zeta)\,d\bar\zeta\wedge d\zeta}{\zeta - z}$$

converges uniformly on D to u, and therefore we have in the sense of distributions $\partial u/\partial\bar z = \lim \partial u_k/\partial\bar z = \lim f_k = f$. ∎

Now we pass to the case of several complex variables.

Notation. Let $\mathbb{C}^n$ $(n = 1, 2, \dots)$ be the space of all n-tuples $z = (z_1, \dots, z_n)$ of complex numbers z_j. The components $z_1, \dots, z_n$ of $z \in \mathbb{C}^n$ will be called the *canonical (complex) coordinates* of z. By $x_1, \dots, x_{2n}$ we denote the *real coordinates in* $\mathbb{C}^n$ such that $z_j = x_j + ix_{j+n}$. For $Y \subseteq \mathbb{C}^n$, let $C^0(Y)$ be the space of continuous complex-valued functions on Y. If $D \subseteq \mathbb{C}^n$ is open, then $C^k(D)$ $(k = 1, 2, \dots, \infty)$ is the space of k-times continuously differentiable (with respect to the real coordinates x_j) complex-valued functions in D, and $C_0^k(D)$ is the subspace of all $f \in C^k(D)$ vanishing outside a compact subset of D. We introduce the differential operators

$$\frac{\partial}{\partial z_j} := \frac{1}{2}\left(\frac{\partial}{\partial x_j} + \frac{1}{i}\frac{\partial}{\partial x_{j+n}}\right) \quad\text{and}\quad \frac{\partial}{\partial\bar z_j} := \frac{1}{2}\left(\frac{\partial}{\partial x_j} - \frac{1}{i}\frac{\partial}{\partial x_{j+n}}\right).$$

These operators $\partial/\partial z_j$ and $\partial/\partial\bar z_j$ will also be denoted by ∂_j and $\bar\partial_j$, respectively. A *multi-order* is an n-tuple $k = (k_1, \dots, k_n)$ of non-negative integers. For every multi-order k we write $z^k := z_1^{k_1} \dots z_n^{k_n}$ if $z \in \mathbb{C}^n$, $k! := k_1! \dots k_n!$, $\partial^k := \partial_1^{k_1} \circ \dots \circ \partial_n^{k_n}$ and $\bar\partial^k := \bar\partial_1^{k_1} \circ \dots \circ \bar\partial_n^{k_n}$. If $D \subseteq \mathbb{C}^n$ is open and $f \in C^0(D)$, then we define

$$\partial f := \sum_{j=1}^n \frac{\partial f}{\partial z_j}dz_j \quad\text{and}\quad \bar\partial f := \sum_{j=1}^n \frac{\partial f}{\partial\bar z_j}d\bar z_j \text{ in } D.$$

Then $df = \partial f + \bar\partial f$ in D.

(Remark. In general, ∂f and $\bar\partial f$ are differential forms whose coefficients are distributions. However, in this book we need ∂f ($\bar\partial f$) only if ∂f ($\bar\partial f$) is continuous.)

A set $P \subseteq \mathbb{C}^n$ is said to be an *open (closed) polydisc* if there are open (closed) discs $P_1, \dots, P_n$ in $\mathbb{C}^1$ such that $P = P_1 \times \dots \times P_n$. If ξ_j is the center of P_j, then the point

$(\xi_1, \dots, \xi_n)$ is called the *center* of P. If r_j is the radius of P_j, then $(r_1, \dots, r_n)$ is called the *multi-radius* of P.

1.1.4. Lemma. *Let a_k be complex numbers defined for all multi-orders k. Suppose that for some $\xi \in \mathbb{C}^n$ ($\xi_j \neq 0$ for all j) the power series $\sum_k a_k \xi^k$ converges. Then the series $\sum_k a_k z^k$ converges normally in the polydisc $|z_j| < |\xi_j|$, $j = 1, \dots, n$, that is, for all $r < 1$*

$$\sum_k \sup_{|z_j| \leqq r|\xi_j|} |a_k z^k| < \infty .$$

Proof. By hypothesis there is $C < \infty$ such that $|a_k \xi^k| \leqq C$ for all k. Consequently,

$$\sum_k \sup_{|z_j| \leqq r|\xi_j|} |a_k z^k| \leqq C \sum_k r^{k_1 + \dots + k_n} < \infty . \blacksquare$$

1.1.5. Theorem. *Let $D \subseteqq \mathbb{C}^n$ be an open set and f a complex-valued function in D. Then the following conditions are equivalent:*

(i) *$f \in C^0(D)$ and $\bar{\partial} f = 0$ in D.*

(ii) *$f \in C^0(D)$ and f is holomorphic in each variable z_j when the other variables are kept fixed.*

(iii) *$f \in C^0(D)$ and, for every polydisc $P = P_1 \times \dots \times P_n \subset\subset D$,*

$$f(z) = \frac{1}{(2\pi i)^n} \int_{\partial P_1 \times \dots \times \partial P_n} \frac{f(\zeta)\, d\zeta_1 \wedge \dots \wedge d\zeta_n}{(\zeta_1 - z_1) \cdot \dots \cdot (\zeta_n - z_n)}, \quad z \in P . \tag{1.1.2}$$

(iv) *f admits local power series expansions, that is, for every point $\xi \in D$ there are complex numbers a_k defined for every multi-order k such that, for all z in some neighbourhood of ξ,*

$$f(z) = \sum_k a_k (z - \xi)^k . \tag{1.1.3}$$

If these equivalent conditions are fulfilled, then, moreover:

(v) *$f \in C^\infty(D)$.*

(vi) *For every polydisc $P = P_1 \times \dots \times P_n \subset\subset D$ and every multi-order k,*

$$\partial^k f(z) = \frac{k!}{(2\pi i)^n} \int_{\partial P_1 \times \dots \times \partial P_n} \frac{f(\zeta)\, d\zeta_1 \wedge \dots \wedge d\zeta_n}{(\zeta_1 - z_1)^{k_1+1} \dots (\zeta_n - z_n)^{k_n+1}}, \quad z \in P . \tag{1.1.4}$$

(vii) *The coefficients in the power series expansion (1.1.3) are uniquely determined, where*

$$a_k = \frac{\partial^k f(\xi)}{k!} . \tag{1.1.5}$$

(viii) *The power series expansion (1.1.3) converges uniformly in each polydisc $P \subset\subset D$ centered at ξ.*

Proof. (i) $\Leftrightarrow$ (ii) according to Corollary 1.1.2. By repeated use of (1.1.1), we obtain the implication (i) $\Rightarrow$ (iii).

We prove that (iii) $\Rightarrow$ (iv). Let $\xi \in D$ and $P = P_1 \times \dots \times P_n \subset\subset D$ be a polydisc centered at ξ. Since for $z \in P$ and $\zeta \in \partial P_1 \times \dots \times \partial P_n$

$$\frac{1}{(\zeta_1 - z_1) \dots (\zeta_n - z_n)} = \sum_{k_1, \dots, k_n = 0}^{\infty} \frac{(z_1 - \xi_1)^{k_1} \dots (z_n - \xi_n)^{k_n}}{(\zeta_1 - \xi_1)^{k_1+1} \dots (\zeta_n - \xi_n)^{k_n+1}}$$

with uniform convergence in ζ, it follows from (1.1.2) that

$$f(z) = \sum_k \left[\frac{1}{(2\pi i)^n} \int\limits_{\partial P_1 \times \ldots \times \partial P_n} \frac{f(\zeta)\, d\zeta_1 \wedge \ldots \wedge d\zeta_n}{(\zeta_1 - \xi_1)^{k_1+1} \ldots (\zeta_n - \xi_n)^{k_n+1}}\right] (z - \xi)^k, \qquad z \in P. \tag{1.1.5'}$$

Now we prove that (iv) $\Rightarrow$ (ii). It follows from (1.1.3) and Lemma 1.1.4 that f is locally the uniform limit of polynomials in $z_1, \ldots, z_n$. Since such polynomials fulfil (ii), and since the uniform limit of continuous functions is continuous and the uniform limit of holomorphic functions of one complex variable is holomorphic, we conclude that f satisfies condition (ii).

Now we suppose that the equivalent conditions (i)—(iv) are fulfilled. Then (v) follows from (iii) and (vi) can be obtained by differentiation under the sign of integration in (1.1.2). Property (vii) follows, having chosen a sufficiently small polydisc $P = P_1 \times \ldots \times P_n$ centered at ξ, by the following computation, which is permissible in view of (1.1.4) and Lemma 1.1.4:

$$\begin{aligned} \partial^k f(\xi) &= \frac{k!}{(2\pi i)^n} \sum_l a_l \int\limits_{\partial P_1 \times \ldots \times \partial P_n} \frac{(\zeta - \xi)^l\, d\zeta_1 \wedge \ldots \wedge d\zeta_n}{(\zeta_1 - \xi_1)^{k_1+1} \ldots (\zeta_n - \xi_n)^{k_n+1}} \\ &= \frac{k!}{(2\pi i)^n} \sum_l a_l \frac{(2\pi i)^n}{l!} \partial_\zeta^k (\zeta - \xi)^l \big|_{\zeta = \xi} = k!\, a_k . \end{aligned}$$

(viii) follows from (vii), (1.1.5') and Lemma 1.1.4. ■

1.1.6. Definition. Let $D \subseteqq \mathbb{C}^n$ be an open set. A complex-valued function f in D is said to be *holomorphic* (or *analytic*) if the equivalent conditions in Theorem 1.1.5 are fulfilled. The set of all holomorphic functions in D will be denoted by $\mathcal{O}(D)$.

1.1.7. Corollary. *For every open set $D \subseteqq \mathbb{C}^n$, $\mathcal{O}(D)$ is a ring, that is, if $f, g \in \mathcal{O}(D)$, then $f + g \in \mathcal{O}(D)$ and $fg \in \mathcal{O}(D)$. Further, if $f \in \mathcal{O}(D)$ and $f(z) \neq 0$ for all $z \in D$, then $1/f \in \mathcal{O}(D)$.*

Proof. This follows from condition (ii) in Theorem 1.1.5 and the corresponding properties of holomorphic functions of one complex variable.

1.1.8. Corollary (Maximum principle). *Let $D \subseteqq \mathbb{C}^n$ be an open set and $f \in \mathcal{O}(D)$. Suppose that there exists a point $\xi \in D$ such that $|f(z)| \leqq |f(\xi)|$ for all $z \in D$. Then f is constant in D if D is connected.*

Proof. This follows from condition (ii) in Theorem 1.1.5 and the maximum principle for holomorphic functions of one complex variable. ■

1.1.9. Corollary. *Let $D \subseteqq \mathbb{C}^n$ be an open set and $f_j \in \mathcal{O}(D)$. If $f_j \to f$ when $j \to \infty$, uniformly on every compact subset of D, then $f \in \mathcal{O}(D)$.*

Proof. This follows from condition (ii) in Theorem 1.1.5 and the corresponding property of holomorphic functions of one complex variable. Clearly, it can also be obtained from condition (iii) in Theorem 1.1.5. ■

1.1.10. Corollary (the uniqueness of holomorphic continuation). *Let $D \subseteqq \mathbb{C}^n$ be an open set and $f \in \mathcal{O}(D)$. If there is a point $\xi \in D$ where $\partial^k f(\xi) = 0$ for all multi-orders k, then $f = 0$ in D if D is connected.*

Proof. This follows from condition (iv) and relation (1.1.5) in Theorem 1.1.5. ■

1.1.11. Corollary. *For every open set $D \subseteqq \mathbb{C}^n$ and all multi-orders k, we have $\partial^k \mathcal{O}(D) \subseteqq \mathcal{O}(D)$.*

Proof. This follows from property (vi) in Theorem 1.1.5. ■

1.1.12. Corollary (Cauchy's inequality). *If f is holomorphic in the polydisc $P := \{z \in \mathbb{C}^n : |z_j| < r_j, j = 1, \dots, n\}$, where $0 < r_j < \infty$, $j = 1, \dots, n$, then for all multi-orders k*

$$|\partial^k f(0)| \leqq k!\, r^{-k} \sup_{z \in P} |f(z)| \, .$$

Proof. This follows from property (vi) in Theorem 1.1.5. ■

1.1.13. Theorem. *Let $D \subseteqq \mathbb{C}^n$ be an open set. Then, for every compact set $K \subset\subset D$ and every multi-order k, there exists a constant $C = C(K, k)$ such that*

$$\max_{z \in K} |\partial^k f(z)| \leqq C \int_D |f| \, d\sigma_{2n} \quad \textit{for all} \quad f \in \mathcal{O}(D) \, ,$$

where $d\sigma_{2n}$ is the Lebesgue measure in $\mathbb{C}^n$.

Proof. First let $n = 1$. Choose $\chi \in C_0^\infty(D)$ such that $\chi = 1$ in a neighbourhood U_K of K. Then, by (1.1.1), for every $f \in \mathcal{O}(D)$

$$\chi(z) f(z) = \frac{-1}{2\pi i} \int_{D \setminus U_K} \frac{f(\zeta)\, \bar{\partial}\chi(\zeta) \wedge d\zeta}{\zeta - z} \, ,$$

and differentiation leads to the required estimate. Repeated use of this gives the theorem for polydiscs D of arbitrary dimension. The general case follows by use of a covering of K by a finite number of polydiscs $\subset\subset D$. ■

1.1.14. Corollary (Stieltjes-Vitali). *Let $D \subseteqq \mathbb{C}^n$ be an open set and let f_k be a sequence of holomorphic functions in D which is uniformly bounded on every compact subset of D. Then there is a subsequence f_{k_j} converging uniformly on every compact subset of D to a limit in $\mathcal{O}(D)$.*

Proof. Since $\bar{\partial} f_k = 0$, it follows from Theorem 1.1.13 that all first-order derivatives (with respect to the real coordinates) of f_k are uniformly bounded on any compact subset of D. Hence the corollary follows from Ascoli's theorem and Corollary 1.1.9. ■

1.1.15. Definition. Let $D \subseteqq \mathbb{C}^n$ be an open set. A map $f = (f_1, \dots, f_m) : D \to \mathbb{C}^m$ is said to be *holomorphic* in D if $f_1, \dots, f_m \in \mathcal{O}(D)$. The set of all holomorphic maps $f: D \to \mathbb{C}^m$ will be denoted by $\mathcal{O}^m(D)$. If $f \in \mathcal{O}^m(D)$ and $\xi \in D$, then the matrix

$$J_f(\xi) := \left(\frac{\partial f_j(\xi)}{\partial \xi_k} \right)_{k=1,\dots,n}^{j=1,\dots,m} \quad (j \text{ is the row index})$$

is called the (*complex*) *Jacobi matrix* of f at ξ. f is called *regular* at $\xi \in D$ if $\operatorname{rank} J_f(\xi) = \min\{n, m\}$. Let $D, G \subseteqq \mathbb{C}^n$ be open sets. A *biholomorphic map from D onto G* is by definition a bijective map f from D onto G such that both $f \in \mathcal{O}^n(D)$ and $f^{-1} \in \mathcal{O}^n(G)$. Then we shall also say that f is *biholomorphic in D*.

1.1.16. Proposition. *Let $D \subseteqq \mathbb{C}^n$ be an open set and $f = (f_1, \dots, f_m) \in \mathcal{O}^m(D)$. Then*

(i) *For every $\xi \in D$,*

$$f(\xi + z) = f(\xi) + J_f(\xi)\, z + O(|z|^2) \quad \textit{when} \quad \mathbb{C}^n \ni z \to 0 \, . \qquad (1.1.6)$$

(ii) *If g is a continuous complex-valued function in some neighbourhood of $f(D)$, then (in the sense of distributions)*

$$\frac{\partial(g \circ f)}{\partial z_j} = \sum_{k=1}^{m} \frac{\partial g}{\partial f_k} \frac{\partial f_k}{\partial z_j} \quad \textit{for} \quad j = 1, \dots, n, \tag{1.1.7}$$

and

$$\frac{\partial(g \circ f)}{\partial \bar{z}_j} = \sum_{k=1}^{m} \frac{\partial g}{\partial \bar{f}_k} \frac{\partial \bar{f}_k}{\partial \bar{z}_j} \quad \textit{for} \quad j = 1, \dots, n. \tag{1.1.8}$$

Proof. Since $\bar{\partial} f_j = 0$ for $j = 1, \dots, n$, and $\mathrm{d} = \partial + \bar{\partial}$, (1.1.6) follows from the Taylor formula. Since $\mathrm{d} = \partial + \bar{\partial}$ and $\bar{\partial} f_k = \partial \bar{f}_k = 0$, we obtain

$$\sum_{j=1}^{n} \left(\frac{\partial(g \circ f)}{\partial z_j} \mathrm{d}z_j + \frac{\partial(g \circ f)}{\partial \bar{z}_j} \mathrm{d}\bar{z}_j \right) = \mathrm{d}(g \circ f)$$

$$= \sum_{k=1}^{m} \left(\frac{\partial g}{\partial f_k} \mathrm{d}f_k + \frac{\partial g}{\partial \bar{f}_k} \mathrm{d}\bar{f}_k \right) = \sum_{j=1}^{n} \sum_{k=1}^{m} \left(\frac{\partial g}{\partial f_k} \frac{\partial f_k}{\partial z_j} \mathrm{d}z_j + \frac{\partial g}{\partial \bar{f}_k} \frac{\partial \bar{f}_k}{\partial \bar{z}_j} \mathrm{d}\bar{z}_j \right),$$

which implies (1.1.7) amd (1.1.8) by comparison of the coefficients of $\mathrm{d}z_1, \dots, \mathrm{d}z_n$, $\mathrm{d}\bar{z}_1, \dots, \mathrm{d}\bar{z}_n$. ■

1.1.17. Corollary. *Let $D \subseteq \mathbb{C}^n$, $G \subseteq \mathbb{C}^m$ be open sets, and let $f \in \mathcal{O}^m(D)$, $g \in \mathcal{O}^k(G)$ such that $f(D) \subseteq G$. Then $g \circ f \in \mathcal{O}^k(D)$ and*

$$J_{g \circ f}(\xi) = J_g(f(\xi))\, J_f(\xi) \quad \textit{for all} \quad \xi \in D. \tag{1.1.9}$$

Proof. That $g \circ f \in \mathcal{O}^k(D)$ follows from (1.1.8). (1.1.9) follows from (1.1.7). ■

1.1.18. Theorem (Implicit function theorem). *Let U be a neighbourhood of $\xi \in \mathbb{C}^n$ and $f \in \mathcal{O}^n(U)$. Then f is biholomorphic in some neighbourhood of ξ if and only if*

$$\det J_f(\xi) \neq 0. \tag{1.1.10}$$

Proof. It follows from (1.1.9) that (1.1.10) is necessary. We prove that condition (1.1.10) is sufficient. Without loss of generality we can assume that $\xi = 0$, $f(\xi) = 0$, and $J_f(\xi)$ is the unit matrix. Denote by id the identity map of $\mathbb{C}^n$ and define

$$\tilde{f} := \mathrm{id} - f \quad \text{on} \quad U.$$

Choose a neighbourhood V of $0 \in \mathbb{C}^n$ such that $V \subset\subset U$. Then, by (1.1.6), for some $C < \infty$

$$|\tilde{f}(z)| \leqq C\, |z|^2 \quad \text{for} \quad z \in V.$$

We use the notation $E_0(\delta) := \{z \in \mathbb{C}^n : |z| < \delta\}$, $\delta > 0$, and choose $\varepsilon > 0$ so small that $\varepsilon < 1/2C$ and $E_0(2\varepsilon) \subseteq V$. Then $\tilde{f}(E_0(\varepsilon/2^k)) \subseteq E_0(\varepsilon/2^{k+1})$ for $k = 0, 1, 2, \dots$. Consequently, the series

$$h := \sum_{k=1}^{\infty} \tilde{f}^k, \quad \text{where} \quad \tilde{f}^k := \underbrace{\tilde{f} \circ \dots \circ \tilde{f}}_{k \text{ times}},$$

converges uniformly on $E_0(\varepsilon)$ and $h(E_0(\varepsilon)) \subseteq E_0(2\varepsilon) \subseteq U$. As the uniform limit of holomorphic maps h is holomorphic. Further, $(\mathrm{id} + h) \circ f = f \circ (\mathrm{id} + h) = \mathrm{id}$. ■

1.1.19. Corollary. *If $X \subseteq \mathbb{C}^n$ and $k \in \{1, 2, \dots, n-1\}$, then the following conditions are equivalent:*

(i) *For every point $\xi \in X$ there exists a biholomorphic map $f = (f_1, \dots, f_n)$ in some neighbourhood U of ξ such that $X \cap U = \{z \in U : f_{k+1}(z) = \dots = f_n(z) = 0\}$.*

(ii) *For every point $\xi \in X$ there exist a neighbourhood V of ξ and a regular holomorphic map $g: V \to \mathbb{C}^{n-k}$ such that $X \cap V = \{z \in V: g(z) = 0\}$.*

Proof. The implication (i) $\Rightarrow$ (ii) is trivial. We prove that (ii) $\Rightarrow$ (i). Let $\xi \in X$ and let V, g as in condition (ii). Denote by G the linear map from $\mathbb{C}^n$ into $\mathbb{C}^{n-k}$ defined by the Jacobi matrix $J_g(\xi)$. Since rank $J_g(\xi) = n - k$, G is onto and we can find a linear map $A: \mathbb{C}^n \to \mathbb{C}^k$ such that the map $A \oplus G: \mathbb{C}^n \to \mathbb{C}^n$ defined by $(A \oplus G)(z) := \big(A(z), G(z)\big)$ is invertible. Define $f(z) := \big(A(z), g(z)\big)$ for $z \in V$. Then $\det J_f(\xi) = \det A \oplus G \neq 0$ and, by Theorem 1.1.18, f is biholomorphic in some neighbourhood $U \subseteq V$ of ξ. Since $X \cap U = \{z \in U: g(z) = 0\}$, and since $(f_{k+1}, \dots, f_n) = g$, it follows that $X \cap U = \{z \in U: f_{k+1}(z) = \dots = f_n(z) = 0\}$. ■

1.1.20. Definition. Let $D \subseteq \mathbb{C}^n$ be an open set. A subset $X \subseteq D$ is said to be a *complex submanifold of* $\mathbb{C}^n$ if the equivalent conditions (i) and (ii) in Corollary 1.1.19 are fulfilled. If, moreover, X is a closed subset of D, then X is called a *closed complex submanifold of* D.

It follows from condition (i) in Corollary 1.1.19 that every complex submanifold of $\mathbb{C}^n$ in a natural way becomes a complex manifold in the following abstract sense:

1.1.21. Definition. A *complex manifold of complex dimension* n is a real $2n$-dimensional C^∞-manifold X together with a family $\{(U_j, \varphi_j)\}_{j \in I}$ satisfying the following conditions:

(i) For every $j \in I$, U_j is an open subset of X, and $\bigcup_{j \in I} U_j = X$.

(ii) For all $j \in I$, φ_j is a C^∞-diffeomorphism from U_j onto some open set in $\mathbb{C}^n$.

(iii) For all $i, j \in I$, $\varphi_j \circ \varphi_i^{-1}$ is a biholomorphic map from $\varphi_i(U_i \cap U_j)$ onto $\varphi_j(U_i \cap U_j)$.

Then a couple (V, ψ) is called a *system of holomorphic coordinates in* V, if the family $\{(U_j, \varphi_j)\}_{j \in I} \cup \{(V, \psi)\}$ fulfils conditions (i)—(iii). A family $(V_\alpha, \psi_\alpha)_{\alpha \in A}$ is called a *holomorphic atlas of* X if every (V_α, ψ_α) is a system of holomorphic coordinates and $X = \bigcup_{\alpha \in A} V_\alpha$. (In particular, the system $\{(U_j, \varphi_j)\}_{j \in I}$ above is a holomorphic atlas of X.) Every open subset of a complex manifold becomes in an obvious sense a complex manifold.

A function f defined on a complex manifold X is said to be *holomorphic* on X if, for every system (V, ψ) of holomorphic coordinates in V, the function $f \circ \psi^{-1}$ is holomorphic on $\psi(V)$. If X, Y are two complex manifolds, then a map $f: X \to Y$ is called *holomorphic* if, for every system (U, φ) of holomorphic coordinates in X and every system (V, ψ) of holomorphic coordinates in Y, the map $\psi \circ f \circ \varphi^{-1}$ is holomorphic in $\varphi\big(U \cap f^{-1}(V)\big)$. If, moreover, f is a bijective map from X onto Y and f^{-1} is also holomorphic, then f is said to be *biholomorphic from X onto Y*. If there exists a biholomorphic map from X onto Y, then X and Y are called *biholomorphically equivalent*.

A subset Z of a complex manifold X is said to be a *complex submanifold of* X if, for every system (U, φ) of holomorphic coordinates in X, $\varphi(U \cap Z)$ is a complex submanifold of $\varphi(U)$. If, moreover, Z is a closed subset of X, then Z is called a *closed complex submanifold of* X. Every complex submanifold of a complex manifold becomes in an obvious sense a complex manifold.

If U is an open set in a complex manifold, then we denote by $\mathcal{O}(U)$ the set of all holomorphic functions in U, and by $\mathcal{O}^k(U)$ we denote the set of all holomorphic maps from U into $\mathbb{C}^k$.

1.2. Application of the Cauchy-Green formula to the $\bar\partial$-equation

In this section we begin the study of the inhomogeneous Cauchy-Riemann equations for several complex variables. This is the system of differential equations

$$\frac{\partial u}{\partial \bar z_j} = f_j\,, \qquad j = 1, \dots, n\,, \tag{1.2.1}$$

which is also called the $\bar\partial$-*equation*. It is clear that the following compatibility condition is necessary for the existence of a solution:

$$\frac{\partial f_j}{\partial \bar z_k} = \frac{\partial f_k}{\partial \bar z_j}\,, \qquad j, k = 1, \dots, n\,. \tag{1.2.2}$$

Using essentially only the Cauchy-Green formula, we shall prove in the present section that condition (1.2.2) is also sufficient for the local solvability of (1.2.1) as well as for global solvability if the functions $f_1, \dots, f_n$ have compact support, where in this case, if $n \geqq 2$, the solution can also be chosen with compact support. In general, global solvability on open sets $D \subseteqq \mathbb{C}^n$ can be obtained only under additional conditions on D.

1.2.1. Theorem. *Let* $n \geqq 2$, $f_1, \dots, f_n \in C^1_0(\mathbb{C}^n)$, *and assume that the compatibility condition* (1.2.2) *is fulfilled. Then there is a* $u \in C^1_0(\mathbb{C}^n)$ *satisfying* (1.2.1).

Proof. Define

$$u(z) := -\frac{1}{2\pi i} \int\limits_{\zeta \in \mathbb{C}^1} \frac{f_1(\zeta, z_2, \dots, z_n)\, d\bar\zeta \wedge d\zeta}{\zeta - z_1}\,. \tag{1.2.3}$$

After a change of variables we can write

$$u(z) = \frac{1}{2\pi i} \int\limits_{\zeta \in \mathbb{C}^1} \frac{f_1(z_1 - \zeta, z_2, \dots, z_n)\, d\bar\zeta \wedge d\zeta}{\zeta}\,.$$

This shows that $u \in C^1(\mathbb{C}^n)$. From Theorem 1.1.3 it follows that $\partial u/\partial \bar z_1 = f_1$ in $\mathbb{C}^n$. Now let $k \in \{2, \dots, n\}$. Then, by differentiating under the sign of integration we obtain

$$\frac{\partial u}{\partial \bar z_k} = -\frac{1}{2\pi i} \int\limits_{\zeta \in \mathbb{C}^1} \frac{(\partial f_1(\zeta, z_2, \dots, z_n)/\partial \bar z_k) d\bar\zeta \wedge d\zeta}{\zeta - z_1}\,.$$

Since $\partial f_1/\partial \bar z_k = \partial f_k/\partial \bar z_1$, and since by (1.1.1),

$$-\frac{1}{2\pi i} \int\limits_{\zeta \in \mathbb{C}^1} \frac{(\partial f_k(\zeta, z_2, \dots, z_n)/\partial \bar\zeta)\, d\bar\zeta \wedge d\zeta}{\zeta - z_1} = f_k(z)\,,$$

we conclude that $\partial u/\partial \bar z_k = f_k$ also for $k = 2, \dots, n$. Hence u satisfies (1.2.1). In particular, u is holomorphic outside $\operatorname{supp} f_1 \cup \dots \cup \operatorname{supp} f_n$. On the other hand it follows from (1.2.3) that $u(z) = 0$ if $|z_2| + \dots + |z_n|$ is large enough. By the uniqueness of holomorphic continuation this implies that $u(z) = 0$ in the unbounded component of $\mathbb{C}^n \setminus (\operatorname{supp} f_1 \cup \dots \cup \operatorname{supp} f_n)$. ■

Note that Theorem 1.2.1 is in contrast with the situation in the case of one complex variable. In fact, if $u, f \in C^1_0(\mathbb{C}^1)$ such that $\partial u/\partial \bar z = f$, then by Stokes' formula

$\int_{\mathbb{C}^1} f(z)\, d\bar{z} \wedge dz = \int_{\mathbb{C}^1} d(u(z)\, dz) = 0$. This contrast becomes even clearer in the following theorem of Hartogs, which is a consequence of Theorem 1.2.1.

1.2.2. Theorem (Hartogs' theorem). *Let $n \geqq 2$, let $D \subseteqq \mathbb{C}^n$ be an open set, and let K be a compact subset of D such that $D \setminus K$ is connected. Then, for every $h \in \mathcal{O}(D \setminus K)$ there exists $H \in \mathcal{O}(D)$ such that $H = h$ in $D \setminus K$.*

Proof. Choose $\chi \in C_0^\infty(D)$ such that $\chi = 1$ in some neighbourhood of K, and set

$$f_j := \begin{cases} -(\partial\chi/\partial\bar{z}_j)\, h & \text{in } D \setminus K\,, \\ 0 & \text{outside } D \setminus K\,. \end{cases}$$

Then the compatibility condition (1.2.2) is fulfilled and Theorem 1.2.1 gives $u \in C_0^1(D)$ such that $\partial u/\partial\bar{z}_j = f_j$. By setting $H := (1-\chi)\, h - u$, we define $H \in \mathcal{O}(D)$. Let U be the unbounded component of the complement of supp χ. Then $u = 0$ in U and, therefore, $H = h$ in $D \cap U$. Since $D \setminus K$ is connected and $(D \setminus K) \cap U \neq \emptyset$, we conclude that $H = h$ in $D \setminus K$. ■

1.2.3. Corollary (Non-existence of isolated singularities). *Let $n \geqq 2$ and let $D \subseteqq \mathbb{C}^n$ be an open set. If $z \in D$, then every holomorphic function in $D \setminus z$ can be holomorphically continued to z.*

1.2.4. Corollary. *Let $n \geqq 2$, and let $D \subseteqq \mathbb{C}^n$ be a connected open set. If $f \in \mathcal{O}(D)$ and $N(f) := \{z \in D : f(z) = 0\}$, then*

(i) *$D \setminus N(f)$ is connected,*

(ii) *$N(f)$ is not compact.*

Proof. (i). We only have to prove that for every open ball $B \subseteqq D$ the set $B \setminus N(f)$ is connected. To do this we consider two arbitrary points $z, w \in B \setminus N(f)$ and show that z and w belong to the same component of $B \setminus N(f)$. Let X be the complex line which contains z and w. Then, by Corollary 1.1.17, the restriction of f to $X \cap B$ can be considered as a holomorphic function of one complex variable. Since $f \not\equiv 0$ on $X \cap B\,(f(z) \neq 0$ and $f(w) \neq 0)$, $X \cap B \cap N(f)$ is discrete in $X \cap B$. Since $X \cap B$ is a disc and therefore connected, this implies that $(B \cap X) \setminus N(f)$ is connected. Consequently, z and w belong to the same component of $B \setminus N(f)$.

(ii). Assume that $N(f)$ is compact. Since, by part (i), $D \setminus N(f)$ is connected, then it follows from Theorem 1.2.2 that $1/f$ can be continued holomorphically to $N(f)$. This is a contradiction, because $f = 0$ on $N(f)$. ■

To solve the inhomogeneous Cauchy-Riemann equations locally without the hypothesis on compact support is slightly more complicated than the proof of Theorem 1.2.1. In Section 2.1 we shall do this by means of an integral formula for strictly convex smooth domains. Here we prove the following result in polydiscs, which also implies local solvability:

1.2.5. Theorem. *Let $D \subseteqq \mathbb{C}^n$ be an open set and let $f_1, \dots, f_n \in C^\infty(D)$ such that the compatibility condition (1.2.2) is fulfilled in D. Then, for every open polydisc $P = P_1 \times \dots \times P_n \subset\subset D$ there is a $u \in C^\infty(P)$ satisfying (1.2.1) in P.*

Proof. We shall prove inductively that the theorem is true if $f_{m+1} = \dots = f_n = 0$. This is trivial if $m = 0$. Assume that it has already been proved for $m - 1$. Let $D' := D_1' \times \dots \times D_n'$ and $D'' := D_1'' \times \dots \times D_n''$ be open polydiscs such that $P \subset\subset D'' \subset\subset D' \subset\subset D$ and choose $\chi \in C_0^\infty(D_m')$ such that $\chi = 1$ in D_m''. Define for $z \in D'$

$$v(z) := -\frac{1}{2\pi i} \int\limits_{\zeta \in D'_m} \frac{\chi(\zeta) f_m(z_1, \dots, z_{m-1}, \zeta, z_{m+1}, \dots, z_n) \, d\bar{\zeta} \wedge d\zeta}{\zeta - z_m}.$$

Since $\chi \in C_0^\infty(D'_m)$, after the change of variables $\zeta - z_m \to \zeta$ we see that $v \in C^\infty(D')$. Since $\chi = 1$ in D''_m it follows from Theorem 1.1.3 that

$$\frac{\partial v}{\partial \bar{z}_m} = f_m \quad \text{in} \quad D''. \tag{1.2.4}$$

Since $\partial f_m / \partial \bar{z}_j = \partial f_j / \partial \bar{z}_m = 0$ for $j = m + 1, \dots, n$, we obtain by differentiation under the sign of integration that

$$\frac{\partial v}{\partial \bar{z}_j} = 0 = f_j \quad \text{in} \quad D' \quad \text{for} \quad j = m + 1, \dots, n. \tag{1.2.5}$$

Then the functions $\tilde{f}_j := f_j - \partial v / \partial \bar{z}_j$, $j = 1, \dots, n$, fulfil the compatibility condition (1.2.2) and it follows from (1.2.4) and (1.2.5) that $\tilde{f}_m = \dots = \tilde{f}_n = 0$ in D''. Therefore, by the inductive hypothesis, we can find $w \in C^\infty(P)$ such that $\partial w / \partial \bar{z}_j = \tilde{f}_j$ in P and $u := v + w$ is the required solution of (1.2.1) in P. ■

1.3. Domains of holomorphy

Hartogs' Theorem 1.2.2 shows that, for $n \geqq 2$, there are examples of open sets $D \subsetneqq \tilde{D} \subseteqq \mathbb{C}^n$ such that every holomorphic function in D admits a holomorphic extension to $\tilde{D}$. In this section we shall examine this phenomenon more closely. Open sets $D \subseteqq \mathbb{C}^n$ with the property that, roughly speaking, there is no part of the boundary across which every holomorphic function in D can be continued holomorphically are of special interest in the theory of functions of several complex variables. We give a precise definition:

1.3.1. Definition. An open set $D \subseteqq \mathbb{C}^n$ is called a *domain of holomorphy* if the following condition is fulfilled: If $U, V \subseteqq \mathbb{C}^n$ are open sets such that

(1) $\emptyset \neq U \subseteqq V \cap D \neq V$,

(2) V is connected,

then there exists an $f \in \mathcal{O}(D)$ such that it is not possible to find $\tilde{f} \in \mathcal{O}(V)$ with $\tilde{f} = f$ in U.

Remark. We do not assume that a domain of holomorphy is connected.

The following definition is closely connected with the notion of a domain of holomorphy:

1.3.2. Definition. Let $D \subseteqq \mathbb{C}^n$ be an open set and K a compact subset of D. Then we define

$$\hat{K}_D^{\mathcal{O}} := \{z \in D : |f(z)| \leqq \sup_{\zeta \in K} |f(\zeta)| \quad \text{for all} \quad f \in \mathcal{O}(D)\}.$$

The set $\hat{K}_D^{\mathcal{O}}$ is called the *$\mathcal{O}(D)$-hull* of K. If $K = \hat{K}_D^{\mathcal{O}}$, then K is called *$\mathcal{O}(D)$-convex.*

1.3.3. Proposition. *For every compact set $K \subset\subset \mathbb{C}^n$, the $\mathcal{O}(\mathbb{C}^n)$-hull is contained in the convex hull of K (convex in the geometric sense of $\mathbb{C}^n = \mathbb{R}^{2n}$).*

Proof. Let $x_j(z)$ be the real coordinates of $z \in \mathbb{C}^n$ such that $z_j = x_j(z) + ix_{j+n}(z)$. If the point $w \in \mathbb{C}^n$ does not belong to the convex hull of K, then we can find real numbers $y_1, \dots, y_{2n}$ such that $\sum_{j=1}^{2n} x_j(w)\, y_j = 0$, but $\sum_{j=1}^{2n} x_j(z)\, y_j < 0$ if $z \in K$. Set $v_j := y_j + iy_{j+n}$. Then $f(z) := \exp\left(\sum_{j=1}^{n} z_j \bar{v}_j\right)$ is a holomorphic function in $\mathbb{C}^n$ such that $|f(w)| = 1$ and $|f(z)| < 1$ for all $z \in K$. ■

1.3.4. Corollary. *For every open set $D \subseteq \mathbb{C}^n$ and every compact subset K of D, $\hat{K}_D^{\mathcal{O}}$ is bounded.*

Proof. By Proposition 1.3.3, $\hat{K}_{\mathbb{C}^n}^{\mathcal{O}}$ is bounded. Clearly, $\hat{K}_D^{\mathcal{O}} \subseteq \hat{K}_{\mathbb{C}^n}^{\mathcal{O}}$. ■

1.3.5. Theorem. *Let $D \subseteq \mathbb{C}^n$ be a domain of holomorphy, $K \subset\subset D$ a compact set, and let P be an open polydisc with centre at 0. If*

$$P + w \subseteq D \quad \textit{for all} \quad w \in K\,, \tag{1.3.1}$$

then

$$P + w \subseteq D \quad \textit{for all} \quad w \in \hat{K}_D^{\mathcal{O}}\,. \tag{1.3.2}$$

Proof. Let r be the multi-radius of P. For $0 < t < 1$ we denote by P_t the polydisc of multi-radius tr with centre at 0. It is sufficient to prove that $P_t + w \subseteq D$ for all $w \in \hat{K}_D^{\mathcal{O}}$ and $0 < t < 1$. To do this we fix some $w_0 \in \hat{K}_D^{\mathcal{O}}$ and $0 < t_0 < 1$. Without loss of generality we can assume that P is bounded. Then it follows from Cauchy's inequality (Corollary 1.1.12) and (1.3.1) that, for all $f \in \mathcal{O}(D)$ and all multi-orders k,

$$\max_{w \in K} \frac{|\partial^k f(w)|}{k!} (t_0 r)^k \leqq C_f\,,$$

where

$$C_f := \max\,\{|f(z)| : z \in (\bar{P}_{t_0} + w) \text{ for some } w \in K\}\,.$$

Since $\partial^k \mathcal{O}(D) \subseteq \mathcal{O}(D)$, and by definition of $\hat{K}_D^{\mathcal{O}}$, this implies that for every $f \in \mathcal{O}(D)$ and all multi-orders k

$$\frac{|\partial^k f(w_0)|}{k!} (t_0 r)^k \leqq C_f\,.$$

Taking into account (1.1.5), the last inequality leads to the conclusion that, for every $f \in \mathcal{O}(D)$, the power series expansion of f at w_0 converges in the polydisc $P_{t_0} + w_0$. Since D is a domain of holomorphy, this implies that $P_{t_0} + w_0 \subseteq D$. ■

1.3.6. Corollary. *Let $D \subseteq \mathbb{C}^n$ be a domain of holomorphy. Then, for every compact set $K \subset\subset D$,*

$$\operatorname{dist}(K, \partial D) = \operatorname{dist}(\hat{K}_D^{\mathcal{O}}, \partial D)\,, \tag{1.3.3}$$

where $\operatorname{dist}(K, \partial D) := \inf\,\{|w - z| : w \in K, z \in \partial D\}$.

1.3.7. Theorem. *For every open set $D \subseteq \mathbb{C}^n$, the following conditions are equivalent:*

(i) *D is a domain of holomorphy.*

(ii) *There exists an $f \in \mathcal{O}(D)$ such that for every couple of open sets $U, V \subseteq \mathbb{C}^n$ satisfying conditions* (1) *and* (2) *in Definition* 1.3.1, *it is not possible to find $\tilde{f} \in \mathcal{O}(V)$ with $\tilde{f} = f$ in U.*

(iii) *For every compact subset $K \subset\subset D$, $\hat{K}_D^{\mathcal{O}}$ is also compact.*

(iv) *For every compact subset $K \subset\subset D$,* (1.3.3) *is valid.*

(v) *For every infinite set $X \subseteqq D$, which is discrete in D, there exists an $f \in \mathcal{O}(D)$ which is unbounded on X.*

Proof. The implication (ii) $\Rightarrow$ (i) is trivial. (i) $\Rightarrow$ (iv) according to Corollary 1.3.6. Since, by Corollary 1.3.4, for every compact $K \subset\subset D$, $\hat{K}_D^{\mathcal{O}}$ is bounded, we obtain that (iv) $\Rightarrow$ (iii).

Now we prove that (iii) $\Rightarrow$ (ii). From condition (iii) we obtain a sequence $\{K_m\}_1^\infty$ of $\mathcal{O}(D)$-convex compact sets $K_m \subset\subset D$ such that every compact subset of D is contained in some K_m. Let X be a countable dense set in D, and let $\{\xi^{(m)}\}_{m=1}^\infty$ be a sequence of points in X which contains every point in X infinitely many times. Denote by B_m the maximal open ball centered at $\xi^{(m)}$ which belongs to D. Fix $\eta^{(m)} \in B_m \setminus K_m$. Since all K_m are $\mathcal{O}(D)$-convex, we can then find functions $f_m \in \mathcal{O}(D)$ such that $f_m(\eta^{(m)}) = 1$, but $|f_m| < 1$ in K_m. By raising these functions to high powers we may arrange that $f_m(\eta^{(m)}) = 1$ and $|f_m| < 1/2^m$ in K_m. Moreover, we can assume that every f_m is not identically 1 in any of the components of D. Then the infinite product $f := \prod_{m=1}^{\infty} (1 - f_m)^m$ converges uniformly on every compact subset of D, $f \in \mathcal{O}(D)$, and f is not identically 0 in any of the components of D.

Now let U, V be open sets satisfying conditions (1) and (2) in Definition 1.3.1. Assume that there is an $\tilde{f} \in \mathcal{O}(V)$ such that $\tilde{f} = f$ in U. Let W be the component of $V \cap D$ containing U. Then $\tilde{f} = f$ in W. Since V is connected and V is not contained in D, the set $\partial W \cap \partial D \cap V$ is not empty. Fix $\zeta \in \partial W \cap \partial D \cap V$. Since $X \cap W$ is dense in W and the sequence $\{\xi^{(m)}\}$ contains every point in $X \cap W$ infinitely many times, we can find a subsequence $\{\xi^{(m_s)}\}_{s=1}^\infty$ of points in $X \cap W$ which tends to ζ. Choosing m_1 sufficiently large, we can assume that $B_{m_s} \subseteqq W$ and, therefore, $\eta^{(m_s)} \in W$ for all s. Since all derivatives of f of order $< m_s$ vanish at $\eta^{(m_s)}$ and since $\tilde{f} = f$ in W, it follows that all derivatives of $\tilde{f}$ at $\zeta (= \lim \eta^{(m_s)})$ vanish. Consequently, $\tilde{f}$ is identically 0 in V, which is a contradiction, because f is not identically 0 in W.

The equivalence of conditions (i)—(iv) is proved. It remains to show that (iii) $\Leftrightarrow$ (v).

First we assume that (iii) is violated. Then there is a compact set $K \subset\subset D$ such that $\hat{K}_D^{\mathcal{O}}$ is not compact. Since $\hat{K}_D^{\mathcal{O}}$ is bounded and relatively closed in D, we can find a sequence $\xi^{(k)} \in \hat{K}_D^{\mathcal{O}}$ tending to some point $\xi^{(\infty)} \in \partial D$. This sequence is discrete in D, but for all $f \in \mathcal{O}(D)$

$$\sup_k |f(\xi^{(k)})| \leqq \sup_{z \in \hat{K}_D^{\mathcal{O}}} |f(z)| = \sup_{z \in K} |f(z)| < \infty,$$

that is, (v) is violated.

Now we assume that (iii) is fulfilled. Then we can find a sequence $\{K_m\}_{m=1}^\infty$ of $\mathcal{O}(D)$-convex subsets of D such that every compact subset of D is contained in some K_m. Let $X \subseteqq D$ be an infinite set which is discrete in D. Then we can choose a subsequence $\{K_{m_j}\}_{j=1}^\infty$ and a sequence $\{\xi^{(j)}\}_{j=1}^\infty$ of points in X such that $\xi^{(j)} \in K_{m_{j+1}} \setminus K_{m_j}$. Since all K_{m_j} are $\mathcal{O}(D)$-convex, we can find $f_j \in \mathcal{O}(D)$ such that $f_j(\xi^{(j)}) > 1$ but $|f_j| < 1$ on K_{m_j}. Replacing f_j by a high power of f_j, we may arrange that $|f_j| < 1/2^j$ on K_{m_j}, but $f_j(\xi^{(j)}) > j$. Then $f := \sum_j f_j$ converges uniformly on every compact subset of D and defines a holomorphic function f on D such that $\lim f(\xi^{(j)}) = \infty$. Hence condition (v) is fulfilled. ■

1.3.8. Corollary. *Every convex open set $D \subseteq \mathbb{C}^n$ is a domain of holomorphy.*

Proof. This follows from condition (iii) in Theorem 1.3.7 and Proposition 1.3.3. ■

1.3.9. Corollary. *Let $\{D_j\}_{j\in J}$ be a system of domains of holomorphy in $\mathbb{C}^n$. Then the interior D of $\bigcap_{j\in J} D_j$ is a domain of holomorphy.*

Proof. Let K be a compact subset of D. Since by Corollary 1.3.6 $\operatorname{dist}(\hat{K}^{\mathcal{O}}_{D_j}, \partial D_j) = \operatorname{dist}(K, \partial D_j)$, and since $\hat{K}^{\mathcal{O}}_D \subseteq \hat{K}^{\mathcal{O}}_{D_j}$ for all $j \in J$ we conclude that

$$\operatorname{dist}(\hat{K}^{\mathcal{O}}_D, \partial D) \geqq \inf_j \operatorname{dist}(\hat{K}^{\mathcal{O}}_{D_j}, \partial D_j) = \operatorname{dist}(K, \partial D),$$

that is, condition (iv) in Theorem 1.3.7 is fulfilled. ■

1.3.10. Corollary. *Let $D \subseteq \mathbb{C}^n$ and $G \subseteq \mathbb{C}^m$ be domains of holomorphy. Then $D \times G$ is a domain of holomorphy.*

Proof. Let $X \subseteq D \times G$ be an infinite set which is discrete in $D \times G$. We can assume that $X_D := \{z \in D : (z, w) \in X \text{ for some } w \in G\}$ is infinite and discrete in D (otherwise X_G is so). Then, by hypothesis, there is $f \in \mathcal{O}(D)$ which is unbounded on X_D. Setting $F(z, w) := f(z)$ for $(z, w) \in D \times G$, we obtain $F \in \mathcal{O}(D \times G)$ which is unbounded on X. Hence, by condition (v) in Theorem 1.3.7, $D \times G$ is a domain of holomorphy. ■

1.3.11. Theorem. *Let $D \subseteq \mathbb{C}^n$ be an open set, $G \subseteq \mathbb{C}^m$ a domain of holomorphy, and let $F: D \to \mathbb{C}^m$ be a holomorphic map. Suppose that at least one of the following conditions is fulfilled:*
(i) *D is a domain of holomorphy.*
(ii) *$F^{-1}(G) \subset\subset D$.*
Then $F^{-1}(G)$ is a domain of holomorphy.

Proof. Let $X \subseteq F^{-1}(G)$ be an infinite set which is discrete in $F^{-1}(G)$. By condition (v) in Theorem 1.3.7, it is sufficient to find $f \in \mathcal{O}(F^{-1}(G))$ which is unbounded on X. Since one of the conditions (i) and (ii) is fulfilled, we only have to consider the case that X is not discrete in D. In this case we can choose a sequence $\{\xi^{(k)}\}_{k=1}^{\infty}$ of points in X which converges to some point $\xi \in D \setminus F^{-1}(G)$. Since F is continuous and $F(\xi) \notin G$ then the set $\{F(\xi^{(k)})\}$ is infinite and discrete in G. Since G is a domain of holomorphy, we can therefore find a $g \in \mathcal{O}(G)$ which is unbounded on $\{F(\xi^{(k)})\}$. It remains to set $f := g \circ F$. ■

1.3.12. Definition. A bounded open set $D \subset\subset \mathbb{C}^n$ is called an *analytic polyhedron* if there are holomorphic functions $F_1, \ldots, F_N$ in some neighbourhood $U_{\bar{D}}$ of $\bar{D}$ such that

$$D = \{z \in U_{\bar{D}} : |F_j(z)| < 1 \quad \text{for} \quad j = 1, \ldots, N\}.$$

1.3.13. Corollary. *Every analytic polyhedron is a domain of holomorphy.*

Proof. First we observe that every open polydisc is a domain of holomorphy (this follows, for example, from Corollary 1.3.8). Let G be the open polydisc of multiradius $(1, \ldots, 1)$ and with centre $0 \in \mathbb{C}^N$. If $F = (F_1, \ldots, F_N)$, then $D = F^{-1}(G)$ and, therefore, the corollary follows from Theorem 1.3.11 (ii). ■

We conclude this section with the following

Remark. Every open set D in the complex plane is a domain of holomorphy. This follows from the Weierstrass product theorem. Another possibility is to check, for example, condition (v) in Theorem 1.3.7: If X is an infinite discrete set in D, then X is unbounded or $\overline{X} \cap \partial D \neq \emptyset$. In the first case the function $f(z) := z$ is unbounded on X and in the second case, for every $w \in \overline{X} \cap \partial D$, the function $f(z) := 1/(z - w)$ is holomorphic on D and unbounded on X.

1.4. Plurisubharmonic functions

In the next section we introduce the concept of pseudoconvexity. Pseudoconvexity will be defined by means of plurisubharmonic functions, which will be investigated in the present section. First we recall the notion of a subharmonic function of one complex variable.

1.4.1. Definition. Let $D \subseteq \mathbb{C}^1$ be an open set. A *continuous subharmonic function in D* is a continuous[1]) function $\varrho: D \to \mathbb{R}^1$ such that the following condition is fulfilled: For every $\xi \in D$ and all $0 < r < \operatorname{dist}(\xi, \partial D)$

$$\varrho(\xi) \leqq \frac{1}{2\pi r} \int\limits_{|z-\xi|=r} \varrho(z)\,|\mathrm{d}z| = \frac{1}{2\pi} \int\limits_0^{2\pi} \varrho(\xi + r\,\mathrm{e}^{i\varphi})\,\mathrm{d}\varphi\,, \tag{1.4.1}$$

where

$$|\mathrm{d}z| := \frac{r}{x_1}\,\mathrm{d}x_2 = -\frac{r}{x_2}\,\mathrm{d}x_1 = r\,\mathrm{d}\varphi \quad \text{if} \quad z = \xi + x_1 + \mathrm{i}x_2 = \xi + r\,\mathrm{e}^{\mathrm{i}\varphi}\,. \tag{1.4.2}$$

Every harmonic function is subharmonic; then (1.4.1) is valid with equality. We give some other equivalent definitions of continuous subharmonic functions.

1.4.2. Theorem. *Let $D \subseteq \mathbb{C}^1$ be an open set, and let $\varrho: D \to \mathbb{R}$ be continuous. Then each of the following conditions is necessary and sufficient for ϱ to be subharmonic:*

(i) *For every $\xi \in D$ there exists an $\varepsilon > 0$ such that* (1.4.1) *holds for all $0 < r < \varepsilon$.*

(ii) *For every compact set $K \subset\subset D$ and every harmonic function h in some neighbourhood of K such that $h \geqq \varrho$ on ∂K, we have $h \geqq \varrho$ on K.*

(iii) *For every $\xi \in D$, all $0 < r < \operatorname{dist}(\xi, \partial D)$, and every harmonic function h in some neighbourhood of the disc $|\zeta - \xi| = r$ such that $h(\zeta) \geqq \varrho(\zeta)$ for $|\zeta - \xi| = r$, we have $h(\zeta) \geqq \varrho(\zeta)$ for all $|\zeta - \xi| \leqq r$.*

Proof. It is trivial that subharmonicity implies condition (i). Further, it is clear that (ii) $\Rightarrow$ (iii). Since for harmonic functions (1.4.1) is valid with equality, (iii) implies subharmonicity. It remains to prove that (i) $\Rightarrow$ (ii). Let $K \subset\subset D$ be a compact set and h a harmonic function in some neighbourhood of K such that $h \geqq \varrho$ on ∂K. We have to prove that

$$C := \max_{z \in K} \big(\varrho(z) - h(z)\big) \leqq 0\,.$$

Assume that $C > 0$. Set $K_C := \{z \in K: \varrho(z) - h(z) = C\}$. Then K_C is a non-empty compact subset of the interior of K. Choose a point $\xi \in K_C$ with $\operatorname{dist}(\xi, \partial K) = \operatorname{dist}(K_C, \partial K)$. Then for all sufficiently small $r > 0$

$$\frac{1}{2\pi r} \int\limits_{|z-\xi|=r} \big(\varrho(z) - h(z)\big)\,|\mathrm{d}z| < C\,.$$

[1]) Usually the notion of subharmonicity is defined for upper semicontinuous functions with values in $\mathbb{R} \cup \{-\infty\}$, but in the present book we need this for continuous functions only.

This contradicts the inequality $C = \varrho(\xi) - h(\xi) \leqq (1/2\pi r) \int\limits_{|z-\xi|=r} (\varrho(z) - h(z))\,|dz|$, which is valid for all sufficiently small $r > 0$, because of condition (i) and harmonicity of h. ■

In this book subharmonic C^2-functions are of special interest. These functions can be characterized as follows:

1.4.3. Theorem. *Let $D \subseteqq \mathbb{C}^1$ be an open set. A C^2-function $\varrho\colon D \to \mathbb{R}^1$ is subharmonic if and only if*

$$\frac{\partial^2 \varrho(z)}{\partial z\, \partial \bar{z}} \geqq 0 \quad \textit{for all} \quad z \in D\,. \tag{1.4.3}$$

This theorem is an immediate consequence of the following lemma:

1.4.4. Lemma. *Let $D \subseteqq \mathbb{C}^1$ be an open set, $\varrho\colon D \to \mathbb{R}^1$ a C^2-function, and $\xi \in D$. If $0 < r < \operatorname{dist}(\xi, \partial D)$, then we define* (see (1.4.2) for notation)

$$M(r) := \frac{1}{2\pi r} \int\limits_{|\xi - z| = r} \varrho(z)\,|dz| = \frac{1}{2\pi} \int\limits_0^{2\pi} \varrho(\xi + r\,e^{i\varphi})\,d\varphi\,. \tag{1.4.4}$$

Then

$$\lim_{r\to 0} M(r) = \varrho(\xi) \tag{1.4.5}$$

and

$$\frac{dM(r)}{dr} = \frac{2}{\pi r} \int\limits_{|\xi - z| < r} \frac{\partial^2 \varrho}{\partial z\, \partial \bar{z}}\, dx_1 \wedge dx_2\,. \tag{1.4.6}$$

Proof. (1.4.5) is trivial. We prove (1.4.6). First observe that

$$\frac{dM(r)}{dr} = \frac{1}{2\pi} \int\limits_0^{2\pi} \frac{d}{dr} \varrho(\xi + r\,e^{i\varphi})\,d\varphi = \frac{1}{2\pi r} \int\limits_{|z-\xi|=r} \frac{\partial \varrho}{\partial r}\,|dz|\,.$$

Since by (1.4.2)

$$\frac{\partial \varrho}{\partial r}\,|dz| = \left(\frac{\partial \varrho}{\partial x_1}\frac{x_1}{r} + \frac{\partial \varrho}{\partial x_2}\frac{x_2}{r}\right)|dz| = \frac{\partial \varrho}{\partial x_1}\,dx_2 - \frac{\partial \varrho}{\partial x_2}\,dx_1\,,$$

and by Stokes' formula, this implies that

$$\frac{dM(r)}{dr} = \frac{1}{2\pi r} \int\limits_{|z-\xi|<r} \left(\frac{\partial^2 \varrho}{\partial x_1^2} + \frac{\partial^2 \varrho}{\partial x_2^2}\right) dx_1 \wedge dx_2\,.$$

Taking into account the relation

$$\frac{\partial^2 \varrho}{\partial x_1^2} + \frac{\partial^2 \varrho}{\partial x_2^2} = 4\,\frac{\partial^2 \varrho}{\partial z\, \partial \bar{z}}\,, \tag{1.4.7}$$

we obtain (1.4.6). ■

1.4.5. Definition. Let $D \subseteqq \mathbb{C}^1$ be an open set. A C^2-function $\varrho\colon D \to \mathbb{R}^1$ is said to be *strictly subharmonic in D* if

$$\frac{\partial^2 \varrho(z)}{\partial z\, \partial \bar{z}} > 0 \quad \text{for all} \quad z \in D\,. \tag{1.4.8}$$

Now we pass to the case of several complex variables.

1.4.6. Definition. Let $D \subseteqq \mathbb{C}^n$ be an open set.

(i) A *continuous plurisubharmonic function in D* is a continuous function $\varrho: D \to \mathbb{R}^1$ such that the following condition is fulfilled: For arbitrary $v, w \in \mathbb{C}^n$, the function $\zeta \to \varrho(v + \zeta w)$ is subharmonic in that part of $\mathbb{C}^1$ where it is defined. The set of all continuous plurisubharmonic functions in D will be denoted by $P^0(D)$.

(ii) A C^2-function $\varrho: D \to \mathbb{R}^1$ is called *strictly plurisubharmonic* if, for arbitrary z, $w \in \mathbb{C}^n$ with $w \neq 0$, the function $\zeta \to \varrho(z + \zeta w)$ is strictly subharmonic in that part of $\mathbb{C}^1$ where it is defined.

1.4.7. Theorem. *Let $D \subseteqq \mathbb{C}^n$ be an open set, and let $\varrho: D \to \mathbb{R}^1$ be a C^2-function. Then*
(i) *ϱ is plurisubharmonic if and only if*

$$\sum_{j,k=1}^{n} \frac{\partial^2 \varrho(z)}{\partial z_j \, \partial \bar{z}_k} \xi_j \bar{\xi}_k \geqq 0 \quad \text{for all} \quad z \in D \quad \text{and} \quad \xi \in \mathbb{C}^n . \tag{1.4.9}$$

(ii) *ϱ is strictly plurisubharmonic if and only if*

$$\sum_{j,k=1}^{n} \frac{\partial^2 \varrho(z)}{\partial z_j \, \partial \bar{z}_k} \xi_j \bar{\xi}_k > 0 \quad \text{if} \quad z \in D \quad \text{and} \quad 0 \neq \xi \in \mathbb{C}^n . \tag{1.4.10}$$

Proof. Let $v, w \in \mathbb{C}^n$ and $\tilde{\varrho}(\zeta) = \varrho(v + \zeta w)$. Then by Proposition 1.1.16

$$\frac{\partial^2 \tilde{\varrho}(\zeta)}{\partial \zeta \, \partial \bar{\zeta}} = \sum_{j,k=1}^{n} \left. \frac{\partial^2 \varrho(z)}{\partial z_j \, \partial \bar{z}_k} \right|_{z = v + \zeta w} w_j \bar{w}_k .$$

Together with Theorem 1.4.3 and Definition 1.4.5 this proves the theorem. ■

Notation. The Hermitian form

$$\sum_{j,k=1}^{n} \frac{\partial^2 \varrho(z)}{\partial z_j \, \partial \bar{z}_k} \xi_j \bar{\xi}_k \tag{1.4.11}$$

is called the *Levi form of ϱ at z*.

1.4.8. Proposition. *Let $D \subseteqq \mathbb{C}^n$ be an open set and $f \in \mathcal{O}(D)$. Then $|f| \in P^0(D)$.*

Proof. We only have to prove this for $n = 1$. Let $\xi \in D$ and $0 < r < \operatorname{dist}(\xi, \partial D)$. Then by Cauchy's formula

$$f(\xi) = \frac{1}{2\pi i} \int_{|z - \xi| = r} \frac{f(z)}{z - \xi} \, dz ,$$

which implies (1.4.1) for $|f|$. ■

In the following proposition we collect some properties of plurisubharmonic functions which are immediate consequences of Definition 1.4.1 of subharmonicity.

1.4.9. Proposition. *Let $D \subseteqq \mathbb{C}^n$ be an open set.*
(i) *If $\varrho, \varphi \in P^0(D)$, then $\varrho + \varphi \in P^0(D)$.*
(ii) *If $\varrho \in P^0(D)$ and $0 \leqq c < \infty$, then $c\varrho \in P^0(D)$.*
(iii) *If $\varrho_j \in P^0(D)$ for all j in some index set J and if $\sup_{j \in J} \varrho_j$ is continuous in D, then $\sup_{j \in J} \varrho_j \in P^0(D)$.*
(iv) *If $\varrho_k \in P^0(D)$ and $\varrho_k \to \varrho$ when $k \to \infty$, uniformly on every compact subset of D, then $\varrho \in P^0(D)$.*
(v) *Every convex function from $C^0(D)$ (strictly convex function from $C^2(D)$) is plurisubharmonic (strictly plurisubharmonic).*

Remark. If both $\varrho \in P^0(D)$ and $-\varrho \in P^0(D)$, then ϱ is pluriharmonic, that is, the functions $\varrho(v + \zeta w)$ in Definition 1.4.6 are harmonic. If $f \in \mathcal{O}(D)$ and $-|f| \in P^0(D)$, then f is a constant.

1.4.10. Theorem. *Let $\varphi \in C_0^\infty(\mathbb{C}^n)$ such that $\varphi \geqq 0$ in $\mathbb{C}^n$, $\varphi(z) = 0$ if $|z| \geqq 1$, and*

$$\int_{\mathbb{C}^n} \varphi(\zeta)\, \mathrm{d}\sigma_{2n}(\zeta) = 1\,, \tag{1.4.12}$$

where $\mathrm{d}\sigma_{2n}(\zeta)$ is the Lebesgue measure in $\mathbb{C}^n$. Let $D \subseteqq \mathbb{C}^n$ be an open set and $\varrho \in P^0(D)$. If $\varepsilon > 0$, then we set $D_\varepsilon := \{z \in D : \operatorname{dist}(z, \partial D) > \varepsilon\}$ and

$$\varrho_\varepsilon(z) := \int_{\mathbb{C}^n} \varrho(z - \varepsilon\zeta)\, \varphi(\zeta)\, \mathrm{d}\sigma_{2n}(\zeta) \quad \text{for} \quad z \in D_\varepsilon\,. \tag{1.4.13}$$

Then $\varrho_\varepsilon \in C^\infty(D_\varepsilon) \cap P^0(D_\varepsilon)$ for all $\varepsilon > 0$ and $\varrho_\varepsilon \to \varrho$ when $\varepsilon \to 0$, uniformly on every compact subset of D.

Proof. We have

$$\varrho_\varepsilon(z) = \int_{\mathbb{C}^n} \varrho(\zeta)\, \varphi\left(\frac{z - \zeta}{\varepsilon}\right) \frac{\mathrm{d}\sigma_{2n}(\zeta)}{\varepsilon^{2n}} \quad \text{for} \quad z \in D_\varepsilon\,. \tag{1.4.14}$$

Together with (1.4.12) this implies that $\varrho_\varepsilon \to \varrho$ uniformly on every compact subset of D. By differentiation under the sign of integration in (1.4.14) we see that $\varrho_\varepsilon \in C^\infty(D_\varepsilon)$. From (1.4.13) it follows that, on every compact subset of D_ε, ϱ_ε is the uniform limit of functions of the form $z \to \sum_{j=1}^{N} \varrho(z - \varepsilon\zeta_j)\, c_j$, where $N < \infty$, $\zeta_j \in \mathbb{C}^n$ and $c_j > 0$. In view of Proposition 1.4.9 this implies that $\varrho_\varepsilon \in P^0(D_\varepsilon)$. ■

1.4.11. Theorem. *Let $G \subseteqq \mathbb{C}^m$ and $D \subseteqq \mathbb{C}^n$ be open sets, and let $f: G \to D$ be a holomorphic map. Then*

(i) *If $\varrho \in P^0(D)$, then $\varrho \circ f \in P^0(G)$.*

(ii) *Suppose that f is biholomorphic from G onto D. If $\varrho \in C^0(D)$, then $\varrho \in P^0(D)$ if and only if $\varrho \circ f \in P^0(G)$. If $\varrho \in C^2(D)$, then g is strictly plurisubharmonic in D if and only if $\varrho \circ f$ is strictly plurisubharmonic in G.*

Proof. If $\varrho \in C^2(D)$, $z \in G$, $w = f(z)$ and $\xi \in \mathbb{C}^m$, then by Proposition 1.1.16 (ii)

$$\sum_{j,k=1}^{m} \frac{\partial^2 \varrho \circ f(z)}{\partial z_j\, \partial \bar{z}_k} \xi_j \bar{\xi}_k = \sum_{\nu,\mu=1}^{n} \frac{\partial^2 \varrho(w)}{\partial w_\nu\, \partial \bar{w}_\mu} \left(\sum_{j=1}^{m} \frac{\partial w_\nu}{\partial z_j} \xi_j\right) \overline{\left(\sum_{k=1}^{m} \frac{\partial w_\mu}{\partial z_k} \xi_k\right)}\,. \tag{1.4.15}$$

Since for biholomorphic f the Jacobi matrix $(\partial w_\nu/\partial z_j)_{\nu,j}$ is invertible, and by Theorem 1.4.7 (ii), this implies part (ii). Further, it follows from (1.4.15) and Theorem 1.4.7 (i) that part (i) is valid for all $\varrho \in P^0(D) \cap C^2(D)$. To complete the proof of part (i), we consider an arbitrary $\varrho \in P^0(D)$. By Theorem 1.4.10 we can then find a sequence of plurisubharmonic C^∞-functions ϱ_l which converges to ϱ uniformly on every compact subset of D. Then all $\varrho_l \circ f \in P^0(G)$ and it follows from Proposition 1.4.9 (iv) that $\varrho \circ f \in P^0(G)$. ■

Remark. In view of part (ii) of this theorem, continuous plurisubharmonic functions and strictly plurisubharmonic C^k-functions can be defined also on complex manifolds (cp. Definition 1.1.21). If Y is a complex submanifold of a complex manifold X and ϱ is a continuous plurisubharmonic (strictly plurisubharmonic C^k) function on X, then it is clear (by Definition 1.4.6) that the restriction of ϱ to Y is plurisubharmonic (strictly plurisubharmonic).

1.4.12. Theorem. *Let $G \subseteqq \mathbb{R}^k$ be an open set, and let $\varphi: G \to \mathbb{R}^1$ be a C^2-function which is convex, that is,*

$$\sum_{\nu,\mu=1}^{k} \frac{\partial^2\varphi(t)}{\partial t_\nu\, \partial t_\mu} y_\nu y_\mu \geqq 0 \quad \textit{for all} \quad t \in G \quad \textit{and} \quad y \in \mathbb{R}^k . \tag{1.4.16}$$

Suppose that $D \subseteqq \mathbb{C}^n$ is open and $\varrho_1, \dots, \varrho_k \in P^0(D)$ such that $(\varrho_1(z), \dots, \varrho_k(z)) \in G$ for all $z \in D$. Define $\varrho(z) := \varphi(\varrho_1(z), \dots, \varrho_k(z))$ for $z \in D$. Then

(i) *If*

$$\min_{\nu=1,\dots,k} \frac{\partial\varphi(t)}{\partial t_\nu} \geqq 0 \quad \textit{for all} \quad t \in G , \tag{1.4.17}$$

then $\varrho \in P^0(D)$.

(ii) *If* (1.4.17) *is fulfilled and, moreover,*

$$\max_{\nu=1,\dots,k} \frac{\partial\varphi(t)}{\partial t_\nu} > 0 \quad \textit{for all} \quad t \in G , \tag{1.4.17'}$$

and if the functions $\varrho_1, \dots, \varrho_k$ are C^2 and strictly plurisubharmonic in D, then ϱ is a strictly plurisubharmonic C^2-function in D.

Proof. We only have to prove this for $n = 1$. In view of the Approximation Theorem 1.4.10 we can also assume that the functions ϱ_j are C^2 (cp. the end of the proof of Theorem 1.4.11). Now let $z = x_1 + ix_2 \in D$, $x_j \in \mathbb{R}^1$, and $t = (\varrho_1(z), \dots, \varrho_k(z))$. Then

$$\frac{\partial^2\varrho(z)}{\partial x_j^2} = \sum_{\nu,\mu=1}^{k} \frac{\partial^2\varphi(t)}{\partial t_\nu\, \partial t_\mu} \frac{\partial\varrho_\nu(z)}{\partial x_j} \frac{\partial\varrho_\mu(z)}{\partial x_j} + \sum_{\nu=1}^{k} \frac{\partial\varphi(t)}{\partial t_\nu} \frac{\partial^2\varrho_\nu(z)}{\partial x_j^2}$$

for $j = 1, 2$. Since by (1.4.16) the first sum is $\geqq 0$, and since $\partial^2/\partial x_1^2 + \partial^2/\partial x_2^2 = 4\partial^2/\partial z\, \partial\bar{z}$, it follows that

$$\frac{\partial^2\varrho(z)}{\partial z\, \partial\bar{z}} \geqq \sum_{\nu=1}^{k} \frac{\partial\varphi(t)}{\partial t_\nu} \frac{\partial^2\varrho_\nu(z)}{\partial z\, \partial\bar{z}} .$$

In view of Theorem 1.4.7 this completes the proof. ∎

Notation. Let $D \subseteqq \mathbb{C}^n$ be an open set and ϱ a real-valued C^2-function in D. Then we define

$$\hat{F}_\varrho(z, \zeta) := -\left[2 \sum_{j=1}^{n} \frac{\partial\varrho(\zeta)}{\partial\zeta_j} (z_j - \zeta_j) + \sum_{j,k=1}^{n} \frac{\partial^2\varrho(\zeta)}{\partial\zeta_j\, \partial\zeta_k} (z_j - \zeta_j)(z_k - \zeta_k)\right]$$

for $\quad z, \zeta \in D.$ (1.4.18)

Remark. $\hat{F}_\varrho(z, \zeta)$ is only continuous in ζ. Later we shall replace the continuous derivatives $\partial^2\varrho(\zeta)/\partial\zeta_j\, \partial\zeta_k$ by sufficiently close C^1-functions. The obtained modification of $\hat{F}_\varrho(z, \zeta)$ will be denoted by $F(z, \zeta) = F_\varrho(z, \zeta)$ then.

The function $\hat{F}_\varrho(z, \zeta)$ is called the *Levi polynomial* of ϱ. It plays an important role for strictly plurisubharmonic C^2-functions. The following lemma describes the connection between the Levi polynomial (1.4.18) and the Levi form (1.4.11).

1.4.13. Lemma. *Let $D \subseteqq \mathbb{C}^n$ be an open set, and let ϱ be a real-valued C^2-function in D. Then, for all $\zeta \in D$ and $z \to \zeta$,*

$$\varrho(z) = \varrho(\zeta) - \operatorname{Re} \hat{F}_\varrho(z, \zeta) + \sum_{j,k=1}^{n} \frac{\partial^2 \varrho(\zeta)}{\partial \zeta_j \, \partial \bar{\zeta}_k} (z_j - \zeta_j)(\bar{z}_k - \bar{\zeta}_k) + o(|\zeta - z|^2), \tag{1.4.19}$$

where $\operatorname{Re} \hat{F}_\varrho$ *is the real part of* $\hat{F}_\varrho$.

Proof. Let $x_j = x_j(\zeta)$ be the real coordinates of $\zeta \in \mathbb{C}^n$ such that $\zeta_j = x_j(\zeta) + i x_{j+n}(\zeta)$. Then a computation gives

$$\sum_{j=1}^{2n} \frac{\partial \varrho(\zeta)}{\partial x_j} (x_j(z) - x_j(\zeta)) = 2 \operatorname{Re} \left[\sum_{j=1}^{n} \frac{\partial \varrho(\zeta)}{\partial \zeta_j} (z_j - \zeta_j) \right]$$

and

$$\frac{1}{2} \sum_{j,k=1}^{2n} \frac{\partial^2 \varrho(\zeta)}{\partial x_j \, \partial x_k} (x_j(z) - x_j(\zeta))(x_k(z) - x_k(\zeta))$$

$$= \sum_{j,k=1}^{n} \frac{\partial^2 \varrho(\zeta)}{\partial \zeta_j \, \partial \bar{\zeta}_k} (z_j - \zeta_j)(\bar{z}_k - \bar{\zeta}_k) + \operatorname{Re} \left[\sum_{j,k=1}^{n} \frac{\partial^2 \varrho(\zeta)}{\partial \zeta_j \, \partial \zeta_k} (z_j - \zeta_j)(z_k - \zeta_k) \right].$$

Consequently, (1.4.19) is the Taylor expansion of ϱ at ζ. ■

It is trivial that every strictly convex C^2-function ϱ is strictly plurisubharmonic (Proposition 1.4.9 (v)). The following theorem shows that if $d\varrho \neq 0$, then, after an appropriate local change of holomorphic coordinates, the converse is also true.

1.4.14. Theorem. *Let ϱ be a strictly plurisubharmonic C^2-function in some neighbourhood of $0 \in \mathbb{C}^n$. If $d\varrho(0) \neq 0$, then there exists a biholomorphic map h from some neighbourhood U of 0 onto a neighbourhood W of 0 such that $\varrho \circ h^{-1}$ is strictly convex in W, that is,*

$$\sum_{j,k=1}^{2n} \frac{\partial^2 \varrho \circ h^{-1}(\zeta)}{\partial x_j \, \partial x_k} t_j t_k > 0 \quad \text{for all} \quad \zeta \in W \quad \text{and} \quad 0 \neq t \in \mathbb{R}^{2n}, \tag{1.4.20}$$

where $x_j = x_j(\zeta)$ are the real coordinates of $\zeta \in \mathbb{C}^n$ such that $\zeta_j = x_j(\zeta) + i x_{j+n}(\zeta)$.

Proof. Since $d\varrho(0) \neq 0$ and ϱ is real, we have $\partial\varrho(0) \neq 0$. Assume that $\partial\varrho(0)/\partial\zeta_1 \neq 0$. Then $h(z) := (\hat{F}_\varrho(z, 0), z_2, \ldots, z_n)$ is biholomorphic from some neighbourhood U onto some neighbourhood V of $0 = h(0)$ (for the definition of $\hat{F}_\varrho$ see (1.4.18)). Let $f(\zeta) = (f_1(\zeta), \ldots, f_n(\zeta)) := h^{-1}(\zeta)$, $\zeta \in V$. Then, by Lemma 1.4.13, for $\zeta \to 0$

$$\varrho \circ f(\zeta) = \varrho(0) - \operatorname{Re} \hat{F}_\varrho(f(\zeta), 0) + \sum_{j,k=1}^{n} \frac{\partial^2 \varrho(0)}{\partial z_j \, \partial \bar{z}_k} f_j(\zeta) \overline{f_k(\zeta)} + o(|f(\zeta)|^2).$$

Since, by definition of $f = h^{-1}$, $\hat{F}_\varrho(f(\zeta), 0) = \zeta_1$, this implies that for $\zeta \to 0$

$$\varrho \circ f(\zeta) = \varrho(0) - \operatorname{Re} \zeta_1 + \sum_{j,k=1}^{n} \frac{\partial^2 \varrho(0)}{\partial z_j \, \partial \bar{z}_k} f_j(\zeta) \overline{f_k(\zeta)} + o(|f(\zeta)|^2).$$

Let $F = (F_1, \ldots, F_n)$ be the linear part of f at 0, that is, the linear map $F\colon \mathbb{C}^n \to \mathbb{C}^n$ such that $f(\zeta) = F(\zeta) + O(|\zeta|^2)$ for $\zeta \to 0$. Then we further obtain that for $\zeta \to 0$

$$\varrho \circ f(\zeta) = \varrho(0) - \operatorname{Re} \zeta_1 + \sum_{j,k=1}^{n} \frac{\partial^2 \varrho(0)}{\partial z_j \, \partial \bar{z}_k} F_j(\zeta) \overline{F_k(\zeta)} + o(|\zeta|^2).$$

Hence, by uniqueness of the Taylor expansion,

$$\sum_{j,k=1}^{2n} \frac{\partial^2 \varrho \circ f(0)}{\partial x_j\, \partial x_k} x_j(\zeta)\, x_k(\zeta) = \sum_{j,k=1}^{n} \frac{\partial^2 \varrho(0)}{\partial z_j\, \partial \bar{z}_k} F_j(\zeta)\overline{F_k(\zeta)} \quad \text{for all} \quad \zeta \in \mathbb{C}^n .$$

Since $F(\zeta) \neq 0$ if $\zeta \neq 0$ (f is biholomorphic at 0), and in view of Theorem 1.4.7 (i), this implies (1.4.20) for $\zeta = 0$ and, therefore, for all ζ in some neighbourhood W of 0. ■

Remark. Theorem 1.4.14 is not valid without the hypothesis $d\varrho(0) \neq 0$. Counterexample: $\varrho(z) = 2x_1^2 - x_2^2$, $z = x_1 + ix_2 \in \mathbb{C}^1$ ($d\varrho(0) = 0$ but ϱ does not have a local minimum at 0). In the general case there is the following result:

1.4.15. Theorem. *Let ϱ be a strictly plurisubharmonic C^2-function in some neighbourhood of $0 \in \mathbb{C}^n$. Then there exist a complex-linear isomorphism $t\colon \mathbb{C}^n \to \mathbb{C}^n$ and numbers $\lambda_j \geqq 0$ ($j = 1, \dots, n$) such that*

$$\left(\frac{\partial^2 \varrho \circ t(0)}{\partial x_j\, \partial x_k}\right)_{j,k=1}^{2n} = \begin{pmatrix} 1+\lambda_1 & & & & & 0 \\ & \ddots & & & & \\ & & 1+\lambda_n & & & \\ & & & 1-\lambda_1 & & \\ & & & & \ddots & \\ 0 & & & & & 1-\lambda_n \end{pmatrix}$$

where $x_j = x_j(\zeta)$ are the real coordinates of $\zeta \in \mathbb{C}^n$ such that $\zeta_j = x_j(\zeta) + ix_{j+n}(\zeta)$.

Proof. Let

$$L := \left(\frac{\partial^2 \varrho(0)}{\partial \zeta_j\, \partial \bar{\zeta}_k}\right)_{j,k=1}^{n} \qquad (j = \text{row index}) .$$

By Theorem 1.4.7 (ii), L is strictly positive-definite. Consequently, there is an invertible complex $n \times n$ matrix $V = (v_{jk})$ (j = row index) such that $V^+ L \bar{V}$ is the unit matrix, where $\bar{V} = (\bar{v}_{jk})$ and $V^+ = (v_{kj})$ (j = row index). If v is the complex-linear isomorphism of $\mathbb{C}^n$ defined by V, then this implies (cp. Proposition 1.1.16)

$$\left(\frac{\partial^2 \varrho \circ v(0)}{\partial \zeta_j\, \partial \bar{\zeta}_k}\right)_{j,k=1}^{n} = V^+ L \bar{V} = \begin{pmatrix} 1 & & 0 \\ & \ddots & \\ 0 & & 1 \end{pmatrix}. \tag{1.4.21}$$

Without loss of generality we can assume that $\varrho(0) = 0$ and $d\varrho(0) = 0$. Then, by (1.4.21) and Lemma 1.4.13, we have the Taylor expansion at $\zeta = 0$

$$\varrho \circ v(\zeta) = |\zeta|^2 + \operatorname{Re} \sum_{j,k=1}^{n} \frac{\partial^2 \varrho(0)}{\partial \zeta_j\, \partial \zeta_k} \zeta_j \zeta_k + o(|\zeta|^2) . \tag{1.4.22}$$

Let

$$\left(\frac{\partial^2 \varrho(0)}{\partial \zeta_j\, \partial \zeta_k}\right)_{j,k=1}^{n} = A + iB ,$$

where A and B are real matrices. Set

$$R := \begin{pmatrix} A & -B \\ -B & -A \end{pmatrix}. \tag{1.4.23}$$

Then R is a real symmetric $2n \times 2n$ matrix and a computation shows that

$$\operatorname{Re} \sum_{j,k=1}^{n} \frac{\partial^2 \varrho(0)}{\partial \zeta_j\, \partial \zeta_k} \zeta_j \zeta_k = \big(Rx(\zeta), x(\zeta)\big) ,$$

where $x(\zeta) = (x_1(\zeta), \dots, x_{2n}(\zeta))$ and $(\cdot,\cdot)$ is the scalar product in $\mathbb{R}^{2n}$. Together with (1.4.22) this implies that

$$\varrho \circ v(\zeta) = (x(\zeta) + Rx(\zeta), x(\zeta)) + o(|\zeta|^2) \quad \text{for} \quad \zeta \to 0\,. \tag{1.4.24}$$

If $e \in \mathbb{C}^n$ such that $x(e)$ is an eigenvector of the matrix R, with eigenvalue λ, then it follows from (1.4.23) that $x(ie)$ is an eigenvector of R, with eigenvalue $-\lambda$. Consequently, since R is symmetric, we can find vectors $e_1, \dots, e_n \in \mathbb{C}^n$ and numbers $\lambda_1, \dots, \lambda_n \geqq 0$ such that the vectors $x(e_1), \dots, x(e_n), x(ie_1), \dots, x(ie_n)$ form an orthonormal basis in $\mathbb{R}^{2n}$ and

$$Rx(e_j) = \lambda_j x(e_j)\,, \qquad Rx(ie_j) = -\lambda_j x(ie_j) \quad \text{for} \quad j = 1, \dots, n\,. \tag{1.4.25}$$

Then the vectors $e_1, \dots, e_n$ form an orthonormal basis in $\mathbb{C}^n$ and the complex-linear map $u\colon \mathbb{C}^n \to \mathbb{C}^n$ defined by $u(\zeta) := \sum_{j=1}^{n} \zeta_j e_j$, $\zeta \in \mathbb{C}^n$, is unitary. The corresponding map of real coordinates $x(\zeta) \to x(u(\zeta))$ is defined by the matrix

$$U := \begin{pmatrix} x_1(e_1) & \dots & x_1(e_n) & x_1(ie_1) & \dots & x_1(ie_n) \\ \cdot & \cdot & \cdot & \cdot & \cdot & \cdot \\ x_{2n}(e_1) & \dots & x_{2n}(e_n) & x_{2n}(ie_1) & \dots & x_{2n}(ie_n) \end{pmatrix}.$$

Since U is orthogonal, and since by (1.4.25)

$$U^{-1}RU = \begin{pmatrix} \lambda_1 & & & & & 0 \\ & \ddots & & & & \\ & & \lambda_n & & & \\ & & & -\lambda_1 & & \\ & & & & \ddots & \\ 0 & & & & & -\lambda_n \end{pmatrix},$$

it follows from (1.4.24) that for $|\zeta| \to 0$

$$\varrho \circ v \circ u(\zeta) = \sum_{j=1}^{n} (1 + \lambda_j)\, |x_j(\zeta)|^2 + \sum_{j=1}^{n} (1 - \lambda_j)\, |x_{j+n}(\zeta)|^2 + o(|\zeta|^2)\,.$$

We conclude the proof by setting $t := v \circ u$. ■

1.4.16. Definition. A subset $T \subseteq \mathbb{C}^n$ is said to be a *real (complex) plane in* $\mathbb{C}^n$ if, for $z \in T$, $T - z$ is a real-(complex-)linear subspace of $\mathbb{C}^n$. Complex planes of complex dimension 1 are called *complex lines*.

A real plane T in $\mathbb{C}^n$ is said to be *totally real* if there is no complex line which is contained in T.

A C^1-submanifold of $\mathbb{C}^n$ is called *totally real* if the real tangent plane at every point of X is totally real.

1.4.17. Proposition. (i) *If X is a totally real C^1-submanifold of $\mathbb{C}^n$, then the real dimension of X is $\leqq n$.*

(ii) *If X is a C^1-submanifold of $\mathbb{C}^n$ and the real tangent plane of X at some point $\xi \in X$ is totally real, then there exists a neighbourhood U of ξ such that $U \cap X$ is totally real.*

The simple proof is left to the reader.

1.4.18. Theorem. *Let X be a totally real C^1-submanifold of $\mathbb{C}^n$. Then there exist a neighbourhood U of X and a non-negative strictly plurisubharmonic C^2-function ϱ in U such that*

$$X = \{z \in U : \varrho(z) = 0\} = \{z \in U : \mathrm{d}\varrho(z) = 0\}\,. \tag{1.4.26}$$

Proof. For $z \in X$, we denote by $t_z(X)$ the real tangent plane of X at z, and set

$$\hat{t}_z(X) := \{\zeta \in \mathbb{C}^n : \zeta + z \in t_z(X)\} .$$

Let P_z be the orthogonal real-linear projection from $\mathbb{C}^n$ onto the real-orthogonal complement of the real-linear subspace $\hat{t}_z(X)$. Since X is C^1, then P_z depends continuously on $z \in X$ and

$$P_z(\zeta - z) = o(|\zeta - z|) \quad \text{for} \quad \zeta, z \in X, |\zeta - z| \to 0 , \tag{1.4.27}$$

uniformly in every compact subset of X. Consequently,

$$(P_z(\zeta - z), \zeta - z)_{\mathbb{R}} = o(|\zeta - z|^2) \quad \text{for} \quad \zeta, z \in X, \qquad |\zeta - z| \to 0 , \tag{1.4.28}$$

uniformly in every compact subset of X, where $(\cdot, \cdot)_{\mathbb{R}}$ denotes the real scalar product in $\mathbb{C}^n = \mathbb{R}^{2n}$. By Whitney's extension theorem (see, for example, MALGRANGE [2]), it follows from (1.4.27) and (1.4.28) that there is a real-valued C^2-function ϱ on $\mathbb{C}^n$ such that at $z \in X$ we have the Taylor expansion

$$\varrho(\zeta) = (P_z(\zeta - z), \zeta - z)_{\mathbb{R}} + o(|\zeta - z|^2) . \tag{1.4.29}$$

This implies that, for some neighbourhood U of X, (1.4.26) holds.

It remains to prove that ϱ is strictly plurisubharmonic in some neighbourhood of X. By Definitions 1.4.6 (ii) and 1.4.5 we have to prove that for every point $z \in X$ and all $0 \neq w \in \mathbb{C}^n$

$$\left.\frac{\partial^2 \varrho(z + \lambda w)}{\partial \lambda\, \partial \bar{\lambda}}\right|_{\lambda=0} > 0 , \qquad \lambda \in \mathbb{C}^1 . \tag{1.4.30}$$

Let $\lambda = y_1 + iy_2$, $y_j \in \mathbb{R}$. Then it follows from (1.4.29) and the relation $\partial^2/\partial\lambda\, \partial\bar{\lambda} = (1/4)\,(\partial^2/\partial y_1^2 + \partial^2/\partial y_2^2)$ that

$$\left.\frac{\partial^2 \varrho(z + \lambda w)}{\partial \lambda\, \partial \bar{\lambda}}\right|_{\lambda=0} = \frac{1}{4} [(P_z(w), w)_{\mathbb{R}} + (P_z(iw), iw)_{\mathbb{R}}] .$$

Since P_z is a real-orthogonal projection, it follows that

$$\left.\frac{\partial^2 \varrho(z + \lambda w)}{\partial \lambda\, \partial \bar{\lambda}}\right|_{\lambda=0} = \frac{1}{4} [|P_z(w)|^2 + |P_z(iw)|^2] . \tag{1.4.31}$$

Since X is totally real and $w \neq 0$, at least one of the following relations is valid: $w \notin \hat{t}_z(X)$ or $iw \notin \hat{t}_z(X)$, that is, $|P_z(w)|^2 + |P_z(iw)|^2 \neq 0$. Together with (1.4.31) this implies (1.4.30). ■

1.4.19. Lemma. *Let ϱ be a strictly plurisubharmonic C^2-function in some neighbourhood of a compact set $K \subset\subset \mathbb{C}^n$. Then there exists an $\varepsilon > 0$ with the following property: If φ is a real-valued C^2-function in a neighbourhood of K such that*

$$\left|\frac{\partial^2 \varphi(z)}{\partial z_j\, \partial \bar{z}_k}\right| < \varepsilon \quad \textit{for all} \quad z \in K ,$$

then $\varrho + \varphi$ is strictly plurisubharmonic in a neighbourhood of K.

Proof. In view of Theorem 1.4.7 (ii), it is sufficient to set

$$\varepsilon := \frac{1}{n^2} \min_{z \in K, \xi \in \mathbb{C}^n, |\xi| = 1} \sum_{j,k=1}^{n} \frac{\partial^2 \varrho(z)}{\partial z_j\, \partial \bar{z}_k} \xi_j \bar{\xi}_k . \; ■$$

1.4.20. Corollary. *Let $X \subseteqq \mathbb{C}^n$ be a totally real C^1-submanifold of $\mathbb{C}^n$ and K a compact subset of X. Then there exists a non-negative strictly plurisubharmonic C^2-function ϱ in some neighbourhood U of K such that*

$$K = \{z \in U_K : \varrho(z) = 0\} . \tag{1.4.32}$$

Proof. By Theorem 1.4.18 there is a non-negative strictly plurisubharmonic C^2-function ϱ_X in some neighbourhood U_X of X such that $X = \{z \in U_X : \varrho_X(z) = 0\}$. Choose a C^∞-function $\varphi \geqq 0$ on $\mathbb{C}^n$ such that $K = \{z \in \mathbb{C}^n : \varphi(z) = 0\}$ (see, for example, Lemma 1.4.13 in Narasimhan [2]). By Lemma 1.4.19 we then can find $\varepsilon > 0$ such that $\varrho := \varrho_X + \varepsilon\varphi$ is strictly plurisubharmonic in some neighbourhood U of K. Clearly, $K = \{z \in U : \varrho(z) = 0\}$. ■

1.4.21. Theorem. *Let $D \subseteqq \mathbb{C}^n$ be an open set and ϱ a strictly plurisubharmonic C^2-function in D. Then, for every $\xi \in D$, there exist a neighbourhood $U \subseteqq D$ of ξ and a totally real C^1-submanifold X of U such that*

$$\{z \in U : \mathrm{d}\varrho(z) = 0\} \subseteqq X . \tag{1.4.33}$$

Proof. Without loss of generality we can assume that $\xi = 0$, $\varrho(0) = 0$ and $\mathrm{d}\varrho(0) = 0$. Then, by Theorem 1.4.15, we can also assume that for $\zeta \to 0$

$$\varrho(\zeta) = \sum_{j=1}^{n} (1 + \lambda_j) |x_j(\zeta)|^2 + \sum_{j=1}^{n} (1 - \lambda_i) |x_{j+n}(\zeta)|^2 + o(|\zeta|^2) , \tag{1.4.34}$$

where $\lambda_j \geqq 0$ and $x_j = x_j(\zeta)$ are the real coordinates of ζ such that $\zeta_j = x_j(\zeta) + ix_{j+n}(\zeta)$. Set $\varphi_j(\zeta) := \partial\varrho(\zeta)/\partial x_j$, $j = 1, \ldots, n$. Then, by (1.4.34), $\mathrm{d}\varphi_1 \wedge \ldots \wedge \mathrm{d}\varphi_n \neq 0$ in some neighbourhood $\tilde{U}$ of 0. Consequently, $\tilde{X} := \{z \in \tilde{U} : \varphi_1(z) = \ldots = \varphi_n(z) = 0\}$ is a C^1-submanifold of U. The real tangent plane of $\tilde{X}$ at 0 is the totally real plane $\{z \in \mathbb{C}^n : x_1(z) = \ldots = x_n(z) = 0\}$. Hence there exists a neighbourhood $U \subseteqq \tilde{U}$ of 0 such that $X = \tilde{X} \cap U$ is totally real. Relation (1.4.33) holds, because if $\mathrm{d}\varrho(z) = 0$, then, in particular, $\varphi_j(z) = \partial\varrho(z)/\partial x_j = 0$, $j = 1, \ldots, n$. ■

By Corollary 1.4.20 every compact subset of a totally real C^1-submanifold is the zero set of some non-negative strictly plurisubharmonic C^2-function. The following corollary shows that locally the converse is also valid.

1.4.22. Corollary. *Let $D \subseteqq \mathbb{C}^n$ be an open set and ϱ a non-negative strictly plurisubharmonic C^2-function in D. Then, for every point $\xi \in D$, there exist a neighbourhood $U \subseteqq D$ of ξ and a totally real C^1-submanifold X of U such that*

$$\{z \in U : \varrho(z) = 0\} \subseteqq \{z \in U : \mathrm{d}\varrho(z) = 0\} \subseteqq X . \tag{1.4.35}$$

Proof. This follows from Theorem 1.4.21, because ϱ is non-negative, and therefore $\mathrm{d}\varrho(z) = 0$ if $\varrho(z) = 0$. ■

1.5. Pseudoconvex sets

If $D \subsetneqq \mathbb{C}^n$ is an open set, then we write $\operatorname{dist}(z, \partial D) := \inf\{|z - \zeta| : \zeta \in \partial D\}$. It is clear that the function $D \ni z \to \operatorname{dist}(z, \partial D)$ is continuous and > 0 in D. Consequently, $-\ln \operatorname{dist}(z, \partial D)$ is a continuous real-valued function in D.

1.5.1. Definition. An open set $D \subsetneqq \mathbb{C}^n$ is said to be *pseudoconvex* if the function $-\ln \operatorname{dist}(z, \partial D)$ is plurisubharmonic in D. $\mathbb{C}^n$ is called *pseudoconvex*.

1.5.2. Corollary. (i) *If* $D \subseteqq \mathbb{C}^n$ *and* $G \subseteqq \mathbb{C}^m$ *are pseudoconvex open sets, then* $D \times G \subseteqq \mathbb{C}^{n+m}$ *is pseudoconvex.*

(ii) *If* $\{D_j\}_{j\in J}$ *is a family of pseudoconvex open sets in* $\mathbb{C}^n$, *then the interior of* $\bigcap_{j\in J} D_j$ *is pseudoconvex.*

Proof. This follows from Proposition 1.4.9 (iii). ■

1.5.3. Definition. Let $D \subseteqq \mathbb{C}^n$ be an open set and K a compact subset of D. Then we define

$$\hat{K}_D^P := \{z \in D : \varrho(z) \leqq \max_{\zeta\in K} \varrho(\zeta) \quad \text{for all} \quad \varrho \in P^0(D)\}\,,$$

where $P^0(D)$ is the set of all continuous plurisubharmonic functions in D. $\hat{K}_D^P$ is called the $P^0(D)$-*hull of* K. If $\hat{K}_D^P = K$, then K is said to be $P^0(D)$-*convex.*

1.5.4. Proposition. *Let* $D \subseteqq \mathbb{C}^n$ *be an open set and* K *a compact subset of* D. *Then*

$$\hat{K}_D^P \subseteqq \hat{K}_D^{\mathcal{O}}\,. \tag{1.5.1}$$

Proof. If $f \in \mathcal{O}(D)$, then, by Proposition 1.4.8, $|f| \in P^0(D)$. ■

Remark. In Section 2.7 we shall prove that, for pseudoconvex D, (1.5.1) is valid with equality. This fact is called the *solution of the Levi problem.*

1.5.5. Theorem. *If* $D \subseteqq \mathbb{C}^n$ *is an open set, then the following conditions are equivalent:*
(i) D *is pseudoconvex.*
(ii) *For every compact subset* K *of* D, $\hat{K}_D^P$ *is also a compact subset of* D.
(iii) *There exists a* $\varrho \in P^0(D)$ *such that, for every* $\alpha \in \mathbb{R}$,

$$D_\alpha := \{z \in D : \varrho(z) < \alpha\} \subset\subset D\,.$$

Proof. The case $D = \mathbb{C}^n$ is trivial. Then (i) holds by definition, (ii) follows from Proposition 1.5.4, and the function $|z|^2$ satisfies condition (iii).

Let $D \subsetneqq \mathbb{C}^n$. If (i) is valid, then the function $|z|^2 - \ln \operatorname{dist}(z, \partial D)$ fulfils condition (iii). It is clear that (iii) $\Rightarrow$ (ii). So we only have to prove that (ii) $\Rightarrow$ (i). Suppose that (ii) is fulfilled. We have to prove that $-\ln \operatorname{dist}(z, \partial D)$ is plurisubharmonic. By Theorem 1.4.2 (iii) it is sufficient to prove the following lemma:

Lemma. *Let* $\xi \in D$, $0 \neq w \in \mathbb{C}^n$ *and* $r > 0$ *so that*

$$\{\xi + \lambda w : \lambda \in \mathbb{C}^1, |\lambda| \leqq r\} \subseteqq D\,, \tag{1.5.2}$$

and let h be a harmonic function in some neighbourhood of the disc $|\lambda| \leqq r$ *in* $\mathbb{C}^1$ *such that*

$$h(\lambda) \geqq -\ln \operatorname{dist}(\xi + \lambda w, \partial D) \quad \textit{for} \quad |\lambda| = r\,. \tag{1.5.3}$$

Then

$$h(\lambda) \geqq -\ln \operatorname{dist}(\xi + \lambda w, \partial D) \quad \textit{for} \quad |\lambda| \leqq r\,. \tag{1.5.4}$$

Proof of the lemma. Let h^* be a harmonic function in some neighbourhood of $|\lambda| \leqq r$ such that $f := h + ih^*$ is holomorphic in a neighbourhood of $|\lambda| \leqq r$. Then (1.5.3) is equivalent to the inequality $\operatorname{dist}(\xi + \lambda w, \partial D) \geqq |e^{-f(\lambda)}|$ for $|\lambda| = r$, that is,

$$\xi + \lambda w + \zeta\, e^{-f(\lambda)} \in D \quad \text{for} \quad |\lambda| = r \quad \text{and} \quad \zeta \in \mathbb{C}^n \quad \text{with} \quad |\zeta| < 1\,. \tag{1.5.5}$$

Relation (1.5.4) is equivalent to

$$\xi + \lambda w + \zeta\, e^{-f(\lambda)} \in D \quad \text{for} \quad |\lambda| \leqq r \quad \text{and} \quad \zeta \in \mathbb{C}^n \quad \text{with} \quad |\zeta| < 1\,. \tag{1.5.6}$$

To prove (1.5.6), we fix some $\zeta \in \mathbb{C}^n$ with $|\zeta| < 1$ and define

$$\Gamma_t := \{\xi + \lambda w + t\zeta\, e^{-f(\lambda)} : |\lambda| \leqq r\} \quad \text{for} \quad 0 \leqq t \leqq 1\,.$$

Let T be the set of all $0 \leqq t \leqq 1$ such that $\Gamma_t \subseteqq D$. Then we have to prove that $1 \in T$. Since, by (1.5.2), $0 \in T$, it is sufficient to show that T is both open and closed in $[0, 1]$. That T is open is clear. To prove that T is closed, we consider the compact set

$$K := \{\xi + \lambda w + t\zeta\, e^{-f(\lambda)} : |\lambda| = r, 0 \leqq t \leqq 1\}\ .$$

By (1.5.5), $K \subseteqq D$ and in view of condition (ii), $\hat{K}_D^P$ is a compact subset of D. To prove that T is closed it is consequently sufficient to show that, for every $t \in T$, $\Gamma_t \subseteqq \hat{K}_D^P$.

Consider some $t \in T$. Then, for every $\varrho \in P^0(D)$, the function $\lambda \to \varrho(\xi + \lambda w + t\zeta\, e^{-f(\lambda)})$ is subharmonic in a neighbourhood of the disc $|\lambda| \leqq r$ (cp. Theorem 1.4.11 (i)). By (1.4.1) this implies that, for every $\varrho \in P^0(D)$ and all $|\lambda_0| \leqq r$,

$$\varrho(\xi + \lambda_0 w + t\zeta\, e^{-f(\lambda_0)}) \leqq \max_{|\lambda| = r} \varrho(\xi + \lambda w + t\zeta\, e^{-f(\lambda)}) \leqq \max_{\tau \in K} \varrho(\tau)\ ,$$

that is, $\Gamma_t \subseteqq \hat{K}_D^P$. ∎

1.5.6. Corollary. *Every domain of holomorphy $D \subseteqq \mathbb{C}^n$ is pseudoconvex.*

Proof. This follows from Theorem 1.5.5 (ii), Theorem 1.3.7 (iii) and Proposition 1.5.4. ∎

The following results shows that pseudoconvexity is a local property of the boundary.

1.5.7. Theorem. *Let $D \subseteqq \mathbb{C}^n$ be an open set. If, for every $\xi \in \partial D$, there is a neighbourhood U of ξ such that $U \cap D$ is pseudoconvex, then D is pseudoconvex.*

Proof. For every $\xi \in \partial D$ and every neighbourhood U of ξ we can find a neighbourhood $V \subseteqq U$ such that $\operatorname{dist}(z, \partial D) = \operatorname{dist}(z, \partial(U \cap D))$ for $z \in V$. Therefore, it follows from the hypothesis of the theorem that there is a neighbourhood W of ∂D such that $-\ln \operatorname{dist}(z, \partial D)$ is plurisubharmonic in $W \cap D$. Choose an increasing convex C^2-function $\varphi: \mathbb{R} \to \mathbb{R}$ such that $\varphi(|z|) > -\ln \operatorname{dist}(z, \partial D)$ for $z \in D \setminus W$ and $\varphi(t) \to \infty$ when $t \to \infty$. By Theorem 1.4.12 (i), $\varphi(|z|)$ is plurisubharmonic for all $z \in \mathbb{C}^n$. By Proposition 1.4.9 (iii), the function $\varrho(z) := \max(\varphi(|z|), -\ln \operatorname{dist}(z, \partial D))$ is plurisubharmonic for $z \in D$. Now the assertion follows from Theorem 1.5.5 (iii). ∎

1.5.8. Theorem. *Let $D \subseteqq \mathbb{C}^n$ be a pseudoconvex open set, and let K be a compact subset of D such that $K = \hat{K}_D^P$. Then, for every neighbourhood U_K of K, there exists a function ϱ satisfying the following conditions:*

(i) *ϱ is a strictly plurisubharmonic C^∞-function in D.*

(ii) *$\varrho < 0$ in K and $\varrho > 0$ in $D \setminus U_K$.*

(iii) *For every $\alpha \in \mathbb{R}$, $\{z \in D : \varrho(z) < \alpha\} \subset\subset D$.*

Proof. Let $\psi \in P^0(D)$ satisfy condition (iii) in Theorem 1.5.5. Adding a constant to ψ, we can assume that $\psi < 0$ in K. Consider the compact set $K' := \{z \in D : \psi(z) \leqq 0\}$. Since K is $P^0(D)$-convex, for every $z \in K' \cap (D \setminus U_K)$, we can find $\varphi_z \in P^0(D)$ such that $\varphi_z(z) > 0$ and $\varphi_z < 0$ in K. Since $K' \cap (D \setminus U_K)$ is compact, we can therefore choose a finite number of functions $\varphi_1, \ldots, \varphi_N \in P^0(D)$ such that $\max(\varphi_1, \ldots, \varphi_N) > 0$ on $K' \cap (D \setminus U_K)$ and $\max(\varphi_1, \ldots, \varphi_N) < 0$ on K. Set $\varphi := \max(\psi, \varphi_1, \ldots, \varphi_N)$. Then

$$\varphi < 0 \quad \text{in } K \quad \text{and} \quad \varphi > 0 \quad \text{in } D \setminus U_K\,, \tag{1.5.7}$$

and, by Proposition 1.4.9 (iii), $\varphi \in P^0(D)$. Clearly, φ satisfies conditions (ii) and (iii).

To obtain a strictly plurisubharmonic C^∞-function ϱ in D satisfying (ii) and (iii), we set $D_j := \{z \in D: \varphi(z) < j\}$, $j = 0, 1, 2, \ldots$, and construct a sequence ϱ_j $(j = 1, 2, \ldots)$ such that for $j = 1, 2, \ldots$ the following conditions are fulfilled:

(a) ϱ_j is a strictly plurisubharmonic C^∞-function in some neighbourhood of $\bar{D}_j$.

(b) $\varrho_j < 0$ in K.

(c) $\varrho_j \geqq \varphi$ in $\bar{D}_j$.

(d) $\varrho_j = \varrho_{j-1}$ in D_{j-2} if $j \geqq 2$.

Then, by setting $\varrho := \varrho_j$ in D_{j-2}, we obtain the required function.

At first we construct ϱ_1. By (1.5.7) we can choose $\varepsilon > 0$ so small that $\varphi + \varepsilon < 0$ in K, and by Theorem 1.4.10 we can find a plurisubharmonic C^∞-function $\tilde{\varrho}_1$ in some neighbourhood of $\bar{D}_1$ such that $|\tilde{\varrho}_1 - \varphi| < \varepsilon/2$ in $\bar{D}_1$. Then, for $\delta > 0$, $\tilde{\varrho}(z) + \delta|z|^2$ is strictly plurisubharmonic in a neighbourhood of $\bar{D}_1$, and if δ is sufficiently small, then the function $\varrho_1(z) := \tilde{\varrho}_1(z) + \delta\,|z^2| + \varepsilon/2$ has the required properties.

Now suppose that the functions $\varrho_1, \ldots, \varrho_{j-1}$ are already constructed, $j \geqq 2$. Since ϱ_{j-1} is C^∞ in some neighbourhood of $\bar{D}_{j-1}$, we can find a C^∞-function $\hat{\varrho}_{j-1}$ in $\mathbb{C}^n$ such that $\hat{\varrho}_{j-1} = \varrho_{j-1}$ in a neighbourhood of $\bar{D}_{j-1}$. Further, by Theorem 1.4.10, we can choose a plurisubharmonic C^∞-function $\tilde{\varrho}_j$ in a neighbourhood of $\bar{D}_j$ such that $|\tilde{\varrho}_j - \varphi| < 1/2$ in $\bar{D}_j$. Let $\delta > 0$ be so small that the strictly plurisubharmonic C^∞-function $\hat{\varrho}_j(z) := \tilde{\varrho}_j(z) + \delta\,|z|^2$ also fulfils the estimate $|\hat{\varrho}_j - \varphi| < 1/2$ on $\bar{D}_j$. Then $\hat{\varrho}_j < j - 3/2$ in $\bar{D}_{j-2}$ and $\hat{\varrho}_j > j - 3/2$ in $\bar{D}_j \setminus D_{j-1}$. Choose a convex C^∞-function $\chi: \mathbb{R} \to \mathbb{R}$ such that $\chi(t) = 0$ if $t \leqq j - 3/2$ and $\mathrm{d}\chi(t)/\mathrm{d}t > 0$ if $t > j - 3/2$. Then $\chi \circ \hat{\varrho}_j = 0$ in $\bar{D}_{j-2}$ and $\chi \circ \hat{\varrho}_j > 0$ in $\bar{D}_j \setminus D_{j-1}$, and it follows from Theorem 1.4.12 that $\chi \circ \hat{\varrho}_j$ is plurisubharmonic in a neighbourhood of $\bar{D}_j$ and strictly plurisubharmonic in a neighbourhood of $\bar{D}_j \setminus D_{j-1}$.

Now we claim that for sufficiently large $C \in \mathbb{R}$ the function $\varrho_j := \hat{\varrho}_{j-1} + C\chi \circ \hat{\varrho}_j$ fulfils conditions (a)—(d). Firstly observe that (d) and (b) are valid for every real C. In fact, since $\chi \circ \hat{\varrho}_j = 0$ and $\hat{\varrho}_{j-1} = \varrho_{j-1}$ in D_{j-2}, we obtain (d), and, since $K \subseteqq D_0 \subseteqq D_{j-2}$, this implies (b), because ϱ_{j-1} satisfies (b). That (c) is fulfilled for sufficiently large C follows from the fact that $\hat{\varrho}_{j-1}\,(= \varrho_{j-1}) \geqq \varphi$ in $\bar{D}_{j-1}$, $\chi \circ \hat{\varrho}_j \geqq 0$ on $\bar{D}_j$, and $\chi \circ \hat{\varrho}_j > 0$ on $\bar{D}_j \setminus D_{j-1}$. It remains to show that (a) is true if C is sufficiently large. Since $\hat{\varrho}_{j-1}\,(= \varrho_{j-1})$ is strictly plurisubharmonic and $\chi \circ \hat{\varrho}_j$ is plurisubharmonic in a neighbourhood of $\bar{D}_{j-1}$, we obtain that, for $C > 0$, ϱ_j is strictly plurisubharmonic in a neighbourhood of $\bar{D}_{j-1}$. Further, since $\chi \circ \varrho_j$ is strictly plurisubharmonic in a neighbourhood of $\bar{D}_j \setminus D_{j-1}$, it follows from Lemma 1.4.19 that ϱ_j is also strictly plurisubharmonic in a neighbourhood of $\bar{D}_j \setminus D_{j-1}$ if C is large enough. ■

Theorem 1.4.21 and Proposition 1.4.17 (i) show that the set of critical points of a strictly plurisubharmonic C^2-function is contained locally in a real C^1-submanifold of real dimension $\leqq n$. (A point z is called a *critical point* of a C^1-function ϱ if $\mathrm{d}\varrho(z) = 0$.) The following lemma of Morse shows that every strictly plurisubharmonic C^2-function can be uniformly approximated on compact sets by strictly plurisubharmonic C^2-functions whose critical points are isolated.

1.5.9. Lemma (Morse's lemma). *Let $D \subseteqq \mathbb{C}^n$ be an open set and ϱ a strictly plurisubharmonic C^2-function in D. Then, for every $\varepsilon > 0$, there exists a real linear map*

$L: \mathbb{C}^n \to \mathbb{R}^1$ such that

$$\max_{z \in \mathbb{C}^n, |z|=1} |L(z)| \leqq \varepsilon, \tag{1.5.8}$$

the set Crit $(\varrho + L) := \{z \in D : d(\varrho + L)(z) = 0\}$ *is discrete in* D[1]), *and* $\varrho + L$ *is strictly plurisubharmonic in* D.

Proof. Since ϱ is C^2, by a lemma of M. Morse (see, for example, § 2, Lemma A in Milnor [1]), for almost all real-linear maps $L: \mathbb{C}^n \to \mathbb{R}^1$ the critical points of $\varrho + L$ are non-degenerate and therefore isolated. In particular, this is the case for arbitrarily small L. That $\varrho + L$ is strictly plurisubharmonic follows from Theorem 1.4.7 (ii) and the fact that the second-order derivatives of L vanish. ■

1.5.10. Corollary. *Under the hypotheses of Theorem* 1.5.8 *there exists a function* ϱ *satisfying conditions* (i)—(iii) *in Theorem* 1.5.8 *and such that the set* Crit $(\varrho) := \{z \in D: d\varrho(z) = 0\}$ *is discrete in* D.

Proof. This follows immediately from Lemma 1.5.9. ■

1.5.11. Corollary. *Let $D \subseteqq \mathbb{C}^n$ be a pseudoconvex open set. Then, for every compact set $K \subset\subset D$, there exists a C^∞ strictly pseudoconvex*[2]) *open set D' such that $K \subset\subset D' \subset\subset D$.*

Proof. By Theorem 1.5.5, $\hat{K}_D^P$ is a compact subset of D. Therefore, by Corollary 1.5.10, we can find a strictly plurisubharmonic C^∞-function ϱ in D such that $\varrho < 0$ in $\hat{K}_D^P$, $\{z \in D : \varrho(z) < \alpha\} \subset\subset D$ for all $\alpha \in \mathbb{R}$, and the set $\{z \in D : d\varrho(z) = 0\}$ is discrete in D. Choose $\alpha' > 0$ so that $d\varrho(z) \neq 0$ for $z \in \partial D_{\alpha'}$ and set $D' := D_{\alpha'}$. ■

1.5.12. Proposition. *Let $D \subseteqq \mathbb{C}^n$ be an open set. If there exists a continuous plurisubharmonic function ϱ in some neighbourhood θ of ∂D such that*

$$D \cap \theta = \{z \in \theta : \varrho(z) < 0\}, \tag{1.5.9}$$

then D is pseudoconvex.

Proof. Let $\xi \in \partial D$ and $\varepsilon > 0$ so small that the ball $E := \{z \in \mathbb{C}^n : |z - \xi| < \varepsilon\}$ is relatively compact in θ. Since pseudoconvexity is a local property of the boundary (Theorem 1.5.7), we have to prove only that $E \cap D$ is pseudoconvex. To do so we consider the continuous plurisubharmonic function $\varphi(z) := \max\left(|z - \xi| - \varepsilon, \varrho(z)\right)$ is some neighbourhood of $\overline{D \cap E}$ (cp. Propositions 1.4.8 and 1.4.9 (iii)). Then $\varphi = 0$ on $\partial(D \cap E)$ but $\varphi < 0$ in $D \cap E$. In view of Theorem 1.5.5 (ii), this implies that $D \cap E$ is pseudoconvex. ■

We shall now examine when an open set with C^2-boundary is pseudoconvex. Firstly we introduce the notion of the complex tangent plane:

1.5.13. The complex tangent plane. Let $D \subseteqq \mathbb{C}^n$ be an open set with C^2-boundary ∂D, and let ϱ be a real-valued C^2-function in some neighbourhood $U_{\bar{D}}$ of $\bar{D}$ such that $D = \{z \in U_{\bar{D}} : \varrho(z) < 0\}$ and $d\varrho(z) \neq 0$ for $z \in \partial D$. For $\xi \in \partial D$ we denote by t_ξ the real tangent plane of ∂D at ξ, that is,

$$t_\xi := \left\{w \in \mathbb{C}^n : \sum_{j=1}^{2n} \frac{\partial \varrho(\xi)}{\partial x_j} x_j(w - \xi) = 0\right\},$$

---

[1]) A subset $M \subseteqq D$ is called *discrete* in D if for every $z \in M$ there is a neighbourhood U of z such that $U \cap M = \{z\}$.

[2]) See Definition 1.5.15.

where $x_j(\zeta)$ are the real coordinates of $\zeta \in \mathbb{C}^n$ such that $\zeta_j = x_j(\zeta) + i x_{j+n}(\zeta)$. The maximal complex plane (cp. Definition 1.4.16) which is contained in t_ξ is called the *complex tangent plane* of ∂D at ξ and will be denoted by T_ξ that is, $T_\xi := \{\xi + i\{t_\xi - \xi\}\} \cap t_\xi$.

Proposition.

$$T_\xi = \left\{ w \in \mathbb{C}^n : \sum_{j=1}^{n} \frac{\partial \varrho(\xi)}{\partial \xi_j} (w_j - \xi_j) = 0 \right\}. \tag{1.5.10}$$

Proof.

$$\sum_{j=1}^{n} \frac{\partial \varrho(\xi)}{\partial \xi_j} (w_j - \xi_j) = \frac{1}{2} \sum_{j=1}^{n} \left(\frac{\partial \varrho(\xi)}{\partial x_j} - i \frac{\partial \varrho(\xi)}{\partial x_{j+n}} \right) (x_j(w - \xi) + i x_{j+n}(w - \xi)) = 0$$

if and only if

$$\sum_{j=1}^{n} \frac{\partial \varrho(\xi)}{\partial x_j} x_j(w - \xi) + \sum_{j=1}^{n} \frac{\partial \varrho(\xi)}{\partial x_{j+n}} x_{j+n}(w - \xi) = 0$$

and

$$\sum_{j=1}^{n} \frac{\partial \varrho(\xi)}{\partial x_j} x_{j+n} (w - \xi) - \sum_{j=1}^{n} \frac{\partial \varrho(\xi)}{\partial x_{j+n}} x_j(w - \xi) = 0 .$$

Since $x_j(i(w - \xi)) = -x_{j+n}(w - \xi)$ and $x_{j+n}(i(w - \xi)) = x_j(w - \xi)$, $j = 1, \dots, n$, this implies that $\sum_{j=1}^{n} (\partial\varrho(\xi)/\partial\xi_j)(w_j - \xi_j) = 0$ if and only if $w \in t_\xi$ and $\xi + i(w - \xi) \in t_\xi$. ■

1.5.14. Theorem. *Let $D \subseteq \mathbb{C}^n$ be an open set with C^2-boundary ∂D, and let ϱ be a real-valued C^2-function in a neighbourhood $U_{\partial D}$ of ∂D such that $U_{\partial D} \cap D = \{z \in U_{\partial D} : \varrho(z) < 0\}$ and $d\varrho(z) \neq 0$ for $z \in \partial D$. Then D is pseudoconvex if and only if*

$$\sum_{j,k=1}^{n} \frac{\partial^2 \varrho(z)}{\partial z_j \partial \bar{z}_k} w_j \bar{w}_k \geqq 0 \quad \textit{for all} \quad z \in \partial D \quad \textit{and} \quad w \in \{T_z - z\} , \tag{1.5.11}$$

where T_z is the complex tangent plane of ∂D at z.

Proof. Let ϱ_1 be another C^2-function on $U_{\partial D}$ such that $U_{\partial D} \cap D = \{z \in U_{\partial D} : \varrho_1(z) < 0\}$ and $d\varrho_1(z) \neq 0$ for $z \in \partial D$. Then $\varrho_1 = \varphi\varrho$, where φ is a C^1-function and > 0 on $U_{\partial D}$. Consequently, for $z \in \partial D$ and $w \in \{T_z - z\}$,

$$\sum_{j,k=1}^{n} \frac{\partial^2 \varrho_1(z)}{\partial z_j \partial \bar{z}_k} w_j \bar{w}_k = \varphi(z) \sum_{j,k=1}^{n} \frac{\partial^2 \varrho(z)}{\partial z_j \partial \bar{z}_k} w_j \bar{w}_k .$$

This implies that (1.5.11) is independent of the choice of ϱ.

To prove that (1.5.11) is necessary for pseudoconvexity of D, we set

$$\varrho(z) := \begin{cases} -\operatorname{dist}(z, \partial D) & \text{if } z \in D, \\ +\operatorname{dist}(z, \partial D) & \text{if } z \in \mathbb{C}^n \setminus D. \end{cases} \tag{1.5.12}$$

Since ∂D is C^2, then ϱ is C^2 in some neighbourhood $U_{\partial D}$ of ∂D. If D is pseudoconvex, then by definition, $-\ln(-\varrho)$ is plurisubharmonic in D. By Theorem 1.4.7 (i), this implies that

$$\sum_{j,k=1}^{n} \left[\frac{1}{\varrho^2} \frac{\partial \varrho(z)}{\partial z_j} \frac{\partial \varrho(z)}{\partial \bar{z}_k} - \frac{1}{\varrho} \frac{\partial^2 \varrho(z)}{\partial z_j \partial \bar{z}_k} \right] w_j \bar{w}_k \geqq 0$$

for all $z \in U_{\partial D} \cap D$ and $w \in \mathbb{C}^n$. This implies that for $z \in U_{\partial D} \cap D$ and $z + w \in T_z$

$$\sum_{j,k=1}^{2} \frac{\partial^2 \varrho(z)}{\partial z_j \partial \bar{z}_k} w_j \bar{w}_k \geqq 0 .$$

A passage to the limit shows that this is also true for $z \in \partial D$.

Now we prove that (1.5.11) is sufficient for pseudoconvexity of D. To do this we assume that D is not pseudoconvex. Let $U_{\partial D}$ be a neighbourhood of ∂D such that the function ϱ defined by (1.5.12) is C^2 in $U_{\partial D}$. By Theorem 1.5.7, then there exists a point $\xi \in U_{\partial D} \cap D$ such that $-\ln(-\varrho)$ is not plurisubharmonic at ξ. This means, by Theorem 1.4.3, that for some $0 \neq w \in \mathbb{C}^n$,

$$\gamma := \frac{\partial^2}{\partial\lambda\,\partial\bar\lambda} \ln\left(-\varrho(\xi + \lambda w)\right)\big|_{\lambda=0} > 0 .$$

Then we have by Taylor's formula (cp. also Lemma 1.4.13) that for $\lambda \to 0$

$$\ln\left(-\varrho(\xi + \lambda w)\right) = \ln|\varrho(\xi)| + \operatorname{Re}(\alpha\lambda + \beta\lambda^2) + \gamma|\lambda|^2 + o(|\lambda|^2) , \tag{1.5.13}$$

where α, β are constants. Now we choose $\eta \in C^n$ such that $|\eta| = |\varrho(\xi)|$ and $\xi + \eta \in \partial D$. Define $\zeta_s(\lambda) := \xi + \lambda w + s\eta\, e^{\alpha\lambda + \beta\lambda^2}$ if $0 < s \leqq 1$. Then by (1.5.13) we can find an $\varepsilon > 0$ such that for $|\lambda| \leqq \varepsilon$ and $0 < s \leqq 1$

$$\operatorname{dist}\left(\zeta_s(\lambda), \partial D\right) \geqq -\varrho(\xi + \lambda w) - s|\eta|\,|e^{\alpha\lambda + \beta\lambda^2}| \geqq |\varrho(\xi)|\,(e^{\gamma|\lambda|^2/2} - s)\,|e^{\alpha\lambda+\beta\lambda^2}| . \tag{1.5.14}$$

This implies that dist $(\zeta_s(\lambda), \partial D) > 0$ if $0 < s < 1$ and $|\lambda| \leqq \varepsilon$. Since $\zeta_s(0) = \xi + s\eta \in D$ for $0 < s < 1$, we conclude that $\zeta_s(\lambda) \in D$ for all $0 < s < 1$ and $|\lambda| \leqq \varepsilon$. Consequently a passage to the limit shows that $\zeta_1(\lambda) \in \bar D$ for all $|\lambda| \leqq \varepsilon$. Since $\xi \in U_{\partial D}$, after shrinking ε, we can assume that $\zeta_1(\lambda) \in \bar D \cap U_{\partial D}$ for $|\lambda| \leqq \varepsilon$. Therefore, $\varrho(\zeta_1(\lambda)) = -\operatorname{dist}(\zeta_1(\lambda), \partial D)$ for $|\lambda| \leqq \varepsilon$, and by (1.5.14)

$$-\varrho(\zeta_1(\lambda)) \geqq |\varrho(\xi)|\,(e^{\gamma|\lambda|^2/2} - 1)\,|e^{\alpha\lambda+\beta\lambda^2}| \quad \text{for} \quad |\lambda| \leqq \varepsilon .$$

The function on the right-hand side is strictly convex and $\geqq 0$ in some neighbourhood of $\lambda = 0$. Since $\varrho(\zeta_1(0)) = 0$, it follows that $-\varrho \circ \zeta_1$ is strictly convex at 0 and $d(\varrho \circ \zeta_1)(0) = 0$. In particular,

$$\frac{\partial^2\varrho(\zeta_1(\lambda))}{\partial\lambda\,\partial\bar\lambda}\bigg|_{\lambda=0} < 0 \quad \text{and} \quad \frac{\partial\varrho(\zeta_1(\lambda))}{\partial\lambda}\bigg|_{\lambda=0} = 0 .$$

Since $\zeta_1(\lambda)$ is an analytic function of λ, we conclude that (cp. Proposition 1.1.16)

$$\sum_{j,k=1}^{n} \frac{\partial^2\varrho(z)}{\partial z_j\,\partial\bar z_k}\bigg|_{z=\xi+\eta} v_j\bar v_k < 0 \quad \text{and} \quad \sum_{j=1}^{n} \frac{\partial\varrho(z)}{\partial z_j}\bigg|_{z=\xi+\eta} v_j = 0 .$$

By (1.5.10), this contradicts (1.5.11) at $\xi + \eta \in \partial D$. ■

In this book, pseudoconvex open sets are of special interest which can be defined by strictly plurisubharmonic C^2-functions:

1.5.15. Definition. Let $D \subset\subset \mathbb{C}^n$ be an open set. D is called *strictly pseudoconvex* if there exists a strictly plurisubharmonic C^2-function ϱ in some neighbourhood θ of the boundary of D such that $D \cap \theta = \{z \in \theta : \varrho(z) < 0\}$. If, moreover, the boundary of D is smooth of class C^k ($k = 2, 3, \ldots,$), then D is said to be a *C^k strictly pseudoconvex open set.*

Remark. The boundary of a strictly pseudoconvex open set $D \subset\subset \mathbb{C}^n$ need not be smooth. Example: $\{z = x_1 + ix_2 \in \mathbb{C}^1 : 2x_1^2 - x_2^2 + x_2^4 < 0\}$. However, it is interesting that, by Theorem 1.4.21 and Proposition 1.4.17 (i), the set of points where ∂D is not smooth is not too „large", namely, it is locally contained in a real C^1-submanifold of real dimension $\leqq n$. Observe also that strictly pseudoconvex open sets $D \subset\subset \mathbb{C}^n$ may consist of infinitely many components. If the function ϱ in Definition 1.5.15 is strictly convex at some point $\xi \in \partial D$, then $d\varrho(\xi) \neq 0$, because ϱ has not a local minimum at ξ. The example above shows that this is not the case when ϱ is strictly plurisubharmonic only. However we have the following

1.5.16. Proposition. *Let $D \subset\subset \mathbb{C}^n$ be a strictly pseudoconvex open set, and let ϱ be a strictly plurisubharmonic C^2-function in a neighbourhood θ of ∂D such that $D \cap \theta = \{z \in \theta : \varrho(z) < 0\}$. If the boundary of D is of class C^2, then $d\varrho(z) \neq 0$ for all $z \in \partial D$.*

Proof. By hypothesis there is a real-valued C^2-function φ in $\mathbb{C}^n$ such that $\varphi < 0$ in D, $\varphi > 0$ in $\mathbb{C}^n \setminus D$, and $d\varphi(z) \neq 0$ for $z \in \partial D$. Then $\psi := \varrho/\varphi$ is a C^1-function in θ and $d\varrho(z) = \psi(z)\, d\varphi(z)$ for $z \in \partial D$. Therefore, we have to prove that $\psi \neq 0$ on ∂D. Assume that $\psi(\xi) = 0$ for some $\xi \in \partial D$. Since $\psi \geqq 0$, then $d\psi(\xi) = 0$, too. Since also $\varphi(\xi) = 0$, we conclude that all second-order derivatives of $\varrho = \psi\varphi$ vanish at ξ. This is not possible, because ϱ is strictly plurisubharmonic at ξ (cp. Theorem 1.4.7 (ii)).[1] ∎

We shall now examine when an open set with C^2-boundary is strictly pseudoconvex.

1.5.17. Theorem. *Let $D \subset\subset \mathbb{C}^n$ be an open set with C^2-boundary and ϱ a real-valued C^2-function in a neighbourhood $U_{\partial D}$ of ∂D such that $U_{\partial D} \cap D = \{z \in U_{\partial D} : \varrho(z) < 0\}$ and $d\varrho(z) \neq 0$ for $z \in \partial D$. Then D is strictly pseudoconvex if and only if*

$$\sum_{j,k=1}^{n} \frac{\partial^2 \varrho(z)}{\partial z_j \partial \bar{z}_k} w_j \bar{w}_k > 0 \quad \textit{for all} \quad z \in \partial D \quad \textit{and} \quad 0 \neq w \in \{T_z - z\}\,, \tag{1.5.15}$$

where T_z is the complex tangent plane of ∂D at z (see Subsection 1.5.13). *If condition* (1.5.15) *is fulfilled, then, for sufficiently large $\lambda > 0$, the function*

$$\tilde{\varrho} := e^{\lambda\varrho} - 1 \tag{1.5.16}$$

is strictly plurisubharmonic in some neighbourhood $V_{\partial D} \subseteqq U_{\partial D}$ of ∂D, and $D \cap V_{\partial D} = \{z \in V_{\partial D} : \tilde{\varrho}(z) < 0\}$.

Proof. As in the beginning of the proof of Theorem 1.5.14 we see that condition (1.5.15) is independent of the choice of ϱ. If D is strictly pseudoconvex, then, by definition and Proposition 1.5.16, ϱ can be chosen to be strictly plurisubharmonic. By Theorem 1.4.7 (ii) this implies that condition (1.5.15) is necessary.

Assume now that (1.5.15) is fulfilled. We only have to prove that, for sufficiently large λ, the function $\tilde{\varrho}$ defined by (1.5.16) is strictly plurisubharmonic in some neighbourhood of ∂D, because, clearly, $D \cap U_{\partial D} = \{z \in U_{\partial D} : \tilde{\varrho}(z) < 0\}$. We have

$$\frac{\partial^2 \tilde{\varrho}(z)}{\partial z_j \partial \bar{z}_k} = \lambda \frac{\partial^2 \varrho(z)}{\partial z_j \partial \bar{z}_k} e^{\lambda\varrho} + \lambda^2 \frac{\partial \varrho(z)}{\partial z_j} \frac{\partial \varrho(z)}{\partial \bar{z}_k} e^{\lambda\varrho}\,.$$

Consequently, for $z \in \partial D$ and all $w \in \mathbb{C}^n$,

$$\sum_{j,k=1}^{n} \frac{\partial^2 \tilde{\varrho}(z)}{\partial z_j \partial \bar{z}_k} w_j \bar{w}_k = \lambda \sum_{j,k=1}^{n} \frac{\partial^2 \varrho(z)}{\partial z_j \partial \bar{z}_k} w_j \bar{w}_k + \lambda^2 \left| \sum_{j=1}^{n} \frac{\partial \varrho(z)}{\partial z_j} w_j \right|^2. \tag{1.5.17}$$

---

[1]) This proof shows that the following more general proposition is true: Let $D \subseteqq \mathbb{R}^m$ be an open set with C^2-boundary, and let ϱ be a real-valued C^2-function in some neighbourhood $U_{\partial D}$ of ∂D such that $D \cap U_{\partial D} = \{x \in U_{\partial D} : \varrho(x) < 0\}$. If, for some $x \in \partial D$, $d\varrho(x) = 0$, then the second-order derivatives of ϱ at x vanish.

Set

$$K := \left\{(w, z) \in \mathbb{C}^n \times \partial D : |w| = 1 \text{ and } \sum_{j,k=1}^{n} \frac{\partial^2 \varrho(z)}{\partial z_j \, \partial \bar{z}_k} w_j \bar{w}_k \leqq 0\right\}.$$

Since K is compact, and since by (1.5.15),

$$\left|\sum_{j=1}^{n} \frac{\partial \varrho(z)}{\partial z_j} w_j\right| > 0 \quad \text{for} \quad (w, z) \in K,$$

we can choose λ so large that

$$\max_{(w,z)\in K} \left|\sum_{j,k=1}^{n} \frac{\partial^2 \varrho(z)}{\partial z_j \, \partial \bar{z}_k} w_j \bar{w}_k\right| < \lambda \min_{(w,z)\in K} \left|\sum_{j=1}^{n} \frac{\partial \varrho(z)}{\partial z_j} w_j\right|.$$

By (1.5.17) and Theorem 1.4.7 (ii), then $\tilde{\varrho}$ is strictly plurisubharmonic in some neighbourhood of ∂D. ■

1.5.18. Corollary. *Let $D \subset\subset \mathbb{C}^n$ be a strictly pseudoconvex open set with C^k-boundary ($k = 2, 3, \dots, \infty$). Then there exists a strictly plurisubharmonic C^k-function ϱ in some neighbourhood $U_{\partial D}$ of ∂D such that $D \cap U_{\partial D} = \{z \in U_{\partial D} : \varrho(z) < 0\}$.*

Proof. By hypothesis the function ϱ in Theorem 1.5.17 can be chosen to be C^k. Then, for sufficiently large λ, the function $\tilde{\varrho}$ defined by (1.5.16) has the required properties. ■

1.5.19. Theorem. *For every strictly pseudoconvex open set $D \subset\subset \mathbb{C}^n$ there exists a strictly plurisubharmonic C^2-function ϱ in some neighbourhood $U_{\bar{D}}$ of $\bar{D}$ such that*

$$D = \{z \in U_{\bar{D}} : \varrho(z) < 0\} \tag{1.5.18}$$

and

$$\partial D = \{z \in U_{\bar{D}} : \varrho(z) = 0\}. \tag{1.5.19}$$

If ∂D is of class C^k ($k = 2, 3, \dots, \infty$), then this function ϱ can be chosen to be C^k in $U_{\bar{D}}$.

Proof. By definition there is a strictly plurisubharmonic C^2-function ϱ_0 in some neighbourhood $U_{\partial D}$ of ∂D such that $D \cap U_{\partial D} = \{z \in U_{\partial D} : \varrho_0(z) < 0\}$. Let $\delta > 0$ so small that $K_\delta := \{z \in U_{\partial D} : -\delta \leqq \varrho_0(z) \leqq 0\}$ is a compact subset of $U_{\partial D}$. Choose a C^∞-function $\chi: \mathbb{R} \to \mathbb{R}$ with the following properties: $\chi(t) = -\delta$ if $t \leqq -\delta$, $\chi(0) = 0$, $d^2\chi(t)/dt^2 \geqq 0$ for all $-\infty < t < \infty$, and $d\chi(t)/dt > 0$ if $-\delta < t < \infty$. Define $\varrho_1 := -\delta$ in $D \setminus K_\delta$ and $\varrho_1 := \chi \circ \varrho_0$ in $U_{\partial D}$. Then

$$D = \{z \in D \cup U_{\partial D} : \varrho_1(z) < 0\} \tag{1.5.20}$$

and, by Theorem 1.4.12, ϱ_1 is a plurisubharmonic C^2-function in $D \cup U_{\partial D}$ that is strictly plurisubharmonic for all $z \in U_{\partial D}$ with $\varrho_0(z) > -\delta$.

By Theorem 1.5.8 there is a strictly plurisubharmonic C^∞-function ϱ_2 in D such that $D_\alpha := \{z \in D : \varrho_2(z) < \alpha\} \subset\subset D$ for all $\alpha \in \mathbb{R}$. Choose $\beta \in \mathbb{R}$ so large that $\varrho_1 > -\delta/2$ on $D \setminus D_\beta$, and let ψ be a real-valued C^∞-function on $\mathbb{C}^n$ such that $\psi = 1$ in some neighbourhood of $\bar{D}_\beta$ and $\psi = 0$ in a neighbourhood of $\mathbb{C}^n \setminus D$. Define $\tilde{\varrho}_2 := \psi\varrho_2$ in D and $\tilde{\varrho}_2 := 0$ in $\mathbb{C}^n \setminus D$. Then $\tilde{\varrho}_2$ is strictly plurisubharmonic in a neighbourhood of $\bar{D}_\beta$. Consequently, for every $c > 0$, $\varrho_1 + c\varrho_2$ is strictly plurisubharmonic in a neighbourhood of $\bar{D}_\beta$. Since ϱ_1 is strictly plurisubharmonic in $U_{\partial D} \setminus D_\beta$, and since $\tilde{\varrho}_2 = 0$ in $U_{\partial D} \setminus D$, it follows from Lemma 1.4.19 that, for sufficiently small $c > 0$, $\varrho := \varrho_1 + c\tilde{\varrho}_2$ is a strictly plurisubharmonic C^2-function in $U_{\bar{D}} := D \cup U_{\partial D}$. Moreover, since $\tilde{\varrho}_2 = 0$ in a neighbourhood of $\mathbb{C}^n \setminus D$, it follows from (1.5.20) that (1.5.18) is fulfilled if $c > 0$ is sufficiently small.

It remains to satisfy (1.5.19). To do this, we choose a C^∞-function $\varphi \geqq 0$ on $\mathbb{C}^n$ such that $\bar{D} = \{z \in \mathbb{C}^n : \varphi(z) = 0\}$ (see, for example, Lemma 1.4.13 in NARASIMHAM [2]). Then, by Lemma 1.4.19, for sufficiently small $\varepsilon > 0$ and after shrinking $U_{\bar{D}}$, $\varrho + \varepsilon\varphi$ is strictly plurisubharmonic in $U_{\bar{D}}$, and (1.5.18) and (1.5.19) are satisfied if we replace ϱ by $\varrho + \varepsilon\varphi$.

Finally, observe that if ∂D is C^k, then, by Corollary 1.5.18, ϱ_0 can be chosen to be C^k, which implies that $\varrho + \varepsilon\varphi$ is C^k. ■

1.5.20. Corollary. *Let $D \subset\subset \mathbb{C}^n$ be a strictly pseudoconvex open set. Then, for every compact set $K \subseteqq \partial D$ and every neighbourhood $U_K \subseteqq \mathbb{C}^n$ of K, there exists a strictly pseudoconvex open set $\tilde{D} \subseteqq \mathbb{C}^n$ such that $D \cup K \subseteqq \tilde{D} \subseteqq D \cup U_K$. If the boundary of D is of class C^k ($k = 2, 3, \ldots, \infty$), then this strictly pseudoconvex open set $\tilde{D}$ can be chosen with C^k-boundary, too.*

Proof. Let ϱ and $U_{\bar{D}}$ be as in Theorem 1.5.19. Choose a non-negative real-valued $\chi \in C_0^\infty(U_K \cap U_{\bar{D}})$ such that $\chi > 0$ on K. Then, by Lemma 1.4.19 and Proposition 1.5.16, for sufficiently small $\varepsilon > 0$, $\tilde{D} := \{z \in U_{\bar{D}} : \varrho(z) - \varepsilon\chi(z) < 0\}$ has the required properties. ■

1.5.21. Theorem. *For every strictly pseudoconvex open set $D \subset\subset \mathbb{C}^n$ and every neighbourhood $U_{\bar{D}}$ of $\bar{D}$, there exists a C^∞ strictly pseudoconvex open set $\tilde{D} \subset\subset \mathbb{C}^n$ such that $\bar{D} \subseteqq \tilde{D} \subseteqq U_{\bar{D}}$.*

Proof. By Theorem 1.5.19, after shrinking $U_{\bar{D}}$, we can find a strictly plurisubharmonic C^2-function ϱ in $U_{\bar{D}}$ satisfying (1.5.18) and (1.5.19). Then there exists $\varepsilon_0 > 0$ so small that $D_{\varepsilon_0} := \{z \in U_{\bar{D}} : \varrho(z) < 3\varepsilon_0\}$ is a relatively compact subset of $U_{\bar{D}}$. In view of the Approximation Theorem 1.4.10, we can choose a plurisubharmonic C^∞-function φ in some neighbourhood of $\bar{D}_{\varepsilon_0}$ which is so close to ϱ on $\bar{D}_{\varepsilon_0}$ that, for some $\delta > 0$, the strictly plurisubharmonic function $\tilde{\varrho}(z) := \varphi(z) + \delta\,|z|^2$ satisfies the condition

$$D \subseteqq \tilde{D}_\varepsilon := \{z \in D_{\varepsilon_0} : \tilde{\varrho}(z) < \varepsilon\} \subseteqq D_{\varepsilon_0} \quad \text{for all} \quad \varepsilon_0 \leqq \varepsilon \leqq 2\varepsilon_0 .$$

In view of Morse's Lemma 1.5.9, we can moreover assume that $d\tilde{\varrho}(z) \neq 0$ for $z \in \bar{D}_{\varepsilon_0}$ except for not more than a finite number of points. Consequently, we can choose $\varepsilon_0 \leqq \varepsilon_1 \leqq 2\varepsilon_0$ so that $d\tilde{\varrho}(z) \neq 0$ for all $z \in \partial\tilde{D}_{\varepsilon_1}$. Setting $\tilde{D} := \tilde{D}_{\varepsilon_1}$ we conclude the proof. ■

Remark. Theorem 1.5.21 is not valid without the hypothesis on strict pseudoconvexity. DIEDERICH and FORNAESS [1] constructed an example of a smooth pseudoconvex open set $D \subset\subset \mathbb{C}^n$ without a basis of pseudoconvex neighbourhoods of $\bar{D}$.

Since every strictly convex C^2-function is strictly plurisubharmonic, we obtain

1.5.22. Proposition. *Every C^2 strictly convex open set $D \subset\subset \mathbb{C}^n$ is strictly pseudoconvex.*

It is interesting that, after an appropriate local change of holomorphic coordinates, the converse is also true. This will be proved below (Theorem 1.5.24). To do this we need part (i) of the following lemma. (Part (ii) will be needed later.)

1.5.23. Lemma. (i) *Let $G \subseteqq \mathbb{C}^n$ be an open set and $\varrho : G \to \mathbb{R}$ a strictly convex C^k-function ($k = 2, 3, \ldots, \infty$). Set $G_- := \{z \in G : \varrho(z) < 0\}$. Then, for every convex compact*

set $K \subseteqq \overline{G}_- \cap G$ and every neighbourhood U_K of K, there exists a strictly convex open set D with C^k-boundary such that $K \subseteqq \overline{D} \subseteqq U_K \cap \overline{G}_-$.

(ii) *Let $G \subseteqq \mathbb{C}^n$ be an open set, and let ϱ be a strictly plurisubharmonic C^k-function in G ($k = 2, 3, \dots, \infty$). Set $G_- := \{z \in G : \varrho(z) < 0\}$. Further, let $K \subseteqq \overline{G}_- \cap G$ be a compact set with a basis of pseudoconvex neighbourhoods. Then, for every neighbourhood U_K of K, there exists a strictly pseudoconvex open set $D \subset\subset \mathbb{C}^n$ such that $K \subseteqq \overline{D} \subseteqq U_K \cap \overline{G}_-$. Moreover, if $\mathrm{d}\varrho(z) \neq 0$ for all $z \in \Gamma := \{z \in G : \varrho(z) = 0\}$, then this set can be chosen with C^k-boundary.*

Proof. (i) Since K is convex, we can find a C^∞ strictly convex open set D_1^0 such that $K \subset\subset D_1^0 \subset\subset U_K \cap G$. Choose a real-valued strictly convex C^∞-function ϱ_1 in some neighbourhood $U_1 \subset\subset U_K \cap G$ of $\overline{D}_1^0$ such that $D_1^0 = \{z \in U_1 : \varrho_1(z) < 0\}$. Since no point in ∂D_1^0 is a local minimum of ϱ_1, and since ϱ_1 is strictly convex, then $\mathrm{d}\varrho_1(\zeta) \neq 0$ for $\zeta \in \partial D_1^0$. Therefore, we can find $\varepsilon > 0$ so small that $D_1^\varepsilon := \{z \in U_1 : \varrho_1(z) < \varepsilon\} \subset\subset U_1$.

Now we choose C^∞-functions $f, g: \mathbb{R} \to \mathbb{R}$ with the following properties: $\mathrm{d}f(t)/\mathrm{d}t > 0$ and $\mathrm{d}^2f(t)/\mathrm{d}t^2 \geqq 0$ for all $t \in \mathbb{R}$, $-1 < f(t) < 0$ if $t < 0$, $f(0) = 0$, $f(t) > 0$ if $t > 0$; $\mathrm{d}g(t)/\mathrm{d}t \geqq 0$ and $\mathrm{d}^2g(t)/\mathrm{d}t^2 \geqq 0$ for all $t \in \mathbb{R}$, $g(t) = 0$ if $t \leqq 0$, $g(t) > 0$ if $t > 0$, $g(t) > 1$ if $t > \varepsilon$.

Set $\varphi(x, y) := f(x) + g(y)$, $x, y \in \mathbb{R}$. Then φ is a convex C^∞-function on $\mathbb{R}^2$, $\partial\varphi(x, y)/\partial x > 0$ and $\partial\varphi(x, y)/\partial y \geqq 0$ on $\mathbb{R}^2$, $\varphi(x, y) > 0$ if $\max(x, y - \varepsilon) > 0$, $\varphi(x, y) < 0$ if $x < 0$ and $y \leqq 0$.

Then $\psi(z) := \varphi(\varrho(z), \varrho_1(z))$ is a strictly convex C^k-function in U_1. Define $D := \{z \in U_1 : \psi(z) < 0\}$. Then $D_1^0 \cap G_- \subseteqq D \subseteqq D_1^\varepsilon \cap G_-$, and therefore $K \subseteqq \overline{D} \subseteqq U_K \cap \overline{G}_-$. Since ψ is strictly convex, and since no point in ∂D is a local minimum of ψ, we have $\mathrm{d}\psi(z) \neq 0$ for $z \in \partial D$. Consequently, D is a C^k strictly convex open set.

(ii) Since K has a basis of pseudoconvex neighbourhoods, by Corollary 1.5.11, we can find a C^∞ strictly pseudoconvex open set D_1^0 such that $K \subset\subset D_1^0 \subset\subset U_K \cap G$. In view of Theorem 1.5.19, there is a strictly plurisubharmonic C^∞-function ϱ_1 in some neighbourhood $U_1 \subset\subset U_K \cap G$ of D_1^0 such that $D_1^0 = \{z \in U_1 : \varrho_1(z) < 0\}$. Choose $\varepsilon > 0$ so small that $D_1^\varepsilon := \{z \in U_1 : \varrho_1(z) < \varepsilon\} \subset\subset U_1$. (This is possible, because, by Proposition 1.5.16, $\mathrm{d}\varrho_1(z) \neq 0$ for all $z \in \partial D_1^0$.)

Define $\psi(z) := \varphi(\varrho(z), \varrho_1(z))$, $z \in U_1$, where $\varphi(x, y) = f(x) + g(y)$ is the same function as in the proof of part (i). Then, by Theorem 1.4.12 (ii), ψ is strictly plurisubharmonic in U_1, and $D := \{z \in U_1 : \psi(z) < 0\}$ is strictly pseudoconvex, where $D_1^0 \cap G_- \subseteqq D \subseteqq D_1^\varepsilon \cap G_-$ and, therefore, $K \subseteqq \overline{D} \subseteqq U_K \cap \overline{G}_-$.

Now we assume that $\mathrm{d}\varrho(z) \neq 0$ for all $z \in \Gamma$. If $z \in \Gamma \cap \partial D$, then $\varrho(z) = 0$ and $\psi(z) = f(\varrho(z)) + g(\varrho_1(z)) = 0$. Since $f(0) = 0$ and $g(t) > 0$ for $t > 0$, this implies that $\varrho(z) = 0$ and $\varrho_1(z) \leqq 0$ for $z \in \Gamma \cap \partial D$. Since $g(t) = 0$ if $t \leqq 0$ and, therefore, $\mathrm{d}g(t)/\mathrm{d}t = 0$ if $t \leqq 0$, we conclude that

$$\mathrm{d}\psi(z) = \left.\frac{\mathrm{d}f(t)}{\mathrm{d}t}\right|_{t=\varrho(z)} \mathrm{d}\varrho(z) \quad \text{for all} \quad z \in \Gamma \cap \partial D\,.$$

Since $\mathrm{d}f(t)/\mathrm{d}t > 0$ for all $t \in \mathbb{R}$, and since $\mathrm{d}\varrho(z) \neq 0$ for all $z \in \Gamma$, it follows that $\mathrm{d}\psi(z) \neq 0$ if $z \in \Gamma \cap \partial D$. Choose a neighbourhood $V \subset\subset U_1$ of $\Gamma \cap \partial D$ such that $\mathrm{d}\psi(z) \neq 0$ for all $z \in \overline{V}$. Let $\chi \geqq 0$ be a C^∞-function on $\mathbb{C}^n$ such that $\chi = 1$ in $\mathbb{C}^n \setminus V$ and $\chi = 0$ in some neighbourhood $W \subset\subset V$ of $\Gamma \cap \partial D$. In view of Morse's Lemma 1.5.9, there is an arbitrarily small real-linear map $L: \mathbb{C}^n \to \mathbb{R}^1$ such that the critical points of $\psi + L$ are isolated. If $\varepsilon_0 := \sup_{z \in U_1} |L(z)|$, then we can choose $\varepsilon_0 \leqq \varepsilon \leqq 2\varepsilon_0$ so that $d(\psi + L + \varepsilon)(z) \neq 0$ if $(\psi + L + \varepsilon)(z) = 0$, $z \in U_1$. Since $V \subset\subset U_1$, $\chi = 1$ on $U_1 \setminus V$ and $\psi + L + \varepsilon$ is strictly plurisubharmonic in U_1, by Lemma 1.4.19, we can choose L (and therefore ε) so small that $\psi + \chi(L + \varepsilon)$ is strictly plurisubharmonic in U_1. Further, for sufficiently small L, $d(\psi + \chi(L + \varepsilon))(z) \neq 0$ if $(\psi + \chi(L + \varepsilon))(z) = 0$, $z \in U_1$. Since, moreover, $\chi(L + \varepsilon) \geqq 0$ on U_1 ($\varepsilon \geqq \varepsilon_0$, $\chi \geqq 0$), we conclude that $D' := \{z \in U_1 : (\psi + \chi(L + \varepsilon))(z) < 0\}$ is a C^k strictly pseudoconvex open set $\subseteqq D$. To complete the proof, we observe that, for sufficiently small L, $\psi + \chi(L + \varepsilon) \leqq 0$ on K, and therefore $K \subseteqq \bar{D}'$, because $\chi = 0$ on W, $\psi \leqq 0$ on K and $\psi < 0$ on $K \setminus W$ ($\varrho < 0$ and $\varrho_1 < 0$ on $K \setminus W$). ■

1.5.24. Theorem. *Let $G \subseteqq \mathbb{C}^n$ be an open set, and let ϱ be a strictly plurisubharmonic C^k-function in G such that $d\varrho(z) \neq 0$ for all $z \in \Gamma := \{z \in G : \varrho(z) = 0\}$, $k = 2, 3, \ldots, \infty$. Set $G_- := \{z \in G : \varrho(z) < 0\}$. Then, for every point $\xi \in \Gamma$, there exist a neighbourhood U_ξ of ξ and a C^k strictly pseudoconvex open set $D \subset\subset \mathbb{C}^n$ such that*

(i) *$U_\xi \cap G_- \subseteqq D \subseteqq G_-$.*

(ii) *There is a biholomorphic map h of some neighbourhood of D such that $h(D)$ is strictly convex.*

Proof. By Theorem 1.4.14, there is a biholomorphic map h from some neighbourhood V_ξ of ξ onto some open set $W \subseteqq \mathbb{C}^n$ such that $\varrho \circ h^{-1}$ is strictly convex in W. Let $U' \subset\subset W$ be an open ball centered at $h(\xi)$. Then, by Lemma 1.5.23 (i), there exists a C^k strictly convex open set D' such that $\bar{U}' \cap h(V_\xi \cap \bar{G}_-) \subseteqq \bar{D}' \subseteqq h(V_\xi \cap \bar{G}_-)$. By setting $U_\xi := h^{-1}(U')$ and $D := h^{-1}(D')$, we complete the proof. ■

1.5.25. Corollary. *An open set $D \subset\subset \mathbb{C}^n$ with C^k-boundary ($k = 2, 3, \ldots, \infty$) is strictly pseudoconvex if and only if, for every point $\xi \in \partial D$, there exist a neighbourhood $U_\xi \subseteqq \mathbb{C}^n$ of ξ and a biholomorphic map h_ξ in some neighbourhood of $\bar{U}_\xi$ such that $h_\xi(U_\xi \cap D)$ is a strictly convex set with C^k-boundary.*

Proof. By Theorem 1.5.24, the condition is necessary. That the condition is sufficient follows from Theorem 1.5.17 and the fact that condition (1.5.15) is invariant under holomorphic changes of coordinates (cp. (1.4.15) and the corresponding computation for $\sum_{j=1}^{n} (\partial\varrho \circ f(z)/\partial z_j)\, \xi_j$). ■

We conclude this section with the following

Remark. Every open set $D \subset\subset \mathbb{C}^1$ with C^2-boundary is strictly pseudoconvex. This follows, for example, from Theorem 1.5.17 and the fact that, for $n = 1$, condition (1.5.15) is trivially fulfilled, because then $T_z = 0$ for all $z \in \partial D$.

1.6. Preliminaries concerning differential forms

We assume that the reader knows the calculus of differential forms on real manifolds as it is developed in standard books. In the present section we collect certain more special material and introduce some notations.

Let X be a real C^1-manifold. Then we denote by $\dim_{\mathbb{R}} X$ the real dimension of X. The notation of a *differential form on X* (*form on X*) will be used for differential forms with complex-valued measurable coefficients. The *degree* of a differential form f on X will be denoted by $\deg f$, the *support* of f (this is the smallest closed subset K in X such that $f = 0$ in $X \setminus K$) will be denoted by $\operatorname{supp} f$. An *s-form* on X is a differential form of degree s on X; a *C^k-form* (if X is a C^{k+1}-manifold) is a differential form on X whose coefficients are k times continuously differentiable ($k = 0, 1, \ldots, \infty$); a *$C^k_{(s)}$-form* on X is a C^k-form of degree s on X. The space of all $C^k_{(s)}$-forms on X will be denoted by $C^k_{(s)}(X)$ ($k = 0, 1, \ldots, \infty$; $s = 1, \ldots, \dim_{\mathbb{R}} X$).

1.6.1. Integration with respect to a part of the variables. Let X and Y be two real C^1-manifolds, and let f be a differential form on $X \times Y$. If $y_1, \ldots, y_n$, are local coordinates on Y, then the form f can uniquely be written

$$f(x, y) = \sum_{|I| \leqq \deg f}{}' f_I(x, y) \wedge dy^I , \tag{1.6.1}$$

where $f_I(x, y)$ are differential forms of degree $\deg f - |I|$ on X depending on $y \in Y$. Here the following notations are used: $I = (i_1, \ldots, i_r)$ is a multi-index of integers $1 \leqq i_s \leqq n$, $dy^I := dy_{i_1} \wedge \ldots \wedge dy_{i_r}$, $|I| = r$ is the length of I, and $\sum'$ means that the summation is over all multi-indices with strictly increasing components. If X is oriented and if the integrals $\int_X f_I(x, y)$ exist for all $|I| = \deg f - \dim_{\mathbb{R}} X$ and $y \in Y$, then we define

$$\int_X f(x, y) = \sum_{|I| = \deg f - \dim_{\mathbb{R}} X} \Big(\int_X f_I(x, y)\Big) \, dy^I . \tag{1.6.2}$$

The result of this integration is a differential form of degree $\deg f - \dim_{\mathbb{R}} X$ on Y and is independent of the choice of the local coordinates $y_1, \ldots, y_n$.

Remark. $\int_X f(x, y) = 0$ if f does not contain monomials of degree $\dim_{\mathbb{R}} X$ in x.

1.6.2. The absolute value of a differential form of maximal degree. Let X be an oriented real C^1-manifold of real dimension m, and let f be a differential form of degree m on X. Then we denote by $|f|$ the differential form on X which is defined as follows:

Let $x_1, \ldots, x_m$ be local coordinates on X such that $dx_1 \wedge \ldots \wedge dx_m$ defines the orientation of X. Then $f = \varphi \, dx_1 \wedge \ldots \wedge dx_m$, where φ is a complex-valued function, and we define $|f| := |\varphi| \, dx_1 \wedge \ldots \wedge dx_m$. It is clear that this definition is independent of the choice of the local coordinates $x_1, \ldots, x_m$. The differential form $|f|$ is called the absolute value of f. If f is integrable, then, obviously,

$$\Big| \int_X f \Big| \leqq \int_X |f| .$$

1.6.3. The norm of differential forms at points. Let $D \subseteqq \mathbb{R}^n$ be an open set, and let f be a differential form of degree s on D. Then, for $x \in D$, we denote by $\|f(x)\|$ the Euclidean length of the vector of coefficients of $f(x)$ with respect to the canonical coordinates $x_1, \ldots, x_n$ in $\mathbb{R}^n$, that is, if

$$f(x) = \sum_{1 \leqq i_1 < \ldots < i_s \leqq n} f_{i_1 \ldots i_s}(x) \, dx_{i_1} \wedge \ldots \wedge dx_{i_s} , \qquad x \in D ,$$

then

$$\|f(x)\| := \Big(\sum_{1 \leqq i_1 < \ldots < i_s \leqq n} |f_{i_1 \ldots i_s}(x)|^2 \Big)^{1/2} , \qquad x \in D .$$

$\|f(x)\|$ is called the *norm* of f at x.

Now let X be a real C^1-manifold. Then, for every differential form f on X, we introduce the norm $\|f(x)\|$, $x \in X$, as follows: Let $\{U_j\}$ be a locally finite open covering of X such that, for every j, we can find C^1-coordinates $x_1^{(j)}, \ldots, x_n^{(j)}$ in U_j, and let $\{\chi_j\}$ be a C^1-partition of unity subordinate to $\{U_j\}$. If

$$f(x) = \sum_{1 \leqq i_1 < \ldots < i_s \leqq n} f^{(j)}_{i_1 \ldots i_s}(x) \, dx^{(j)}_{i_1} \wedge \ldots \wedge dx^{(j)}_{i_s} , \qquad x \in U_j ,$$

then we define

$$||f(x)|| := \sum_j \chi_j(x) \Big(\sum_{1 \leq i_1 < \dots < i_s \leq n} |f^{(j)}_{i_1 \dots i_s}(x)|^2 \Big)^{1/2}, \qquad x \in X .$$

Clearly, this norm depends on the choice of the local coordinates and the partition of unity. However two such norms differ by a positive continous factor only.

1.6.4. Differential forms in $\mathbb{C}^n$. Let D be an open set in $\mathbb{C}^n$. We denote by $z_1, \dots, z_n$ the canonical complex coordinates in $\mathbb{C}^n$. Then every differential form f on D can be written uniquely

$$f = \sum_{|I|+|J|=\deg f}{}' f_{IJ} \, dz^I \wedge d\bar{z}^J , \tag{1.6.3}$$

where f_{IJ} are functions on D. Here the following notations are used: $I = (i_1, \dots, i_r)$ and $J = (j_1, \dots, j_s)$ are multi-indices of integers $1 \leq i_\nu, j_\mu \leq n$, $dz^I := dz_{i_1} \wedge \dots \wedge dz_{i_r}$, $d\bar{z}^J := d\bar{z}_{j_1} \wedge \dots \wedge d\bar{z}_{j_s}$, $|I|$ is the length of I, and $\sum'$ means that the summation is over all multi-indices with strictly increasing components.

A differential form f on D is called a *(p, q)-form* ($0 \leq p, q \leq n$) or a form of bidegree (p, q) if in the representation (1.6.3) $f_{IJ} = 0$ for all I, J with $(|I|, |J|) \neq (p, q)$, that is

$$f = \sum_{|I|=p, |J|=q}{}' f_{IJ} \, dz^I \wedge d\bar{z}^J .$$

A (p, q)-form with coefficients in $C^k(D)$, $k = 0, 1, \dots, \infty$, is called a $C^k_{(p,q)}$-form or a *C^k-form of bidegree* (p, q) on D. The space of all $C^k_{(p,q)}$-forms on D will be denoted by $C^k_{(p,q)}(D)$. We write $C^k_{(p,q)}(D) = 0$ for $\max(p, q) > n$. It follows from the uniqueness of the representation (1.6.3) that every differential form f on D can be written uniquely

$$f = \sum_{p+q=\deg f} f_{(p,q)} , \tag{1.6.4}$$

where $f_{(p,q)}$ is a (p, q)-form.

Now we want to generalize the operators ∂ and $\bar\partial$ defined in Section 1.1 for functions to differential forms of arbitrary degree. If f is a continuous differential form represented in the form (1.6.3), then we define (in the sense of distributions)

$$\partial f := \sum_{|I|+|J|=\deg f}{}' \partial f_{IJ} \wedge dz^I \wedge d\bar{z}^J \tag{1.6.5}$$

and

$$\bar\partial f := \sum_{|I|+|J|=\deg f}{}' \bar\partial f_{IJ} \wedge dz^I \wedge d\bar{z}^J . \tag{1.6.6}$$

Then $df = \partial f + \bar\partial f$, and if f is of bidegree (p, q), then ∂f is of bidegree $(p + 1, q)$ and $\bar\partial f$ is of bidegree $(p, q + 1)$.

1.6.4.1. Proposition. *For every continuous differential form f on D, we have (in the sense of distributions)*

$$\partial^2 f = (\partial\bar\partial + \bar\partial\partial) f = \bar\partial^2 f = 0 . \tag{1.6.7}$$

Proof. In view of (1.6.4), we only have to prove this for (p, q)-forms f. Since $d^2 = 0$ and $d = \partial + \bar\partial$, we have

$$\partial^2 f + (\partial\bar\partial + \bar\partial\partial) f + \bar\partial^2 f = d^2 f = 0 .$$

This implies (1.6.7), because the three forms on the left-hand side are of different bidegrees. ■

1.6.4.2. Proposition. *Let $h = (h_1, \dots, h_m)\colon D \to \mathbb{C}^m$ be a holomorphic map and $G \subseteq \mathbb{C}^m$ an open set such that $h(D) \subseteq G$.* (If f is a differential form on G, then we denote by h^*f the pull-back of f with respect to h, that is, the form defined on D by (1.6.9) below.) *Then*

(i) *For every (p, q)-form f on G, h^*f is a (p, q)-form on D.*

(ii) *For every continuous differential form f on G, we have (in the sense of distributions)*

$$\partial h^* f = h^* \partial f \quad \text{and} \quad \bar\partial h^* f = h^* \bar\partial f . \tag{1.6.8}$$

Proof. If

$$f = \sum_{\substack{1 \le i_1 < \dots < i_p \le m \\ 1 \le j_1 < \dots < j_q \le m}} f^{i_1 \dots i_p}_{j_1 \dots j_q} \, dz_{i_1} \wedge \dots \wedge dz_{i_p} \wedge d\bar z_{j_1} \wedge \dots \wedge d\bar z_{j_q} ,$$

then, by definition of h^*f,

$$h^* f = \sum_{\substack{1 \le i_1 < \dots < i_p \le m \\ 1 \le j_1 < \dots < j_q \le m}} f^{i_1 \dots i_p}_{j_1 \dots j_q} \circ h \, dh_{i_1} \wedge \dots \wedge dh_{i_p} \wedge d\bar h_{j_1} \wedge \dots \wedge d\bar h_{j_q} . \tag{1.6.9}$$

This is a (p, q)-form, because the functions h_j are holomorphic, and therefore $dh_j = \partial h_j$ is of bidegree $(1, 0)$, and $d\bar h_j = \bar\partial \bar h_j$ is of bidegree $(0, 1)$. This proves part (i).

To prove part (ii), by (1.6.4), we can assume that f is a (p, q)-form. Since $dh^* f = h^* df$ and $d = \partial + \bar\partial$, we get $(\partial h^* - h^* \partial) f = (h^* \bar\partial - \bar\partial h^*) f$. This implies (1.6.8), because the form on the left-hand side is of bidegree $(p + 1, q)$ whereas the form on the right-hand side is of bidegree $(p, q + 1)$. ■

1.6.5. Differential forms on complex manifolds. If f is a differential form on a complex manifold, then f is called of *bidegree* (p, q) (a *(p, q)-form*), if it is of bidegree (p, q) with respect to local holomorphic coordinates. It follows from Proposition 1.6.4.2 (i) that this definition is independent of the choice of local holomorphic coordinates. The unique representation (1.6.4) is valid on complex manifolds, too. The differential operators ∂ and $\bar\partial$ can be defined on complex manifolds with respect to local holomorphic coordinates. By Proposition 1.6.4.2 (ii) these definitions are independent of the choice of local holomorphic coordinates. Proposition 1.6.4.1 holds also for complex manifolds. For complex manifolds X, we denote by $C^k_{(p,q)}(X)$ the space of all $C^k_{(p,q)}$-forms on X.

1.7. The differential forms $\omega'(v)$ and $\omega(u)$

Let X be a real C^1-manifold, and let $u = (u_1, \dots, u_n)\colon X \to \mathbb{C}^n$ and $v = (v_1, \dots, v_n)\colon X \to \mathbb{C}^n$ be C^1-maps. Then we define the continuous differential forms $\omega(u) := du_1 \wedge \dots \wedge du_n$ and

$$\omega'(v) := \sum_{j=1}^{n} (-1)^{j+1} v_j \, dv_1 \wedge \dots \hat{j} \dots \wedge dv_n , \tag{1.7.1}$$

where $\hat{j}$ means that dv_j must be omitted. Further, we use the notation

$$\langle v, u \rangle := \sum_{j=1}^{n} v_j u_j .$$

If there is a second C^1-manifold Y, the maps u, v are defined for $(x, y) \in X \times Y$, and u, v are C^1 with respect to x, then we write $\omega_x(u) := \mathrm{d}_x u_1 \wedge \ldots \wedge \mathrm{d}_x u_n$ and

$$\omega_x'(v) := \sum_{j=1}^{n} (-1)^{j+1} v_j \mathrm{d}_x v_1 \wedge \ldots \hat{j} \ldots \wedge \mathrm{d}_x v_n\,,$$

where d_x is the exterior differentiation with respect to x. If u and v are C^1 with respect to (x, y), then we also write $\omega_{x,y}'(v)$ and $\omega_{x,y}(u)$ instead of $\omega'(v)$ and $\omega(u)$.

1.7.1. Proposition. *Let X be a real C^∞-manifold, and let $u, v\colon X \to \mathbb{C}^n$ be C^1-maps. Then the differential form*

$$\frac{\omega'(v) \wedge \omega(u)}{\langle v, u\rangle^n}$$

is closed (in the sense of distributions) for all $x \in X$ with $\langle v(x), u(x)\rangle \neq 0$.

Proof. Since $\mathrm{d}\omega(u) = 0$, we have

$$\mathrm{d}\frac{\omega'(v) \wedge \omega(u)}{\langle v, u\rangle^n} = \frac{\mathrm{d}\omega'(v) \wedge \omega(u)}{\langle v, u\rangle^n} - \frac{\mathrm{d}\langle v, u\rangle^n \wedge \omega'(v) \wedge \omega(u)}{\langle v, u\rangle^{2n}}\,.$$

But $\mathrm{d}\omega'(v) = n\omega(v)$ and

$$\begin{aligned}\mathrm{d}\langle v, u\rangle^n \wedge \omega'(v) \wedge \omega(u) &= n\langle v, u\rangle^{n-1} \sum_{j=1}^{n} (v_j\, \mathrm{d}u_j + u_j\, \mathrm{d}v_j) \wedge \omega'(v) \wedge \omega(u)\\ &= n\langle v, u\rangle^n\, \omega(v) \wedge \omega(u).\ \blacksquare\end{aligned}$$

1.7.2. Proposition. *Let $x_j = x_j(\zeta), j = 1, \ldots, 2n$, be the real coordinates of $\zeta \in \mathbb{C}^n$ such that $\zeta_j = x_j(\zeta) + \mathrm{i}x_{j+n}(\zeta)$. Then for every $z \in \mathbb{C}^n$,*

$$\mathrm{d}_\zeta\big(\omega'(\bar\zeta - \bar z) \wedge \omega(\zeta)\big) = n(2\mathrm{i})^n\, \mathrm{d}x_1 \wedge \ldots \wedge \mathrm{d}x_{2n}\,.$$

Proof.

$$\begin{aligned}&\mathrm{d}_\zeta\Big(\sum_j (-1)^{j+1} (\bar\zeta_j - \bar z_j) \bigwedge_{k\neq j} \mathrm{d}\bar\zeta_k \bigwedge_l \mathrm{d}\zeta_l\Big) = n \bigwedge_k \mathrm{d}\bar\zeta_k \bigwedge_l \mathrm{d}\zeta_l\\ &= n \bigwedge_k (\mathrm{d}x_k - \mathrm{i}\mathrm{d}x_{k+n}) \bigwedge_l (\mathrm{d}x_l + \mathrm{i}\mathrm{d}x_{l+n}) = n(2\mathrm{i})^n \bigwedge_j \mathrm{d}x_j\,.\ \blacksquare\end{aligned}$$

1.7.3. Proposition. *For every C^1-function ψ on X, we have*

$$\omega'(\psi v) = \psi^n \omega'(v)\,, \tag{1.7.2}$$

and consequently

$$\frac{\omega'(\psi v) \wedge \omega(u)}{\langle \psi v, u\rangle^n} = \frac{\omega'(v) \wedge \omega(u)}{\langle v, u\rangle^n}\,. \tag{1.7.3}$$

This proposition can be proved by direct computation. We will give a proof, which uses the fact that $\omega'(v)$ can be written as a determinant (this interpretation will be especially useful later). First we need some preparations. For that purpose we consider an arbitrary algebra $\mathfrak{A}$ (for example, the algebra of exterior differential forms). Let $A = (a_{ij})_{i,j=1}^n$ be a matrix with elements $a_{ij} \in \mathfrak{A}$. Then we define

$$\det A = \sum_\sigma \operatorname{sgn}(\sigma)\, a_{\sigma(1),1} \cdots a_{\sigma(n),n}\,, \tag{1.7.4}$$

where the summation is over all permutations σ of $\{1, \ldots, n\}$, and $\operatorname{sgn}(\sigma)$ is the signature of σ. Under this definition the usual relations between row operations (the first index is the row index) and determinants hold. Therefore, as in the commutative case it can be proved.

1.7.4. Proposition. *Let Z be an $n \times n$ matrix with elements in the centre*[1]) *of $\mathfrak{A}$. Then* $\det ZA = \det Z \det A$.

Remark that, if $\mathfrak{A}$ is non-commutative, then in general $\det AZ \neq \det A \det Z$. Further, it is possible that some of the columns of A are equal but $\det A \neq 0$ (cp., for example, (1.7.5) below). However, some of the properties with respect to columns which are true for usual determinants remain valid also for the non-commutative case. It is easy (and left to the reader) to prove the following

1.7.5. Proposition. (i) $\det A$ *is a multi-linear map of the columns with respect to linear combinations with coefficients from the centre of* $\mathfrak{A}$.

(ii) *If, for some* $1 \leqq k < l \leqq n$, $a_{ik} = b_k z_i$ *and* $a_{il} = b_l z_i$ *for all* i, *where* b_k, b_l *are arbitrary elements from* $\mathfrak{A}$ *but all* z_i *belong to the centre of* $\mathfrak{A}$, *then* $\det A = 0$.

The differential form $\omega'(v)$ can be written

$$\omega'(v) = \frac{1}{(n-1)!} \det \begin{pmatrix} v_1 & \underbrace{dv_1 \dots dv_1} \\ \cdot & \cdots\cdots\cdots \\ v_n & \underbrace{dv_n \dots dv_n}_{n-1} \end{pmatrix}. \tag{1.7.5}$$

Proof of Proposition 1.7.3. It follows from Proposition 1.7.5 that

$$\det(\psi v, d(\psi v), \dots, d(\psi v)) = \det(\psi v, \psi\, dv + v\, d\psi, \dots, \psi\, dv + v\, d\psi)$$
$$= \det(\psi v, \psi\, dv, \dots, \psi\, dv) = \psi^n \det(v, dv, \dots, dv).$$

1.7.6. Definition. *Let* $1 \leqq m \leqq n$, *let* $a_1, \dots, a_m$ *be column vectors of length* n *of differential forms, and let* $s_1, \dots, s_m$ *be non-negative integers such that* $s_1 + \dots + s_m = n$. *Then*

$$\det_{s_1 \dots s_m}(a_1, \dots, a_m) := \det(\underbrace{a_1, \dots, a_1}_{s_1}, \dots, \underbrace{a_m, \dots, a_m}_{s_m})$$

1.8. Leray maps and the operators $B_{\partial D}$, B_D, $L^w_{\partial D}$ and $R^w_{\partial D}$

In this section we give the definitions and some simple properties of the integral operators used in the integral representation formulas, which will be proved in Sections 1.9–1.12.

1.8.1. Orientation of $\mathbb{C}^n$. In this book we use the following orientation in $\mathbb{C}^n$: If $x_j = x_j(\zeta)$, $j = 1, \dots, 2n$, are the real coordinates of $\zeta \in \mathbb{C}^n$ such that $\zeta_j = x_j(\zeta) + i x_{j+n}(\zeta)$, then the differential form $dx_1 \wedge \dots \wedge dx_{2n}$ defines the orientation of $\mathbb{C}^n$. For open sets $D \subseteq \mathbb{C}^n$ we use the same orientation. If $D \subseteq \mathbb{C}^n$ is an open set and M is a relatively open subset of ∂D which is smooth of class C^1, then M is oriented by the orientation which is induced from D.

Remark. In the literature the orientation of $\mathbb{C}^n$ is also defined by $dx_1 \wedge dx_{1+n} \wedge \dots \wedge dx_n \wedge dx_{2n} = (-1)^{n(n-1)/2} dx_1 \wedge \dots \wedge dx_{2n}$. Then we obtain a corresponding change of the sign in the integral formulas given in this book.

1.8.2. Open sets with piecewise C^1-boundary. Let $D \subset\subset \mathbb{C}^n$ be an open set. The boundary ∂D of D is said to be *piecewise* C^1 if there is a finite number of real-valued

[1]) The centre of $\mathfrak{A}$ is the set of all $a \in \mathfrak{A}$ such that $ab = ba$ for every $b \in \mathfrak{A}$. If $\mathfrak{A}$ is an algebra of differential forms, then this is the subalgebra of forms of even degree.

C^1-functions $\varrho_1, \dots, \varrho_k$ on $\mathbb{C}^n$ such that $D = \{z \in \mathbb{C}^n : \varrho_j(z) < 0 \text{ for } j = 1, \dots, k\}$ and $d\varrho_{j_1}(z) \wedge \dots \wedge d\varrho_{j_s}(z) \neq 0$ for every collection of indices $1 \leqq j_1 < \dots < j_s \leqq k$ and all points $z \in \partial D$ with $\varrho_{j_1}(z) = \dots = \varrho_{j_s}(z) = 0$.

Remark. For open sets with piecewise C^1-boundary it is easy to find a sequence of open sets $D_m \subset\subset D$ with C^∞-boundary such that the following two conditions are fulfilled: (i) For every compact set $K \subset\subset D$ there is a number m_K such that $K \subseteq D_m$ for all $m \geqq m_K$. (ii) If f and g are continuous differential forms on D of bidegree $2n$ and $2n-1$, respectively, then $\int_D f = \lim \int_{D_m} f$ and $\int_{\partial D} g = \lim \int_{\partial D_m} g$. In the proofs of the integral representations in Sections 1.9–1.12 the existence of such a sequence D_m is used only. Therefore, these formulas are valid also for more general open sets. In the present book we need these formulas only for open sets with piecewise C^1-boundary.

1.8.3. B_D for 1-forms and $B_{\partial D}$ for functions. We recall the notations

$$\omega(\zeta) := d\zeta_1 \wedge \dots \wedge d\zeta_n \quad \text{and} \quad \omega'(\bar\zeta - \bar z) := \sum_{j=1}^n (-1)^{j+1} (\bar\zeta_j - \bar z_j) \bigvee_{k \neq j} d\bar\zeta_k$$

from Section 1.7. Since, for fixed $z \in \mathbb{C}^n$, the differential form $(\omega'(\bar\zeta - \bar z) \wedge \omega(\zeta))/|\zeta - z|^{2n}$ is smooth in $\mathbb{C}^n \setminus z$ and has an integrable singularity at $\zeta = z$ (the order of this singularity is $2n-1$), we can define the following two integral operators:

If $D \subset\subset \mathbb{C}^n$ is an open set and f is a bounded 1-form on D, then we define

$$(B_D f)(z) := \frac{(n-1)!}{(2\pi i)^n} \int_{\zeta \in D} f(\zeta) \wedge \frac{\omega'_\zeta(\bar\zeta - \bar z) \wedge \omega(\zeta)}{|\zeta - z|^{2n}}, \qquad z \in D. \tag{1.8.1}$$

If $D \subset\subset \mathbb{C}^n$ is an open set with piecewise C^1-boundary and f is a bounded measurable function on ∂D, then we define

$$(B_{\partial D} f)(z) := \frac{(n-1)!}{(2\pi i)^n} \int_{\zeta \in \partial D} f(\zeta) \frac{\omega'_\zeta(\bar\zeta - \bar z) \wedge \omega(\zeta)}{|\zeta - z|^{2n}}, \qquad z \in D. \tag{1.8.2}$$

Remark. For $n = 1$, $B_D f(z) = (1/2\pi i) \int_D f(\zeta)\, d\zeta/(\zeta - z)$ and $B_{\partial D} f(z) = (1/2\pi i) \times \int_{\partial D} f(\zeta)\, d\zeta/(\zeta - z)$ are the operators from the Cauchy-Green formula (1.1.1).

1.8.4. B_D and $B_{\partial D}$ for differential forms of abritrary degree. For $(z, \zeta) \in \mathbb{C}^n \times \mathbb{C}^n$ consider the differential form

$$\frac{\omega'_{z,\zeta}(\bar\zeta - \bar z) \wedge \omega(\zeta)}{|\zeta - z|^{2n}} = \sum_{j=1}^n (-1)^{j+1} \frac{\bar\zeta_j - \bar z_j}{|\zeta - z|^{2n}} \bigwedge_{k \neq j} (d\bar\zeta_k - d\bar z_k) \wedge \omega(\zeta). \tag{1.8.3}$$

This differential form is C^∞ for $\zeta \neq z$ and has a singularity of order $2n-1$ at $\zeta = z$. Since this singularity is integrable with respect to ζ, we can give the following definition (see Subsection 1.6.1 for the definition of integration with respect to ζ):

If $D \subset\subset \mathbb{C}^n$ is an open set and f is a bounded differential form on D, then we define

$$(B_D f)(z) := \frac{(n-1)!}{(2\pi i)^n} \int_{\zeta \in D} f(\zeta) \wedge \frac{\omega'_{z,\zeta}(\bar\zeta - \bar z) \wedge \omega(\zeta)}{|\zeta - z|^{2n}}, \qquad z \in D. \tag{1.8.4}$$

If $D \subset\subset \mathbb{C}^n$ is an open set with piecewise C^1-boundary, and f is a bounded differential form on ∂D, then we define

$$(B_{\partial D}f)(z) := \frac{(n-1)!}{(2\pi i)^n} \int_{\zeta \in \partial D} f(\zeta) \wedge \frac{\omega'_{z,\zeta}(\bar{\zeta} - \bar{z}) \wedge \omega(\zeta)}{|\zeta - z|^{2n}}, \qquad z \in D. \tag{1.8.5}$$

Definition (1.8.4) coincides with (1.8.1) when f is a 1-form, and definition (1.8.5) coincides with (1.8.2) when f is a function. To see this we consider the unique representation

$$\frac{(n-1)!}{(2\pi i)^n} \frac{\omega'_{z,\zeta}(\bar{\zeta} - \bar{z}) \wedge \omega(\zeta)}{|\zeta - z|^{2n}} = \sum_{q=0}^{n-1} \Omega_q(z, \zeta),$$

where $\Omega_q(z, \zeta)$ is of bidegree $(0, q)$ in z and of bidegree $(n, n-q-1)$ in ζ. Let f be a bounded differential form on D and let

$$f = \sum_{r+s=\deg f} f_{(r,s)}$$

be its unique representation as a sum of (r, s)-forms. Since $f_{(r,s)}(\zeta) \wedge \omega(\zeta) = 0$ if $r > 0$, then

$$B_D f = B_D f_{(0,\deg f)}. \tag{1.8.6}$$

Further, by degree reasons, $f(\zeta) \wedge \Omega_q(z, \zeta) = 0$ if $\deg f > q + 1$, and, by definition of integration with respect to ζ, we have $\int_{\zeta \in D} f(\zeta) \wedge \Omega_q(z, \zeta) = 0$ when $\deg f < q + 1$. Consequently,

$$B_D f = \begin{cases} \int_{\zeta \in D} f_{(0,\deg f)}(\zeta) \wedge \Omega_{(\deg f)-1}(z, \zeta) & \text{if} \quad 1 \leqq \deg f \leqq n, \\ 0 \quad \text{otherwise}. \end{cases} \tag{1.8.7}$$

Similarly we obtain that

$$B_{\partial D} f(z) = \begin{cases} \int_{\zeta \in \partial D} f_{(0,\deg f)}(\zeta) \wedge \Omega_{\deg f}(z, \zeta) & \text{if} \quad 0 \leqq \deg f \leqq n-1, \\ 0 \quad \text{otherwise}. \end{cases} \tag{1.8.8}$$

Since

$$\Omega(z, \zeta) = \frac{(n-1)!}{(2\pi i)^n} \frac{\omega'_{\zeta}(\bar{\zeta} - \bar{z}) \wedge \omega(\zeta)}{|\zeta - z|^{2n}},$$

this implies that (1.8.4) coincides with (1.8.1) when f is a 1-form, and (1.8.5) coincides with (1.8.2) when f is a function. Further, it follows from (1.8.7) and (1.8.8) that $B_D f$ is of bidegree $(0, (\deg f) - 1)$ and $B_{\partial D} f$ is of bidegree $(0, \deg f)$.

Notation. Let $Y \subseteqq \mathbb{C}^n$ be a measurable set. Then, for every measurable complex-valued function f on Y, we define

$$\|f\|_{0,Y} := \sup_{z \in Y} |f(z)|, \tag{1.8.9}$$

and we denote by $L^\infty(Y)$ the Banach space of all measurable complex-valued functions f on Y with $\|f\|_{0,Y} < \infty$. The space of all continuous complex-valued functions f on Y will be denoted by $C^0(Y)$. We set $BC^0(Y) := C^0(Y) \cap L^\infty(Y)$. $BC^0(Y)$ forms a closed subspace of the Banach space $L^\infty(Y)$. If Y is a compact set, then $BC^0(Y) = C^0(Y)$.

For $f \in C^0(Y)$ and $0 < \alpha < 1$, we define the *α-Hölder norm* $\|f\|_{\alpha,Y}$ by

$$\|f\|_{\alpha,Y} := \|f\|_{0,Y} + \sup_{z,\zeta\in Y} \frac{|f(z) - f(\zeta)|}{|\zeta - z|^\alpha}. \tag{1.8.10}$$

Set $H^\alpha(Y) := \{f \in C^0(Y) : \|f\|_{\alpha,Y} < \infty\}$. Endowed with the norm $\|\cdot\|_{\alpha,Y}$, $H^\alpha(Y)$ forms a Banach space, which is called the *space of α-Hölder continuous functions* (*Hölder space*).

Now suppose that $D \subseteq \mathbb{C}^n$ is an open set. Then for $0 < \alpha < 1$ we denote by $C^\alpha(D)$ the space of all $f \in C^0(D)$ such that $\|f\|_{\alpha,K} < \infty$ for every compact set $K \subset\subset D$ (the functions in $C^\alpha(D)$ need not be bounded). If $k \geqq 0$ is an integer, then by $C^{k+\alpha}(D)$ will be denoted the space of all k times continuously differentiable complex-valued functions on D such that all derivatives of order k belong to $C^\alpha(D)$. The functions in $C^{k+\alpha}(D)$ will be called *$C^{k+\alpha}$-functions* in D.

If $Y \subseteq \mathbb{C}^n$ is a measurable set, then the notations $L^\infty_{(p,q)}(Y)$, $C^0_{(p,q)}(Y)$, $BC^0_{(p,q)}(Y)$, $H^\alpha_{(p,q)}(Y)$ [$L^\infty_{(s)}(Y)$, $C^0_{(s)}(Y)$, $BC^0_{(s)}(Y)$, $H^\alpha_{(s)}(Y)$] will be used for the spaces of differential forms of bidegree (p, q) [degree s] and with coefficients in $L^\infty(Y)$, $C^0(Y)$, $BC^0(Y)$, $H^\alpha(Y)$, respectively. The spaces $L^\infty_{(p,q)}(Y)$, $BC^0_{(p,q)}(Y)$, $L^\infty_{(s)}(Y)$ and $BC^0_{(s)}(Y)$ become Banach spaces if we introduce the norm

$$\|f\|_{0,Y} := \sup_{z\in Y} \|f(z)\|,$$

where the norm $\|f(z)\|$ was defined in Subsection 1.6.3.

If $D \subseteq \mathbb{C}^n$ is an open set, then we denote by $C^{k+\alpha}_{(p,q)}(D)$ [$C^{k+\alpha}_{(s)}(D)$] the space of differential forms of bidegree (p, q) [degree s] in D whose coefficients belong to $C^{k+\alpha}(D)$. The forms in $C^{k+\alpha}_{(p,q)}(D)$ [$C^{k+\alpha}_{(s)}(D)$] are called $C^{k+\alpha}_{(p,q)}$-forms [$C^{k+\alpha}_{(s)}$-forms] or $C^{k+\alpha}$-forms of bidegree (p, q) [degree s].

1.8.5. Lemma. *Let D be a bounded open set in $\mathbb{C}^n$. Then*

(i) *For every bounded differential form f on D, $B_D f$ is a C^α-form in D for all $0 < \alpha < 1$. Moreover, for every $0 < \alpha < 1$, there is a constant $C_\alpha < \infty$ such that, for every bounded differential form f on D, $\|B_D f\|_{\alpha,D} \leqq C_\alpha \|f\|_{0,D}$.*

(ii) *If f is a bounded C^k-form on D, $1 \leqq k \leqq \infty$, then $B_D f$ is a $C^{k+\alpha}$-form in D for all $0 < \alpha < 1$.*

Proof. (i). It follows immediately from the definition of B_D that, for some constant $C_1 < \infty$ and all $z, \xi \in D$,

$$\|B_D f(z) - B_D f(\xi)\| \leqq C_1 \|f\|_{0,D} \sum_{j=1}^n \int_{\zeta\in D} \left| \frac{\bar\zeta_j - \bar z_j}{|\zeta - z|^{2n}} - \frac{\bar\zeta_j - \bar\xi_j}{|\zeta - \xi|^{2n}} \right| d\sigma_{2n},$$

where $d\sigma_{2n}$ is the Lebesgue measure in $\mathbb{C}^n$. In view of Proposition 1, Appendix 1, it follows that, for some $C_1 < \infty$,

$$\|B_D f(z) - B_D f(\xi)\| \leqq C_2 \|f\|_{0,D} |z - \xi| \left|\ln |z - \xi|\right|.$$

Since, for all $0 < \alpha < 1$, $\sup\limits_{z,\xi\in D} |z - \xi|^{1-\alpha} \left|\ln |z - \xi|\right| < \infty$, we obtain assertion (i).

(ii). We fix $z \in D$ and prove that $B_D f$ is $C^{k+\alpha}$ in some neighbourhood of z. Choose a function $\chi \in C_0^\infty(D)$ such that $\chi = 1$ in a neighbourhood U_z of z. Then it is clear that $B_D(1 - \chi) f$ is C^∞ in U_z. It remains to prove that $B_D \chi f$ is $C^{k+\alpha}$ in U_z. By (1.8.7), we can assume that f is a $(0, q)$-form and

$$B_D \chi f(\xi) = \int_{\zeta\in D} \chi(\zeta) f(\zeta) \wedge \Omega_{q-1}(\xi, \zeta).$$

Let $f(\zeta) = \sum'_{|I|=q} f_I(\zeta)\, d\bar{\zeta}^I$ and

$$\Omega_{q-1}(\xi,\zeta) = \sum'_{\substack{|J|=q-1\\|K|=n-q-2}} \varphi_{JK}(\xi,\zeta)\, d\bar{\zeta}^K \wedge \omega(\zeta) \wedge d\bar{\xi}^J .$$

Then $B_D\chi f(\xi) = \sum'_J h_J(\xi)\, d\bar{\xi}^J$, where

$$\begin{aligned} h_J(\xi) &= \sum'_{I,K} \int_{\zeta\in D} \chi(\zeta) f_I(\zeta)\, \varphi_{JK}(\xi,\zeta)\, d\bar{\zeta}^I \wedge d\bar{\zeta}^K \wedge \omega(\zeta) \\ &= \sum'_{I,K} \int_{\zeta\in \mathbb{C}^n} \chi(\zeta+\xi) f_I(\zeta+\xi)\, \varphi_{JK}(\xi,\zeta+\xi)\, d\bar{\zeta}^I \wedge d\bar{\zeta}^K \wedge \omega(\zeta). \end{aligned}$$

The functions $\varphi_{JK}(\xi,\zeta+\xi)$ are independent of ξ and locally integrable with respect to ζ. Therefore, by differentiation under the sign of integration, we obtain that $B_D\chi f$ is a C^k-form. If $\xi_j = x_j + ix_{j+n}$ ($x_j \in \mathbb{R}$), $\nu = (\nu_1, \dots, \nu_{2n})$ is an arbitrary multi-index of non-negative integers with $\nu_1 + \dots + \nu_{2n} = k$, and $D^\nu := \partial^k/\partial x_1^{\nu_1} \dots \partial x_{2n}^{\nu_{2n}}$, then

$$\begin{aligned} D^\nu h_J(\xi) &= \sum'_{I,K} \int_{\zeta\in \mathbb{C}^n} \big(D^\nu(\chi f_I)\big)(\zeta+\xi)\, \varphi_{JK}(\xi,\zeta+\xi)\, d\bar{\zeta}^I \wedge d\bar{\zeta}^K \wedge \omega(\zeta) \\ &= \sum'_{I,K} \int_{\zeta\in D} \big(D^\nu(\chi f_I)\big)(\zeta)\, \varphi_{JK}(\xi,\zeta)\, d\bar{\zeta}^I \wedge d\bar{\zeta}^K \wedge \omega(\zeta) . \end{aligned}$$

It follows immediately from the definition of φ_{JK} that, for some constant $C < \infty$ and all $\xi, w \in D$,

$$\begin{aligned} &|D^\nu h_J(\xi) - D^\nu h_J(w)| \leqq \\ &\leqq C \sup_{\zeta\in D} \sum'_I |(D^\nu(\chi f_I)(\zeta))| \sum_{j=1}^n \int_{\zeta\in D} \left| \frac{\bar{\zeta}_j - \bar{\xi}_j}{|\zeta-\xi|^{2n}} - \frac{\bar{\zeta}_j - \bar{w}_j}{|\zeta-w|^{2n}} \right| d\sigma_{2n} . \end{aligned}$$

As in the proof of part (i), this implies that the functions $D^\nu h_J$ are C^α-functions for all $0 < \alpha < 1$. ■

1.8.6. $L^w_{\partial D}$ for functions and $R^w_{\partial D}$ for 1-forms. Let D be a bounded open set in $\mathbb{C}^n$. A $\mathbb{C}^n$-valued C^1-function $w(z,\zeta) = (w_1(z,\zeta), \dots, w_n(z,\zeta))$ defined for ζ in some neighbourhood of ∂D and $z \in D$ is called a *Leray map for D* if

$$\langle w(z,\zeta), \zeta - z\rangle \neq 0 \quad \text{for all} \quad (z,\zeta) \in D \times \partial D . \tag{1.8.11}$$

Let $w(z,\zeta)$ be a Leray map for D and suppose that the boundary of D is piecewise C^1. Then we define

$$\eta^w(z,\zeta,\lambda) := (1-\lambda)\frac{w(z,\zeta)}{\langle w(z,\zeta), \zeta - z\rangle} + \lambda \frac{\bar{\zeta}-\bar{z}}{\langle \bar{\zeta}-\bar{z}, \zeta - z\rangle} \tag{1.8.12}$$

for all $z \in D$, $0 \leqq \lambda \leqq 1$, and ζ in some neighbourhood of ∂D. Then, by (1.8.11), for every fixed $z \in D$, the differential form

$$\frac{\omega'_\zeta(w(z,\zeta)) \wedge \omega(\zeta)}{\langle w(z,\zeta), \zeta - z\rangle^n} := \sum_{j=1}^n (-1)^{j+1} \frac{w_j(z,\zeta)}{\langle w(z,\zeta), \zeta - z\rangle^n} \bigwedge_{k\neq j} d_\zeta w_k(z,\zeta) \wedge \omega(\zeta)$$

is continuous for ζ in some neighbourhood of ∂D, and the differential form

$$\omega'_{\zeta,\lambda}(\eta^w(z,\zeta,\lambda)) \wedge \omega(\zeta) := \sum_{j=1}^n (-1)^{j+1} \eta_j^w(z,\zeta,\lambda) \bigwedge_{k\neq j} d_{\zeta,\lambda}\eta_k^w(z,\zeta,\lambda) \wedge \omega(\zeta)$$

is continuous for ζ in some neighbourhood of ∂D and for $0 \leqq \lambda \leqq 1$. Therefore, we can define:

If f is a bounded measurable complex-valued function on ∂D, then we set

$$(L^w_{\partial D} f)(z) := \frac{(n-1)!}{(2\pi i)^n} \int_{\zeta \in \partial D} f(\zeta) \frac{\omega'_\zeta(w(z,\zeta)) \wedge \omega(\zeta)}{\langle w(z,\zeta), \zeta - z\rangle^n}, \qquad z \in D. \tag{1.8.13}$$

If f is a bounded 1-form on ∂D, then we define

$$(R^w_{\partial D} f)(z) := \frac{(n-1)!}{(2\pi i)^n} \int_{\substack{\zeta \in \partial D\\ 0 \leq \lambda \leq 1}} f(\zeta) \wedge \omega'_{\zeta,\lambda}(\eta^w(z,\zeta,\lambda)) \wedge \omega(\zeta), \qquad z \in D. \tag{1.8.14}$$

Remark. For every open set $D \subset\subset \mathbb{C}^n$ with piecewise C^1-boundary, $w(z,\zeta) := \bar\zeta - \bar z$ is a Leray map. Then $L^w_{\partial D} = B_{\partial D}$ and $R^w_{\partial D} = 0$.

1.8.7. $L^w_{\partial D}$ and $R^w_{\partial D}$ for differential forms of arbitrary degree. Let $D \subset\subset \mathbb{C}^n$ be an open set with piecewise C^1-boundary, let $w(z,\zeta)$ be a Leray map for D, and let $\eta^w(z,\zeta,\lambda)$ be defined by (1.8.12). We define

$$\omega'(w(z,\zeta)) := \sum_{j=1}^n (-1)^{j+1} w_j(z,\zeta) \bigwedge_{k \neq j} \bar\partial_{z,\zeta} w_k(z,\zeta)$$

and

$$\bar\omega'(\eta^w(z,\zeta,\lambda)) := \sum_{j=1}^n (-1)^{j+1} \eta^w_j(z,\zeta,\lambda) \bigwedge_{k \neq j} (\bar\partial_{z,\zeta} + d_\lambda)\, \eta^w_k(z,\zeta,\lambda).$$

By (1.8.11) the differential forms

$$\frac{\bar\omega'(w(z,\zeta)) \wedge \omega(\zeta)}{\langle w(z,\zeta), \zeta - z\rangle^n} \quad \text{and} \quad \bar\omega'(\eta^w(z,\zeta,\lambda)) \wedge \omega(\zeta)$$

are continuous for $(z,\zeta,\lambda) \in D \times \partial D \times [0,1]$. (Moreover, for every relatively compact open set $\tilde D \subset\subset D$ there is a neighbourhood $\tilde U_{\partial D}$ of ∂D such that these forms are continuous on $\tilde D \times \tilde U_{\partial D} \times [0,1]$.) Therefore, we can give the following definition (for the definition of integration with respect to ζ and (ζ,λ), respectively, see Subsection 1.6.1):

If f is a bounded differential form on ∂D, then we define

$$(L^w_{\partial D} f)(z) := \frac{(n-1)!}{(2\pi i)^n} \int_{\zeta \in \partial D} f(\zeta) \wedge \frac{\bar\omega'(w(z,\zeta)) \wedge \omega(\zeta)}{\langle w(z,\zeta), \zeta - z\rangle^n}, \qquad z \in D, \tag{1.8.15}$$

and

$$(R^w_{\partial D} f)(z) := \frac{(n-1)!}{(2\pi i)^n} \int_{\substack{\zeta \in \partial D\\ 0 \leq \lambda \leq 1}} f(\zeta) \wedge \bar\omega'(\eta^w(z,\zeta,\lambda)) \wedge \omega(\zeta), \qquad z \in D. \tag{1.8.16}$$

Now we shall prove that (1.8.15) coincides with (1.8.13) if f is a function, and (1.8.16) coincides with (1.8.14) if f is a 1-form. Consider the unique representation

$$\frac{(n-1)!}{(2\pi i)^n} \bar\omega'(\eta^w(z,\zeta,\lambda)) \wedge \omega(\zeta) = \sum_{q=0}^{n-1} \Omega^w_q(z,\zeta,\lambda),$$

where $\Omega^w_q(z,\zeta,\lambda)$ is of bidegree $(0,q)$ in z and of bidegree $(n, n-q-1)$ in (ζ,λ). Let f be a bounded differential form on ∂D and $f = \sum f_{(r,s)}$ its unique decomposition in (r,s)-forms. Since $f_{(r,s)}(\zeta) \wedge \omega(\zeta) = 0$ if $r > 0$, then

$$R^w_{\partial D} f = R^w_{\partial D} f_{(0,\deg f)}. \tag{1.8.17}$$

Further, by degree reasons, $f(\zeta) \wedge \Omega_q^w(z, \zeta, \lambda) = 0$ if $\deg f > q + 1$, and, by the definition of integration with respect to (ζ, λ),

$$\int_{\substack{\zeta \in \partial D \\ 0 \leqq \lambda \leqq 1}} f(\zeta) \wedge \Omega_q^w(z, \zeta, \lambda) = 0 \quad \text{if} \quad \deg f < q + 1 .$$

Hence

$$R_{\partial D}^w f = \begin{cases} \int\limits_{\substack{\zeta \in \partial D \\ 0 \leqq \lambda \leqq 1}} f_{(0, \deg f)}(\zeta) \wedge \Omega_{(\deg f)-1}^w(z, \zeta, \lambda) & \text{if} \quad 1 \leqq \deg f \leqq n , \\ 0 \quad \text{otherwise} . \end{cases} \tag{1.8.18}$$

Similarly, using the unique representation

$$\frac{(n-1)!}{(2\pi i)^n} \frac{\bar{\omega}'(w(z, \zeta)) \wedge \omega(\zeta)}{\langle w(z, \zeta), \zeta - z\rangle^n} = \sum_{q=0}^{n-1} \Omega_q^w(z, \zeta)$$

where $\Omega_q^w(z, \zeta)$ is of bidegree $(0, q)$ in z and of bidegree $(n, n - q - 1)$ in ζ, we obtain that

$$L_{\partial D}^w f = \begin{cases} \int\limits_{\zeta \in \partial D} f_{(0, \deg f)}(\zeta) \wedge \Omega_{\deg f}^w(z, \zeta) & \text{if} \quad \deg f \leqq n - 1 , \\ 0 \quad \text{otherwise} . \end{cases} \tag{1.8.19}$$

We have

$$\Omega_0^w(z, \zeta, \lambda) = \frac{(n-1)!}{(2\pi i)^n} \sum_{j=1}^{n} (-1)^{j+1} \eta_j^w(z, \zeta, \lambda) \bigwedge_{k \neq j} (\bar{\partial}_\zeta + d_\lambda) \eta_k^w(z, \zeta, \lambda) \wedge \omega(\zeta) .$$

In view of the factor $\omega(\zeta)$, the operator $\bar{\partial}_\zeta$ on the right-hand side can be replaced by d_ζ. Hence

$$\Omega_0^w(z, \zeta, \lambda) = \frac{(n-1)!}{(2\pi i)^n} \omega'_{\zeta, \lambda}(\eta^w(z, \zeta, \lambda)) \wedge \omega(\zeta)$$

and it follows from (1.8.19) that (1.8.16) coincides with (1.8.14) if f is a 1-form. Similarly, it follows from (1.8.18) that (1.8.15) coincides with (1.8.13) if f is a funtion.

Remark. If $w(z, \zeta) = \bar{\zeta} - \bar{z}$, then $L_{\partial D}^w f = B_{\partial D} f$ and $R_{\partial D}^w f = 0$ for every bounded differential form f on ∂D.

1.9. The Martinelli-Bochner formula

In this section we prove the Martinelli-Bochner formula. This is a first generalization of the Cauchy-Green formula (1.1.1) to several complex variables.

1.9.1. Theorem (Martinelli-Bochner formula). *Let $D \subset\subset \mathbb{C}^n$ be an open set with piecewise C^1-boundary, and let f be a continuous function on $\bar{D}$ such that $\bar{\partial} f$ is also continuous on $\bar{D}$.*[1]) *Then*

$$f = B_{\partial D} f - B_D \bar{\partial} f \quad \text{in} \quad D , \tag{1.9.1}$$

where $B_{\partial D}$ and B_D are the integral operators defined in Subsection 1.8.3.

[1]) This means, $\bar{\partial} f$ which is defined in D in the sense of distributions is continuous in D and admits a continuous continuation to $\bar{D}$.

Proof. Fix $z \in D$ and set

$$\theta(\zeta) := \frac{(n-1)!}{(2\pi i)^n} \frac{\omega'_\zeta(\bar\zeta - \bar z) \wedge \omega(\zeta)}{\langle \bar\zeta - \bar z, \zeta - z\rangle^n}.$$

Since, by Proposition 1.7.1, $d\theta(\zeta) = 0$ in $D \setminus z$, and since $\partial f(\zeta) \wedge \omega(\zeta) = 0$, it follows that $d(f(\zeta)\,\theta(\zeta)) = \bar\partial f(\zeta) \wedge \theta(\zeta)$ in $D \setminus z$. Consequently, for every sufficiently small $\varepsilon > 0$, Stokes' formula gives

$$\int_{|\zeta - z| = \varepsilon} f(\zeta)\,\theta(\zeta) = \int_{\partial D} f(\zeta)\,\theta(\zeta) - \int_{D_\varepsilon} \bar\partial f(\zeta) \wedge \theta(\zeta),$$

where $D_\varepsilon := \{\zeta \in D : |\zeta - z| > \varepsilon\}$. Therefore, we have to prove that the left-hand side of this equality tends to $f(z)$ when $\varepsilon \to 0$. We have

$$\int_{|\zeta - z| = \varepsilon} f(\zeta)\,\theta(\zeta) = f(z) \int_{|\zeta - z| = \varepsilon} \theta(\zeta) + \int_{|\zeta - z| = \varepsilon} (f(\zeta) - f(z))\,\theta(\zeta).$$

By Proposition 1.7.2, Stokes' formula gives

$$\int_{|\zeta - z| = \varepsilon} \theta(\zeta) = \frac{(n-1)!}{(2\pi i)^n} \int_{|\zeta - z| = \varepsilon} \frac{\omega'_\zeta(\bar\zeta - \bar z) \wedge \omega(\zeta)}{\varepsilon^{2n}}$$

$$= \frac{(n-1)!}{\varepsilon^{2n}(2\pi i)^n} \int_{|\zeta - z| < \varepsilon} d\big(\omega'_\zeta(\bar\zeta - \bar z) \wedge \omega(\zeta)\big)$$

$$= \frac{n!}{\varepsilon^{2n}\pi^n} \int_{|\zeta - z| < \varepsilon} dx_1 \wedge \ldots \wedge dx_{2n} = 1.$$

Now we consider the integral

$$\int_{|\zeta - z| = \varepsilon} (f(\zeta) - f(z))\,\theta(\zeta) = \frac{(n-1)!}{(2\pi i)^n \varepsilon^{2n-1}} \int_{|\zeta - z| = \varepsilon} (f(\zeta) - f(z)) \frac{\omega'_\zeta(\bar\zeta - \bar z) \wedge \omega(\zeta)}{|\zeta - z|}.$$

Since the form $\big(\omega'_\zeta(\bar\zeta - \bar z) \wedge \omega(\zeta)\big)/|\zeta - z|$ is bounded in D, this implies the estimate

$$\Big| \int_{|\zeta - z| = \varepsilon} (f(\zeta) - f(z))\,\theta(\zeta) \Big| \leqq C \max_{|\zeta - z| \leqq \varepsilon} |f(\zeta) - f(z)|,$$

where $C < \infty$ is a constant which is independent of ε. This completes the proof. ∎

1.9.2. Corollary. *Let $D \subset\subset \mathbb{C}^n$ be an open set, and let f be a continuous function on $\bar D$ that is holomorphic in D. Then*

$$f = B_{\partial D} f \quad \text{in} \quad D, \tag{1.9.2}$$

where $B_{\partial D}$ is the integral operator defined in Subsection 1.8.3.

Remark. For $n = 1$, $\omega'_\zeta(\bar\zeta - \bar z) = \bar\zeta - \bar z$ and the Martinelli-Bochner formula is the Cauchy-Green formula (1.1.1).

1.10. The Leray formula

For $n \geqq 2$ the kernel in the Martinelli-Bochner formula does not depend holomorphically on z, because then $(\bar\zeta - \bar z)/|\zeta - z|^{2n}$ is not holomorphic in z. Therefore, in the present section we give a generalization of the Martinelli-Bochner formula which is obtained by replacing the map $\bar{\zeta} - \bar{z}$ by a more general Leray map $w(z, \zeta)$. This generalization was found in 1956 by J. Leray and forms the basic idea in the constructions of formulas for solving the inhomogeneous Cauchy-Riemann equations presented in this book. For important classes of open sets (for example, for strictly pseudoconvex open sets) this Leray map can be chosen to be holomorphic in z and, moreover, to satisfy appropriate growth conditions.

1.10.1 Theorem (Leray formula). *Let D be a bounded open set in $\mathbb{C}^n$ with piecewise C^1-boundary, and let $w(z, \zeta)$ be a Leray map for D. Then, for every continuous function f on $\bar{D}$ such that $\bar{\partial} f$ is also continuous on $\bar{D}$, we have*

$$f = L^w_{\partial D} f - R^w_{\partial D}\, \bar{\partial} f - B_D\, \bar{\partial} f \quad in \quad D\,, \tag{1.10.1}$$

where $L^w_{\partial D}$, $R^w_{\partial D}$ and B_D are the integral operators defined in Subsections 1.8.6 *and* 1.8.3.

Proof. In view of the Martinelli-Bochner formula (1.9.1) we only have to prove that

$$R^w_{\partial D}\, \bar{\partial} f = L^w_{\partial D} f - B_{\partial D} f \quad \text{in} \quad D\,. \tag{1.10.2}$$

Fix $z \in D$. Since $\langle \eta^w(z, \zeta, \lambda), \zeta - z\rangle = 1$ for $0 \leqq \lambda \leqq 1$ and ζ in a neighbourhood of ∂D, Proposition 1.7.1 gives $d_{\zeta,\lambda}[\omega'_{\zeta,\lambda}(\eta^w(z, \zeta, \lambda)) \wedge \omega(\zeta)] = 0$. Since $\partial f(\zeta) \wedge \omega(\zeta) = 0$, this implies that

$$d_{\zeta,\lambda}[f(\zeta)\, \omega'_{\zeta,\lambda}(\eta^w(z, \zeta, \lambda)) \wedge \omega(\zeta)] = \bar{\partial} f(\zeta) \wedge \omega'_{\zeta,\lambda}(\eta^w(z, \zeta, \lambda)) \wedge \omega(\zeta)\,.$$

Now we apply Stokes' formula to $f(\zeta)\, \omega'_{\zeta,\lambda}(\eta^w(z, \zeta, \lambda)) \wedge \omega(\zeta)$ on $\partial D \times [0, 1]$. This implies (1.10.2) if we take into account that, by Proposition 1.7.3,

$$\omega'_{\zeta,\lambda}(\eta^w(z, \zeta, \lambda)) \wedge \omega(\zeta)|_{\lambda=0} = \omega'_\zeta\left(\frac{w(z, \zeta)}{\langle w(z, \zeta), \zeta - z\rangle}\right) \wedge \omega(\zeta) = \frac{\omega'_\zeta(w(z, \zeta)) \wedge \omega(\zeta)}{\langle w(z, \zeta), \zeta - z\rangle^n}$$

and

$$\omega'_{\zeta,\lambda}(\eta^w(z, \zeta, \lambda)) \wedge \omega(\zeta)|_{\lambda=1} = \omega'_\zeta\left(\frac{\bar{\zeta} - \bar{z}}{|\zeta - z|^2}\right) \wedge \omega(\zeta) = \frac{\omega'_\zeta(\bar{\zeta} - \bar{z}) \wedge \omega(\zeta)}{|\zeta - z|^{2n}}\,. \blacksquare$$

1.10.2 Corollary. *Let D be a bounded open set in $\mathbb{C}^n$ with piecewise C^1-boundary, and let $w(z, \zeta)$ be a Leray map for D. Then, for every continuous function f on $\bar{D}$ that is holomorphic in D,*

$$f = L^w_{\partial D} f \quad in \quad D\,, \tag{1.10.3}$$

where $L^w_{\partial D}$ is the integral operator defined in Subsection 1.8.6.

Remark. If $w(z, \zeta) = \bar{\zeta} - \bar{z}$, then $L^w_{\partial D} = B_{\partial D}$ and $R^w_{\partial D} = 0$. Therefore, in this case the Leray formula (1.10.1) is the Martinelli-Bochner formula (1.9.1).

1.10.3. Remark. In the proof of Theorem 1.10.1 we did not use any information about the kind of dependence of $w(z, \zeta)$ on z, that is, we proved the following theorem: *Let D and f be as in Theorem* 1.10.1, *let $z_0 \in D$, and let $w(\zeta)$ be a $\mathbb{C}^n$-valued C^1-map in a neighbourhood of ∂D such that $\langle w(\zeta), \zeta - z_0\rangle \neq 0$ for all $\zeta \in \partial D$. Then $f(z_0)$ $= (L^w_{\partial D} f)(z_0) - (R^w_{\partial D}\, \bar{\partial} f)(z_0) - (B_D\, \bar{\partial} f)(z_0)$.* (Here the numbers $(L^w_{\partial D} f)(z_0)$ and $(R^w_{\partial D}\, \bar{\partial} f)(z_0)$ are defined by (1.8.12)—(1.8.14), where instead of $w(z, \zeta)$ the map $w(\zeta)$ is used.)

1.11. The Koppelman formula

In the case of one complex variable we need only integral representation formulas for functions. For several complex variables we also need such formulas for differential forms. Here we can restrict ourselves to $(0, q)$-forms, because every (p, q)-form can be understood as a $(0, q)$-form with a values in certain holomorphic vector bundle, which will be considered in Sections 2.10—2.13 and 4.12. In the present section we prove the Koppelman formula, which is a generalization of the Martinelli-Bochner formula to differential forms. The Koppelman-Leray formula, which is a generalization of the Leray formula as well as of the Koppelman formula, will be proved in the next section.

1.11.1. Theorem (Koppelman formula). *Let $D \subset\subset \mathbb{C}^n$ be an open set with piecewise C^2-boundary, and let f be a continuous $(0, q)$-form on $\bar{D}$ such that $\bar\partial f$ is also continuous on $\bar{D}$, $0 \leqq q \leqq n$. Then the forms $B_{\partial D}f$, $B_D\, \bar\partial f$, $B_D f$ and $\bar\partial B_D f$ are continuous in D, and we have*

$$(-1)^q f = B_{\partial D} f - B_D\, \bar\partial f + \bar\partial B_D f \quad in \quad D\,. \tag{1.11.1}$$

(Here $B_{\partial D}$ and B_D are the integral operators defined in Subsection 1.8.4.)

Proof. If $q = 0$, then, by (1.8.7), $B_D f = 0$ and (1.11.1) is the Martinelli-Bochner formula (1.9.1). Therefore, we must prove the theorem for $1 \leqq q \leqq n$ only. It is trivial that $B_{\partial D}f$ is continuous in D, and it follows from Lemma 1.8.5 that $B_D\, \bar\partial f$ and $B_D f$ are continuous in D. Therefore, it remains to prove that in the sense of distributions $\bar\partial B_D f = (-1)^q f - B_{\partial D} f + B_D\, \bar\partial f$ in D. That is, we have to prove that, for every C^∞-form v with compact support in D,

$$(-1)^q \int_D B_D f \wedge \bar\partial v = (-1)^q \int_D f \wedge v - \int_D B_{\partial D} f \wedge v + \int_D B_D\, \bar\partial f \wedge v\,. \tag{1.11.2}$$

Since, by (1.8.7) and (1.8.8), $B_{\partial D}f$ and $B_D\, \bar\partial f$ are of bidegree $(0, q)$ and $B_D f$ is of bidegree $(0, q-1)$, we can assume that v is of bidegree $(n, n-q)$. Using the notation

$$\theta(z, \zeta) := \frac{(n-1)!}{(2\pi i)^n} \frac{\omega'_{z,\zeta}(\bar\zeta - \bar z) \wedge \omega(\zeta)}{|\zeta - z|^{2n}}$$

and the definitions of $B_{\partial D}$ and B_D, (1.11.2) can be written

$$(-1)^q \int_{(\zeta,z)\in D\times D} f(\zeta) \wedge \theta(z,\zeta) \wedge \bar\partial v(z) = (-1)^q \int_{z\in D} f(z) \wedge v(z)$$
$$- \int_{(\zeta,z)\in\partial D\times D} f(\zeta) \wedge \theta(z,\zeta) \wedge v(z) + \int_{(\zeta,z)\in D\times D} \bar\partial f(\zeta) \wedge \theta(z,\zeta) \wedge v(z)\,. \tag{1.11.3}$$

Set $\tilde\theta(z, \zeta) := \left((n-1)!/(2\pi i)^n\right) \omega'_{z,\zeta}(\bar\zeta - \bar z) \wedge \omega_{z,\zeta}(\zeta - z)/|\zeta - z|^{2n}$. Then, by Proposition 1.7.1, $d_{z,\zeta}\tilde\theta(z, \zeta) = 0$ for $\zeta \neq z$. Since all monomials in the difference $\tilde\theta(z, \zeta) - \theta(z, \zeta)$ contain at least one of the differentials $dz_1, \dots, dz_n$ and $v(z)$ contains the factor $dz_1 \wedge \dots \wedge dz_n$, it follows that

$$d_{z,\zeta}\left(\theta(z,\zeta) \wedge v(z)\right) = d_{z,\zeta}\left(\tilde\theta(z,\zeta) \wedge v(z)\right)$$
$$= (-1)^{2n-1}\, \tilde\theta(z,\zeta) \wedge dv(z) = -\theta(z,\zeta) \wedge \bar\partial v(z) \quad \text{for} \quad \zeta \neq z\,. \tag{1.11.4}$$

Moreover, $df(\zeta) \wedge \theta(z,\zeta) = \bar\partial f(\zeta) \wedge \theta(z,\zeta)$ for $\zeta \neq z$, because $\theta(z,\zeta)$ contains the factor $d\zeta_1 \wedge \dots \wedge d\zeta_n$. Hence

$$d_{z,\zeta}\left(f(\zeta) \wedge \theta(z,\zeta) \wedge v(z)\right)$$
$$= \bar\partial f(\zeta) \wedge \theta(z,\zeta) \wedge v(z) - (-1)^q f(\zeta) \wedge \theta(z,\zeta) \wedge \bar\partial v(z) \quad \text{for} \quad \zeta \neq z\,. \tag{1.11.5}$$

We consider the open sets

$$U_\varepsilon := \{(\zeta, z) \in \mathbb{C}^n \times \mathbb{C}^n : |\zeta - z| < \varepsilon\}, \qquad \varepsilon > 0 .$$

Since supp $v \subset\subset D$, then, for sufficiently small $\varepsilon > 0$,

$$\partial(D \times D \setminus U_\varepsilon) \cap (\mathbb{C}^n \times \operatorname{supp} v) = (\partial D \times D \cup \partial U_\varepsilon) \cap (\mathbb{C}^n \times \operatorname{supp} v) .$$

This is a relation between sets only without orientation. Taking into account the orientation, we obtain

$$\partial(D \times D \setminus U_\varepsilon) \cap (\mathbb{C}^n \times \operatorname{supp} v) = (\partial D \times D - \partial U_\varepsilon) \cap (\mathbb{C}^n_\circ \times \operatorname{supp} v) .$$

By Stokes' formula, this and (1.11.5) imply that

$$\int_{\partial D \times D} f(\zeta) \wedge \theta(z, \zeta) \wedge v(z) - \int_{\partial U_\varepsilon} f(\zeta) \wedge \theta(z, \zeta) \wedge v(z)$$
$$= \int_{D \times D \setminus U_\varepsilon} \bar\partial f(\zeta) \wedge \theta(z, \zeta) \wedge v(z) - (-1)^q \int_{D \times D \setminus U_\varepsilon} f(\zeta) \wedge \theta(z, \zeta) \wedge \bar\partial v(z) .$$

It is clear that the integrals on the right-hand side tend to the corresponding integrals in (1.11.3) when $\varepsilon \to 0$. Therefore, it remains to prove that

$$\lim_{\varepsilon \to 0} \int_{\partial U_\varepsilon} f(\zeta) \wedge \theta(z, \zeta) \wedge v(z) = (-1)^q \int_D f(z) \wedge v(z) . \tag{1.11.6}$$

Consider the map $T(\xi, z) := (z + \xi, z)$ from $\mathbb{C}^n \times \mathbb{C}^n$ onto itself. If $S_\varepsilon := \{\xi \in \mathbb{C}^n : |\xi| = \varepsilon\}$, then $T(S_\varepsilon \times \mathbb{C}^n) = \partial U_\varepsilon$. Let T^* be the pull-back of differential forms defined by T, and let $f(\zeta) = \sum'_{|I|=q} f_I(\zeta)\, d\bar\zeta^I$. Since $v(z)$ contains the factor $dz_1 \wedge \ldots \wedge dz_n$ and, therefore, $\omega_{z,\xi}(z + \xi) \wedge v(z) = \omega(\xi) \wedge v(z)$, it follows that

$$T^*(f(\zeta) \wedge \theta(z, \zeta) \wedge v(z))$$
$$= \sum_{|I|=q} f_I(z + \xi)\, d(\bar z + \bar\xi)^I \frac{(n-1)!}{(2\pi i)^n} \frac{\omega'(\bar\xi) \wedge \omega(\xi)}{|\xi|^{2n}} \wedge v(z) .$$

Since the degree of $\omega'(\bar\xi) \wedge \omega(\xi)$ is $2n - 1 = \dim_{\mathbb{R}} S_\varepsilon$, we obtain that

$$d(\bar z + \bar\xi)^I \wedge \omega'(\bar\xi) \wedge \omega(\xi)|_{S_\varepsilon \times \mathbb{C}^n} = d\bar z^I \wedge \omega'(\bar\xi) \wedge \omega(\xi)|_{S_\varepsilon \times \mathbb{C}^n} .$$

Moreover, $d\bar z^I \wedge \omega'(\bar\xi) \wedge \omega(\xi) = (-1)^q \omega'(\bar\xi) \wedge \omega(\xi) \wedge d\bar z^I$. Therefore,

$$\int_{\partial U_\varepsilon} f(\zeta) \wedge \theta(z, \zeta) \wedge v(z)$$
$$= (-1)^q \int_{z \in \mathbb{C}^n} \sum_{|I|=q} \left[\frac{(n-1)!}{(2\pi i)^n} \int_{\xi \in S_\varepsilon} f_I(z + \xi) \frac{\omega'(\bar\xi) \wedge \omega(\xi)}{|\xi|^{2n}}\right] d\bar z^I \wedge v(z) .$$

This implies (1.11.6). In fact, by the Martinelli-Bochner formula (1.9.2), the term in brackets is equal to

$$f_I(z) + \frac{(n-1)!}{(2\pi i)^n} \int_{\xi \in S_\varepsilon} (f_I(z + \xi) - f_I(z)) \frac{\omega'(\bar\xi) \wedge \omega(\xi)}{|\xi|^{2n}},$$

which tends to $f_I(z)$ when $\varepsilon \to 0$. ■

1.12. The Koppelman-Leray formula

In this section we prove an integral representation formula, which contains the Koppelman formula (1.11.1) as well as the Leray formula (1.10.1) as special cases. As a corollary we obtain a formula for solving the inhomogeneous Cauchy-Riemann equations for the case that there exists a Leray map $w(z, \zeta)$ which is holomorphic in z.

1.12.1. Theorem (Koppelman-Leray formula). *Let $D \subset\subset \mathbb{C}^n$ be an open set with piecewise C^1-boundary D and let $w(z, \zeta)$ be a Leray map for D. Suppose, in addition, that all derivatives of $w(z, \zeta)$ which are of order $\leqq 2$ in z and of order $\leqq 1$ in ζ are continuous for all $z \in D$ and ζ in some neighbourhood of ∂D.*[1] *Then, for every continuous $(0, q)$-form f on $\overline{D}$ such that $\bar\partial f$ is also continuous on $\overline{D}$, $0 \leqq q \leqq n$, the forms $L^w_{\partial D}f$, $R^w_{\partial D}f$, $B_D\,\bar\partial f$, $R^w_{\partial D}\bar\partial f$, $B_D f$, $\bar\partial R^w_{\partial D} f$ and $\bar\partial B_D f$ are continuous in D, and we have*

$$(-1)^q f = L^w_{\partial D}f - (R^w_{\partial D} + B_D)\,\bar\partial f + \bar\partial(R^w_{\partial D} + B_D)\,f\,. \tag{1.12.1}$$

(Here $L^w_{\partial D}$, $R^w_{\partial D}$ and B_D are the integral operators defined in Section 1.8.)

Remark. If $w(z, \zeta) = \bar\zeta - \bar z$, then $L^w_{\partial D} = B_{\partial D}$, $R^w_{\partial D} = 0$ and, therefore, (1.12.1) is the Koppelman formula (1.11.1). If $q = 0$, then, by (1.8.7) and (1.8.18), $B_D f = 0$ and $R^w_{\partial D} f = 0$, and therefore (1.12.1) is the Leray formula (1.10.1).

Proof of Theorem 1.12.1. By the remark above we can assume that $1 \leqq q \leqq n$. It is clear that $L^w_{\partial D}f$ and $R^w_{\partial D}f$ are continuous in D, and, by Theorem 1.11.1, $B_D\,\bar\partial f$, $B_D f$ and $\bar\partial B_D f$ are continuous in D. Since, by hypothesis of the theorem, $w(z, \zeta)$ is C^2 in z, by differentiation under the sign of integration, we obtain that $\bar\partial R^w_{\partial D} f$ is continuous in D. It remains to prove (1.12.1). In view of the Koppelman formula (1.11.1) we only have to prove that

$$\bar\partial R^w_{\partial D} f = B_{\partial D} f - L^w_{\partial D} f + R^w_{\partial D}\,\bar\partial f \quad \text{in} \quad D\,. \tag{1.12.2}$$

To do this we introduce the following abbreviations: $\omega := \omega(\zeta)$,

$$\theta := \frac{(n-1)!}{(2\pi i)^n} \sum_{j=1}^{n} (-1)^{j+1}\, \eta^w_j(z, \zeta, \lambda) \bigwedge_{k \neq j} d_{z,\zeta,\lambda}\, \eta^w_k(z, \zeta, \lambda)\,,$$

$$\bar\theta := \frac{(n-1)!}{(2\pi i)^n} \sum_{j=1}^{n} (-1)^{j+1}\, \eta^w_j(z, \zeta, \lambda) \bigwedge_{k \neq j} (\bar\partial_{z,\zeta} + d_\lambda)\, \eta^w_k(z, \zeta, \lambda)\,.$$

These forms are continuous in some neighbourhood $W \subseteq D \times \mathbb{C}^n \times [0, 1]$ of $D \times \partial D \times [0, 1]$. Since $\langle \eta^w(z, \zeta, \lambda), \zeta - z\rangle = 1$ in W, it follows from Proposition 1.7.1 that (in the sense of distributions) $d_{z,\zeta,\lambda}(\theta \wedge \omega) = 0$ in W. Since $\partial_\zeta(\theta \wedge \omega) = 0$, this implies that $(\bar\partial_{z,\zeta} + d_\lambda + \partial_z)\,(\theta \wedge \omega) = 0$. Hence

$$(\bar\partial_{z,\zeta} + d_\lambda)\,(\bar\theta \wedge \omega) + \partial_z(\bar\theta \wedge \omega) + (\bar\partial_{z,\zeta} + d_\lambda + \partial_z)\,\big((\theta - \bar\theta) \wedge \omega\big) = 0 \quad \text{in} \quad W\,. \tag{1.12.3}$$

The monomials in $\theta - \bar\theta$ (with respect to dz_j, $d\bar z_j$, $d\zeta_j$, $d\bar\zeta_j$ and $d\lambda$) and, therefore, in $(\bar\partial_{z,\zeta} + d_\lambda + \partial_z)\,\big((\theta - \bar\theta) \wedge \omega\big)$ contain at least one of the differentials $dz_1, \dots, dz_n$ as a factor. The same is true for $\partial_z(\bar\theta \wedge \omega)$, whereas the monomials in $(\bar\partial_{z,\zeta} + d_\lambda)\,(\bar\theta \wedge \omega)$

[1] The theorem is valid also without this assumption, but then its proof becomes technically more complicated. In this book we need this theorem only for the case that $w(z, \zeta)$ is even holomorphic in z.

do not contain any of the differentials dz_j. Therefore, it follows from (1.12.3) that $(\bar\partial_{z,\zeta} + d_\lambda)(\bar\theta \wedge \omega) = 0$ in W, that is,

$$(\bar\partial_\zeta + d_\lambda)(\bar\theta \wedge \omega) = -\bar\partial_z(\bar\theta \wedge \omega)\,.$$

By hypothesis of the theorem, both $\bar\partial f$ and the form on the right-hand side of the last relation are continuous in W (after shrinking W if necessary). Therefore, the following computation is correct (that is, there is no problem concerning products of distributions)

$$d_{\zeta,\lambda}(f \wedge \bar\theta \wedge \omega) = (\bar\partial_\zeta + d_\lambda)(f \wedge \bar\theta \wedge \omega) = (\bar\partial f) \wedge \bar\theta \wedge \omega + (-1)^q f \wedge (\bar\partial_\zeta + d_\lambda)(\bar\theta \wedge \omega) = (\bar\partial f) \wedge \bar\theta \wedge \omega - \bar\partial_z(f \wedge \bar\theta \wedge \omega)\,, \tag{1.12.4}$$

and the forms on the right and the left-hand side of (1.12.4) are continuous in W. Consequently, we can apply Stokes' formula to the form $f \wedge \bar\theta \wedge \omega$ on $\partial D \times [0, 1]$ for fixed $z \in D$. This gives

$$\int\limits_{\partial D \times [0,1]} (\bar\partial f) \wedge \bar\theta \wedge \omega - \bar\partial_z \int\limits_{\partial D \times [0,1]} f \wedge \bar\theta \wedge \omega = \int\limits_{\partial D \times 0} f \wedge \bar\theta \wedge \omega - \int\limits_{\partial D \times 1} f \wedge \bar\theta \wedge \omega\,. \tag{1.12.5}$$

Using the proof of Proposition (1.7.3), one obtains

$$\bar\theta \wedge \omega|_{\lambda=0} = \frac{(n-1)!}{(2\pi i)^n}\,\overline{\omega}'\left(\frac{w(z,\zeta)}{\langle w(z,\zeta), \zeta - z\rangle}\right) \wedge \omega(\zeta) = \frac{(n-1)!}{(2\pi i)^n}\,\frac{\overline{\omega}'(w(z,\zeta)) \wedge \omega(\zeta)}{\langle w(z,\zeta), \zeta - z\rangle^n}$$

and

$$\bar\theta \wedge \omega|_{\lambda=1} = \frac{(n-1)!}{(2\pi i)^n}\,\overline{\omega}'\left(\frac{\bar\zeta - \bar z}{|\zeta - z|^2}\right) \wedge \omega(\zeta) = \frac{(n-1)!}{(2\pi i)^n}\,\frac{\omega'(\bar\zeta - \bar z) \wedge \omega(\zeta)}{|\zeta - z|^{2n}}\,.$$

Consequently, (1.12.5) coincides with (1.12.2). ■

1.12.2. Corollary. *Let $D \subset\subset \mathbb{C}^n$ be an open set with piecewise C^1-boundary, and let $w(z,\zeta)$ be a Leray map for D, which depends holomorphically on $z \in D$. For $q = 1, \dots, n$, we introduce the abbreviation*

$$T_q := (-1)^q (R^w_{\partial D} + B_D)\,. \tag{1.12.6}$$

Then, for every continuous $(0, q)$-form f on $\bar D$ such that $\bar\partial f$ is also continuous on $\bar D$, $1 \leq q \leq n$, we have

$$f = \bar\partial T_q f + T_{q+1}\bar\partial f\,. \tag{1.12.7}$$

If $\bar\partial f = 0$, then $u := T_q f$ is a continuous solution of the equation $\bar\partial u = f$ in D. Moreover, this solution u is C^α in D for all $0 < \alpha < 1$, and, if f is C^k in D, $k = 1, 2, \dots, \infty$, then u is $C^{k+\alpha}$ in D for all $0 < \alpha < 1$.

Proof. By (1.8.19), $L^w_{\partial D} f = \int\limits_{\partial D} f \wedge \Omega^w_q$, where Ω^w_q is the sum of all monomials in

$$\frac{(n-1)!}{(2\pi i)^n} \sum_{j=1}^n (-1)^{j+1} \frac{w_j(z,\zeta)}{\langle w(z,\zeta), \zeta - z\rangle^n} \bigwedge_{k \neq j} (\bar\partial_z + \bar\partial_\zeta)\, w_k(z,\zeta) \wedge \omega(\zeta)$$

which are of bidegree $(0, q)$ in z. However, since $w(z,\zeta)$ is holomorphic in z, these monomials vanish if $q \geq 1$. Therefore, $L^w_{\partial D} f = 0$ and (1.12.7) follows from the Koppelman-Leray formula (1.12.1). It is clear from (1.12.7) that $u = T_q f$ is a solution of $\bar{\partial}u = f$ if $\bar{\partial}f = 0$. The assertion about the smoothness of u follows from Lemma 1.8.5 (i) and the obvious fact that $R^w_{\partial D}f$ is C^∞ in D. ■

Notes

The Cauchy-Green formula for the first time appeared in Pompeiju [1] (1904). This formula is one of the basic tools in the theory of functions of one complex variable. Examples for profound applications of this formula can be found in Carleson [1] and Vitruškin [1]. Theorem 1.2.1 was proved in 1951 by Grothendieck (see Dolbeault [1]). Theorem 1.2.2 was found in 1906 by Hartogs [1]. The proof given in Section 1.2 is due to Ehrenpreis [1] (see also Hörmander [1]). The characterization of domains of holomorphy in Theorem 1.3.7 is due to Cartan/Thullen [1] (1932). Plurisubharmonic functions were introduced by Oka [1] (1941) and Lelong [1] (1945). The presentation in Section 1.4 partially follows Hörmander [1]. Theorem 1.4.14 is due to Narasimhan [1]. Theorems 1.4.15 and 1.4.18 are proved in Harvey/Wells [1, 2]. Theorem 1.4.21 is due to Harvey/Wells [1] and Narasimhan [1]. The presentation of pseudoconvex sets in Section 1.5 partially follows Hörmander [1]. Corollary 1.5.6 was proved in 1910 by E. Levi [1], who also stated the converse problem (the Levi problem). The Martinelli-Bochner formula was found by Martinelli [1, 2] (1938, 1943) and Bochner [1] (1943). Its generalization to differential forms, the Koppelman formula, was obtained in 1967 by Koppelman [1]. The Leray formula for holomorphic functions (Corollary 1.10.2) was proved in 1956 by Leray [1] using an idea of Fantappiè [1] (1943). Its generalization to smooth functions (Theorem 1.10.1) was obtained in Henkin [1] (1970) (see also Grauert/Lieb [1] (1970)). The Leray-Koppelman formula for the first time appeared in Lieb [1] (1970) and Øvrelid [1] (1971) (see also Poljakov [1] (1971)).

Exercises, remarks and problems (one complex variable)

Here we give exercises, remarks and problems concerning the case of one complex variable which are of interest from the point of view of several complex variables.

1. Let $D \subset\subset \mathbb{C}^1$ be an open set. Prove that, for every $f \in C^\infty(D)$, the equation $\partial u/\partial\bar{z} = f$ has a solution $u \in C^\infty(D)$. Hint. Use Theorem 1.1.3 and the Runge approximation theorem.

2. (Solvability of the Cousin problems). Let $U_j \subseteqq \mathbb{C}^1$, $j \in J$, be open sets and $D = \bigcup_j U_j$, and let $f_{ij} \in \mathcal{O}(U_i \cap U_j)$, $i, j \in J$, such that $f_{ij} = f_{ik} + f_{kj}$ in $U_i \cap U_j \cap U_k$ for all $i, j, k \in J$. Prove that there exist $f_j \in \mathcal{O}(U_j)$, $j \in J$, such that $f_{ij} = f_i - f_j$ in $U_i \cap U_j$ for all $i, j \in J$. Hint. Consider the case $J = \{1, 2\}$. If $c_j \in C^\infty(U_j)$, $j = 1, 2$, such that $f_{12} = c_1 - c_2$ in $U_1 \cap U_2$, then, by setting $\varphi := \partial c_j/\partial\bar{z}$ on U_j, we obtain $\varphi \in C^\infty(U_1 \cup U_2)$. If $\partial u/\partial\bar{z} = \varphi$, where $u \in C^\infty(U_1 \cup U_2)$, then $f_j := c_j - u$, $j = 1, 2$, solves the problem.

3. Let $R_1, R_2 \subset\subset \mathbb{C}^1$ be open rectangles such that $R_1 \cup R_2$ is also a rectangle.

a) Prove that, for every bounded holomorphic function f in $R_1 \cap R_2$, there are bounded holomorphic functions f_j in R_j such that $f = f_1 - f_2$ in $R_1 \cap R_2$.

b) Prove that, for every $0 < \alpha < 1$ and every $f \in H^\alpha(\overline{R_1 \cap R_2}) \cap \mathcal{O}(R_1 \cap R_2)$ there are $f_j \in H^\alpha(\overline{R_j}) \cap \mathcal{O}(R_j)$ such that $f = f_1 - f_2$ in $\overline{R_1 \cap R_2}$.

c) Prove that for every $f \in C^0(\overline{R_1 \cap R_2}) \cap \mathcal{O}(R_1 \cap R_2)$ there exist $f_j \in C^0(\overline{R_j}) \cap \mathcal{O}(R_j)$ such that $f = f_1 - f_2$ in $\overline{R_1 \cap R_2}$.

Hint. Use the fact that, by Theorem 1.1.3 and Lemma 1.8.5 (i), for every bounded continuous function φ on $R_1 \cup R_2$ there exists a solution of $\partial u/\partial\bar{z} = \varphi$ which belongs to $H^\alpha(R_1 \cup R_2)$ for all $0 < \alpha < 1$.

4. Let $D \subseteqq \mathbb{C}^1$ be open and $f_1, f_2 \in \mathcal{O}(D)$ such that $|f_1(z)| + |f_2(z)| \neq 0$ for all $z \in D$. Prove that there exist $g_1, g_2 \in \mathcal{O}(D)$ such that $g_1 f_1 + g_2 f_2 \equiv 1$ on D.

5. Let $D \subset\subset \mathbb{C}^1$ be an open set with rectifiable boundary. For $f \in C^0(\partial D)$, we define $K_D f(z) := \int_{\partial D} f(\zeta)\, d\zeta/(\zeta - z)$, $z \in D$.

a) (Plemelj-Privalov theorem, see, for example, MUSCHELIŠVILI [1]). Prove that, if D is the unit disc, then K_D is bounded as an operator from $H^\alpha(\partial D)$ into $H^\alpha(\bar{D})$, $0 < \alpha < 1$.

b) It is easy to see that Exercise 3b) can be solved by means of the Plemelj-Privalov theorem. Prove that this is possible also for Exercise 3c).

c) (SALAEV [1]). Suppose there is a constant $C < \infty$ such that, for every $\xi \in \partial D$ and all $\delta > 0$, the length of the curve $\{\zeta \in \partial D : |\zeta - \xi| < \delta\}$ is $\leqq C\delta$. Prove that, for every $0 < \alpha < 1$, K_D is bounded as an operator from $H^\alpha(\partial D)$ into $H^\alpha(\bar{D})$.

d) Remark. DYNKIN [1] constructed an example of a domain $D \subset\subset \mathbb{C}^1$ with rectifiable boundary such that K_D is not bounded as an operator from $H^{1/2}(\partial D)$ into $C^0(\bar{D})$.

6 (RUDIN [1], see also HOFFMAN [1]). Let $D \subset\subset \mathbb{C}^1$ be the unit disc. Prove that there does not exist a bounded linear projection from $H^\alpha(\partial D)$ onto $H^\alpha(\bar{D}) \cap \mathcal{O}(D)$ if $\alpha = 0, 1$.

Hint. Assume there exists such a projection P; prove that then $\tilde{P}f := (1/2\pi) \int_{-\pi}^{\pi} (Pf_t)_{-t}\, dt$ is also bounded, where $f_t(s) := f(s + t)$; prove that $\tilde{P} = K_D$ (Exercise 5); this is not possible, because K_D is not bounded between these spaces.

7. Let $D \subset\subset \mathbb{C}^1$ be the unit disc. Prove that there does not exist a bounded linear projection from $C^1(\bar{D})$ onto $C^1(\bar{D}) \cap \mathcal{O}(D)$ ($C^1(\bar{D})$ is the space of all $f \in C^1(D)$ whose first-order derivatives are continuous on $\bar{D}$).

8. Remark. Let $D \subseteqq \mathbb{C}^1$ be the unit disc. A complex measure μ on D is called a *Carleson measure* if there is a constant $C < \infty$ such that, for all $\delta > 0$ and $z \in \partial D$,

$$\int_{\zeta \in D, |\zeta - z| < \delta} |d\mu| (\zeta) \leqq C\delta .$$

For $p > 0$ we denote by $\mathfrak{H}^p(D)$ the space of holomorphic functions f in D such that

$$\sup_{0<r<1} \int_{|\zeta|=1} |f(r\zeta)|^p \, |d\zeta| < \infty .$$

Theorem (CARLESON [2], for a simpler proof see HAVIN/VINOGRADOV [1]). *Let μ be a complex measure on D and let $p > 0$. Then $\mathfrak{H}^p(D)$ is continuously imbedded in $L^p(|d\mu|, D)$ (by the natural map) if and only if μ is a Carleson measure.*

In particular, this implies that every Carleson measure μ on D defines a continuous linear functional $h \to \int_D h(\zeta)\, d\mu(\zeta)$ on $\mathfrak{H}^1(D)$. Conversely, it was proved by VAROPOULOS [1] that every continuous linear functional on $\mathfrak{H}^1(D)$ can be defined by a Carleson measure.

One of the important results in the theory of functions of one complex variable, whose generalization to the case of several complex variables is an open problem (cp. point 14b) in Exercises, remarks and problems at the end of Chapter 2) is the following Corona theorem of CARLESON [2, 3]:

Corona theorem. *Let $f_1, \dots, f_n$ be bounded holomorphic functions in D such that $\inf_{z \in D} |f(z)| > 0$, where $|f|^2 := \sum_{j=1}^n |f_j|^2$. Then there exist bounded holomorphic functions h_j in D such that $\sum_{j=1}^n f_j h_j = 1$ in D.*

We give an outline of CARLESON's [2, 3] original proof as well as of WOLFF's recent more elementary proof (see KOOSIS [1] and GAMELIN [1]).

Carleson's proof. 1. The „soft" part. From CARLESON's imbedding theorem above and the Hahn-Banach theorem we obtain that, for every Carleson measure μ, there is a function $u \in L^\infty(\partial D)$ such that

$$\frac{1}{2\pi i} \int_D \frac{d\mu(\zeta)}{\zeta - z} + \frac{1}{2\pi i} \int_{\partial D} \frac{u(\zeta)\, d\zeta}{\zeta - z} = 0 \quad \text{for} \quad z \in \mathbb{C}^1 \setminus D . \tag{*}$$

For $z \in D$ the left-hand side of (*) defines a function u in D that is bounded on ∂D and such that $\partial u/\partial \bar{z} = \mu$ in D.

2. The „hard" part. For simplicity let $n = 2$. This part amounts to finding $\varepsilon > 0$ and functions φ_1, φ_2 in D such that $\varphi_1 + \varphi_2 \equiv 1$, $0 \leqq \varphi_j \leqq 1$, $\varphi_j(z) = 0$ if $|f_j(z)| \leqq \varepsilon$, $(\partial\varphi_j/\partial\bar{z})\, d\bar{z} \wedge dz$ is a Carleson measure ($j = 1, 2$). Then, by the „soft" part we can find a function

u in D which is bounded on ∂D such that

$$\frac{\partial u}{\partial \bar z} = \frac{1}{|f|^2}\left(\frac{\bar f_1}{f_2}\frac{\partial \varphi_2}{\partial \bar z} - \frac{\bar f_2}{f_1}\frac{\partial \varphi_1}{\partial \bar z}\right),$$

and it remains to set $h_1 := (\varphi_1/f_1) + uf_2$ and $h_2 := (\varphi_2/f_2) - uf_1$. ■

Wolff's proof. For simplicity let $n = 2$. Set $\psi_j := \bar f_j/|f|^2$. Then $f_1\psi_1 + f_2\psi_2 = 1$. Set $U_j := \{z \in D : f_j(z) \neq 0\}$, $\varphi := \partial(\psi_1/f_2)/\partial \bar z$ in U_2 and $\varphi := -\partial(\psi_2/f_1)/\partial \bar z$ in U_1. It remains to find a function u in D such that u is bounded on ∂D and $\partial u/\partial \bar z = \varphi$, because then we can set $h_1 := \psi_1 - uf_2$ and $h_2 := \psi_2 - f_1 u$. To find u we take the solution u_0 of $\partial u_0/\partial \bar z = \varphi$ which is given by the Cauchy-Green formula and look for a holomorphic function g in D such that $u_0 + g$ is bounded on ∂D. This means, essentially, we have to estimate the norm of $u_0|_{\partial D}$ as a linear functional on $\mathfrak{H}^1(D)$. This can be reduced to the proof of the fact that $(1 - |z|)\,|\varphi|^2\, d\bar z \wedge dz$ and $(1 - |z|)\,|\partial\varphi/\partial z|\, d\bar z \wedge dz$ are Carleson measures and therefore define continuous linear functionals on $\mathfrak{H}^1(D)$. ■

9. (Poljakov [2]). Let $D_1, D_2 \subset\subset \mathbb{C}^1$ be open discs. Prove that, for every bounded holomorphic function f in $D_1 \cap D_2$, there are bounded holomorphic functions f_j in D_j such that $f = f_1 - f_2$ in $D_1 \cap D_2$.

10. Let $D \subseteq \mathbb{C}^1$ be an open set and let A be an $n \times n$ matrix whose entries are continuous complex-valued functions in D. Consider the equation

$$\partial u/\partial \bar z + Au = f, \qquad (**)$$

where f is a given continuous $\mathbb{C}^n$-valued function in D. Prove that (**) can be solved locally. Hint. Use that the operator for solving $\partial u/\partial \bar z = f$ given in Theorem 1.1.3 is „small" if D is „small".

11. Let $R_1, R_2 \subset\subset \mathbb{C}^1$ be open rectangles such that $R_1 \cup R_2$ also is a rectangle. We denote by $L(n, \mathbb{C})$ the space of all complex $n \times n$ matrices, and we denote by $GL(n, \mathbb{C})$ the group of invertible matrices in $L(n, \mathbb{C})$.

a) (Cartan's lemma, Cartan [1]). Prove that, for every continuous function $A: \overline{R_1 \cap R_2} \to GL(n, \mathbb{C})$ that is holomorphic in $R_1 \cap R_2$, there are continuous functions $A_j: \overline{R}_j \to GL(n, \mathbb{C})$ which are holomorphic in R_j such that $A = A_1A_2$. Hint. The group of all continuous $GL(n, \mathbb{C})$-valued functions on $\overline{R_1 \cap R_2}$ which are holomorphic in $R_1 \cap R_2$ is connected; consequently, $A = (\exp M_1) \dots (\exp M_k)$, where $M_j: \overline{R_1 \cap R_2} \to L(n, \mathbb{C})$ are continuous on $\overline{R_1 \cap R_2}$ and holomorphic in $R_1 \cap R_2$; consequently, A can be approximated uniformly on $\overline{R_1 \cap R_2}$ by holomorphic functions in $\mathbb{C}^1$ with values in $GL(n, \mathbb{C})$; consequently, we have to prove the assertion only for the case that A is sufficiently close to the unit matrix; this can be done by means of the implicit function theorem in Banach spaces and Exercise 3c).

b) Let $0 < \alpha < 1$. Prove that, for every α-Hölder continuous function $A: \overline{R_1 \cap R_2} \to GL(n, \mathbb{C})$ that is holomorphic in $R_1 \cap R_2$, there are α-Hölder continuous functions $A_j: \overline{R}_j \to GL(n, \mathbb{C})$ which are holomorphic in R_j such that $A = A_1A_2$. Warning. The set of holomorphic functions in a neighbourhood of $\overline{R_1 \cap R_2}$ is not dense in $H^\alpha(\overline{R_1 \cap R_2}) \cap \mathcal{O}(R_1 \cap R_2)$. Hint. Prove and use a version of the Implicit function theorem in Banach spaces with two norms. (For another proof, see, for example, Leiterer [1].)

12 (Röhrl [1]). Let $U_j \subseteq \mathbb{C}^1$, $j \in J$, be open sets, let $D := \bigcup_{j \in J} U_j$, and let $A_{ij}: U_i \cap U_j \to GL(n, \mathbb{C})$ (see Exercise 11) be a system of holomorphic functions such that $A_{ij}A_{jk} = A_{ik}$ in $U_i \cap U_j \cap U_k$ for all $i, j, k \in J$. Prove that there exists a system of holomorphic functions $A_j: U_j \to GL(n, \mathbb{C})$ such that $A_{ij} = A_iA_j^{-1}$ in $U_i \cap U_j$ for all $i, j \in J$. Remark. This statement admits a profound generalization to the case of several complex variables (Grauert [1, 2, 3], see also Cartan [2]). For simple direct proofs of the special case above see, for example, Grauert/Remmert [1] and Gohberg/Leiterer [1, 2, 3].

13. Let $L(n, \mathbb{C})$, $GL(n, \mathbb{C})$ be as in Exercise 11. Prove that, for every open $D \subseteq \mathbb{C}^1$ and every continuous $A: D \to L(n, \mathbb{C})$, there exists a continuous solution $U: D \to GL(n, \mathbb{C})$ of

$$\frac{\partial U}{\partial \bar{z}} = UA \quad \text{in} \quad D\,. \tag{***}$$

Hint. Solve (***) locally; if U_1, U_2 are two local solutions, then $U_1 U_2^{-1}$ is holomorphic; use Exercise 12.

14. Prove that (**) can be solved globally. Hint. Solve $\partial v/\partial z = Uf$, where U is a solution of (***), and set $u := U^{-1}v$.

15. Remark. Let $D \subseteq \mathbb{C}^1$ be the unit disc, and let $z_k \in D$ $(k = 1, 2, \ldots)$ be a sequence such that

$$\inf_k |B_k(z_k)| > 0\,, \quad \text{where} \quad B_k(z) := \prod_{\substack{j=1\\ j \neq k}}^{\infty} \frac{\bar{z}_j}{|z_j|} \frac{z_j - z}{1 - \bar{z}_j z}\,.$$

Theorem (Carleson [2], see also Havin/Vinogradov [1]). *For every bounded sequence w_k of complex numbers, there exists a bounded holomorphic function h in D such that $h(z_k) = w_k$ $(k = 1, 2, \ldots)$.*

This theorem can be proved by the following scheme: Let B be the Blaschke product of the sequence $\{z_k\}$; prove that $(\partial(1/B)/\partial\bar{z})\, d\bar{z} \wedge dz$ is a Carleson measure (Remark 8); let $W(z) := 0$ if $z \notin \{z_1, z_2, \ldots\}$ and $W(z_k) := w_k$ $(k = 1, 2, \ldots)$; using the result stated in the „soft" part of Carleson's proof of the Corona theorem (Remark 8), we obtain a solution of $\partial u/\partial\bar{z} = WB$ that is bounded on ∂D; set $h := Bu$. For the details of this simplified version of Carleson's [2] proof see Amar [2].

Remark also that, by a recent result of Jones (see Gorin/Hruščev/Vinogradov [1]), this theorem can be proved by means of the following explicit formula:

$$h(z) := \sum_{k=1}^{\infty} w_k \frac{B_k(z)}{B_k(z_k)} \frac{F_k(z)}{F_k(z_k)}\,,$$

where

$$F_k(z) := \left(\frac{1 - |z_k|^2}{1 - \bar{z}_k}\right)^2 \exp\left(-\frac{1}{2 \log(e/\delta^2)} \sum_{j=k}^{\infty} \frac{1 + \bar{z}_k z}{1 - \bar{z}_k z}(1 - |z_k|^2)\right).$$

16. Let R_1, R_2 and $GL(n, \mathbb{C})$ be as in Exercise 11, and let $A: R_1 \cap R_2 \to GL(n, \mathbb{C})$ be a bounded holomorphic function such that A^{-1} also is bounded in $R_1 \cap R_2$.

a) **Open problem.** Do there exist bounded holomorphic functions $A_j: R_j \to GL(n, \mathbb{C})$ such that $A = A_1 A_2$ in $R_1 \cap R_2$?

b) Prove that for $n = 1$ the answer is affirmative.

Exercises, remarks and problems (several complex variables)

17. Prove that for every analytic polyhedron $D \subset\subset \mathbb{C}^n$ there exist an open polydisc $P \subset\subset \mathbb{C}^N$ (for some $N > n$) and a biholomorphic map h from some neighbourhood of $\bar{D}$ onto some closed submanifold Y in some neighbourhood of $\bar{P}$ such that $h(D) = P \cap Y$.

18 (Čirka/Henkin [1], Fornaess [1]). Prove that for every strictly pseudoconvex C^2-domain $D \subset\subset \mathbb{C}^n$ there exist a strictly convex C^2-domain $G \subset\subset \mathbb{C}^N$ (for some $N > n$), a closed complex submanifold Y in some neighbourhood of G such that the intersection $Y \cap \partial G$ is transversal, and a biholomorphic map h from some neighbourhood of $\bar{D}$ onto Y such that $h(D) = G \cap Y$.

19. Open problem. Is the open unit ball in $\mathbb{C}^n$ $(n \geqq 2)$ biholomorphically equivalent to some closed complex submanifold in a bounded open polydisc?

20. Open problem. Let $D \subset\subset \mathbb{C}^n$ $(n \geqq 2)$ be a pseudoconvex domain with C^∞-boundary. Does there exist a closed complex submanifold X in some convex domain $G \subset\subset \mathbb{C}^N$ (for some $N > n$) such that D is biholomorphically equivalent to X? Remark. Kohn/Nirenberg [1] constructed a pseudoconvex domain $D \subset\subset \mathbb{C}^n$ with real analytic boundary and with the following property: There does not exist a closed C^∞-submanifold Y in a neighbourhood of some convex C^∞ domain $G \subset\subset \mathbb{C}^N$ such that $G \cap Y$ is a complex submanifold of G which is biholomorphically equivalent to D.

21. Subharmonicity and plurisubharmonicity can be defined also for upper continuous functions with values in $[-\infty, +\infty)$ (by the same Definitions 1.4.1 and 1.4.6). Prove that, for every holomorphic function f in the open set $D \subseteq \mathbb{C}^n$, $\ln |f|$ is plurisubharmonic in D; if $f \neq 0$, then $\ln |f|$ is even pluriharmonic.

22 (Poincaré-Lelong equality). Let f be a holomorphic function in the open set $D \subseteq \mathbb{C}^n$, and $M_f := \{z \in D : f(z) = 0\}$.

a) Suppose, additionally, that M_f is a smooth submanifold of D. Prove that: 1. M_f is a complex submanifold of D; 2. $\dim_{\mathbb{C}} M_f = n - 1$; 3. there is an integer $1 \leq k < \infty$ such that, for every fixed $w \in M_f$, $0 < \lim\limits_{z \to w} |f(z)|/|z - w|^k < \infty$; 4. for every C^∞-function φ with compact support in D, we have

$$\int_D \varphi \, \Delta \ln |f| \, d\sigma_{2n} = 2\pi k \int_{M_f} \varphi \, d\sigma_{2n-2} ,$$

where Δ is the Laplacian and $d\sigma_{2n}$, $d\sigma_{2n-2}$ are the Euclidean volume forms in $\mathbb{C}^n$ and M_f, respectively.

b) Consider the general case (without the condition that M_f is smooth). Let M_j be the regular components of M_f (for properties of zero sets of holomorphic functions (analytic sets) see, for example, GUNNING/ROSSI [1]), and let k_j be the order of vanishing of f on M_j. Prove that then

$$\int_D \varphi \, \Delta \ln |f| \, d\sigma_{2n} = 2\pi \sum k_j \int_{M_j} \varphi \, d\sigma_{2n-2} .$$

23. Let $u: [0, 1] \to \mathbb{C}^n$ be a continuous map, and let $D \subset\subset \mathbb{C}^m$ be an open set. Then every holomorphic function in some neighbourhood of $(u(0) \times \partial D) \cup \bigcup\limits_{0<t\leq 1} (u(t) \times \overline{D}) \subseteq \mathbb{C}^n \times \mathbb{C}^m$ admits a holomorphic extension to a neighbourhood of $\bigcup\limits_{0\leq t\leq 1} (u(t) \times D)$.
Hint. Use the proof of Theorem 1.3.5.

24 (KNESER [1]). Let $D_1, D_2 \subseteq \mathbb{C}^n$ be pseudoconvex open sets with C^2-boundary. Suppose that the intersection $\partial D_1 \cap \partial D_2$ is transversal. Prove that $D_1 \cup D_2$ is pseudoconvex if and only if $\partial D_1 \cap \partial D_2$ is a complex submanifold of $\mathbb{C}^n$.

25. Let $D_n \subset\subset \mathbb{C}^{n^2}$ be the set of all complex $n \times n$ matrices Z such that $ZZ^* - I$ is strictly negative-definite. Here I is the unit matrix and $Z^* = (\bar{z}_{ji})$ if $Z = (z_{ij})$ (i = row index).

a) Prove that D_n is convex.

b) Prove that D_n is not strictly pseudoconvex if $n > 1$.

26. Let $D \subset\subset \mathbb{C}^n$ be an analytic polyhedron, and let $f_1, \dots, f_N$ be holomorphic functions in a neighbourhood U of $\overline{D}$ such that $D = \{z \in U : |f_j(z)| < 1, j = 1, \dots, N\}$. Set $\Gamma_{n-1} := \{z \in U :$ there exists $1 \leq j_1 < \dots < j_{n-1} \leq N$ such that $|f_{j_1}(z)| = \dots = |f_{j_{n-1}}(z)| = 1\}$. Prove that every holomorphic function f in some neighbourhood of Γ_{n-1} admits a holomorphic extension to D.

27. Let $D \subset\subset \mathbb{C}^n$ be an open set with C^1-boundary, and let $f = (f_1, \dots, f_n)$ be a collection of holomorphic functions in some neighbourhood of $\overline{D}$ satisfying the following conditions: (i) $f \neq 0$ on ∂D. (ii) f is biholomorphic in some neighbourhood of every point $z \in D$ where $f(z) = 0$. Then

$$\frac{(n-1)!}{(2\pi i)^n} \int_{\partial D} \frac{\omega'(f) \wedge \omega(f)}{|f|^{2n}}$$

is the number of zeros of f in D.

28 (BOCHNER [1], HARVEY/LAWSON [1], DAUTOV/KYTMANOV [1], WEINSTOCK [1]). Let $D \subset\subset \mathbb{C}^n$ be a domain with C^2-boundary, and let f be a continuous function on ∂D. Set $f_+(z) := (B_{\partial D} f)(z)$ for $z \in D$ and $f_-(z) := (B_{\partial D} f)(z)$ for $z \in \mathbb{C}^n \setminus \overline{D}$. (For $z \in \mathbb{C}^n \setminus \overline{D}$ we define $(B_{\partial D} f)(z)$ also by (1.8.2)).

a) Let $\nu(z)$, $z \in \partial D$, be the unit exterior normal vector of ∂D at z. Prove that

$$\lim_{\varepsilon \to 0} [f_+(z - \varepsilon \nu(z)) - f_-(z + \varepsilon \nu(z))] = f(z) ,$$

uniformly in $z \in \partial D$.

b) If $\int_{\partial D} f \wedge \bar{\partial} g = 0$ for every $C^\infty_{(n,n-2)}$-form g in some neighbourhood of ∂D, then f_+ is holomorphic in D and f_- is holomorphic in $\mathbb{C}^n \setminus \overline{D}$. If, moreover, $\mathbb{C}^n \setminus \overline{D}$ is connected, then $f = f_+$.

29 (E. Levi [1], H. Lewy [1, 2]). Let $D \subset\subset \mathbb{C}^2$ be a strictly pseudoconvex open set with C^2-boundary, and let ϱ be a strictly plurisubharmonic C^2-function in some neighbourhood θ of $\overline{D}$ such that $D = \{z \in \theta : \varrho(z) < 0\}$ and $d\varrho(z) \neq 0$ for $z \in \partial D$. Let f be a C^1-function in some neighbourhood U of $\xi \in \partial D$. Define

$$Lf := \frac{\partial \varrho}{\partial \bar{z}_1} \frac{\partial f}{\partial \bar{z}_2} - \frac{\partial \varrho}{\partial \bar{z}_2} \frac{\partial f}{\partial \bar{z}_1} \quad \text{in} \quad U .$$

a) Prove that $Lf|_{\partial D} = 0$ if and only if $\int_{\partial D \cap U} f \wedge \bar{\partial} g = 0$ for every $C^\infty_{(2,0)}$-form g with compact support in U. (This is true also without the hypothesis that ϱ is strictly plurisubharmonic.)

b) Let $\delta > 0$ and $\hat{H}_\xi(\delta) := \{z \in \theta : |\hat{F}_\varrho(z, \xi)| < \delta^2\}$ (for the definition of $\hat{F}_\varrho(z, \xi)$ see (1.4.18)). Prove that, if $Lf = 0$ on $\hat{H}_\xi(\delta) \cap \partial D$, then f admits a C^1-continuation to $\hat{H}_\xi(\delta) \cap \overline{D}$ that is holomorphic in $\hat{H}_\xi(\delta) \cap D$.

30 (Agranovskij/Val'skij [1], Stout [2]). Let $D \subset\subset \mathbb{C}^n$ be an open set with C^∞-boundary, and let f be a continuous function on ∂D. Prove that, if, for every complex line $X \subseteq \mathbb{C}^n$, f admits a continuous continuation to $\overline{D} \cap X$ that is holomorphic in $D \cap X$, then f admits a continuous continuation to $\overline{D}$ that is holomorphic in D.

2. The $\bar{\partial}$-equation and the „fundamental problems" of the theory of functions on Stein manifolds

Summary. In Section 2.1 we construct a formula for solving the $\bar{\partial}$-equation in strictly convex open sets in $\mathbb{C}^n$ with C^2-boundary. In Section 2.2 we prove that the solution obtained by means of this formula admits 1/2-Hölder estimates. In Section 2.3 we prove solvability of the $\bar{\partial}$-equation with 1/2-Hölder estimates in C^2 strictly pseudoconvex open sets in $\mathbb{C}^n$. Here the main idea is that, after an appropriate choice of local holomorphic coordinates, the boundary of such sets is strictly convex. This makes it possible to use locally the formula from Section 2.1 and, by means of the Hölder estimates, to obtain Fredholm solvability. Some elementary additional arguments then complete the proof. In Sections 2.4 and 2.5, for C^2 strictly pseudoconvex open sets, we construct a Leray map $w(z, \zeta)$ which is holomorphic in z. In Section 2.6 we prove 1/2-Hölder estimates for the corresponding integral operators. This makes it possible to obtain a more explicit solution of the $\bar{\partial}$-equation with 1/2-Hölder estimates. In Section 2.7 by means of the Leray map from Section 2.5 we prove that, if $D \subseteqq \mathbb{C}^n$ is a pseudoconvex open set, then every holomorphic function in a neighbourhood of a $P^0(D)$-convex compact subset K of D can be approximated uniformly on K by holomorphic functions in D. Then we show that every pseudoconvex open set in $\mathbb{C}^n$ is a domain of holomorphy (solution of the Levi problem) and, in Section 2.8, we prove that the $\bar{\partial}$-equation can be solved in arbitrary pseudoconvex open sets in $\mathbb{C}^n$. In Section 2.9 we use the Leray map from Section 2.5 to show that, for every C^2 strictly pseudoconvex open set $D \subset\subset \mathbb{C}^n$, every continuous function on $\bar{D}$ that is holomorphic in D can be approximated uniformly on $\bar{D}$ by holomorphic functions in a neighbourhood of $\bar{D}$.

In Sections 2.10—2.13 we investigate the $\bar{\partial}$-equation in a more general situation: We consider $(0, q)$-forms with values in holomorphic vector bundles over complex manifolds. First, in the same way as in Section 2.3, we prove Fredholm solvability (with 1/2-Hölder estimates) of the $\bar{\partial}$-equation on complex manifolds with strictly pseudoconvex C^2-boundary (Section 2.11). Then, in Section 2.12, we prove that the $\bar{\partial}$-equation can be solved on complex manifolds which admit a strictly plurisubharmonic exhausting C^2-function — if D is a C^2 strictly pseudoconvex open set in such a manifold, then 1/2-Hölder estimates are obtained. Here the main idea is an inductive procedure with respect to the levels of a strictly plurisubharmonic exhausting C^2-function whose critical points are isolated (using the results in $\mathbb{C}^n$). In Section 2.13 we give the definition of a Stein manifold and prove that a complex manifold is Stein if and only if it admits a strictly plurisubharmonic exhausting C^2-function (solution of the Levi problem for manifolds).

2.1. Formula for solving the $\bar{\partial}$-equation in C^2 strictly convex open sets

Let $D \subset\subset \mathbb{C}^n$ be a strictly convex open set with C^2-boundary, that is, let

$$D = \{z \in \mathbb{C}^n : \varrho(z) < 0\} , \qquad (2.1.1)$$

where ϱ is a real-valued C^2-function in $\mathbb{C}^n$ such that, for some $\alpha > 0$,

$$\sum_{j,k=1}^{2n} \frac{\partial^2 \varrho(z)}{\partial x_j \, \partial x_k} t_j t_k \geqq \alpha \, |t|^2 \quad \text{for all} \quad z \in \partial D \quad \text{and} \quad t \in \mathbb{R}^{2n}, \tag{2.1.2}$$

where $x_j = x_j(z)$ are the real coordinates of $z \in \mathbb{C}^n$ such that $z_j = x_j(z) + \mathrm{i}x_{j+n}(z)$.

2.1.1. Remark. Since no point of ∂D is a local minimum of ϱ, it follows from (2.1.2) that $\mathrm{d}\varrho(z) \neq 0$ for all $z \in \partial D$

2.1.2. Definition

$$w_\varrho(\zeta) := 2\left(\frac{\partial \varrho(\zeta)}{\partial \zeta_1}, \ldots, \frac{\partial \varrho(\zeta)}{\partial \zeta_n}\right).$$

2.1.3. Lemma. *There exist a neighbourhood $U_{\partial D}$ of ∂D and numbers $\varepsilon, \beta > 0$ such that for all $\zeta \in U_{\partial D}$ and $z \in \mathbb{C}^n$* with $|\zeta - z| \leqq \varepsilon$

$$\operatorname{Re} \langle w_\varrho(\zeta), \zeta - z \rangle \geqq \varrho(\zeta) - \varrho(z) + \beta \, |\zeta - z|^2 \tag{2.1.3}$$

(Re := *real part of*).

Proof. Let $x_j = x_j(\zeta)$ be the real coordinates of $\zeta \in \mathbb{C}_n$ such that $\zeta_j = x_j(\zeta) + \mathrm{i}x_{j+n}(\zeta)$. Then

$$\begin{aligned} \operatorname{Re} \langle w_\varrho(\zeta), \zeta - z \rangle &= \operatorname{Re} \sum_{j=1}^{n} \left(\frac{\partial \varrho(\zeta)}{\partial x_j} - \mathrm{i} \frac{\partial \varrho(\zeta)}{\partial x_{j+n}}\right) (x_j(\zeta - z) + \mathrm{i}x_{j+n}(\zeta - z)) \\ &= \sum_{j=1}^{2n} \frac{\partial \varrho(\zeta)}{\partial x_j} x_j(\zeta - z) \, . \end{aligned}$$

Hence, Taylor's theorem gives

$$\begin{aligned} \varrho(z) &= \varrho(\zeta) - \operatorname{Re} \langle w_\varrho(\zeta), \zeta - z \rangle \\ &\quad + \frac{1}{2} \sum_{j,k=1}^{2n} \frac{\partial^2 \varrho(\zeta)}{\partial x_j \, \partial x_k} x_j(\zeta - z) \, x_k(\zeta - z) + o(|\zeta - z|^2) \, . \end{aligned}$$

By (2.1.2) this implies that, for a sufficiently small neighbourhood $U_{\partial D}$ of ∂D and sufficiently small $\varepsilon > 0$,

$$\operatorname{Re} \langle w_\varrho(\zeta), \zeta - z \rangle \geqq \varrho(\zeta) - \varrho(z) + \frac{\alpha}{4} |\zeta - z|^2 \quad \text{if} \quad \zeta \in U_{\partial D} \quad \text{and}$$
$$|\zeta - z| \leqq \varepsilon \, . \; \blacksquare$$

2.1.4. Corollary. *$w_\varrho(\zeta)$ is a Leray map for D* (cp. Subsection 1.8.6).

Proof. Let $z \in D$ and $\zeta \in \partial D$, and let $\varepsilon > 0$ as in Lemma 2.1.3. If $|\zeta - z| \leqq \varepsilon$, then, by (2.1.3), $\operatorname{Re} \langle w_\varrho(\zeta), \zeta - z \rangle \geqq -\varrho(z) > 0$. If $|\zeta - z| > \varepsilon$, then we set

$$z_\varepsilon := \left(1 - \frac{\varepsilon}{|\zeta - z|}\right)\zeta + \frac{\varepsilon}{|\zeta - z|} z \, .$$

Since $|\zeta - z_\varepsilon| = \varepsilon$, (2.1.3) gives

$$\operatorname{Re} \langle w_\varrho(\zeta), \zeta - z \rangle = \operatorname{Re} \frac{|\zeta - z|}{\varepsilon} \langle w_\varrho(\zeta), \zeta - z_\varepsilon \rangle \geqq -\varrho(z_\varepsilon) \, .$$

Since D is convex and, therefore, $z_\varepsilon \in D$, this completes the proof. $\blacksquare$

2.1.5. Theorem. *Let $D \subset\subset \mathbb{C}^n$ be a C^2 strictly convex open set defined by* (2.1.1), *and let f be a continuous $(0, q)$-form on $\bar{D}$ such that $\bar{\partial} f = 0$ in D, $1 \leqq q \leqq n$. Then*

$$u := (-1)^q \left(R_{\partial D}^{w_\varrho} f + B_D f\right) \tag{2.1.4}$$

is a solution of $\bar\partial u = f$ *in* D (for the definitions of $R^{w_\varrho}_{\partial D}$ and B_D see Section 1.8). *This solution belongs to* $C^\alpha_{(0,q-1)}(D)$ *for all* $0 < \alpha < 1$ (for the definition of $C^\alpha_{(0,q-1)}(D)$ see Subsection 1.8.4). *If, moreover, for some* $k = 1, 2, \dots, \infty$, $f \in C^k_{(0,q)}(D)$, *then* $u \in C^{k+\alpha}_{(0,q-1)}(D)$ *for all* $0 < \alpha < 1$.

Proof. Since $w_\varrho(\zeta)$ is independent of z (and, therefore, holomorphic in z), the theorem follows from Corollary 1.12.2 and Corollary 2.1.4. ■

2.1.6. Corollary. *Let* $D \subseteq \mathbb{C}^n$ *be an open set and* f *a continuous function in* D. *If* $\bar\partial f$ *is a* C^k*-form in* D, $k = 0, 1, \dots, \infty$, *then* f *is a* $C^{k+\alpha}$*-function in* D *for all* $0 < \alpha < 1$.

Proof. By Theorem 2.1.5, the equation $\bar\partial u = \bar\partial f$ can be solved locally with a $C^{k+\alpha}$-function u when $\bar\partial f$ is C^k. By condition (i) in Theorem 1.1.5, then $u - f$ is holomorphic and, therefore, f is also $C^{k+\alpha}$. ■

2.2. Hölder estimates

In this section we prove 1/2-Hölder estimates for the solution of $\bar\partial u = f$ given by (2.1.4) in strictly convex open sets with C^2-boundary. The estimation of the operator B_D is simple and was already given in Lemma 1.8.5. The estimation of the operator $R^{w_\varrho}_{\partial D}$ is more complicated. Later we need such an estimate also for $R^w_{\partial D}$, where $D \subset\subset \mathbb{C}^n$ is a C^2 strictly pseudoconvex open set and w is a specially constructed Leray map for D. Therefore, in the following lemma, we give the needed estimate for a general Leray map satisfying some conditions (which will be fulfilled in both cases).

2.2.1. Lemma. *Let* $D \subset\subset \mathbb{C}^n$ *be an open set with* C^1*-boundary, and let* $U_{\partial D}$ *and* $U_{\bar D}$ *be neighbourhoods of* ∂D *and* $\bar D$, *respectively. Suppose* $w = w(z, \zeta) = (w_1(z,\zeta), \dots, w_n(z,\zeta))$ *is a collection of* C^1*-functions defined for* $(z, \zeta) \in U_{\bar D} \times U_{\partial D}$ *with the following properties:*

(i) $w(z,\zeta)$ *depends holomorphically on* $z \in U_{\bar D}$ *and (consequently)* $d_\zeta w(z,\zeta)$ *depends holomorphically on* $z \in U_{\bar D}$.

(ii) $w(z,\zeta)$ *is a Leray map for* D (cp. Subsection 1.8.6).

(iii) *For every point* $\xi \in \partial D$ *there are a neighbourhood* U_ξ *of* ξ *and real* C^1*-functions* $t_1(z,\zeta), \dots, t_{2n-1}(z,\zeta)$ *defined for* $z, \zeta \in U_\xi$ *such that the following conditions are fulfilled:* (a) *For every fixed* $z \in U_\xi$, $t_1(z,\cdot), \dots, t_{2n-1}(z,\cdot)$ *are real coordinates on* $\partial D \cap U_\xi$. (b) *For some* $\delta > 0$ *and all* $z \in D \cap U_\xi$, $\zeta \in \partial D \cap U_\xi$, *we have the estimate*

$$|\langle w(z,\zeta), \zeta - z\rangle| \geqq \delta\big(|t_1(z,\zeta)| + |t(z,\zeta)|^2 + \operatorname{dist}(z, \partial D)\big), \tag{2.2.1}$$

where $|t(z,\zeta)|^2 = \sum_{j=1}^{2n-1} |t_j(z,\zeta)|^2$ *and* $\operatorname{dist}(z,\partial D) = \inf\{|z - v| : v \in \partial D\}$.

Then there is a constant $C < \infty$ *such that, for every continuous differential form* f *on* $\bar D$,

$$\|R^w_{\partial D} f\|_{1/2, D} \leqq C\, \|f\|_{0,D}$$

(for the definitions of $\|\cdot\|_{1/2,D}$ and $\|\cdot\|_{0,D}$ see Section 1.8).

Proof. If $n = 1$, then $R^w_{\partial D} = 0$. We can therefore assume that $n \geqq 2$. For simplicity we write $\Phi := \langle w(z,\zeta), \zeta - z\rangle$. Then by definition of $R^w_{\partial D}$ (see Definition 1.7.6 for the notation $\det_{1,n-1}$)

$$(2\pi i)^n (R^w_{\partial D} f)(z) = \int\limits_{\substack{\zeta\in\partial D\\ 0\leqq\lambda\leqq 1}} f(\zeta) \wedge \det\nolimits_{1,n-1}\big(\eta^w(z,\zeta,\lambda), (d_\lambda + \bar\partial_{z,\zeta})\,\eta^w(z,\zeta,\lambda)\big) \wedge \omega(\zeta), \tag{2.2.2}$$

where
$$\eta^w(z,\zeta,\lambda) = (1-\lambda)\frac{w(z,\zeta)}{\Phi(z,\zeta)} + \lambda\frac{\bar\zeta - \bar z}{|\zeta - z|^2}. \tag{2.2.3}$$
Since $\bar\partial_z \Phi = 0$, $\bar\partial_z w = 0$, we have
$$(\mathrm{d}_\lambda + \bar\partial_{z,\zeta})\,\eta^w(z,\zeta,\lambda) = \left(\frac{\bar\zeta - \bar z}{|\zeta - z|^2} - \frac{w}{\Phi}\right)\mathrm{d}\lambda + (1-\lambda)\left(\frac{\bar\partial_\zeta w}{\Phi} - \frac{w}{\Phi}\frac{\bar\partial_\zeta \Phi}{\Phi}\right)$$
$$+ \lambda\left(\frac{\mathrm{d}\bar\zeta - \mathrm{d}\bar z}{|\zeta - z|^2} - \frac{\bar\zeta - \bar z}{|\zeta - z|^2}\frac{\bar\partial_{z,\zeta}|\zeta - z|^2}{|\zeta - z|^2}\right). \tag{2.2.4}$$

Now we want to expand the determinant in (2.2.2). Observe that in addition to the rules given in Proposition 1.7.5 we also have the following rule: If in a matrix of differential forms every column contains only forms of the same degree, then after interchanging two columns the determinant of this matrix changes only by a factor ± 1. Using these rules and taking into account that only monomials of degree 1 in λ contribute to the integral (2.2.2), we obtain
$$(R^w_{\partial D} f)(z)$$
$$= \int\limits_{\partial D\times[0,1]} f \wedge \sum_{s=0}^{n-2} p_s \det{}_{1,1,n-s-2,s}\left(\frac{w}{\Phi}, \frac{\bar\zeta - \bar z}{|\zeta - z|^2}, \frac{\bar\partial_\zeta w}{\Phi}, \frac{\mathrm{d}\bar\zeta - \mathrm{d}\bar z}{|\zeta - z|^2}\right) \wedge \mathrm{d}\lambda \wedge \omega(\zeta), \tag{2.2.5}$$
where p_s are polynomials in λ. Further, by multi-linearity of the determinant,
$$(R^w_{\partial D} f)(z) = \int\limits_{\partial D\times[0,1]} f \wedge \sum_{s=0}^{n-2} p_s \frac{\det_{1,1,n-s-2,s}(w, \bar\zeta - \bar z, \bar\partial_\zeta w, \mathrm{d}\bar\zeta - \mathrm{d}\bar z) \wedge \mathrm{d}\lambda \wedge \omega(\zeta)}{\Phi^{n-s-1}|\zeta - z|^{2s+2}}.$$
Integrating with respect to λ, we obtain
$$(R^w_{\partial D} f)(z) = \sum_{s=0}^{n-2} A_s \int\limits_{\partial D} \frac{f \wedge \det_{1,1,n-s-2,s}(w, \bar\zeta - \bar z, \bar\partial_\zeta w, \mathrm{d}\bar\zeta - \mathrm{d}\bar z) \wedge \omega(\zeta)}{\Phi^{n-s-1}|\zeta - z|^{2s+2}}, \tag{2.2.6}$$
where $A_s = \int_0^1 p_s(\lambda)\,\mathrm{d}\lambda$. It follows that the coefficients of the form $R^w_{\partial D} f$ are linear combinations of integrals of the following type
$$E(z) = \int\limits_{\partial D} \frac{f_I \psi}{\Phi^{n-s-1}|\zeta - z|^{2s+2}} \bigwedge_{j\neq m} \mathrm{d}\bar\zeta_j \wedge \omega(\zeta), \tag{2.2.7}$$
where $0 \leqq s \leqq n-2$, $1 \leqq m \leqq n$, f_I is some coefficient of the form f, and ψ is the product of some of the functions w_j, $\bar\zeta_j - \bar z_j$ and $\partial w_j/\partial\bar\zeta_k$ $(j, k = 1, \ldots, n)$. Since ψ contains at least one of the functions $\bar\zeta_j - \bar z_j$ as a factor, we have for some $C_1 < \infty$
$$|\psi| \leqq C_1 |\zeta - z|. \tag{2.2.8}$$
To estimate the integrals (2.2.7) we apply Proposition 2 in Appendix 1. In view of this Proposition, it is sufficient to prove that, for $j = 1, \ldots, n$,
$$\left|\frac{\partial E(z)}{\partial z_j}\right|, \left|\frac{\partial E(z)}{\partial \bar z_j}\right| \leqq C\,\|f\|_{0,D}\,[\mathrm{dist}\,(z, \partial D)]^{-1/2}, \tag{2.2.9}$$

where $C < \infty$ is a constant which is independent of $z \in D$ and f. We have

$$\frac{\partial}{\partial z_j} \frac{\psi}{\Phi^{n-s-1} |\zeta - z|^{2s+2}} = \frac{\partial \psi/\partial z_j}{\Phi^{n-s-1} |\zeta - z|^{2s+2}} - \frac{(n-s-1)(\partial \Phi/\partial z_j)\psi}{\Phi^{n-s} |\zeta - z|^{2s+2}} + \frac{(s+1)(\bar\zeta_j - \bar z_j)\psi}{\Phi^{n-s-1} |\zeta - z|^{2s+4}}$$

and, since $\partial \Phi/\partial \bar z_j = 0$,

$$\frac{\partial}{\partial \bar z_j} \frac{\psi}{\Phi^{n-s-1} |\zeta - z|^{2s+2}} = \frac{\partial \psi/\partial \bar z_j}{\Phi^{n-s-1} |\zeta - z|^{2s+2}} + \frac{(s+1)(\zeta_j - z_j)\psi}{\Phi^{n-s-1} |\zeta - z|^{2s+4}}.$$

Since $\partial\psi/\partial z_j$, $\partial\Phi/\partial z_j$ and $\partial\psi/\partial \bar z_j$ are bounded for $(z, \zeta) \in D \times \partial D$ and (2.2.8) holds, this implies that for some $C_2 < \infty$

$$\left| \frac{\partial}{\partial z_j} \frac{\psi}{\Phi^{n-s-1} |\zeta - z|^{2s+2}} \right| \leqq \frac{C_2}{|\Phi|^{n-s-1} |\zeta - z|^{2s+2}} + \frac{C_2}{|\Phi|^{n-s} |\zeta - z|^{2s+1}}$$

and

$$\left| \frac{\partial}{\partial \bar z_j} \frac{\psi}{\Phi^{n-s-1} |\zeta - z|^{2s+2}} \right| \leqq \frac{C_2}{|\Phi|^{n-s-1} |\zeta - z|^{2s+2}}.$$

Hence we can find $C_3 < \infty$ such that

$$\left| \frac{\partial E(z)}{\partial z_j} \right|, \left| \frac{\partial E(z)}{\partial \bar z_j} \right| \leqq C_3 \|f\|_{0,D} \left[\int_{\partial D} \frac{d\sigma_{2n-1}}{|\Phi|^{n-s-1} |\zeta - z|^{2s+2}} + \int_{\partial D} \frac{d\sigma_{2n-1}}{|\Phi|^{n-s} |\zeta - z|^{2s+1}} \right],$$

where $d\sigma_{2n-1}$ is the volume form on ∂D. Therefore, it remains to prove that, for every $\xi \in \partial D$, there are a neighbourhood U_ξ of ξ and a constant $C_\xi < \infty$ such that

$$\int_{\partial D \cap U_\xi} \frac{d\sigma_{2n-1}}{|\Phi|^{n-s-1} |\zeta - z|^{2s+2}} \leqq C_\xi [\operatorname{dist}(z, \partial D)]^{-1/2} \tag{2.2.10}$$

and

$$\int_{\partial D \cap U_\xi} \frac{d\sigma_{2n-1}}{|\Phi|^{n-s} |\zeta - z|^{2s+1}} \leqq C_\xi [\operatorname{dist}(z, \partial D)]^{-1/2}. \tag{2.2.11}$$

Consider (2.2.11). Let U_ξ and $t_1(z, \zeta), \ldots, t_{2n-1}(z, \zeta)$ be as in hypothesis (iii) of the lemma, and let $\varphi(z, \zeta)$ be the function defined for $z \in U_\xi$, $\zeta \in \partial D \cap U_\xi$ such that, for every fixed $z \in U_\xi$,

$$d\sigma_{2n-1} = \varphi(z, \zeta)\, d_\zeta t_1(z, \zeta) \wedge \ldots \wedge d_\zeta t_{2n-1}(z, \zeta) \quad \text{on} \quad \partial D \cap U_\xi.$$

Then, for some $\gamma > 0$ and all $z \in U_\xi, \zeta \in \partial D \cap U_\xi$, we have the estimate $|\zeta - z| \geqq \gamma\, |t(z, \zeta)|$, and it follows from (2.2.1) that, for $z \in D \cap U_\xi$,

$$\int_{\partial D \cap U_\xi} \frac{d\sigma_{2n-1}}{|\Phi|^{n-s} |\zeta - z|^{2s+1}} \leqq \frac{1}{\delta\gamma} \int_{\partial D \cap U_\xi} \frac{\varphi(z, \zeta)\, d_\zeta t_1(z, \zeta) \wedge \ldots \wedge d_\zeta t_{2n-1}(z, \zeta)}{(|t_1(z, \zeta)| + |t(z, \zeta)|^2 + \operatorname{dist}(z, \partial D))^{n-s}\; t(z, \zeta)|^{2s+1}}. \tag{2.2.12}$$

After shrinking U_ξ if necessary, we can find constants Γ, $R < \infty$ such that $|\varphi(z, \zeta)| \leq \Gamma$ for $z \in U_\xi$ and $\zeta \in \partial D \cap U_\xi$, and, for every $z \in U_\xi$, the surface $\partial D \cap U_\xi$ is diffeomorphically mapped by $\zeta \to (t_1(z, \zeta), \dots, t_{2n-1}(z, \zeta))$ onto an open subset of the ball with radius R centered at $0 \in \mathbb{R}^{2n-1}$. Therefore, it follows from (2.2.12) that

$$\int_{\partial D \cap U_\xi} \frac{d\sigma_{2n-1}}{|\Phi|^{n-s} |\zeta - z|^{2s+1}} \leq \frac{\Gamma}{\delta\gamma} \int_{\substack{x \in \mathbb{R}^{2n-1} \\ |x| < R}} \frac{dx_1 \wedge \dots \wedge dx_{2n-1}}{(|x_1| + |x|^2 + \operatorname{dist}(z, \partial D))^{n-s} |x|^{2s+1}}.$$

Since

$$(|x_1| + |x|^2 + \operatorname{dist}(z, \partial D))^{n-s} |x|^{2s+1} \geq (|x_1| + |x|^2 + \operatorname{dist}(z, \partial D))^2 |x|^{2n-3},$$

estimate (2.2.11) now follows from Proposition 3 (ii), Appendix 1. In the same way we obtain estimate (2.2.10) if we use part (i) of Proposition 3, Appendix 1. ■

2.2.2. Theorem. *Let $D \subset\subset \mathbb{C}^n$ be a C^2 strictly convex open set defined by* (2.1.1). *Then there exists a constant $C < \infty$ such that, for every continuous differential form f on $\bar{D}$,*

$$||R^{w_\varrho}_{\partial D} f + B_D f||_{1/2, D} \leq C\, ||f||_{0, D}.$$

In particular, if f is a continuous $(0, q)$-form on $\bar{D}$ such that $\bar{\partial} f = 0$ in D, $1 \leq q \leq n$, then the solution of $\bar{\partial} u = f$ given by (2.1.4) *admits the estimate*

$$||u||_{1/2, D} \leq C\, ||f||_{0, D}$$

(for the definition of $||\cdot||_{1/2, D}$ and $||\cdot||_{0, D}$ see Section 1.8).

Proof. By Lemma 1.8.5 (i), for every $0 < \alpha < 1$, there is a constant $C_\alpha < \infty$ such that $||B_D f||_{\alpha, D} \leq C_\alpha ||f||_{0, D}$. Therefore, we only have to estimate $||R^{w_\varrho}_{\partial D} f||_{1/2, D}$. To do this we prove that $w_\varrho(\zeta) = 2(\partial\varrho/\partial\zeta_1, \dots, \partial\varrho/\partial\zeta_n)$ fulfills conditions (i)—(iii) in Lemma 2.2.1. (i) is clear, because $w_\varrho(\zeta)$ is independent of z. (ii) is Corollary 2.1.4.

It remains to show condition (iii). Let $x_j = x_j(\zeta)$ be the real coordinates of $\zeta \in \mathbb{C}^n$ such that $\zeta_j = x_j(\zeta) + i x_{j+n}(\zeta)$. Set $t_1(z, \zeta) := \operatorname{Im} \langle w_\varrho(\zeta), \zeta - z \rangle$ (Im := imaginary part of). Then

$$t_1(z, \zeta) = \sum_{j=1}^{n} \left[\frac{\partial\varrho(\zeta)}{\partial x_j} x_{j+n}(\zeta - z) - \frac{\partial\varrho(\zeta)}{\partial x_{j+n}} x_j(\zeta - z) \right]$$

and, therefore,

$$d_\zeta t_1(z, \zeta)|_{\zeta = z} = \sum_{j=1}^{n} \left[\frac{\partial\varrho(z)}{\partial x_j} dx_{j+n}(z) - \frac{\partial\varrho(z)}{\partial x_{j+n}} dx_j(z) \right]. \tag{2.2.13}$$

Hence

$$d_\zeta t_1(z, \zeta)|_{\zeta = z} \wedge d\varrho(z) \neq 0 \quad \text{for all} \quad z \in \partial D; \tag{2.2.14}$$

in fact, by (2.2.13), the coefficient of $dx_j(z) \wedge dx_{j+n}(z)$ is $-((\partial\varrho(z)/\partial x_j)^2 + (\partial\varrho(z)/\partial x_{j+n})^2)$, which cannot be zero for all $1 \leq j \leq n$, because $d\varrho(z) \neq 0$.

Fix $\xi \in \partial D$. Then, by (2.2.14), we can find a neighbourhood U_ξ of ξ and C^1-functions $t_2, \dots, t_{2n-1}$ on U_ξ such that, for every fixed $z \in U_\xi$, $t_1(z, \cdot), t_2, \dots, t_{2n-1}$ are coordinates on $\partial D \cap U_\xi$. Set $t_j(z, \zeta) = t_j(\zeta) - t_j(z)$ for $z, \zeta \in U_\xi$, $j = 2, \dots, 2n - 1$. Then, for every fixed $z \in U_\xi$, $t_1(z, \cdot), \dots, t_{2n-1}(z, \cdot)$ are coordinates on $\partial D \cap U_\xi$. It remains to prove estimate (2.2.1).

Since $t_j(z, z) = 0$, after shrinking U_ξ, it follows from Taylor's formula that for some $\delta_1 > 0$

$$|\zeta - z| \geq \delta_1 |t(z, \zeta)| \quad \text{for} \quad \zeta, z \in U_\xi.$$

By Lemma 2.1.3, this implies that, after shrinking U_ξ, for some $\delta_2 > 0$,

$$|\langle w_\varrho(\zeta), \zeta - z\rangle| \geqq \delta_2(|t_1(z,\zeta)| + \varrho(\zeta) - \varrho(z) + |t(z,\zeta)|^2), \zeta, z \in U_\xi .$$

This implies (2.2.1), because $\varrho = 0$ on ∂D, and since, after shrinking U_ξ, we can find $\delta_3 > 0$ such that $-\varrho(z) \geqq \delta_3 \operatorname{dist}(z, \partial D)$ for all $z \in D \cap U_\xi$. ■

2.2.3. The Hölder exponent 1/2 is the best one possible. In this subsection we present an example of E. M. Stein which shows that for the case that D is a ball in $\mathbb{C}^2$, the Hölder exponent 1/2 in Theorem 2.2.2 is the best one possible in the following sense: There exist continuous (0, 1)-forms f on $\bar{D}$ which are $\bar\partial$-closed in D such that, for every $\alpha > 1/2$, there does not exist a solution of $\bar\partial u = f$ satisfying the estimate $||u||_{\alpha, D} < \infty$.

Example. Let $D := \{(z_1, z_2) \in \mathbb{C}^2 : |z_1|^2 + |z_2|^2 < 1\}$ be the unit ball in $\mathbb{C}^2$. Let $\ln(z_1 - 1)$, $z_1 \notin [1, +\infty)$, be the branch of the logarithm with $0 <$ argument of $\ln(z_1 - 1) < 2\pi$. Define by setting

$$f(z_1, z_2) := \begin{cases} \dfrac{d\bar{z}_2}{\ln(z_1 - 1)} & \text{if } (z_1, z_2) \in \bar{D} \setminus \{\sigma(1, 0)\}, \\ 0 & \text{if } (z_1, z_2) = (1, 0) \end{cases}$$

a (0, 1)-form on $\bar{D}$. Since $\ln(z_1 - 1) \neq 0$ for $z_1 \notin [1, +\infty)$, f is a C^∞-form on $\bar{D} \setminus \{(1, 0)\}$. Since $\ln(z_1 - 1) \to \infty$ when $z_1 \to 1$, f is continuous on $\bar{D}$. Since $1/\ln(z_1 - 1)$ is holomorphic in D, we have $\bar\partial f = 0$ in D. However:

Proposition. *If $\alpha > 1/2$, then there does not exist a function u in D such that $\bar\partial u = f$ and $||u||_{\alpha, D} < \infty$.*

Proof. Suppose u is a solution of $\bar\partial u = f$ and, for some $\alpha > 1/2$, $||u||_{\alpha, D} < \infty$. Since $\bar\partial(\bar{z}_2/\ln(z_1 - 1)) = f$, the function $u - \bar{z}_2/\ln(z_1 - 1)$ is holomorphic in D. Let $0 < 2\varepsilon < 1$. Then the circles $\{(z_1, z_2) \in \mathbb{C}^2 : z_1 = 1 - \varepsilon, |z_2| = \sqrt{\varepsilon}\}$ and $\{(z_1, z_2) \in \mathbb{C}^2 : z_1 = 1 - 2\varepsilon, |z_2| = \sqrt{\varepsilon}\}$ are contained in D. Since $u - \bar{z}_2/\ln(z_1 - 1)$ is holomorphic, Cauchy's formula gives

$$\int_{|z_2| = \sqrt{\varepsilon}} u(1 - \varepsilon, z_2)\, dz_2 = \int_{|z_2| = \sqrt{\varepsilon}} \frac{\bar{z}_2\, dz_2}{\ln(-\varepsilon)} = \frac{2\pi i \varepsilon}{\ln(-\varepsilon)}$$

and

$$\int_{|z_2| = \sqrt{\varepsilon}} u(1 - 2\varepsilon, z_2)\, dz_2 = \frac{2\pi i \varepsilon}{\ln(-2\varepsilon)}.$$

Since $||u||_{\alpha, D} < \infty$, this implies that for some constant $C < \infty$ and all $0 < 2\varepsilon < 1$

$$\left| \frac{1}{\ln(-\varepsilon)} - \frac{1}{\ln(-2\varepsilon)} \right| \leq C\varepsilon^{\alpha - 1/2}.$$

Hence, for all $0 < 2\varepsilon < 1$, $\ln 2 = |\ln(-2\varepsilon) - \ln(-\varepsilon)| \leq C\varepsilon^{\alpha - 1/2} |\ln(-\varepsilon) \ln(-2\varepsilon)|$. This is impossible. ■

2.3. Solution of the $\bar\partial$-equation with Hölder estimates in C^2 strictly pseudoconvex open sets in $\mathbb{C}^n$

First we prove Fredholm solvability (Lemma 2.3.1 below). To give the precise formulation we need some notations.

Notation. Let $D \subset\subset \mathbb{C}^n$ be an open set, and let $1 \leqq q \leqq n$. We denote by $F^{1/2}_{(0,q-1)}(\bar D)$ the domain of definition of $\bar\partial$ as an operator from $H^{1/2}_{(0,q-1)}(\bar D)$ into $C^0_{(0,q)}(\bar D)$ (for the definitions of $H^{1/2}_{(0,q-1)}(\bar D)$ and $C^0_{(0,q)}(\bar D)$ see Section 1.8), that is, $F^{1/2}_{(0,q-1)}(\bar D)$ is the space of all $f \in H^{1/2}_{(0,q-1)}(\bar D)$ such that $\bar\partial f$ (which is defined in D in the sense of distributions) is continuous in D and admits a continuous continuation to $\bar D$. Set $E^{1/2}_{(0,q)}(\bar D) := \bar\partial F^{1/2}_{(0,q-1)}(\bar D)$. We denote by $Z^0_{(0,q)}(\bar D)$ the space of all $f \in C^0_{(0,q)}(\bar D)$ such that $\bar\partial f = 0$ in D. $Z^0_{(0,q)}(\bar D)$ is a closed subspace of the Banach space $C^0_{(0,q)}(\bar D)$.

2.3.1. Lemma. *Let $D \subset\subset \mathbb{C}^n$ be a strictly pseudoconvex open set with C^2-boundary, and let $1 \leqq q \leqq n$. Then*

(i) *there exists a bounded linear operator S from the Banach space $Z^0_{(0,q)}(\bar D)$ (endowed with the norm $\|\cdot\|_{0,D}$) into the Banach space $H^{1/2}_{(0,q-1)}(\bar D)$ (endowed with the norm $\|\cdot\|_{1/2,D}$) such that $S(Z^0_{(0,q)}(\bar D)) \subseteqq F^{1/2}_{(0,q-1)}(\bar D)$ and $\bar\partial \circ S = \mathrm{id} + K$, where* id *is the identity operator in $Z^0_{(0,q)}(\bar D)$ and K is a compact operator in $Z^0_{(0,q)}(D)$.*[1]

(ii) *$E^{1/2}_{(0,q)}(\bar D)$ is a closed (with respect to the norm $\|\cdot\|_{0,D}$) subspace of $Z^0_{(0,q)}(\bar D)$ of finite codimension.*

Proof. By Corollary 1.5.25, there is a finite covering of $\bar D$ by relatively open sets $U_1, \dots, U_m \subseteqq \bar D$ such that, for every j, there is a biholomorphic map h_j in a neighbourhood of $\bar U_j$ such that $h_j(U_j \cap D)$ is a strictly convex set with C^2-boundary. Therefore, by Theorem 2.2.2, we can find bounded linear operators S_j from $Z^0_{(0,q)}(\bar D)$ into $H^{1/2}_{(0,q-1)}(\bar U_j)$ such that $\bar\partial S_j f = f$ in U_j for all $f \in Z^0_{(0,q)}(\bar D)$. Choose real-valued C^∞-functions φ_j on $\mathbb{C}^n$ such that $D \cap (\operatorname{supp} \varphi_j) \subseteqq U_j$ and $\sum_{j=1}^m \varphi_j = 1$ on $\bar D$, and define

$$Sf := \sum_{j=1}^m \varphi_j S_j f \quad \text{for} \quad f \in Z^0_{(0,q)}(\bar D)\,.$$

Then, for every $f \in Z^0_{(0,q)}(\bar D)$, $Sf \in F^{1/2}_{(0,q-1)}(\bar D)$, and

$$\bar\partial S f = f + \sum_{j=1}^m \bar\partial\varphi_j \wedge S_j f\,.$$

Since the operator K defined by the sum on the right-hand side is bounded as an operator from $Z^0_{(0,q)}(\bar D)$ into $H^{1/2}_{(0,q)}(\bar D)$, it follows from Ascoli's theorem that K is compact as an operator from $Z^0_{(0,q)}(\bar D)$ into itself. This proves part (i) of the lemma.

Since K is compact, $(\mathrm{id} + K)(Z^0_{(0,q)}(\bar D))$ is a subspace of $Z^0_{(0,q)}(\bar D)$ of finite codimension. Since $\bar\partial \circ S = \mathrm{id} + K$ and, therefore, $E^{1/2}_{(0,q)}(\bar D)$ contains this subspace, it follows that $E^{1/2}_{(0,q)}(\bar D)$ has finite codimension in $Z^0_{(0,q)}(\bar D)$. Since, moreover, $E^{1/2}_{(0,q)}(\bar D)$ is the image of a closed operator (namely, of $\bar\partial$ as an operator with the domain of definition $F^{1/2}_{(0,q)}(\bar D)$), this implies that $E^{1/2}_{(0,q)}(\bar D)$ is a closed subspace of $Z^0_{(0,q)}(\bar D)$ (cp. Proposition 4, Appendix 2). ■

The remainder of this section is devoted to the proof that the finite codimension of $E^{1/2}_{(0,q)}(\bar D)$ stated in Lemma 2.3.1 is zero.

---

[1]) Recall that a linear operator between Banach spaces is called compact if every bounded set is mapped into a relatively compact set.

2.3.2. Lemma. *Let $D \subset\subset \mathbb{C}^n$ be a C^2 strictly pseudoconvex open set, and let $z_1, \dots, z_n$ be the canonical coordinates of $z \in \mathbb{C}^n$. Then there exists a holomorphic polynomial p which depends only on z_1 and which is not identically zero such that, for every $f \in Z^0_{(0,q)}(\bar{D})$, $1 \leqq q \leqq n$,*

$$pf \in E^{1/2}_{(0,q)}(\bar{D}) .$$

Proof. By Lemma 2.3.1, there is a finite number of forms $f_1, \dots, f_N \in Z^0_{(0,q)}(\bar{D})$ such that

$$\text{linear span } (E^{1/2}_{(0,q)}(\bar{D}) \cup \{f_1, \dots, f_N\}) = Z^0_{(0,q)}(\bar{D}) . \tag{2.3.1}$$

For $j = 1, \dots, N$, the system $f_j, z_1 f_j, \dots, z_1^N f_j$ is linearly independent (we can assume that every f_j is $\neq 0$) and of length $N + 1$. Since the codimension of $E^{1/2}_{(0,q)}(\bar{D})$ is $\leqq N$, it follows that, for every $j = 1, \dots, N$, we can find a vector $(c_0^{(j)}, \dots, c_N^{(j)}) \neq 0$ of complex numbers such that

$$\sum_{k=0}^{N} c_k^{(j)} z_1^k f_j \in E^{1/2}_{(0,q)}(\bar{D}) . \tag{2.3.2}$$

Set

$$p_j(z) := \sum_{k=0}^{N} c_k^{(j)} z_1^k \quad \text{and} \quad p := \prod_{k=1}^{N} p_k .$$

Now let f be an arbitrary form in $Z^0_{(0,q)}(\bar{D})$. By (2.3.1), f can be written as

$$f = \bar{\partial} v + \sum_{j=1}^{N} \lambda_j f_j ,$$

where λ_j are complex numbers and $v \in H^{1/2}_{(0,q-1)}(\bar{D})$. Further, by (2.3.2), we can find forms $u_j \in H^{1/2}_{(0,q-1)}(\bar{D})$ such that $\bar{\partial} u_j = p_j f_j$. Setting $u := pv + \sum_{j=1}^{N} \lambda_j (\prod_{k \neq j} p_k) u_j$, we obtain $u \in H^{1/2}_{(0,q-1)}(\bar{D})$ such that $\bar{\partial} u = pf$. Hence p is a polynomial with the required properties. ∎

2.3.3. Lemma. *Let p_1 and p_2 be two polynomials in one complex variable without joint zeros. Then there exist polynomials q_1 and q_2 in one complex variable such that $q_1 p_1 + q_2 p_2 \equiv 1$.*

Proof. The assertion is trivial when p_1 is of degree zero. Assume that the lemma is already proved for the case that the degree of p_1 is k. Consider the case that the degree of p_1 is $k + 1$. Let z_0 be a zero of p_1. Then, by inductive hypothesis, there are polynomials Q_1 and Q_2 such that $Q_1 p_1/(z - z_0) + Q_2 p_2 \equiv 1$, and, therefore,

$$\left[\frac{Q_1(z) - Q_1(z_0)\, p_2(z)/p_2(z_0)}{z - z_0}\right] p_1(z) + \left[Q_2(z) + \frac{p_1(z)\, Q_1(z_0)}{(z - z_0)\, p_2(z_0)}\right] p_2(z) \equiv 1. \ ∎$$

2.3.4. Lemma. *Let $D \subset\subset \mathbb{C}^n$ be a strictly pseudoconvex open set, and let f be a continuous $(0, q)$-form in some neighbourhood $U_{\bar{D}}$ of $\bar{D}$ such that $\bar{\partial} f = 0$ in $U_{\bar{D}}$, $1 \leqq q \leqq n$. Then there exists a $u \in H^{1/2}_{(0,q-1)}(\bar{D})$ such that $\bar{\partial} u = f$ in D.*

Proof. By Theorem 1.5.21 we can find a strictly pseudoconvex open set G with C^∞-boundary such that $D \subset\subset G \subset\subset U_{\bar{D}}$. Set $G_\varepsilon := \{z \in \mathbb{C}^n : (z_1 + \varepsilon, z_2, \dots, z_n) \in G\}$, $\varepsilon > 0$. Choose $\varepsilon > 0$ so small that $D \subset\subset G_\varepsilon \subset\subset U_{\bar{D}}$, and the form $f_\varepsilon(z) := f(z_1 - \varepsilon, z_2, \dots, z_n)$ belongs to $Z^0_{(0,q)}(G)$. By Lemma 2.3.2, we can find a polynomial p in z_1 and forms $u_0 \in H^{1/2}_{(0,q-1)}(G)$, $u_\varepsilon \in H^{1/2}_{(0,q-1)}(G)$ such that

$$p(z_1) f(z) = \bar{\partial} u_0(z) \quad \text{and} \quad p(z_1) f_\varepsilon(z) = \bar{\partial} u_\varepsilon(z) \quad \text{for} \quad z \in G\,.$$

Then $p(z_1 + \varepsilon) f(z) = \bar{\partial} u_\varepsilon(z_1 + \varepsilon, z_2, \dots, z_n)$ for $z \in G_\varepsilon$. Choose $\varepsilon > 0$ so small that the polynomials $p(z_1)$ and $p(z_1 + \varepsilon)$ do not have joint zeros. Then, by Lemma 2.3.3, we can find polynomials q and q_ε in z_1 such that $q(z_1) p(z_1) + q_\varepsilon(z_1) p(z_1 + \varepsilon) \equiv 1$. Setting $u(z) := q(z_1) u_0(z) + q_\varepsilon(z_1) u_\varepsilon(z_1 + \varepsilon, z_2, \dots, z_n)$ we complete the proof. ∎

2.3.5. Theorem. *Let $D \subset\subset \mathbb{C}^n$ be a C^2 strictly pseudoconvex open set, and let $1 \leqq q \leqq n$. Then there exists a constant $C < \infty$ such that, for every continuous $(0, q)$-form f on $\bar{D}$ such that $\bar{\partial} f = 0$ in D, there exists a form $u \in H^{1/2}_{(0,q-1)}(\bar{D})$ such that*

$$\bar{\partial} u = f \quad \text{in} \quad D \quad \text{and} \quad \|u\|_{1/2,D} \leqq C \|f\|_{0,D}$$

(for the definitions of $\|\cdot\|_{1/2,D}$ and $\|\cdot\|_{0,D}$ see Section 1.8). *Moreover, there is a bounded linear operator T from $Z^0_{(0,q)}(\bar{D})$ into $H^{1/2}_{(0,q-1)}(\bar{D})$ such that $T(Z^0_{(0,q)}(\bar{D})) \subseteq F^{1/2}_{(0,q-1)}(\bar{D})$ and $\bar{\partial} \circ T = \mathrm{id}$, where id is the identity operator.*

Remark. In this section we give a first proof of this theorem. This proof can be bypassed without loss of continuity, because, by means of Lemma 2.3.4 only, in Sections 2.4–2.6 we construct a global integral representation formula, which gives (among other things) solutions of the $\bar{\partial}$-equation as stated in Theorem 2.3.5 (cp. Theorem 2.6.1).

Proof of Theorem 2.3.5. We only have to prove that, for every $f \in Z^0_{(0,q)}(\bar{D})$, there is $u \in H^{1/2}_{(0,q-1)}(\bar{D})$ with $\bar{\partial} u = f$ in D, because the rest of the theorem then follows from Banach's open mapping theorem (which gives the estimates) and Lemma 2.3.1 (which gives the operator T — cp. Proposition 3, Appendix 2).

If $K \subseteq \partial D$ is a compact set, then we denote by (K) the following assertion: If U_K is a neighbourhood of K and f is a continuous $(0, q)$-form on $\bar{D} \cup U_K$ such that $\bar{\partial} f = 0$ in $D \cup U_K$, then $f \in E^{1/2}_{(0,q)}(\bar{D})$.

Let ϱ be a strictly plurisubharmonic C^2-function in some neighbourhood $U_{\partial D}$ of ∂D such that $U_{\partial D} \cap D = \{z \in U_{\partial D} : \varrho(z) < 0\}$. For $r < 0$ and $\xi \in \mathbb{C}^n$ we denote by $E_\xi(r)$ the open ball of radius r centered at ξ. By Proposition 1.5.16, $d\varrho(z) \neq 0$ for $z \in \partial D$. Therefore, by Theorem 1.4.14, we can find $\varepsilon > 0$ such that the following condition is fulfilled: For every $\xi \in \partial D$, there exist a neighbourhood $U_\xi \subseteq U_{\partial D}$ of $\overline{E_\xi(2\varepsilon)}$ and a biholomorphic map h_ξ in some neighbourhood of $\bar{U}_\xi$ such that $\varrho \circ h^{-1}$ is strictly convex in $h_\xi(\bar{U}_\xi)$ and $h_\xi(U_\xi)$ is a ball.

By Lemma 2.3.4, (∂D) is true. We have to prove that $(\emptyset)$ is true. To do this it is enough to show that, for every compact set $K \subseteq \partial D$ and every point $\xi \in \partial D$, the implication $(K) \Rightarrow (K \setminus E_\xi(\varepsilon))$ is valid.

Suppose $K \subseteq \partial D$ is a compact set such that (K) is true. Let $\xi \in \partial D$, let $\theta \subseteq U_{\partial D}$ be a neighbourhood of $K \setminus E_\xi(\varepsilon)$, and let f be a continuous $(0, q)$-form in $\bar{D} \cup \theta$ such that $\bar{\partial} f = 0$ in $D \cup \theta$. We have to prove that $f \in E^{1/2}_{(0,q)}(\bar{D})$.

Choose a real-valued C^∞-function $\chi \geqq 0$ in $\mathbb{C}^n$ with the following properties:

(i) $\chi > 0$ on $K \setminus E_\xi(\varepsilon)$,

(ii) $\operatorname{supp} \chi \subset\subset \theta$,

(iii) χ and the first- and second-order derivatives of χ are so small that $(\varrho - \chi) \circ h^{-1}$ is strictly convex on $h_\xi(\bar{U}_\xi)$.

Define $G := D \cup \{z \in U_{\partial D} : \varrho(z) < \chi(z)\}$. Then by (ii)

$$\overline{G} \subseteqq \overline{D} \cup \theta .$$

Since $(\varrho - \chi) \circ h^{-1}$ is strictly convex on the ball $h_\xi(\overline{U}_\xi)$, and since the set $h_\xi(G \cap U_\xi)$ is defined in $h_\xi(U_\xi)$ by $(\varrho - \chi) \circ h^{-1} < 0$, it follows from Lemma 1.5.23 (i) that there is a strictly convex open set W with C^2-boundary such that

$$h_\xi(E_\xi(2\varepsilon) \cap G) \subseteqq W \subseteqq h_\xi(U_\xi \cap G) . \tag{2.3.3}$$

Since $h_\xi^{-1}(\overline{W}) \subseteqq \overline{D} \cup \theta$, it follows from Theorem 2.2.2 that there exists $u_0 \in H^{1/2}_{(0,q-1)}(h_\xi^{-1}(\overline{W}))$ such that $f = \bar{\partial} u_0$ in $h_\xi^{-1}(W)$. Choose $\varphi \in C_0^\infty(E_\xi(2\varepsilon))$ such that $\varphi = 1$ on $E_\xi(\varepsilon)$. Since, by (2.3.3), $E_\xi(2\varepsilon) \cap \overline{G} \subseteqq h_\xi^{-1}(\overline{W})$, then, after continuation by zero, φu_0 belongs to $H^{1/2}_{(0,q-1)}(\overline{G})$, $f - \bar{\partial}(\varphi u_0) \in Z^0_{(0,q)}(\overline{G})$ and $f - \bar{\partial}(\varphi u_0) = 0$ in $E_\xi(\varepsilon) \cap \overline{G}$. Hence, $f - \bar{\partial}(\varphi u_0)$ can be continued by zero to $\overline{G} \cup E_\xi(\varepsilon) \supseteqq \overline{D} \cup (G \cup E_\xi(\varepsilon))$. Since $G \cup E_\xi(\varepsilon)$, is a neighbourhood of K and, by hypothesis, (K) is true, it follows that $f - \bar{\partial}(\varphi u_0) = \bar{\partial} u_1$ in D for some $u_1 \in H^{1/2}_{(0,q-1)}(\overline{D})$. Setting $u = u_1 + \varphi u_0$, we obtain $u \in H^{1/2}_{(0,q-1)}(\overline{D})$ such that $\bar{\partial} u = f$ in D. ■

2.4. The support function $\Phi(z, \zeta)$

It is the aim of the present and the next section to construct a Leray map $w(z, \zeta)$ for strictly pseudoconvex open sets in $\mathbb{C}^n$ which depends holomorphically on z and satisfies appropriate growth conditions. The corresponding integral representation formulas will be used in Sections 2.6—2.9. The first step of this construction is the construction of the function $\Phi(z, \zeta) = \langle w(z, \zeta), \zeta - z \rangle$ and will be carried out in the present section. Locally the Levi polynomial (cp. (1.4.18)) can be used as $\Phi(z, \zeta)$. To obtain $\Phi(z, \zeta)$ globally, we have to solve some $\bar{\partial}$-equation which depends continuously differentiably on a parameter. This can be done by means of the bounded linear operator T for solving the $\bar{\partial}$-equation stated in Theorem 2.3.5. However, in our opinion, it is interesting to show that for our purpose it is sufficient to use only Lemma 2.3.4 and certain general arguments which follow from Banach's open mapping theorem. Therefore, we begin with Lemma 2.4.1 below. First we give some notations.

Notation. If D is an open set in $\mathbb{C}^n$, then we denote by $C^\infty(D)$ the Fréchet space of all complex-valued C^∞-functions in D endowed with the topology of uniform convergence on compact sets together with all derivatives. By $Z^\infty_{(0,1)}(D)$ will be denoted the Fréchet space (endowed with the same topology) of all $C^\infty_{(0,1)}$-forms f in D such that $\bar{\partial} f = 0$.

2.4.1. Lemma. *Let $D \subset\subset \mathbb{C}^n$ be a strictly pseudoconvex open set, and let $U_{\overline{D}}$ be a neighbourhood of $\overline{D}$. Then there exists a continuous linear operator $T : Z^\infty_{(0,1)}(U_{\overline{D}}) \to C^\infty(D)$ such that*

$$\bar{\partial} T f = f \text{ in } D \quad \text{for all} \quad f \in Z^\infty_{(0,1)}(U_{\overline{D}}) .$$

Proof. Choose a strictly pseudoconvex neighbourhood $V_{\overline{D}}$ of $\overline{D}$ such that $V_{\overline{D}} \subset\subset U_{\overline{D}}$. Consider the commutative diagram

$$\begin{array}{ccccc} & & C^\infty(V_{\overline{D}}) & \longrightarrow & C^\infty(D) \\ & & \downarrow \bar{\partial} & & \downarrow \bar{\partial} \\ Z^\infty_{(0,1)}(U_{\overline{D}}) & \to & Z^\infty_{(0,1)}(V_{\overline{D}}) & \longrightarrow & Z^\infty_{(0,1)}(D) \end{array}$$

where the horizontal arrows denote the canonical restriction maps. We would like to apply Proposition 1 from Appendix 2. According to Lemma 2.3.4 and Corollary 2.1.6, $Z^{\infty}_{(0,1)}(U_{\bar{D}}) \subseteqq \bar{\partial}(C^{\infty}(V_{\bar{D}}))$, that is, condition (i) in Proposition 1, Appendix 2, is fulfilled.

We prove condition (ii) in this proposition. Let $L_2(D)$ be the Hilbert space of square integrable functions on D, and let P be the orthogonal projection from $L_2(D)$ onto the subspace $L_2^{\text{hol}}(D)$ of all holomorphic functions in $L_2(D)$. Since the natural maps $r_1: C^{\infty}(V_{\bar{D}}) \to L_2(D)$ and $r_2: L_2^{\text{hol}}(D) \to C^{\infty}(D)$ are continuous (see Theorem 1.1.13 for r_2), it follows that $r_2 P r_1$ can be taken as P_{21} in condition (ii) of Proposition 1, Appendix 2. Now the lemma follows from this proposition. ■

2.4.2. Lemma. *Let $\theta \subset\subset \mathbb{C}^n$ be an open set, and let ϱ be a C^2 strictly plurisubharmonic function in some neighbourhood of $\bar{\theta}$. Set*

$$\beta := \frac{1}{3} \min_{\zeta \in \bar{\theta},\, \xi \in \mathbb{C}^n, |\xi| = 1} \sum_{j,k=1}^{n} \frac{\partial^2 \varrho(\zeta)}{\partial \zeta_j \partial \bar{\zeta}_k} \xi_j \bar{\xi}_k . \tag{2.4.1}$$

Let a_{jk} be C^1-functions in a neighbourhood of $\bar{\theta}$ such that

$$\max_{\zeta \in \bar{\theta}} \left| a_{jk}(\zeta) - \frac{\partial^2 \varrho(\zeta)}{\partial \zeta_j \partial \zeta_k} \right| < \frac{\beta}{n^2} \quad \text{for} \quad j, k = 1, \dots, n . \tag{2.4.2}$$

Let $\varepsilon > 0$ be so small that

$$\max_{\zeta, z \in \bar{\theta},\, |\zeta - z| \leqq \varepsilon} \left| \frac{\partial^2 \varrho(\zeta)}{\partial x_j \partial x_k} - \frac{\partial^2 \varrho(z)}{\partial x_j \partial x_k} \right| < \frac{\beta}{2n^2} \quad \text{for} \quad j, k = 1, \dots, 2n , \tag{2.4.3}$$

where $x_j = x_j(\xi)$ are the real coordinates of $\xi \in \mathbb{C}^n$ such that $\xi_j = x_j(\xi) + i x_{j+n}(\xi)$.[1] *For $z, \zeta \in \bar{\theta}$ we define*

$$F(z, \zeta) := -\left[2 \sum_{j=1}^{n} \frac{\partial \varrho(\zeta)}{\partial \zeta_j} (z_j - \zeta_j) + \sum_{j,k=1}^{n} a_{jk}(\zeta) (z_j - \zeta_j)(z_k - \zeta_k) \right] . \tag{2.4.4}$$

Then, for all $z, \zeta \in \bar{\theta}$ with $|\zeta - z| \leqq \varepsilon$ we have the estimate

$$\operatorname{Re} F(z, \zeta) \geqq \varrho(\zeta) - \varrho(z) + \beta |\zeta - z|^2 \tag{2.4.5}$$

(Re := *real part of*).

Proof. Let $\zeta, z \in \bar{\theta}$ and $|\zeta - z| \leqq \varepsilon$. Then it follows from Lemma 1.4.13 and Taylor's theorem that

$$\operatorname{Re} \hat{F}_\varrho(z, \zeta) = \varrho(\zeta) - \varrho(z) + \sum_{j,k=1}^{n} \frac{\partial^2 \varrho(\zeta)}{\partial \zeta_j \partial \bar{\zeta}_k} (z_j - \zeta_j)(\bar{z}_k - \bar{\zeta}_k) + R(z, \zeta) ,$$

where, by (2.4.3), $|R(z, \zeta)| \leqq \beta |\zeta - z|^2$. By (2.4.1) this implies that $\operatorname{Re} \hat{F}_\varrho(z, \zeta) \geqq \varrho(\zeta) - \varrho(z) + 2\beta |\zeta - z|^2$. Since, by (2.4.2), $|\hat{F}_\varrho(z, \zeta) - F(z, \zeta)| \leqq \beta |\zeta - z|^2$, we obtain (2.4.5). ■

2.4.3. Theorem. *Let $D \subset\subset \mathbb{C}^n$ be a strictly pseudoconvex open set, let θ be a neighbourhood of ∂D, and let ϱ be a strictly plurisubharmonic C^2-function defined in some neighbourhood of $\bar{\theta}$ such that*

$$D \cap \theta = \{ z \in \theta : \varrho(z) < 0 \} .$$

[1] Since ϱ is a C^2-function, and since $\beta > 0$, such a_{jk} and ε exist.

Let ε, β, and $F(z, \zeta)$ be as in Lemma 2.4.2, where, moreover, ε is chosen so small that

$$\{z \in \mathbb{C}^n : |\zeta - z| \leqq 2\varepsilon\} \subseteqq \theta \quad \text{for every} \quad \zeta \in \partial D . \tag{2.4.6}$$

Then there exists a C^1-function $\Phi(z, \zeta)$ defined for ζ in some neighbourhood $U_{\partial D} \subseteqq \theta$ of ∂D and $z \in U_{\bar{D}} := D \cup U_{\partial D}$ such that:

(i) *$\Phi(z, \zeta)$ depends holomorphically on $z \in U_{\bar{D}}$.*

(ii) $\Phi(z, \zeta) \neq 0$ *for all* $z \in U_{\bar{D}}, \zeta \in U_{\partial D}$ *with* $|\zeta - z| \geqq \varepsilon$. (2.4.7)

(iii) *There is a C^1-function $M(z, \zeta) \neq 0$ defined for all points $z \in U_{\bar{D}}, \zeta \in U_{\partial D}$ with $|\zeta - z| \leqq \varepsilon$ such that*

$$\Phi(z, \zeta) = F(z, \zeta) M(z, \zeta) \quad \text{for} \quad z \in U_{\bar{D}}, \zeta \in U_{\partial D}, |\zeta - z| \leqq \varepsilon . \tag{2.4.8}$$

Proof. If follows from (2.4.5) that

$$\operatorname{Re} F(z, \zeta) \geqq \varrho(\zeta) - \varrho(z) + \beta\varepsilon^2 \quad \text{for} \quad z, \zeta \in \theta, \epsilon \leqq |\zeta - z| \leqq 2\varepsilon . \tag{2.4.9}$$

Since $\varrho = 0$ on ∂D and by (2.4.6), we can choose a neighbourhood $V_{\partial D} \subseteqq \theta$ of ∂D so small that $|\varrho| \leqq \beta\varepsilon^2/3$ on $V_{\partial D}$ and, for every $\zeta \in V_{\partial D}$, the ball $|\zeta - z| \leqq 2\varepsilon$ is contained in θ. Set $V_{\bar{D}} := D \cup V_{\partial D}$. Then, for every $(z, \zeta) \in V_{\bar{D}} \times V_{\partial D}$ with $|\zeta - z| \leqq 2\varepsilon$, both ζ and z belong to θ and it follows from (2.4.9) that $\operatorname{Re} F(z, \zeta) \geqq \beta\varepsilon^2/3$ for all $z \in V_{\bar{D}}$ and $\zeta \in V_{\partial D}$ with $\varepsilon \leqq |\zeta - z| \leqq 2\varepsilon$. Therefore, we can define $\ln F(z, \zeta)$ for $z \in V_{\bar{D}}$ and $\zeta \in V_{\partial D}$ with $\varepsilon \leqq |\zeta - z| \leqq 2\varepsilon$. Choose a C^∞-function χ on $\mathbb{C}^n$ such that $\chi(\xi) = 1$ for $|\xi| \leqq \varepsilon + \varepsilon/4$ and $\chi(\xi) = 0$ for $|\xi| \geqq 2\varepsilon - \varepsilon/4$. For $z \in V_{\bar{D}}$ and $\zeta \in V_{\partial D}$, we define

$$f(z, \zeta) := \begin{cases} \bar{\partial}_z[\chi(\zeta - z) \ln F(z, \zeta)] & \text{if} \quad \varepsilon \leqq |\zeta - z| \leqq 2\varepsilon , \\ 0 & \text{otherwise} . \end{cases}$$

Then the map $V_{\partial D} \ni \zeta \to f(\cdot, \zeta)$ is continuously differentiable with values in the Fréchet space $Z^\infty_{(0,1)}(V_{\bar{D}})$. Now we choose a neighbourhood $U_{\partial D} \subset\subset V_{\partial D}$ such that $U_{\bar{D}} := D \cup U_{\partial D}$ is strictly pseudoconvex. Then, by Lemma 2.4.1, there is a continuous linear operator $T : Z^\infty_{(0,1)}(V_{\bar{D}}) \to C^\infty(U_{\bar{D}})$ such that $\bar{\partial} T \varphi = \varphi$ on $U_{\bar{D}}$ for all $\varphi \in Z^\infty_{(0,1)}(V_{\bar{D}})$. For $z \in U_{\bar{D}}$ and $\zeta \in U_{\partial D}$, we define

$$u(z, \zeta) := \big(Tf(\cdot, \zeta)\big)(z) , \qquad M(z, \zeta) := \exp\big(-u(z, \zeta)\big)$$

and

$$\Phi(z, \zeta) := \begin{cases} F(z, \zeta) M(z, \zeta) & \text{if} \quad |\zeta - z| \leqq \varepsilon , \\ \exp[\chi(\zeta - z) \ln F(z, \zeta) - u(z, \zeta)] & \text{if} \quad |\zeta - z| \geqq \varepsilon . \end{cases}$$

This completes the proof. ■

Remark. We point out that in Theorem 2.4.3 we do not assume that ∂D is smooth.

Remark. It follows from Theorem 2.4.3 that every strictly pseudoconvex open set $D \subset\subset \mathbb{C}^n$ is a domain of holomorphy. In fact, for every fixed $\zeta_0 \in \partial D$, the function $f(z) := 1/\Phi(z, \zeta_0)$ is holomorphic in D (this follows from properties (i)—(iii) of Φ and estimate (2.4.5)) and $f(z) \to \infty$ for $D \ni z \to \zeta_0$.

2.5. The Oka-Hefer lemma and solution of $\langle w(z, \zeta), \zeta - z\rangle = \Phi(z, \zeta)$

In this section we construct a Leray map $w(z, \zeta)$ for strictly pseudoconvex open sets in $\mathbb{C}^n$ which is holomorphic in z and such that $\langle w(z, \zeta), \zeta - z\rangle = \Phi(z, \zeta)$, where $\Phi(z, \zeta)$ is the function from Theorem 2.4.3. Locally this is simple. To do this globally, we use again the $\bar{\partial}$-equation.

2.5.1. Lemma. *Let $D \subset\subset \mathbb{C}^n$ be a strictly pseudoconvex open set, let $M_1 = \{z \in \mathbb{C}^n: z_1 = 0\}$, and let $U_{\bar D}$ be a neighbourhood of $\bar D$. Then, for every holomorphic function f in $M_1 \cap U_{\bar D}$, there exists a holomorphic function $\tilde f$ in D such that $\tilde f = f$ in $M_1 \cap D$.*

Proof. If U_1 is a sufficiently small neighbourhood of $M_1 \cap \bar D$, then, by setting $F(z) := f(0, z_2, \dots, z_n)$, we obtain a holomorphic continuation of f to U_1. Choose neighbourhoods U_1', U_1'' of $M_1 \cap \bar D$ such that $U_1' \subset\subset U_1'' \subset\subset U_1$. Let χ be a C^∞-function on $\mathbb{C}^n$ such that $\chi = 1$ in U_1' and $\chi = 0$ in $\mathbb{C}^n \setminus U_1''$. Define

$$\varphi(z) := \begin{cases} \dfrac{F(z)\,\bar\partial\chi(z)}{z_1} & \text{if} \quad z \in U_1\,, \\ 0 & \text{if} \quad z \in \mathbb{C}^n \setminus U_1\,. \end{cases}$$

Then φ is a $C^\infty_{(0,1)}$-form on $\mathbb{C}^n$ such that $\bar\partial\varphi = 0$. Therefore, by Lemma 2.3.4, there exists a continuous function u on $\bar D$ such that $\bar\partial u = \varphi$ in D, that is, $\bar\partial(F\chi - z_1 u) = 0$ in D. Hence $\tilde f := F\chi - z_1 u$ is a holomorphic function in D. Since $\chi = 1$ in $D \cap M_1$, $\tilde f = F = f$ in $D \cap M_1$. ■

2.5.2. Lemma. *Let $D \subset\subset \mathbb{C}^n$ be a strictly pseudoconvex open set, let*

$$M_k := \{z \in \mathbb{C}^n : z_1 = \dots = z_k = 0\}\,, \qquad 1 \leqq k \leqq n\,,$$

and let $U_{\bar D}$ be a neighbourhood of $\bar D$. Then, for every holomorphic function f in $U_{\bar D}$ with $f = 0$ on $M_k \cap U_{\bar D}$, there exist holomorphic functions $f_1, \dots, f_k$ in D such that

$$f(z) = \sum_{j=1}^{k} z_j f_j(z) \quad \textit{for all} \quad z \in D\,.$$

Proof. For $k = 1$, we can set $f_1(z) = f(z)/z_1$. Suppose the lemma is already proved for $k - 1$, and let f be a holomorphic function in $U_{\bar D}$ such that $f(z) = 0$ for $z \in M_k \cap U_{\bar D}$. Choose a strictly pseudoconvex open set $\tilde D$ such that $D \subset\subset \tilde D \subset\subset U_{\bar D}$. Then $\tilde D \cap M_1$ is a strictly pseudoconvex open set in M_1 $(=\mathbb{C}^{n-1})$, and, by hypothesis, there are holomorphic functions $\tilde f_j(z_2, \dots, z_n)$ in $\tilde D \cap M_1$ such that

$$f(z) = \sum_{j=2}^{k} z_j \tilde f_j(z_2, \dots, z_n) \quad \text{for} \quad z \in \tilde D \cap M_1\,.$$

By Lemma 2.5.1, we can find holomorphic functions f_j in D such that $f_j(z) = \tilde f_j(z_2, \dots, z_n)$ for $z \in D \cap M_1$. Setting

$$f_1(z) := \frac{1}{z_1}\Big(f(z) - \sum_{j=2}^{k} z_j f_j(z)\Big) \quad \text{for} \quad z \in D\,,$$

we conclude the proof. ■

We need an extension of Lemma 2.5.2 to the case when f depends continuously differentiably on a parameter. Then we want that the coefficients f_j also depend continuously differentiably on this parameter. To obtain this, we prove that the coefficients f_j can be given by continuous linear operators. First we introduce some notations.

Notation. If $U \subseteqq \mathbb{C}^n$ is an open set, then we denote by $\mathcal{O}(U)$ the Fréchet space of holomorphic functions on U endowed with the topology of uniform convergence on compact sets in U. Set $\mathcal{O}^k(U) := \mathcal{O}(U) \oplus \dots \oplus \mathcal{O}(U)$ (k times). If $Y \subseteqq \mathbb{C}^n$, then we denote by $\mathcal{O}_Y(U)$ the subspace of all $f \in \mathcal{O}(U)$ such that $f = 0$ on $Y \cap U$.

2.5.3. Lemma. *Under the hypotheses of Lemma 2.5.2 there are continuous linear operators* $T_j\colon \mathcal{O}_{M_k}(U_{\overline{D}}) \to \mathcal{O}(D)$ *such that, for every* $f \in \mathcal{O}_{M_k}(U_{\overline{D}})$,

$$f(z) = \sum_{j=1}^{k} z_j (T_j f)(z) \quad \textit{for all} \quad z \in D\,.$$

Proof. Choose a strictly pseudoconvex open set D' such that $D \subset\subset D' \subset\subset U_{\overline{D}}$. Denote by A the continuous linear map from $\mathcal{O}^k(D')$ into $\mathcal{O}_{M_k}(D')$ and $\mathcal{O}^k(D)$ into $\mathcal{O}_{M_k}(D)$ which is defined by

$$A(f_1, \ldots, f_k) := \sum_{j=1}^{k} z_j f_j\,.$$

Consider the commutative diagram

$$\begin{array}{ccccc} & & \mathcal{O}^k(D') & \longrightarrow & \mathcal{O}^k(D) \\ & & \downarrow A & & \downarrow A \\ \mathcal{O}_{M_k}(U_{\overline{D}}) & \to & \mathcal{O}_{M_k}(D') & \longrightarrow & \mathcal{O}_{M_k}(D)\,, \end{array}$$

where the horizontal arrows denote the canonical restriction maps. By Lemma 2.5.2, $\mathcal{O}_{M_k}(U_{\overline{D}}) \subseteqq A(\mathcal{O}^k(D'))$. Further, if $L_2^k(D)$ is the Hilbert space of square integrable vector functions in $\mathcal{O}^k(D)$, then the natural maps

$$\mathcal{O}^k(D') \to L_2^k(D) \quad \text{and} \quad L_2^k(D) \to \mathcal{O}^k(D)$$

are continuous (see Theorem 1.1.13 for the second map). Therefore, the statement follows from Propositions 1 and 2, Appendix 2. ■

2.5.4. Theorem (the Oka-Hefer lemma). *Let* $D \subset\subset \mathbb{C}^n$ *be a strictly pseudoconvex open set, and let* $U_{\overline{D}}$ *be a neighbourhood of* $\overline{D}$. *Then there are continuous linear maps* $T_j\colon \mathcal{O}(U_{\overline{D}}) \to \mathcal{O}(D \times D)$ *such that, for every* $f \in \mathcal{O}(U_{\overline{D}})$,

$$f(\zeta) - f(z) = \sum_{j=1}^{n} (\zeta_j - z_j)(T_j f)(z, \zeta) \quad \textit{for all} \quad z, \zeta \in D\,.$$

Proof. Choose a strictly pseudoconvex open set $D' \subset\subset \mathbb{C}^{2n}$ such that $D \times D \subset\subset D' \subset\subset U_{\overline{D}} \times U_{\overline{D}}$. Set $\xi_j := \zeta_j - z_j$ and $\xi_{j+n} := z_j$ for $j = 1, \ldots, n$. Define $h(z, \zeta) := \xi$ for $z, \zeta \in U_{\overline{D}}$. Then $h(D')$ also is strictly pseudoconvex. Define

$$Sf(\xi) := f(\xi_1 + \xi_{n+1}, \ldots, \xi_n + \xi_{2n}) - f(\xi_{n+1}, \ldots, \xi_{2n})\,.$$

S is a continuous linear map from $\mathcal{O}(U_{\overline{D}})$ into $\mathcal{O}_{M_n}(h(U_{\overline{D}} \times U_{\overline{D}}))$, where $M_n := \{\xi \in \mathbb{C}^{2n} : \xi_1 = \ldots = \xi_n = 0\}$. In view of Lemma 2.5.3, there are continuous linear operators $\widetilde{T}_j\colon \mathcal{O}_{M_n}(h(U_{\overline{D}} \times U_{\overline{D}})) \to \mathcal{O}(h(D'))$ such that

$$Sf(\xi) = \sum_{j=1}^{n} \xi_j (\widetilde{T}_j Sf)(\xi) \quad \text{for all} \quad \xi \in h(D') \quad \text{and} \quad f \in \mathcal{O}(U_{\overline{D}})\,.$$

Hence

$$f(\zeta) - f(z) = (Sf)(h(z, \zeta)) = \sum_{j=1}^{n} (\zeta_j - z_j)(\widetilde{T}_j Sf)(h(z, \zeta)) \quad \text{for} \quad z, \zeta \in D\,.$$

Setting $(T_j f)(z, \zeta) := (\widetilde{T}_j Sf)(h(z, \zeta))$, we conclude the proof. ■

2.5.5. Theorem. *Let* $D \subset\subset \mathbb{C}^n$ *be a strictly pseudoconvex open set and let* $U_{\partial D}$, $U_{\overline{D}}$, $\Phi(z, \zeta)$ *be as in Theorem 2.4.3. Suppose* $V_{\partial D}$ *is a neighbourhood of* ∂D *such that* $V_{\partial D} \subset\subset U_{\partial D}$ *and* $V_{\overline{D}} := V_{\partial D} \cup D$ *is strictly pseudoconvex. Then there exist continuous* *linear maps* $T_j\colon \mathcal{O}(U_{\bar D}) \to \mathcal{O}(V_{\bar D} \times V_{\bar D})$ *such that the* C^1*-map* $w = (w_1, \ldots, w_n)$ *defined by*

$$w_j(z, \zeta) := T_j(\Phi(\cdot, \zeta))\,(z, \zeta)\,, \qquad \zeta \in V_{\partial D}\,, \qquad z \in V_{\bar D}\,, \tag{2.5.1}$$

has the following properties:

(i) $\langle w(z, \zeta), \zeta - z\rangle = \Phi(z, \zeta)$ *for all* $z \in V_{\bar D}$ *and* $\zeta \in V_{\partial D}$, (2.5.2)

(ii) $w(z, \zeta)$ *is holomorphic in* $z \in V_{\bar D}$,

(iii) $w(z, \zeta)$ *is a Leray map for* D.

Proof. By Theorem 2.5.4 there are continuous linear operators

$$T_j\colon \mathcal{O}(U_{\bar D}) \to \mathcal{O}(V_{\bar D} \times V_{\bar D}^{*})$$

such that

$$\Phi(\xi, \zeta) - \Phi(z, \zeta) = -\sum_{j=1}^{n} (\xi_j - z_j)\,\big(T_j(\Phi(\cdot, \zeta))\big)\,(z, \xi) \quad \text{for} \quad z, \xi \in V_{\bar D}, \zeta \in U_{\partial D}\,.$$

Define $w_j(z, \zeta) := \big(T_j(\Phi(\cdot, \zeta))\big)\,(z, \zeta)$. Since $\Phi(\zeta, \zeta) = 0$, then we obtain (2.5.2). It is clear that $w(z, \zeta)$ is holomorphic in z. That $w(z, \zeta)$ is a Leray map for D follows from (2.5.2) and the fact that, by (2.4.5), $\Phi(z, \zeta) \neq 0$ for $z \in D$ and $\zeta \in \partial D$. ■

2.5.6. Remark. The map $w(z, \zeta)$ from Theorem 2.5.5 is a Leray map not only for D but for each of the strictly pseudoconvex open sets $(D \setminus V_{\partial D}) \cup \{z \in V_{\partial D} : \varrho(z) < -\delta\}$ if $\delta > 0$ is sufficiently small. This follows from (2.5.2) and the fact that, by (2.4.5), $\Phi(z, \zeta) \neq 0$ if $\varrho(z) < \varrho(\zeta)$. We remark also that in Theorem 2.5.5 we do not assume that ∂D is smooth.

2.6. Formula for solving the $\bar\partial$-equation with Hölder estimates in C^2 strictly pseudoconvex open sets in $\mathbb{C}^n$

In the preceding section, for strictly pseudoconvex open sets $D \subset\subset \mathbb{C}^n$, we constructed a Leray map $w(z, \zeta)$ which is holomorphic in z. If ∂D is smooth, then, by Corollary 1.12.2, this Leray map gives an integral operator for solving the $\bar\partial$-equation in D. In the present section, we prove that this integral operator admits 1/2-Hölder estimates.

2.6.1. Theorem. *Let* $D \subset\subset \mathbb{C}^n$ *be a strictly pseudoconvex open set with* C^2*-boundary, let* $w(z, \zeta)$ *be the Leray map from Theorem 2.5.5 for* D. *Then there exists a constant* $C < \infty$ *such that* (for notation see Section 1.8):

(i) *For every continuous differential form* f *on* $\bar D$,

$$\|R_{\partial D}^{w} f\|_{1/2, D} \leq C\, \|f\|_{0, D}\,.$$

(ii) *For every continuous* $(0, q)$*-form* f *on* $\bar D$ *such that* $\bar\partial f = 0$ *in* D, $1 \leq q \leq n$,

$$u := (-1)^q\,(R_{\partial D}^{w} f + B_D f) \tag{2.6.1}$$

is a continuous solution of $\bar\partial u = f$ *in* D *such that* $\|u\|_{1/2, D} \leq C\, \|f\|_{0, D}$. *Moreover,* $u \in C_{(0,q-1)}^{\alpha}(D)$ *for all* $0 < \alpha < 1$ *and, if* f *is* C^k *in* D $(k = 1, 2, \ldots, \infty)$, *then* $u \in C_{(0,q-1)}^{k+\alpha}(D)$ *for all* $0 < \alpha < 1$.

Proof. By Lemma 1.8.5 and Corollary 1.12.2 we only have to prove part (i). To do this, we show that $w(z, \zeta)$ satisfies condition (iii) in Lemma 2.2.1. Fix $\xi \in D$. By

Theorem 2.5.5 (i) and Theorem 2.4.3, $\langle w(z,\zeta), \zeta - z\rangle = F(z,\zeta)\, M(z,\zeta)$, and we can choose a neighbourhood U_ξ of ξ so small that, for some $\delta_1 > 0$,

$$|\langle w(z,\zeta), \zeta - z\rangle| \geqq \delta_1\, |F(z,\zeta)| \quad \text{for all} \quad z, \zeta \in U_\xi . \tag{2.6.2}$$

Set $t_1(z,\zeta) := \operatorname{Im} F(z,\zeta)$ (Im := imaginary part of). Let $x_j = x_j(\zeta)$ be the real coordinates of $\zeta \in \mathbb{C}^n$ such that $\zeta_j = x_j(\zeta) + ix_{j+n}(\zeta)$. Then

$$t_1(z,\zeta) = \sum_{j=1}^{n} \left[\frac{\partial\varrho(\zeta)}{\partial x_j} x_{j+n}(\zeta - z) - \frac{\partial\varrho(\zeta)}{\partial x_{j+n}} x_j(\zeta - z)\right]$$
$$- \operatorname{Im} \sum_{j,k=1}^{n} a_{jk}(\zeta)\,(\zeta_j - z_j)\,(\zeta_k - z_k)\,,$$

and, consequently,

$$\mathrm{d}_\zeta t_1(z,\zeta)|_{\zeta=z} = \sum_{j=1}^{n} \left[\frac{\partial\varrho(z)}{\partial x_j} \mathrm{d}x_{j+n}(z) - \frac{\partial\varrho(z)}{\partial x_{j+n}} \mathrm{d}x_j(z)\right]. \tag{2.6.3}$$

Hence

$$n\|\mathrm{d}_\zeta t_1(z,\zeta)|_{\zeta=z} \wedge \mathrm{d}\varrho(z)\| \geqq \|\mathrm{d}\varrho(z)\|^2 \quad \text{for all} \quad z \in \partial D\,, \tag{2.6.4}$$

because, by (2.6.3), the coefficient of $\mathrm{d}x_j(z) \wedge \mathrm{d}x_{j+n}(z)$ in this form is $-\ [(\partial\varrho(z)/\partial x_j)^2 + (\partial\varrho(z)/\partial x_{j+n})^2]$. Since $\mathrm{d}\varrho(z) \neq 0$ for $z \in \partial D$ (see Proposition 1.5.16), it follows that, after shrinking U_ξ, we can find C^1-functions $\tilde{t}_2, \dots, \tilde{t}_{2n-1}$ on U_ξ such that, for every $z \in U_\xi$, $t_1(z, \cdot), \tilde{t}_2, \dots, \tilde{t}_{2n-1}$ are coordinates on $U_\xi \cap \partial D$. Set $t_j(z,\zeta) := \tilde{t}_j(\zeta) - \tilde{t}_j(z)$ for $z, \zeta \in U_\xi$, $j = 2, \dots, 2n-1$. Then, for every fixed $z \in U_\xi$, the functions $t_1(z,\cdot), \dots, t_{2n-1}(z,\cdot)$ are also coordinates on $\partial D \cap U_\xi$.

To complete the proof, we show estimate (2.2.1) for these coordinates. Since $t_j(z,z) = 0$, after shrinking U_ξ, Taylor's formula gives that there is a $\delta_2 > 0$ such that $|\zeta - z| \geqq \delta_2\, |t(z,\zeta)|$ for all $z, \zeta \in U_\xi$. Therefore, it follows from estimate (2.4.5) in Lemma 2.4.2 that, after shrinking U_ξ, we can find a $\delta_3 > 0$ such that

$$|F(z,\zeta)| \geqq \delta_3(|t_1(z,\zeta)| + \varrho(\zeta) - \varrho(z) + |t(z,\zeta)|^2) \quad \text{for} \quad z, \zeta \in U_\xi\,.$$

Since $\varrho(\zeta) = 0$ for $\zeta \in \partial D$, and since, after shrinking U_ξ, there is $\delta_4 > 0$ such that $-\varrho(z) \geqq \delta_4 \operatorname{dist}(z, \partial D)$ for $z \in D \cap U_\xi$, this implies that

$$|F(z,\zeta)| \geqq \delta_3\delta_4\left(|t_1(z,\zeta)| + |t(z,\zeta)|^2 + \operatorname{dist}(z, \partial D)\right)$$

for $z \in D \cap U_\xi$ and $\zeta \in \partial D \cap U_\xi$. According to (2.6.2) this gives (2.2.1). ∎

Remark. We point out that Theorem 2.6.1 (ii) contains more information than Theorem 2.3.5, because formula (2.6.1) depends more explicitly on D than the operator T from Theorem 2.3.5. For example, one can prove that the constant C in Theorem 2.6.1 (ii) depends continuously (in some sense) on D.

2.7. Oka-Weil approximation

Let $D \subseteqq \mathbb{C}^n$ be a pseudoconvex open set. In this section we prove that every holomorphic function in a neighbourhood of a $P^0(D)$-convex compact subset K of D can be approximated uniformly on K by holomorphic functions in D (Theorem 2.7.1). By means of the support function $\Phi(z,\zeta)$ from Theorem 2.4.3, this makes it possible to prove that every pseudoconvex open set in $\mathbb{C}^n$ is a domain of holomorphy (Corollary 2.7.3). In Section 2.8 we shall use this approximation theorem to solve the

$\bar{\partial}$-equation in arbitrary pseudoconvex open sets in $\mathbb{C}^n$. The basic tool in the proof of Theorem 2.7.1 is the Leray formula for holomorphic functions which is defined by the Leray map from Theorem 2.5.5.

2.7.1. Theorem. *Let $D \subseteqq \mathbb{C}^n$ be a pseudoconvex open set, and let $K \subset\subset D$ be a $P^0(D)$-convex compact set* (cp. Definition 1.5.3). *Then every holomorphic function in a neighbourhood of K can be approximated uniformly on K by holomorphic functions in D.*

Proof. Let h be a holomorphic function in a neighbourhood U_K of K. By Corollary 1.5.10, there is a strictly plurisubharmonic C^∞-function $\varrho: D \to \mathbb{R}$ such that

(i) $D_\alpha := \{z \in D : \varrho(z) < \alpha\} \subset\subset D$ for all $\alpha \in \mathbb{R}$.

(ii) The set Crit $(\varrho) := \{z \in D : d\varrho(z) = 0\}$ is discrete in D.

(iii) $\varrho < 0$ on K and $\varrho > 1$ on $D \setminus U_K$.

Denote by α_0 the supremum of all $\alpha \in \mathbb{R}$ such that h can be approximated uniformly on K by functions in $\mathcal{O}(D_\alpha)$. Since $h \in \mathcal{O}(U_K)$ and by (iii), $\alpha_0 > 0$. First we prove that $\alpha_0 = \infty$. Assume that $\alpha_0 < \infty$.

By Theorems 2.4.3 and 2.5.5 and by estimate (2.4.5), then there are a number $\delta > 0$, a C^1-function $\Phi(z, \zeta)$ and a $\mathbb{C}^n$-valued C^1-map $w(z, \zeta)$ defined for z in some neighbourhood of $\bar{D}_{\alpha_0+\delta}$ and ζ in a neighbourhood of $\bar{D}_{\alpha_0+\delta} \setminus D_{\alpha_0-\delta}$ with the following properties:

(a) $\Phi(z, \zeta)$ and $w(z, \zeta)$ are holomorphic in z.

(b) $\Phi(z, \zeta) = \langle w(z, \zeta), \zeta - z\rangle$.

(c) $\Phi(z, \zeta) \neq 0$ if $\alpha_0 - \delta \leqq \varrho(\zeta) \leqq \alpha_0 + \delta$ and $\varrho(z) < \varrho(\zeta)$.

(d) $w(z, \zeta)$ is a Leray map for $D_{\alpha_0-\delta}$.

By definition of α_0, h can be approximated uniformly on K by holomorphic functions in a neighbourhood of $\bar{D}_{\alpha_0-\delta}$. To obtain a contradiction (with $\alpha_0 < \infty$), it is therefore sufficient to prove that every holomorphic function f in some neighbourhood of $\bar{D}_{\alpha_0-\delta}$ can be approximated uniformly on K by functions in $\mathcal{O}(D_{\alpha_0+\delta})$. In view of property (ii) of ϱ, we can assume that $d\varrho(z) \neq 0$ for $z \in \partial D_{\alpha_0-\delta}$. Then, in view of the Leray formula (1.10.3), we have

$$f(z) = \frac{(n-1)!}{(2\pi i)^n} \int_{\partial D_{\alpha_0-\delta}} f(\zeta) \frac{\omega'_\zeta(w(z, \zeta)) \wedge \omega(\zeta)}{\Phi^n(z, \zeta)}, \quad z \in D_{\alpha_0-\delta}.$$

Since $w(z, \zeta)$ is holomorphic for $z \in D_{\alpha_0+\delta}$, it is therefore sufficient to prove that, for every fixed $\xi \in \partial D_{\alpha_0-\delta}$ the function $1/\Phi(z, \xi)$ can be approximated uniformly on K by functions in $\mathcal{O}(D_{\alpha_0+\delta})$. To do this we fix $\xi_0 \in \partial D_{\alpha_0-\delta}$ and choose a finite number of points $\xi_1, \ldots, \xi_k$ such that $\alpha_0 - \delta < \varrho(\xi_1) < \ldots < \varrho(\xi_k) = \alpha_0 + \delta$ and

$$\sup_{z \in K} \left| 1 - \frac{\Phi(z, \xi_{j-1})}{\Phi(z, \xi_j)} \right| < 1 \quad \text{for} \quad j = 1, \ldots, k.$$

Then for $j = 1, \ldots, k$

$$\frac{1}{\Phi(z, \xi_{j-1})} = \frac{1}{\Phi(z, \xi_j)} \sum_{s=0}^{\infty} \left[1 - \frac{\Phi(z, \xi_{j-1})}{\Phi(z, \xi_j)} \right]^s,$$

where the convergence is uniform in $z \in K$. Therefore, for $j = 1, \ldots, k$, the function $1/\Phi(z, \xi_{j-1})$ can be approximated uniformly on K by polynomials in $1/\Phi(z, \xi_j)$ and $\Phi(z, \xi_{j-1})$. We conclude that $1/\Phi(z, \xi_0)$ can be approximated uniformly on K by polynomials in $1/\Phi(z, \xi_k)$ and $\Phi(z, \xi_0), \Phi(z, \xi_1), \ldots, \Phi(z, \xi_{k-1})$, which are holomorphic functions in $D_{\alpha_0+\delta}$. Hence it is proved that $\alpha_0 = \infty$.

To complete the proof, it is sufficient to show that, for every $\varepsilon > 0$, there is a sequence $g_j \in \mathcal{O}(D_{j+1})$, $j = 0, 1, 2, \ldots$, such that $g_0 := h$ (by (iii), $D_1 \subseteqq U_K$) and

$$\sup_{z \in D_j} |g_j(z) - g_{j+1}(z)| < \frac{\varepsilon}{2^{j+1}} \quad \text{for} \quad j = 0, 1, \ldots,$$

because then $g_j \to g \in \mathcal{O}(D)$ uniformly on every compact subset of D and $\sup_K |h - g| \leqq \varepsilon$. To find this sequence, we suppose that the functions $g_0, \ldots, g_k$ are already constructed (for $k = 0$ this is trivial). Since $\overline{D}_k$ is P^0(D)-convex, then by what is proved above, there is $g_{k+1} \in \mathcal{O}(D_{k+2})$ such that $|g_k(z) - g_{k+1}(z)| < \varepsilon/2^{k+1}$ for all $z \in \overline{D}_k$. ■

2.7.2. Theorem. *Let $D \subseteqq \mathbb{C}^n$ be a pseudoconvex open set. Then, for every compact set $K \subset\subset D$, $\hat{K}_D^P = \hat{K}_D^{\mathcal{O}}$* (cp. Definitions 1.3.2 and 1.5.3).

Proof. By Proposition 1.5.4, $\hat{K}_D^P \subseteqq \hat{K}_D^{\mathcal{O}}$. To prove the converse we fix a point $\xi \in D \setminus \hat{K}_D^P$ and construct a holomorphic function h in D such that

$$|h(\xi)| > \max_{z \in K} |h(z)| . \tag{2.7.1}$$

It follows from Corollary 1.5.10 (or from Theorem 1.5.8 and Lemma 1.4.19) that there is a strictly pseudoconvex open set $G \subset\subset D$ such that $K \subset\subset G$, $\xi \in \partial G$ and $\overline{G}$ is $P^0(D)$-convex. By Theorem 2.4.3 and estimate (2.4.5), there is a C^1-function $\Phi(z, \zeta)$ defined for ζ in some neighbourhood $U_{\partial G}$ of ∂G and z in some neighbourhood $U_{\overline{G}}$ of $\overline{G}$ such that $\Phi(z, \zeta)$ is holomorphic in $z \in U_{\overline{G}}$, $\Phi(\zeta, \zeta) = 0$, and $\Phi(z, \zeta) \neq 0$ for all $\zeta \in U_{\partial G} \setminus \overline{G}$ and $z \in \overline{G}$. Then, for every fixed $\zeta_0 \in U_{\partial G} \setminus \overline{G}$, $1/\Phi(z, \zeta_0)$ is holomorphic for z in some neighbourhood of $\overline{G}$. Since $K \subset\subset G$, $\xi \in \partial G$, and $\Phi(\xi, \xi) = 0$, we can choose $\zeta_0 \in U_{\partial G} \setminus \overline{G}$ so close to ξ that

$$\left| \frac{1}{\Phi(\xi, \zeta_0)} \right| > 1 + \max_{z \in K} \left| \frac{1}{\Phi(z, \zeta_0)} \right| . \tag{2.7.2}$$

Since $\overline{G}$ is $P^0(D)$-convex, by Theorem 2.7.1 we can find a holomorphic function h in D such that $|h(z) - 1/\Phi(z, \zeta_0)| < 1/2$ for all $z \in \overline{G}$. According to (2.7.2) this function h fulfils (2.7.1). ■

2.7.3. Corollary. *An open set in $\mathbb{C}^n$ is a domain of holomorphy if and only if it is pseudoconvex.*

Proof. This follows from Corollary 1.5.6 and Theorems 2.7.2, 1.3.7 (iii) and 1.5.5 (ii).

2.8. Solution of the $\bar{\partial}$-equation in pseudoconvex open sets in $\mathbb{C}^n$

2.8.1. Theorem. *Let $D \subseteqq \mathbb{C}^n$ be a pseudoconvex open set, $1 \leqq q \leqq n$ and $k = 0, 1, 2, \ldots, \infty$. Then, for every $C^k_{(0,q)}$-form f in D such that $\bar{\partial} f = 0$ in D, there exists a solution u of*

$$\bar{\partial} u = f \quad \text{in} \quad D ,$$

which belongs to $C^{k+\alpha}_{(0,q-1)}(D)$ *for all* $0 < \alpha < 1$ (for the definition of $C^{k+\alpha}_{(0,q-1)}(D)$ see Section 1.8).[1])

Proof. By Corollary 1.5.10, there exists a strictly plurisubharmonic C^∞-function ϱ in D such that

(i) $D_\alpha := \{z \in D : \varrho(z) < \alpha\} \subset\subset D$ for all $\alpha \in \mathbb{R}$,

(ii) the set Crit $(\varrho) := \{z \in D : d\varrho(z) = 0\}$ is discrete in D.

In view of (ii), we can find a sequence of numbers $\alpha_s \in \mathbb{R}$, $s = 1, 2, \ldots$, tending to infinity such that $d\varrho(z) \neq 0$ for $z \in \partial D_{\alpha_s}$, that is, D_{α_s} is a C^∞ strictly pseudoconvex open set. Then, by Theorem 2.6.1 (ii), there are solutions u_s of $\bar\partial u_s = f$ in D_{α_s}, which belong to $C^{k+\alpha}_{(0,q-1)}(D_{\alpha_s})$ for all $0 < \alpha < 1$.

To construct a global solution in D, we first consider the case $q \geqq 2$. It is sufficient to find a new sequence $\{v_s\}_{s=3}^\infty$ of solutions of $\bar\partial v_s = f$ in D_{α_s} such that $v_s \in C^{k+\alpha}_{(0,q-1)}(D_{\alpha_s})$ for all $0 < \alpha < 1$ and, in addition, $v_s = v_{s+1}$ in $D_{\alpha_{s-2}}$. To do this we set $v_3 := u_3$ and suppose that, for some $k \geqq 3$, the functions $v_3, \ldots, v_k$ are already constructed. Then $v_k - u_{k+1}$ is $\bar\partial$-close in D_{α_k} and Theorem 2.6.1 (ii) gives a solution φ of $\bar\partial\varphi = v_k - u_{k+1}$ in $D_{\alpha_{k-1}}$ which belongs to $C^{k+\alpha}_{(0,q-1)}(D_{\alpha_{k-1}})$ for all $0 < \alpha < 1$. Choose a C^∞-function χ with compact support in $D_{\alpha_{k-1}}$ such that $\chi = 1$ in $D_{\alpha_{k-2}}$. Set $v_{k+1} := u_{k+1} + \bar\partial(\chi\varphi)$. This completes the proof for $q \geqq 2$.

Now we consider the case $q = 1$. We will construct a sequence $\{v_s\}_{s=2}^\infty$ of solutions of $\bar\partial v_s = f$ in D_{α_s} which belong to $C^{k+\alpha}(D_{\alpha_s})$ for all $0 < \alpha < 1$ such that, moreover, $|v_s(z) - v_{s+1}(z)| < 2^{-s}$ for $z \in D_{\alpha_{s-1}}$. Then this sequence tends uniformly on the compact sets in D to some function u in D, which has the required properties, because the differences $u - v_s$ are holomorphic.

To construct this sequence, we set $v_2 := u_2$ and suppose that the functions $v_2, \ldots, v_k$ are already constructed. Then the difference $v_k - u_{k+1}$ is holomorphic in $D_{\alpha k}$ and, by Theorem 2.7.1, we can find $w \in \mathcal{O}(D)$ such that $|v_k(z) - u_{k+1}(z) - w(z)| < 2^{-k}$ for $z \in D_{\alpha_{k-1}}$. Setting $v_{k+1} := u_{k+1} + w$ we complete the proof. ■

2.9. Uniform approximation

Let $K \subset\subset \mathbb{C}^n$ be a compact set and let f be a continuous function on K that is holomorphic in the inner points of K. Consider the following problem: Can f be approximated uniformly on K by functions holomorphic in some neighbourhood of K? In distinction to the case of one complex variable, for $n \geqq 2$, this problem is solved only in special cases. Diederich and Fornaess [1] constructed a pseudoconvex open set $D \subset\subset \mathbb{C}^n$ with smooth boundary such that, for $K = \bar D$, the answer is negative. In Section 3.5 we shall prove that the answer is positive if $K = \{z \in U_K : \varrho(z) \leqq 0\}$, where ϱ is a strictly plurisubharmonic C^2-function in some neighbourhood U_K of K. (By Corollary 1.4.20 and Theorem 1.5.19, this is the case for every compact subset of a totally real C^1-submanifold of $\mathbb{C}^n$ and for the closure of every strictly pseudoconvex open set in $\mathbb{C}^n$.) In the present section we prove this result for the special case that K is the closure of a strictly pseudoconvex open set $D \subset\subset \mathbb{C}^n$ with smooth boundary. In this case the proof is a simple application of the Leray formula which corresponds

[1]) For $q = 1$, by Corollary 2.1.6, every continuous solution of $\bar\partial u = f$ is a $C^{k+\alpha}$-function in D for all $0 < \alpha < 1$. For $q \geqq 2$ this is not true.

to the Leray map w from Theorem 2.5.5 as well as of the Hölder estimates for $R^w_{\partial D}$ given in Theorem 2.6.1 (i). First we prove the following theorem on decomposition of singularities:

2.9.1. Theorem. *Let $D \subset\subset \mathbb{C}^n$ be a strictly pseudoconvex open set with C^2-boundary and let $U_1, \dots, U_N \subseteq \mathbb{C}^n$ be open sets such that $D \subseteq U_1 \cup \dots \cup U_N$. Then every continuous function f on $\bar{D}$ that is holomorphic in D admits a decomposition*

$$f = \sum_{j=1}^{N} f_j \quad on \quad \bar{D}, \tag{2.9.1}$$

where, for $j = 1, \dots, N$, f_j is continuous on $\bar{D}$ and holomorphic in some neighbourhood of $\bar{D} \setminus (\partial D \cap U_j)$.

Proof. Choose $\chi_j \in C_0^\infty(U_j)$ such that $\sum_{j=1}^{N} \chi_j = 1$ on $\bar{D}$. Define $f_j := L^w_{\partial D}\chi_j f$, where w is the Leray map from Theorem 2.5.5 and $L^w_{\partial D}$ is the corresponding operator introduced in Section 1.8.6. Since, in view of the Leray formula (1.10.3) for holomorphic functions, $f = L^w_{\partial D} f$, then (2.9.1) holds. Since $w(z, \zeta)$ is holomorphic in z and $\chi_j = 0$ in $\mathbb{C}^n \setminus U_j$, f_j is holomorphic in some neighbourhood of $\bar{D} \setminus (\partial D \cap U_j)$. Since, by the Leray formula (1.10.1) for smooth functions, $f_j = \chi_j f + R^w_{\partial D} f\, \bar\partial\chi_j + B_D f\, \bar\partial\chi_j$, and since, by Theorem 2.6.1 (i) and Lemma 1.8.5 (i), $R^w_{\partial D} f\, \bar\partial\chi_j$ and $B_D f\, \bar\partial\chi_j$ are continuous on $\bar{D}$, it follows that the functions f_j are continuous on $\bar{D}$. ∎

2.9.2. Theorem. *Let $D \subset\subset \mathbb{C}^n$ be a strictly pseudoconvex open set with C^2-boundary. Then every continuous function on $\bar{D}$ that is holomorphic in D can be approximated uniformly on $\bar{D}$ by functions holomorphic in a neighbourhood of $\bar{D}$.*

Proof. In the special case that f is holomorphic in some neighbourhood of $\bar{D} \setminus S$, where $S \subseteq \partial D$ is a set of sufficiently small diameter, the required approximation can be obtained by a shift in the direction of the normal vector of ∂D at some point in S (∂D is smooth!). In view of Theorem 2.9.1, the general case follows from this special case. ∎

2.10. The $\bar\partial$-equation and Cousin problems in holomorphic vector bundles over complex manifolds

Until this section we have studied the theory of functions in open subsets of $\mathbb{C}^n$ only. The remainder of this chapter is devoted to a more general situation: Instead of open sets in $\mathbb{C}^n$ we consider complex manifolds, and instead of functions we consider sections in holomorphic vector bundles. In the present section we give definitions and prove a simple special case of Dolbeault's theorem on the connection between the $\bar\partial$-equation and Cousin problems.

2.10.1. Definition. Let X be a complex manifold (Definition 1.1.21), and let $GL(N, \mathbb{C})$ be the group of invertible complex $N \times N$ matrices.

(i) A *holomorphic vector bundle* of complex dimension N over X is a C^∞ vector bundle B over X with the characteristic fibre $\mathbb{C}^N$ together with a *holomorphic atlas of local trivializations of B*, where a *holomorphic atlas of local trivializations of B* is a family $\{U_j, h_j\}$ such that the following conditions are fulfilled: $\{U_j\}$ is an open covering of X; for every j, h_j is a C^∞-bundle isomorphism from $B|_{U_j}$ (the restriction of B to U_j) onto the product bundle $U_j \times \mathbb{C}^N$; the corresponding transition functions $g_{ij}: U_i \cap U_j \to GL(N, \mathbb{C})$ (defined by $(z, g_{ij}(z)\, v) = h_i h_j^{-1}(z, v)$, $z \in U_i \cap U_j$, $v \in \mathbb{C}^N$) are holomorphic. Equipped with the atlas $\{B|_{U_j}, h_j\}$, B becomes a complex manifold.

(ii) A C^∞-bundle homomorphism between holomorphic vector bundles B_1 and B_2 is called *holomorphic* if it is holomorphic as a map between the complex manifolds B_1 and B_2. Similar we introduce the concept of a holomorphic section of a holomorphic vector bundle.

(iii) A holomorphic vector bundle B over X is called *holomorphically trivial* if there exists a holomorphic bundle isomorphism from B onto $X \times \mathbb{C}^N$. B is called holomorphically trivial over an open set $U \subseteq X$ if the restriction $B|_U$ is holomorphically trivial. A *holomorphic trivialization* of B (over U) is a holomorphic bundle isomorphism from B onto $X \times \mathbb{C}^N$ ($B|_U$ onto $U \times \mathbb{C}^N$).

(iv) Recall that a complex-valued differential form of degree r over X can be defined as a section of the vector bundle $\bigwedge^r T^*(X)_c$, where $T^*(X)_c$ is the complexified cotangent bundle of X. A *differential form of degree r with values in a holomorphic vector bundle B* (or a *B-valued differential form*) *over* X is a section of the bundle $\bigwedge^r (T^*(X)_c) \underset{\mathbb{C}}{\otimes} B$. In other words, if $\{U_j\}_{j\in I}$ is an open covering of X such that B is holomorphically trivial over each U_j, and if $\{g_{ij}\}_{i,j\in I}$ is the corresponding system of transition functions, then a *differential form with values in B* can be identified with a system $\{f_j\}$ of N-tuples of differential forms on U_j such that $f_i = g_{ij} f_j$ over $U_i \cap U_j$ for all $i, j \in I$.

A differential form f with values in B is called a *(0, q)-form*, *$C^k_{(0,q)}$-form* etc. if for every open set $U \subseteq X$, where B is holomorphically trivial, the corresponding N-tuple of differential forms on U consists of $(0, q)$-forms, $C^k_{(0,q)}$-forms etc.

Remark. We do not consider (p, q)-forms, because every (p, q)-form with values in a holomorphic vector bundle can be identified with some $(0, q)$-form with values in some other holomorphic vector bundle.

By means of local holomorphic trivializations the $\bar\partial$-operator can be defined locally for differential forms with values in a holomorphic vector bundle. Since the transition functions are holomorphic, this definition is independent of the choice of local trivializations. Therefore, the $\bar\partial$-operator is well-defined for differential forms with values in holomorphic vector bundles.

2.10.2. Definition. Let X be a complex manifold, B a holomorphic vector bundle over X, and $\{U_j\}_{j\in I}$ an open covering of X. *Holomorphic Cousin data* in B means a system $\{f_{ij}\}_{i,j\in I}$ of holomorphic sections $f_{ij}: U_i \cap U_j \to B$ such that

$$f_{ij} + f_{jk} = f_{ik} \quad \text{in} \quad U_i \cap U_j \cap U_k \quad \text{for all} \quad i, j, k \in I\,.$$

The corresponding *Cousin problem* consists in finding a system $\{f_j\}_{j\in I}$ of holomorphic sections $f_j: U_j \to B$ such that

$$f_{ij} = f_i - f_j \quad \text{in} \quad U_i \cap U_j \quad \text{for all} \quad i, j \in I\,.$$

There is a theorem of Dolbeault [1,2] (see also, for example, Gunning/Rossi [1] and Wells [1]), which connects the solvability of the $\bar\partial$-equation with the solvability of Cousin problems (for (0, 1)-forms) and „higher“ Cousin problems (for $(0, q)$-forms, $q \geqq 2$). We need only the following special case of Dolbeault's theorem:

2.10.3. Theorem. *Let X be a complex manifold, and B a holomorphic vector bundle over X. Then the following two conditions are equivalent:*

(i) *Every holomorphic Cousin problem in B has a solution.*

(ii) *For every B-valued $C^\infty_{(0,1)}$-form f on X such that $\bar\partial f = 0$ in X, there exists a C^∞-section $u: X \to B$ such that $\bar\partial u = f$ in X.*

Proof of (i) $\Rightarrow$ (ii). Since B is locally holomorphically trivial, and since the $\bar\partial$-equation is locally solvable (see Theorem 1.2.5), we can find an open covering $\{U_j\}$ of X and C^∞-sections $u_j: U_j \to B$ such that $\bar\partial u_j = f$ in U_j. Then the differences $u_i - u_j$ are holomorphic in $U_i \cap U_j$ and form holomorphic Cousin data in B. According to (i), there are holomorphic sections $h_j: U_j \to B$ such that $u_i - u_j = h_i - h_j$ in $U_i \cap U_j$. Setting $u := u_i - h_i$ in U_i, we conclude the proof.

(ii) $\Rightarrow$ (i). Let $\{U_j\}$ be an open covering of X, and $f_{ij}: U_i \cap U_j \to B$ a holomorphic Cousin data. Choosing a C^∞-partition of unity $\{\chi_j\}$ subordinate to $\{U_j\}$ and setting

$$c_j := -\sum_k \chi_k f_{kj} \quad \text{in} \quad U_j\,,$$

we obtain a „C^∞-solution" of the Cousin problem:

$$f_{ij} = \sum_k \chi_k(f_{ik} + f_{kj}) = c_i - c_j \quad \text{in} \quad U_i \cap U_j\,.$$

$\bar\partial c_i = \bar\partial c_j$ in $U_i \cap U_j$. Therefore, by (ii), we can find a C^∞-section $u: X \to B$ such that $\bar\partial u = \bar\partial c_j$ in U_j. Setting $h_j := c_j - u$ in U_j, we complete the proof. ■

2.11. Fredholm solvability of the $\bar\partial$-equation on complex manifolds with strictly pseudoconvex C^2-boundary

The notion of strict plurisubharmonicity is well-defined for functions on complex manifolds since it is invariant under holomorphic changes of coordinates (see Theorem 1.4.11 (ii)).

2.11.1. Definition. Let X be a complex manifold, and $D \subset\subset X$ an open set. We shall say that D is *strictly pseudoconvex* (or that *D has a strictly pseudoconvex boundary*) if there exists a strictly plurisubharmonic C^2-function ϱ in some neighbourhood $\theta_{\partial D}$ of ∂D such that $D \cap \theta_{\partial D} = \{z \in \theta_{\partial D} : \varrho(z) < 0\}$.

In Section 2.8 we saw that in strictly pseudoconvex open sets in $\mathbb{C}^n$ the $\bar\partial$-equation can be solved. This is not true for strictly pseudoconvex open sets in general complex manifolds. However, Lemma 2.3.1 together with its proof holds also for strictly pseudoconvex open sets in general complex manifolds as well as for forms with values in holomorphic vector bundles. To give the precise formulation of this result we introduce some notations.

2.11.2. The spaces $Z^0_{(0,q)}(\bar D, B)$, $E^{1/2}_{(0,q)}(\bar D, B)$, $H^\alpha_{(0,q)}(\bar D, B)$ and $F^{1/2}_{(0,q)}(\bar D, B)$. Let X be a complex manifold of complex dimension n, and let B be a holomorphic vector bundle of dimension N over X. Choose a locally finite open covering $\{U_j\}$ of X such that, for every j, there are holomorphic coordinates $\varphi_j: U_j \to \mathbb{C}^n$ as well as a holomorphic trivialization $\psi_j: B|_{U_j} \to U_j \times \mathbb{C}^N$. Further, let $\{\chi_j\}$ be a C^∞-partition of unity subordinate to $\{U_j\}$. Then every B-valued form f over a set $Y \subseteq X$ can be identified with a system $\{f^{(j)}\}$ of vectors $f^{(j)} = (f_1^{(j)}, \dots, f_N^{(j)})$ of differential forms $f_s^{(j)}$

over $\varphi_j(U_j \cap Y) \subseteqq \mathbb{C}^n$. We define

$$||f(z)|| := \sum_j \chi_j(z) \sum_{s=1}^{N} ||f_s^{(j)}(\varphi_j(z))|| \quad \text{for} \quad z \in Y,$$

where $|f_s^{(j)}(\varphi_j(z))|$ is the Euclidean length of the vector of coefficients of $f_s^{(j)}(\varphi_j(z))$ (cp. Subsection 1.6.3).

Let $D \subset\subset X$ be an open set. We denote by $C^0_{(0,q)}(\bar D, B)$, $0 \leqq q \leqq n$, the Banach space of continuous B-valued $(0, q)$-forms on $\bar D$ endowed with the norm

$$||f||_{0,D} := \max_{z \in \bar D} ||f(z)|| .$$

The norm $||\cdot||_{0,D}$ depends on the choice of the local coordinates φ_j, the local trivializations ψ_j and the partition of unity $\{\chi_j\}$. However, since $\bar D$ is compact, different choices give equivalent norms. For $0 < \alpha < 1$ and $0 \leqq q \leqq n$, we denote by $H^\alpha_{(0,q)}(\bar D, B)$ the Banach space of α-Hölder continuous B-valued $(0, q)$-forms on $\bar D$ endowed with the norm

$$||f||_{\alpha,D} := \sum_j \sum_{s=1}^{N} ||f_s^{(j)}||_{\alpha, \varphi_j(U_j \cap D)}$$

where $||f_s^{(j)}||_{\alpha,\varphi_j(D \cap U_j)}$ was defined in Section 1.8. Observe that the sum over j is finite since $\bar D$ is compact and the covering $\{U_j\}$ is locally finite. We also write $C^0(\bar D, B)$ instead of $C^0_{(0,0)}(\bar D, B)$ and $H^\alpha(\bar D, B)$ instead of $H^\alpha_{(0,0)}(\bar D, B)$. We denote by $Z^0_{(0,q)}(\bar D, B)$ the subspace of all forms $f \in C^0_{(0,q)}(\bar D, B)$ such that $\bar\partial f = 0$ in D. Further, we denote by $F^{1/2}_{(0,q-1)}(\bar D, B)$ the domain of definition of $\bar\partial$ as an operator from $H^{1/2}_{(0,q-1)}(\bar D, B)$ into $Z^0_{(0,q)}(\bar D, B)$, that is, $F^{1/2}_{(0,q-1)}(\bar D, B)$ is the subspace of all forms $u \in H^{1/2}_{(0,q-1)}(\bar D, B)$ such that $\bar\partial u$ (which is defined in D in the sense of distributions) is continuous in D and admits a continuous continuation to $\bar D$. Finally, we set $E^{1/2}_{(0,q)}(\bar D, B) := \bar\partial F^{1/2}_{(0,q-1)}(\bar D, B)$, $1 \leqq q \leqq n$.

2.11.3. Theorem. *Let B be a holomorphic vector bundle over a complex manifold X, let $D \subset\subset X$ be a strictly pseudoconvex open set with C^2-boundary, and let $1 \leqq q \leqq n$. Then*

(i) *there exists a bounded linear operator S from $Z^0_{(0,q)}(\bar D, B)$ into $H^{1/2}_{(0,q-1)}(\bar D, B)$ such that $S(Z^0_{(0,q)}(\bar D, B)) \subseteqq F^{1/2}_{(0,q-1)}(\bar D, B)$ and*

$$\bar\partial \circ S = \mathrm{id} + K ,$$

where id *is the identity operator in $Z^0_{(0,q)}(\bar D, B)$ and K is a compact operator in $Z^0_{(0,q)}(\bar D, B)$,*

(ii) *$E^{1/2}_{(0,q)}(\bar D, B)$ is a closed subspace of $Z^0_{(0,q)}(\bar D, B)$ of finite codimension.*

Proof. Repeat the proof of Lemma 2.3.1. ■

In the next section we show that $E^{1/2}_{(0,q)}(\bar D, B) = Z^0_{(0,q)}(\bar D, B)$ if $D \subset\subset X$ is a C^2 strictly pseudoconvex open set with the additional property that the strictly plurisubharmonic function ϱ in Definition 2.11.1 can be chosen to be strictly plurisubharmonic in a neighbourhood of $\bar D$ (and not only of ∂D). This result will be proved by means of an inductive procedure with respect to the levels of ϱ. The basic step in this proof is given by the following lemma.

2.11.4. Lemma. *Let B be a holomorphic vector bundle over a complex manifold X, and let $D \subset\subset X$ be a strictly pseudoconvex open set with C^2-boundary such that*

$$E^{1/2}_{(0,q)}(\bar{D}, B) = Z^0_{(0,q)}(\bar{D}, B) \quad \text{for} \quad 1 \leqq q \leqq n\,.$$

Further suppose that $U \subset\subset X$ is an open set with the following properties:

(1) *U is biholomorphically equivalent to the open unit ball in $\mathbb{C}^n$.*

(2) *B is holomorphically trivial over U.*

(3) *There is a strictly plurisubharmonic C^2-function ϱ in U such that $D \cap U = \{z \in U : \varrho(z) < 0\}$ and $\mathrm{d}\varrho(z) \neq 0$ for $z \in \partial D \cap U$.*

Then, for every strictly pseudoconvex open set $G \subset\subset X$ with C^2-boundary such that

$$D \subseteqq G \subseteqq D \cup V\,, \quad \text{where} \quad V \subset\subset U\,, \tag{2.11.1}$$

the following two statements hold:

(i) *$E^{1/2}_{(0,q)}(\bar{G}, B) = Z^0_{(0,q)}(\bar{G}, B)$ for $1 \leqq q \leqq n$.*

(ii) *Every continuous section of B over $\bar{D}$ that is holomorphic in D can be approximated uniformly on $\bar{D}$ by continuous sections of B over $\bar{G}$ which are holomorphic in G.*

Proof of (i). By hypothesis there is a biholomorphic map T from U onto the unit ball in $\mathbb{C}^n$. In view of (2.11.1), we can choose open sets U', U'' such that $U'' \subset\subset U' \subset\subset U$, $\bar{D} \subseteqq \bar{G} \subseteqq \bar{D} \cup U''$ and $T(U')$, $T(U'')$ are also balls. Then it follows from property (3) of U that $T(\bar{U}' \cap \bar{D})$ is $P^0(T(U))$-convex (Definition 1.5.3). By Theorem 2.7.1 this implies the following statement:

$$\begin{gathered}\text{Every holomorphic function in some neighbourhood of } \bar{U}' \cap \bar{D} \\ \text{can be approximated uniformly on } \bar{U}' \cap \bar{D} \text{ by holomorphic} \\ \text{functions in } U.\end{gathered} \tag{2.11.2}$$

By Lemma 1.5.23 (ii) we can find a strictly pseudoconvex open set W with C^2-boundary such that $\bar{G} \cap \bar{U}' \subseteqq \bar{W} \subseteqq \bar{G} \cap U$. Then $W \cap D$ has C^2-boundary, and since, for C^2 open sets, strict pseudoconvexity is a local property of the boundary (this follows from Theorem 1.5.17 or from Theorem 1.5.25), we obtain that $W \cap D$ is a C^2-strictly pseudoconvex open set.

Now let $f \in Z^0_{(0,q)}(\bar{G}, B)$. We have to prove that $f = \bar{\partial} u$ for some $u \in H^{1/2}_{(0,q-1)}(\bar{G}, B)$. Since B is holomorphically trivial over U $\supseteqq \bar{W}$, and since $T(W)$ is a C^2 strictly pseudoconvex open set in $\mathbb{C}^n$, we obtain from Theorem 2.3.5 (or Theorem 2.6.1 (ii)) a form $u_W \in H^{1/2}_{(0,q-1)}(\bar{W}, B)$ such that

$$\bar{\partial} u_W = f \quad \text{in} \quad W\,. \tag{2.11.3}$$

Moreover, by hypothesis of the lemma, there is $u_D \in H^{1/2}_{(0,q-1)}(\bar{D}, B)$ such that

$$\bar{\partial} u_D = f \quad \text{in} \quad D\,. \tag{2.11.4}$$

Now we distinguish the cases $q = 1$ and $2 \leqq q \leqq n$.

First let $q = 1$. Then $u_W - u_D$ is continuous on $\bar{W} \cap \bar{D}$ and holomorphic in $W \cap D$. Since B is holomorphically trivial over U, it follows from Theorem 2.9.2 that $u_W - u_D$ can be approximated uniformly on $\bar{W} \cap \bar{D}$ by holomorphic sections of B over some neighbourhood of $\bar{W} \cap \bar{D}$. In view of (2.11.2), this implies that there are holomorphic sections $\varphi_k \colon U \to B$ such that

$$\lim_{k\to\infty} \|\varphi_k + u_W - u_D\|_{0,\bar{U}'\cap\bar{D}} = 0\,. \tag{2.11.5}$$

Choose a C^∞-function χ on X such that $\chi = 1$ in some neighbourhood of $\bar U''$ and $\chi = 0$ in some neighbourhood of $X \setminus U'$, and define

$$u_k := \begin{cases} u_D & \text{in } \bar G \setminus U, \\ (1-\chi)\, u_D + \chi(\varphi_k + u_W) & \text{in } \bar G \cap U. \end{cases}$$

Then $u_k \in H^{1/2}(\bar G, B)$, $\bar\partial u_k \in E^{1/2}_{(0,1)}(\bar G, B)$ and it follows from (2.11.3)—(2.11.5) that

$$\lim_{k\to\infty} \|\bar\partial u_k - f\|_{0,G} = \lim_{k\to\infty} \| - u_D\, \bar\partial\chi + (\varphi_k + u_W)\, \bar\partial\chi\|_{0,U'\cap D} = 0\,.$$

Since, by Theorem 2.11.3, $E^{1/2}_{(0,1)}(\bar G, B)$ is a closed subspace of $Z^0_{(0,1)}(\bar G, B)$, this implies that $f \in E^{1/2}_{(0,1)}(\bar G, B)$.

Now consider the case $2 \leqq q \leqq n$. Then $u_W - u_D \in Z^0_{(0,q-1)}(\bar W \cap \bar D, B)$. Since $T(W \cap D)$ is a C^2 strictly pseudoconvex open set in $\mathbb{C}^n$, it follows from Theorem 2.3.5 (or Theorem 2.6.1 (ii)) that there is $v \in H^{1/2}_{(0,q-2)}(\bar W \cap \bar D, B)$ such that $u_W - u_D = \bar\partial v$ in $W \cap D$. If χ is as above, then the required solution $u \in H^{1/2}_{(0,q-1)}(\bar G, B)$ of $\bar\partial u = f$ can be obtained by setting $u := u_D$ in $\bar G \setminus U'$ and $u := u_W - \bar\partial((1-\chi)\, v)$ in $\bar G \cap U'$.

(ii). Let U', U'', W and χ be as in the proof of part (i), and let $f\colon \bar D \to B$ be a continuous section that is holomorphic in D. Since B is holomorphically trivial over U, and since $U' \cap D \subseteqq W \cap D$, it follows from (2.11.2) and Theorem 2.9.2 that there are holomorphic sections $u_k\colon U \to B$ such that

$$\lim_{k\to\infty} \|f - u_k\|_{0,U'\cap D} = 0\,. \tag{2.11.6}$$

Then $\bar\partial(\chi(f-u_k)) = (f-u_k)\, \bar\partial\chi \in Z^0_{(0,1)}(\bar G, B)$, and, by part (i) of the lemma, we can find a sequence $\varphi_k \in H^{1/2}(\bar G, B)$ such that $\bar\partial\varphi_k = \bar\partial(\chi(f-u_k))$. In view of (2.11.6) and Banach's open mapping theorem, this sequence can be chosen so that

$$\lim_{k\to\infty} \|\varphi_k\|_{1/2,G} = 0\,. \tag{2.11.7}$$

Setting

$$f_k := \begin{cases} f - \chi(f-u_k) + \varphi_k & \text{on } \bar D, \\ u_k + \varphi_k & \text{on } \bar G \setminus \bar D, \end{cases}$$

we obtain continuous sections $f_k\colon \bar G \to B$ which are holomorphic in G. It follows from (2.11.6) and (2.11.7) that $\lim_{k\to\infty} \|f - f_k\|_{0,D} = 0$. ■

2.12. Solvability of the $\bar\partial$-equation on complex manifolds with strictly plurisubharmonic exhausting C^2-function

In the preceding section we saw that, for strictly pseudoconvex open sets D with C^2-boundary in complex manifolds, the operator $\bar\partial$ has a finitely codimensional closed image as an operator from the space of 1/2-Hölder continuous $(0, q-1)$-forms on $\bar D$ into the space of $\bar\partial$-closed continuous $(0, q)$-forms on $\bar D$. In the present section we shall prove that this operator is onto if D can be defined by a strictly plurisubharmonic C^2-function which is defined not only in a neighbourhood of ∂D but in a neighbourhood of $\bar D$. Moreover, we shall prove Oka-Weil approximation and solvability of the $\bar\partial$-equation on complex manifolds which admit a strictly plurisubharmonic exhausting C^2-function.

2.12.1. Definition. Let X be a complex manifold. A *strictly plurisubharmonic exhausting C^k-function for X, $k = 2, 3, \ldots, \infty$,* is a strictly plurisubharmonic C^k-function ϱ on X such that, for every $\alpha \in \mathbb{R}$, the set $\{z \in X : \varrho(z) < \alpha\}$ is relatively compact in X.

Remark. Let X be a complex manifold, $\alpha_0 \in \mathbb{R}$, and let ϱ be a strictly plurisubharmonic C^k-function on X with values in the interval $(-\infty, \alpha_0)$ such that, for every $-\infty < \alpha < \alpha_0$, the set $\{z \in X : \varrho(z) < \alpha\}$ is relatively compact in X, $k = 2, 3, \ldots, \infty$. Then X admits a strictly plurisubharmonic exhausting C^k-function. In fact, choose a strictly convex and strictly increasing C^∞-function χ on $(-\infty, \alpha_0)$ such that $\chi(\alpha) \to \infty$ for $\alpha \to \alpha_0$. Then, by Theorem 1.4.12 (ii), $\chi \circ \varrho$ is a strictly plurisubharmonic exhausting C^k-function for X.

In particular, the open sets $\{z \in X : \varrho(z) < \alpha\}$ in Definition 2.12.1 are also manifolds with strictly plurisubharmonic exhausting C^k-function.

By Theorem 1.5.8, every pseudoconvex open set in $\mathbb{C}^n$ admits a strictly plurisubharmonic exhausting C^∞-function. Observe that a strictly pseudoconvex open set in a general complex manifold (Definition 2.11.1) does not necessarily admit a strictly plurisubharmonic exhausting C^2-function.

Finally we observe that every complex manifold with strictly plurisubharmonic exhausting C^2-function admits a strictly plurisubharmonic exhausting C^∞-function. We do not use this fact and its simple direct proof is left to the reader. In the next section, we obtain this as a corollary of Theorems 2.13.4 and 2.13.5.

2.12.2. Lemma (Morse's lemma). *Let X be a complex manifold which is countable at infinity, and let ϱ be a strictly plurisubharmonic C^2-function on X. Further, let $K \subset\subset X$ be a compact set such that $d\varrho(z) \neq 0$ for all $z \in K$. Then, for every $\varepsilon > 0$, there exists a strictly plurisubharmonic C^2-function ϱ_ε on X such that the following conditions are fulfilled:*

(i) *$\varrho - \varrho_\varepsilon$ together with its first- and second-order derivatives is $\leqq \varepsilon$ on X.*

(ii) *The set* Crit $(\varrho_\varepsilon) := \{z \in X : d\varrho_\varepsilon(z) = 0\}$ *is discrete in X.*

(iii) *$\varrho_\varepsilon = \varrho$ on K.*

Proof. Let U_K be a neighbourhood of K such that $d\varrho \neq 0$ in U_K. Since X is countable at infinity, we can find a sequence of open sets $U_j \subset\subset V_j \subset\subset X$, $j = 1, 2, \ldots$, such that the following conditions are fulfilled:

(1) $X \setminus U_K \subseteqq \bigcup_{j=1}^{\infty} U_j$.

(2) For every j, V_j is biholomorphically equivalent to some bounded open set in $\mathbb{C}^n$.

(3) $V_j \cap K = \emptyset$ for all j.

(4) For every compact set $Y \subset\subset X$, there is only a finite number of indices j such that $Y \cap V_j \neq \emptyset$.

Now, by means of Lemma 1.4.19 and a lemma of Morse (cp. the proof of Lemma 1.5.9), we can inductively construct a sequence of C^∞-functions χ_j on X such that, for every j, the following conditions are satisfied:

(a) $\chi_j = 0$ in a neighbourhood of $X \setminus V_j$, and consequently, $\chi_j = 0$ on K.

(b) The function $\varrho + \chi_1 + \ldots + \chi_j$ is strictly plurisubharmonic on X and has only non-degenerate critical points in $K \cup \bar{U}_1 \cup \ldots \cup \bar{U}_j$.

(c) The function χ_j together with its first- and second-order derivatives is $\leqq \varepsilon/2^j$ on X.

Then $\varrho_\varepsilon := \varrho + \sum \chi_j$ (this sum is locally finite) has the required properties. ■

The main result of the present section is the following

2.12.3. Theorem. *Let X be a complex manifold with strictly plurisubharmonic exhausting C^2-function ϱ, and let B be a holomorphic vector bundle over X. Set $D_\alpha := \{z \in X : \varrho(z) < \alpha\}$ for $\alpha \in \mathbb{R}$. Then*

(i) *Let $\alpha \in \mathbb{R}$ such that $d\varrho(z) \neq 0$ for all $z \in \partial D_\alpha$, and let $1 \leqq q \leqq n$. Then there exists a bounded linear operator T from $Z^0_{(0,q)}(\bar{D}_\alpha, B)$ into $H^{1/2}_{(0,q-1)}(\bar{D}_\alpha, B)$* (see Subsection 2.11.2) *such that*

$$\bar{\partial} T f = f \quad \textit{in } D_\alpha \textit{ for every } \quad f \in Z^0_{(0,q)}(\bar{D}_\alpha, B) .$$

(ii) *Let $\alpha \in \mathbb{R}$ such that $d\varrho(z) \neq 0$ for $z \in \partial D_\alpha$. Then every continuous section $f : \bar{D}_\alpha \to B$ that is holomorphic in D_α can be approximated uniformly on $\bar{D}_\alpha$ by holomorphic sections of B over X.*

(iii) *Let $\alpha \in \mathbb{R}$ (not necessarily $d\varrho(z) \neq 0$ for $z \in \partial D_\alpha$). Then every holomorphic section of B over some neighbourhood of $\bar{D}_\alpha$ can be approximated uniformly on $\bar{D}_\alpha$ by holomorphic sections of B over X.*

(iv) *Let $1 \leqq q \leqq n$. Then, for every B-valued continuous $(0, q)$-form f on X such that $\bar{\partial} f = 0$ on X, there exists a B-valued continuous $(0, q-1)$-form u on X such that*

$$\bar{\partial} u = f \quad \textit{on} \quad X .$$

This theorem will be proved by means of Lemma 2.11.4 and the following

2.12.4. Lemma. *Let X be a complex manifold with strictly plurisubharmonic exhausting C^2-function ϱ such that* Crit $(\varrho) := \{z \in X : d\varrho(z) = 0\}$ *is discrete in X, and let B be a holomorphic vector bundle over X. Set $D_\alpha := \{z \in X : \varrho(z) < \alpha\}$ for $\alpha \in \mathbb{R}$.*

Then, for every $\alpha_0 \in \mathbb{R}$, there exist $\varepsilon_0 > 0$ and a finite number of open sets $U_1, \dots, U_M \subset\subset X$ such that the following conditions are fulfilled:

(1) *For $j = 1, \dots, M$, U_j is biholomorphically equivalent to the unit ball in $\mathbb{C}^n$.*

(2) *For $j = 1, \dots, M$, B is holomorphically trivial over U_j.*

(3) $\bar{D}_{\alpha_0+\varepsilon_0} \subseteqq U_1 \cup \dots \cup U_M$.

(4) *If $0 \leqq \varepsilon_1, \varepsilon_2 \leqq \varepsilon_0$ such that $d\varrho(z) \neq 0$ for $z \in \partial D_{\alpha_0-\varepsilon_1} \cup \partial D_{\alpha_0+\varepsilon_2}$, then there are strictly plurisubharmonic exhausting C^2-functions $\varrho_0, \varrho_1, \dots, \varrho_M$ for X such that $d\varrho_j(z) \neq 0$ for $\varrho_j(z) = 0$, $j = 0, 1, \dots, M$, and, for the C^2 strictly pseudoconvex open sets $D_j := \{z \in X : \varrho_j(z) < 0\}$, we have the relations $D_0 = D_{\alpha_0-\varepsilon_1}$, $D_M = D_{\alpha_0+\varepsilon_2}$ and $D_{j-1} \subseteqq D_j \subseteqq D_{j-1} \cup V_j$, where $V_j \subset\subset U_j$, $j = 0, 1, \dots, M$.*

Proof. Clearly, we can find open sets $U_1, \dots, U_M \subset\subset X$ such that (1), (2) are fulfilled and $\bar{D}_{\alpha_0+1} \subseteqq U_1 \cup \dots \cup U_M$. It remains to prove that there is $0 < \varepsilon_0 \leqq 1$ so small that condition (4) is satisfied. Choose C^∞-functions $\chi_1, \dots, \chi_M$ on X such that $0 \leqq \chi_j \leqq 1$, supp $\chi_j \subset\subset U_j$ and $\sum_{j=1}^{M} \chi_j = 1$ on $\bar{D}_{\alpha_0+1}$. Since Crit $(\varrho) \cap \bar{D}_{\alpha_0+1}$ is finite, we can choose these functions so that, moreover, the following condition is fulfilled:

There exists a neighbourhood U of Crit $(\varrho) \cap \partial D_{\alpha_0}$ such that, for all $z \in U$, there is $j(z) \in \{1, \dots, M\}$ such that $\chi_{j(z)}(z) = 1$ and $\chi_j(z) = 0$ for $j \neq j(z)$. (2.12.1)

Choose open sets $V_j \subset\subset U_j$ such that supp $\chi_j \subset\subset V_j$. In view of condition (1), we can find holomorphic coordinates $z_{1j}, \dots, z_{nj}$ on U_j. Set

$$C := \sum_{j=1}^{M} \sup_{z \in V_j} \sum_{r,s=1}^{n} \left| \frac{\partial^2 \chi_j(z)}{\partial z_{rj} \, \partial \bar{z}_{sj}} \right| .$$

Since ϱ is strictly plurisubharmonic, there is $\beta > 0$ such that, for all $\xi \in \mathbb{C}^n$ and $j = 1, \ldots, M$,

$$\inf_{z \in V_j} \sum_{r,s=1}^{n} \frac{\partial^2 \varrho(z)}{\partial z_{rj} \partial \bar{z}_{sj}} \xi_s \bar{\xi}_r \geqq \beta |\xi|^2 .$$

Since Crit $(\varrho) \cap D_{\alpha_0+1}$ is finite, and since (2.12.1) holds, we can choose $\varepsilon_0' > 0$ so small that Crit $(\varrho) \cap (\bar{D}_{\alpha_0+\varepsilon_0'} \setminus D_{\alpha_0-\varepsilon_0'}) \subseteqq \partial D_{\alpha_0}$ and

$$||d\varrho(z)|| \geqq 3\varepsilon_0' \sum_{j=1}^{M} ||d\chi_j(z)|| \quad \text{for} \quad z \in \bar{D}_{\alpha_0+\varepsilon_0'} \setminus D_{\alpha_0-\varepsilon_0'} . \tag{2.12.2}$$

To complete the proof, we shall prove that for $\varepsilon_0 := \min (\varepsilon_0', \beta/3C)$ condition (4) is fulfilled. Let $0 \leqq \varepsilon_1, \varepsilon_2 \leqq \varepsilon_0$ such that

$$d\varrho(z) \neq 0 \quad \text{for} \quad z \in \partial D_{\alpha_0-\varepsilon_1} \cup \partial D_{\alpha_0+\varepsilon_2} . \tag{2.12.3}$$

Define $\varrho_0 := \varrho + \varepsilon_1 - \alpha_0$, and, for $j = 1, \ldots, M$, we set

$$\varrho_j := \varrho + \varepsilon_1 - \alpha_0 - (\varepsilon_1 + \varepsilon_2) \sum_{s=1}^{j} \chi_s .$$

Then it follows from (2.12.2) that for $j = 0, 1, \ldots, M$

$$d\varrho_j(z) \neq 0 \quad \text{for} \quad z \in (\bar{D}_{\alpha_0+\varepsilon_0} \setminus D_{\alpha_0-\varepsilon_0}) \setminus \text{Crit}\,(\varrho) . \tag{2.12.4}$$

Further,

$$\text{Crit}\,(\varrho) \cap (\partial D_0 \cup \ldots \cup \partial D_M) = \emptyset . \tag{2.12.5}$$

In fact, assume that $z \in$ Crit $(\varrho) \cap \partial D_j$. Then $d\varrho(z) = 0$ and, by (2.12.1), either $0 = \varrho_j(z) = \varrho(z) - \alpha_0 - \varepsilon_2$ or $0 = \varrho_j(z) = \varrho(z) + \varepsilon_1 - \alpha_0$, that is, $d\varrho(z) = 0$ and $z \in \partial D_{\alpha_0-\varepsilon_1} \cup \partial D_{\alpha_0+\varepsilon_2}$, which is not possible because of (2.12.3).

Since $D_{\alpha_0-\varepsilon_0} \subseteqq D_j \subseteqq D_{\alpha_0+\varepsilon_0}$, it follows from (2.12.4) and (2.12.5) that $d\varrho_j(z) \neq 0$ for $z \in \partial D_j$ for $j = 0, 1, \ldots, M$. Further, it is clear from the definition of C and β as well as from the relations $\varrho_j = \varrho + \varepsilon_1 - \alpha_0$ on $X \setminus (V_1 \cup \ldots \cup V_M)$ that $\varrho_0, \ldots, \varrho_M$ are strictly plurisubharmonic on X. Finally, it follows immediately from the definition of ϱ_j that $D_0 = D_{\alpha_0-\varepsilon_1}$, $D_M = D_{\alpha_0+\varepsilon_2}$ and $D_{j-1} \subseteqq D_j \subseteqq D_{j-1} \cup \text{supp}\, \chi_j \subseteqq D_{j-1} \cup V_j$. ∎

Proof of Theorem 2.12.3. (i). By Lemma 2.12.2, we can assume that Crit (ϱ) $:= \{z \in X : d\varrho(z) = 0\}$ is discrete in X. By Theorem 2.11.3 and Proposition 3, Appendix 2, it is sufficient to prove that for all $\alpha \in \mathbb{R}$ the following statement holds:

(i)$_\alpha$ If $d\varrho(z) \neq 0$ for $z \in \partial D_\alpha$, then $E^{1/2}_{(0,q)}(\bar{D}_\alpha, B) = Z^0_{(0,q)}(\bar{D}_\alpha, B)$ for all $1 \leqq q \leqq n$.

Set $\alpha_{\min} := \min_{z \in X} \varrho(z)$. We denote by α_0 the supremum of all $\alpha \geqq \alpha_{\min}$ such that (i)$_\beta$ is true for all $\alpha_{\min} \leqq \beta < \alpha$. We have to prove that $\alpha_0 = \infty$.

Since the relation $\varrho(z) = \alpha_{\min}$ implies that $z \in$ Crit (ϱ), there is only a finite number of points $z \in X$ with $\varrho(z) = \alpha_{\min}$. Therefore, if $\varepsilon > 0$ is sufficiently small and $d\varrho(z) \neq 0$ for $z \in \partial D_{\alpha_{\min}+\varepsilon}$, then $D_{\alpha_{\min}+\varepsilon}$ is biholomorphically equivalent to some strictly pseudoconvex open set with C^2-boundary in $\mathbb{C}^n$. By Theorem 2.3.5 (or Theorem 2.6.1 (ii)), this implies that (i)$_{\alpha_{\min}+\varepsilon}$ is true for all sufficiently small $\varepsilon > 0$, that is, $\alpha_0 > \alpha_{\min}$.

Assume that $\alpha_0 < \infty$. Then, by Lemma 2.12.4, there are a number $\varepsilon_0 > 0$ and open sets $U_1, \ldots, U_M$ such that conditions (1)–(4) in Lemma 2.12.4 are fulfilled. Since $\alpha_0 > \alpha_{\min}$, and since Crit (ϱ) is discrete, we can find $0 < \varepsilon_1 < \min (\varepsilon_0, \alpha_0 - \alpha_{\min})$ such that $d\varrho(z) \neq 0$ for $z \in \partial D_{\alpha_0-\varepsilon_1}$.

Now let ε_2 be an arbitrary number such that $0 \leqq \varepsilon_2 \leqq \varepsilon_0$ and $d\varrho(z) \neq 0$ for $z \in \partial D_{\alpha_0+\varepsilon_2}$. Let $D_0, \ldots, D_M$ be the corresponding sets from condition (4) in Lemma 2.12.4. Since (i)$_{\alpha_0-\varepsilon_1}$ is true and $D_0 = D_{\alpha_0-\varepsilon_1}$, then $E^{1/2}_{(0,q)}(\bar{D}_0, B) = Z^0_{(0,q)}(\bar{D}_0, B)$ for $1 \leqq q \leqq n$.

Applying M times Lemma 2.11.4 (i), we obtain that $E^{1/2}_{(0,q)}(\bar D_M, B) = Z^0_{(0,q)}(\bar D_M, B)$ for $1 \leqq q \leqq n$. Since $D_M = D_{\alpha_0+\varepsilon_2}$, this is in contradiction to the definition of α_0.

(ii) By Lemma 2.12.2, we can assume that Crit $(\varrho) := \{z \in X : d\varrho(z) = 0\}$ is discrete in X. Let $\beta \in \mathbb{R}$ such that $d\varrho(z) \neq 0$ for $z \in \partial D_\beta$, and let $f: \bar D_\beta \to B$ be a continuous section that is holomorphic in D. We first prove that for all $\beta \leqq \alpha < \infty$ the following statement holds:

(ii)$_\alpha$ If $d\varrho(z) \neq 0$ for $z \in \partial D_\alpha$, then f can be approximated uniformly on $\bar D_\beta$ by continuous sections over $\bar D_\alpha$ which are holomorphic in D_α.

Clearly, (ii)$_\beta$ is true. Denote by α_0 the supremum of all $\alpha \geqq \beta$ such that (ii)$_\alpha$ is true. We must prove that $\alpha_0 = \infty$. Assume that $\alpha_0 < \infty$. Then, by Lemma 2.12.4, there are $\varepsilon_0 > 0$ and open sets $U_1, \ldots, U_N$ such that conditions (1)—(4) in Lemma 2.12.4 are fulfilled. We can find a number $0 \leqq \varepsilon_1 \leqq \min(\varepsilon_0, \alpha_0 - \beta)$ such that $d\varrho(z) \neq 0$ for $z \in \partial D_{\alpha_0-\varepsilon_1}$ and (ii)$_{\alpha_0-\varepsilon_1}$ is true. In fact, if $\alpha_0 = \beta$, then we set $\varepsilon_1 := 0$. If $\alpha_0 > \beta$, then since Crit (ϱ) is discrete we can choose $0 < \varepsilon_1 \leqq \min(\varepsilon_0, \alpha_0 - \beta)$ so that $d\varrho(z) \neq 0$ for $z \in \partial D_{\alpha_0-\varepsilon_1}$. Further, let $0 < \varepsilon_2 < \varepsilon_0$ be such that $d\varrho(z) \neq 0$ for $z \in \partial D_{\alpha_0+\varepsilon_2}$, and $D_0, \ldots, D_M$ the corresponding sets from condition (4) in Lemma 2.12.4. Since $D_0 = D_{\alpha_0-\varepsilon_1}$, then f can be approximated uniformly on $\bar D_\beta$ by continuous sections over $\bar D_0$ which are holomorphic in D_0. By Theorem 2.12.3 (i), we can apply M times Lemma 2.11.4 (ii) and obtain that f can be approximated uniformly on $\bar D_\beta$ by continuous sections over $\bar D_M$ which are holomorphic in D_M. Since $D_M = D_{\alpha_0+\varepsilon_2}$, this is in contradiction to the definition of α_0.

To complete the proof, we choose a sequence $\beta < \alpha_1 < \alpha_2 < \ldots$ tending to infinity such that $d\varrho(z) \neq 0$ for $z \in \partial D_{\alpha_j}$. This is possible since Crit (ϱ) is discrete. By what was proved, for every $\varepsilon > 0$ we then can find continuous sections $f_j: \bar D_{\alpha_j} \to B$ which are holomorphic in D_{α_j} such that

$$\|f_1 - f\|_{0, D_\beta} < \frac{\varepsilon}{2} \quad \text{and} \quad \|f_{j+1} - f_j\|_{0, D_{\alpha_j}} < \frac{\varepsilon}{2^{j+1}} \quad \text{for} \quad j = 1, 2, \ldots .$$

Then the sequence f_j tends to some holomorphic section $\tilde f: X \to B$, uniformly on each compact subset of X, and $\|f - \tilde f\|_{0, D_\beta} < \varepsilon$.

(iii) Let $\alpha \in \mathbb{R}$ and let f be a holomorphic section of B over some neighbourhood U_f of $\bar D_\alpha$. Choose a C^∞-function χ on X such that $\chi = 0$ on $\bar D_\alpha \cup (X \setminus D_{\alpha+1})$ and $\chi > 0$ on $D_{\alpha+1} \setminus \bar D_\alpha$ (see, for example, Lemma 1.4.13 in Narasimhan [2]). Then by Lemma 1.4.19, for sufficiently small $\varepsilon > 0$, $\varrho_1 := \varrho + \varepsilon\chi$ is also a strictly plurisubharmonic exhausting C^2-function for X. We have $D_\alpha = \{z \in X : \varrho_1(z) < \alpha\}$ and, in addition, $\varrho_1 > \alpha$ on $X \setminus \bar D_\alpha$. (Observe that since we do not assume that $d\varrho(z) \neq 0$ for $z \in \partial D_\alpha$, it is possible that $\varrho(z) = \alpha$ for some points $z \in X \setminus \bar D_\alpha$.) Therefore, by Lemma 2.12.2, we can find a strictly plurisubharmonic exhausting C^2-function ϱ_2 for X such that Crit (ϱ_2) is discrete in X and which is so close to ϱ_1 that, for an appropriate $\delta > 0$,

$$D_\alpha \subset\subset G_\delta := \{z \in X : \varrho_2(z) < \alpha\delta\} \subset\subset U_f .$$

Since Crit (ϱ_2) is discrete, we can assume that, moreover, $d\varrho_2(z) \neq 0$ for $z \in \partial G_\delta$. Now we can apply Theorem 2.12.3 (ii).

(iv) By Lemma 2.12.2 we can assume that Crit (ϱ) is discrete in X. Then we can find a sequence $\alpha_1 < \alpha_2 < \ldots$ tending to infinity such that $d\varrho(z) \neq 0$ for $z \in \partial D_{\alpha_j}$. By

Theorem 2.12.3 (i), then there is a sequence of continuous B-valued $(0, q-1)$-forms on $\bar{D}_{\alpha_j}$ such that

$$\bar{\partial} u_j = f \quad \text{in} \quad D_{\alpha_j}.$$

Now we distinguish the cases $q = 1$ and $2 \leqq q \leqq n$. First let $q = 1$. Then $u_j - u_{j+1}$ is holomorphic in D_{α_j} and, by Theorem 2.13.3 (ii), we can find holomorphic sections $v_j: X \to B$ such that

$$\|u_j - u_{j+1} - v_j\|_{0, D_{\alpha_j}} < \frac{1}{2^j}.$$

Then the sequence $\tilde{u}_j$ defined by

$$\tilde{u}_1 := u_1 \quad \text{and} \quad \tilde{u}_{j+1} := u_{j+1} + \sum_{k=1}^{j} v_k \quad \text{for} \quad j = 1, 2, \ldots$$

tends uniformly on each compact subset of X to some continuous section $u: X \to B$ such that $\bar{\partial}\tilde{u} = f$ in X.

Now we consider the case $2 \leqq q \leqq n$. Then $u_j - u_{j+1} \in Z^0_{(0,q-1)}(\bar{D}_{\alpha_j}, B)$ and, by Theorem 2.12.3 (i), we can find $w_j \in H^{1/2}_{(0,q-2)}(\bar{D}_{\alpha_j}, B)$ such that $u_j - u_{j+1} = \bar{\partial} w_j$ in D_{α_j}. Choose C^∞-functions χ_j on X such that $\chi_j = 1$ in $D_{\alpha_{j-1}}$ and supp $\chi_j \subset\subset D_{\alpha_j}$. Then by setting

$$u := u_{j+1} + \sum_{k=1}^{j} \bar{\partial}(\chi_k w_k) \quad \text{in} \quad D_{\alpha_{j-1}} \quad (j = 2, 3, \ldots)$$

we obtain a continuous B-valued $(0, q-1)$-form u over X such that $\bar{\partial} u = f$ on X. ■

2.12.5. Remark. Parts (i) and (iv) of Theorem 2.12.3 can be completed, respectively, by the following statements:

(i)′ If $k \in \{0, 1, \ldots, \infty\}$ and f is C^k in D_α, then Tf is $C^{k+\alpha}$ in D_α for all $0 < \alpha < 1$.

(iv)′ If $k \in \{0, 1, \ldots, \infty)$ and f is C^k on X, then u can be chosen to be $C^{k+\alpha}$ on X for all $0 < \alpha < 1$.

For $q = 1$ this follows from Corollary 2.1.6. For $2 \leqq q \leqq n$ this can be proved, for example, in the same way as the corresponding statements in Theorem 2.6.1 (ii) and 2.8.1 by means of the global integral formula for solving the $\bar{\partial}$-equation on C^2 strictly pseudoconvex open sets in Stein manifolds which will be constructed in Chapter 4.

2.13. Solution of the Levi problem for complex manifolds

In this section we introduce the notion of a Stein manifold, whose definition is modeled on the properties of domains of holomorphy in $\mathbb{C}^n$. It is the aim of this section to prove that a complex manifold is Stein if and only if it admits a strictly plurisubharmonic exhausting C^2-function (solution of the Levi problem for manifolds). It is not difficult to construct a strictly plurisubharmonic exhausting C^∞-function on Stein manifolds. As in the case of domains in $\mathbb{C}^n$, the proof of the opposite direction is more difficult. This proof is based on the solvability of the $\bar{\partial}$-equation on complex manifolds with strictly plurisubharmonic exhausting C^2-function proved in the preceding section.

2.13.1. Definition. Let X be a complex manifold. For every compact set $K \subset\subset X$ we define

$$\hat{K}^{\mathcal{O}}_X := \{z \in X : |f(z)| \leqq \sup_{\zeta \in K} |f(\zeta)| \quad \text{for all} \quad f \in \mathcal{O}(X)\}.$$

$\hat{K}^{\mathcal{O}}_X$ is called the $\mathcal{O}(X)$-*hull* of K. If $K = \hat{K}^{\mathcal{O}}_X$, then K is called $\mathcal{O}(X)$-*convex*.

2.13.2. Definition. A complex manifold X is called *holomorphically convex* if, for every compact set $K \subset\subset X$, $\hat{K}_X^{\mathcal{O}}$ is also compact.

2.13.3. Definition. A complex manifold X of complex dimension n which is countable at infinity is said to be a *Stein manifold* if

(i) X is holomorphically convex,

(ii) for every point $z \in X$, one can find n functions $f_1, \dots, f_n \in \mathcal{O}(X)$ which form holomorphic coordinates at z (that is, there is a neighbourhood U of z such that the map $U \ni \zeta \to (f_1(\zeta), \dots, f_n(\zeta)) \in \mathbb{C}^n$ is biholomorphic).

Remark. Sometimes the following condition is included in the definition of a Stein manifold:

(iii) If z and ζ are different points in X, then there exists an $f \in \mathcal{O}(X)$ such that $f(z) \neq f(\zeta)$.

We shall see (Corollary 2.13.7) that (iii) is a consequence of (i) and (ii).

Examples. 1. By Theorem 1.3.7 (iii), every domain of holomorphy in $\mathbb{C}^n$ is a Stein manifold.

2. Every closed complex submanifold of a Stein manifold is a Stein manifold.

3. If X, Y are Stein manifolds, then $X \times Y$ is a Stein manifold.

The simple proofs of 2. and 3. are left to the reader.

Remark. In particular, every closed complex submanifold of $\mathbb{C}^n$ is a Stein manifold. Conversely, every Stein manifold is biholomorphically equivalent to some closed complex submanifold of some $\mathbb{C}^n$ (see, for example, Hörmander [1] and Gunning/Rossi [1]). We do not use this result and omit the proof.

2.13.4. Theorem. *Let X be a Stein manifold. Then, for every $\mathcal{O}(X)$-convex compact set $K \subset\subset X$ and every neighbourhood U_K of K, there exists a strictly plurisubharmonic exhausting C^∞-function ϱ for X* (Definition 2.12.1) *such that $\varrho < 0$ on K and $\varrho > 0$ on $X \setminus U_K$.*

Proof. Since K is $\mathcal{O}(X)$-convex, and since X is holomorphically convex and countable at infinity, we can find a sequence of $\mathcal{O}(X)$-convex sets K_j, $j = 1, 2, \dots$, such that $K_1 = K$, $K_j \subseteqq \operatorname{int} K_{j+1}$ (int := interior of), and $X = \bigcup_{j=1}^{\infty} K_j$. Set $U_1 := U_K$ and $U_j := \operatorname{int} K_{j+1}$ for $j = 2, 3, \dots$. Since all K_j are $\mathcal{O}(X)$-convex, for every j we can find a finite number of functions $f_j^1, \dots, f_j^{N(j)} \in \mathcal{O}(X)$ such that

$$\sum_{k=1}^{N(j)} |f_j^k(z)|^2 < \frac{1}{2^j} \quad \text{for} \quad z \in K_j\,, \tag{2.13.1}$$

$$\sum_{k=1}^{N(j)} |f_j^k(z)|^2 > j \quad \text{for} \quad z \in K_{j+2} \setminus U_j\,. \tag{2.13.2}$$

By condition (ii) in Definition 2.13.3, we can assume that, moreover,

$$\operatorname{rank}\left[\left(\frac{\partial f_j^k}{\partial z_l}\right)_{l=1,\dots,n}^{k=1,\dots,N(j)}\right] = n \quad \text{for} \quad z \in K_j\,. \tag{2.13.3}$$

By (2.13.1) we can define

$$\varrho(z) := -1 + \sum_{j=1}^{\infty} \sum_{k=1}^{N(j)} |f_j^k(z)|^2 \quad \text{for} \quad z \in X\,. \tag{2.13.4}$$

It remains to prove that this function ϱ has the required properties. By (2.13.1) and (2.13.2), $\varrho < 0$ on K and $\varrho > 0$ in $X \setminus U_K$. The series

$$r(z, \zeta) := \sum_{j,k} f_j^k(z) \overline{f_j^k(\zeta)}, \qquad z, \zeta \in X, \tag{2.13.5}$$

converges uniformly on each compact subset in $X \times X$. Therefore, $r(z, \zeta)$ is holomorphic in z and $\overline{r(z, \zeta)}$ is holomorphic in ζ. Consequently, $\varrho(z) = (z, z) - 1$ is C^∞ on X. It follows from (2.13.2) that $\{z \in X: \varrho(z) < \alpha\}$ is relatively compact in X for all $\alpha \in \mathbb{R}$. Clearly, ϱ is plurisubharmonic. To prove strict plurisubharmonicity, we assume that for some $z^0 \in X$ and $\xi \in \mathbb{C}^n$

$$\sum_{r,s=1}^{n} \frac{\partial^2 \varrho(z^0)}{\partial z_r \, \partial \bar{z}_s} \xi_r \bar{\xi}_s = 0,$$

where $z_1, \dots, z_n$ are holomorphic coordinates in some neighbourhood of z^0. Then for all j

$$\sum_{k=1}^{N(j)} \left| \sum_{r=1}^{n} \frac{\partial f_j^k(z^0)}{\partial z_r} \xi_r \right|^2 = \sum_{k=1}^{N(j)} \sum_{r,s=1}^{n} \frac{\partial^2 |f_j^k(z)|^2}{\partial z_r \, \partial \bar{z}_s} \xi_r \bar{\xi}_s = 0.$$

By (2.13.3) this implies that $\xi = 0$. ■

2.13.5. Theorem. *A complex manifold X is a Stein manifold if and only if there exists a strictly plurisubharmonic exhausting C^2-function ϱ for X* (Definition 2.12.1). *Then, for every $\alpha \in \mathbb{R}$, the set $\{z \in X: \varrho(z) \leqq \alpha\}$ is $\mathcal{O}(X)$-convex.*

That the condition is necessary we know from Theorem 2.13.4. The remainder of the theorem will be obtained as a consequence of the following

2.13.6. Lemma. *Let X be a complex manifold with strictly plurisubharmonic exhausting C^2-function ϱ. Then*

(i) *Let $\xi \in X$, $\alpha := \varrho(\xi)$, $D_\alpha := \{z \in X: \varrho(z) < \alpha\}$ and suppose that $d\varrho(z) \neq 0$ for all $z \in \partial D_\alpha$. Then there exist a sequence $f_k \in \mathcal{O}(X)$, $k = 1, 2, \dots$, and a constant $C < \infty$ such that the following conditions are fulfilled:*

a) *$f_k(\xi) = 1$ for $k = 1, 2, \dots$,*

b) *$\|f_k\|_{0, D_\alpha} \leqq C$ for $k = 1, 2, \dots$,*

c) *for every compact set $K \subseteqq \bar{D}_\alpha \setminus \xi$, $\lim_{k \to \infty} \|f_k\|_{0,K} = 0$.*

(ii) *For every point $\xi \in X$ and each holomorphic function f in some neighbourhood of ξ, there exists a sequence $\{f_k\}_1^\infty$ of holomorphic functions on X such that $\lim_{k \to \infty} \partial f_k(\xi) = \partial f(\xi)$.*

(iii) *For every $\alpha \in \mathbb{R}$ and all $\xi \in \partial D_\alpha$, there exists a holomorphic function f in some neighbourhood of $\bar{D}_\alpha$ such that $f(\xi) = 0$ and $f(z) \neq 0$ for all $\xi \neq z \in \bar{D}_\alpha$.*

Proof. (i). Let $\lambda = (\lambda_1, \dots, \lambda_n)$ be holomorphic coordinates in some neighbourhood V_ξ of ξ. Set

$$u(z) := 2 \sum_{j=1}^{n} \frac{\partial \varrho(\xi)}{\partial \lambda_j} (\lambda_j(z) - \lambda_j(\xi)) + \sum_{j,k=1}^{n} \frac{\partial^2 \varrho(\xi)}{\partial \lambda_j \, \partial \lambda_k} (\lambda_j(z) - \lambda_j(\xi)) (\lambda_k(z) - \lambda_k(\xi)).$$

Then u is holomorphic in V_ξ,

$$u(\xi) = 0 \tag{2.13.6}$$

and, by Lemma 1.4.13,

$$\operatorname{Re} u(z) = \varrho(z) - \varrho(\xi) - \sum_{j,k=1}^{n} \frac{\partial^2 \varrho(\xi)}{\partial \lambda_j \, \partial \bar{\lambda}_k} (\lambda_j(z) - \lambda_j(\xi)) \overline{(\lambda_k(z) - \lambda_k(\xi))}$$
$$+ o(|\lambda(\xi) - \lambda(z)|^2).$$

Since ϱ is strictly plurisubharmonic, this implies that, after shrinking V_ξ, we can find $\beta > 0$ such that

$$\operatorname{Re} u(z) < \varrho(z) - \varrho(\xi) - \beta\,|\lambda(z) - \lambda(\xi)|^2 \quad \text{for all} \quad z \in V_\xi\,. \tag{2.13.7}$$

Then

$$e^{u(\xi)} = 1 \quad \text{and} \quad |e^{u(z)}| < 1 \quad \text{for all} \quad \xi \neq z \in \bar{D}_\alpha \cap V_\xi\,. \tag{2.13.8}$$

Choose a neighbourhood $W_\xi \subset\subset V_\xi$ of ξ and a C^∞-function χ on X such that $\chi = 1$ on W_ξ and $\operatorname{supp}\chi \subset\subset V_\xi$. Then $e^{ku}\,\bar{\partial}\chi$, $k = 1, 2, \ldots$, is a sequence of $\bar{\partial}$-closed $C^\infty_{(0,1)}$-forms on X and

$$\lim_{k\to\infty} \|e^{ku}\,\bar{\partial}\chi\|_{0,D_\alpha} = 0\,.$$

Since $d\varrho(z) \neq 0$ for $z \in \partial D_\alpha$, we can apply Theorem 2.12.3 (i) and obtain continuous functions v_k on $\bar{D}_\alpha$ such that $\bar{\partial}v_k = e^{ku}\,\bar{\partial}\chi$ in D_α and

$$\lim_{k\to\infty} \|v_k\|_{0,D_\alpha} = 0\,. \tag{2.13.9}$$

Setting $\tilde{f}_k := \chi\, e^{ku} - v_k + v_k(\xi)$, one obtains continuous functions $\tilde{f}_k$ on $\bar{D}_\alpha$, which are holomorphic in D_α. Since $\operatorname{supp}\chi \subset\subset V_\xi$, it follows from (2.13.8) and (2.13.9) that $\tilde{f}_k(\xi) = 1$, $k = 1, 2, \ldots$, $\sup_k \|\tilde{f}_k\|_{0,D_\alpha} < \infty$ and, for every compact set $K \subseteqq \bar{D}_\alpha \setminus \xi$, $\lim_k \|\tilde{f}_k\|_{0,K} = 0$. The required functions $f_k \in \mathcal{O}(X)$ we now obtain by means of the Approximation Theorem 2.12.3 (ii).

(ii) We can assume that

$$f(\xi) = 0\,. \tag{2.13.10}$$

Set $\alpha := \varrho(\xi)$ and suppose that W_ξ, V_ξ, u, χ are as in the proof of part (i). After shrinking W_ξ and V_ξ, we can assume that f is holomorphic on V_ξ. Define by $\varphi_k := f\,e^{ku}\,\bar{\partial}\chi$, $k = 1, 2, \ldots$, a sequence of $\bar{\partial}$-closed $C^\infty_{(0,1)}$-forms on X with $\operatorname{supp}\varphi_k \subset\subset V_\xi \setminus W_\xi$. Then it follows from (2.13.7) that, for some $\delta > 0$, $\lim_k \|\varphi_k\|_{0,D_{\alpha+\delta}} = 0$.

By Lemma 2.12.2 we can assume that $\operatorname{Crit}(\varrho) := \{z \in X : d\varrho(z) = 0\}$ is discrete in X. Choose $0 < \varepsilon < \delta$ such that $d\varrho(z) \neq 0$ for $z \in \partial D_{\alpha+\varepsilon}$. Then, by Theorem 2.12.3 (i), we can find continuous functions v_k on $\bar{D}_{\alpha+\varepsilon}$ such that

$$\bar{\partial}v_k = \varphi_k \quad \text{on} \quad D_{\alpha+\varepsilon} \tag{2.13.11}$$

and

$$\lim_{k\to\infty} \|v_k\|_{0,D_{\alpha+\varepsilon}} = 0\,. \tag{2.13.12}$$

Since $\varphi_k = 0$ in W_ξ, the functions v_k are holomorphic in W_ξ. Together with (2.13.12) this implies that

$$\lim_{k\to\infty} \partial v_k(\xi) = 0\,. \tag{2.13.13}$$

Since $\bar{\partial}v_k = \varphi_k$, by setting $\tilde{f}_k := \chi f\,e^{ku} - v_k$, we obtain continuous functions $\tilde{f}_k$ on $\bar{D}_{\alpha+\varepsilon}$, which are holomorphic in $D_{\alpha+\varepsilon}$. Since, by (2.13.6) and (2.13.10), $f(\xi) = u(\xi) = 0$ and $\chi = 1$ in W_ξ, we have

$$\partial\tilde{f}_k(\xi) = \partial f(\xi) - \partial v_k(\xi)\,. \tag{2.13.14}$$

By Theorem 2.12.3 (ii), we can find $f_k \in \mathcal{O}(X)$ such that $\|f_k - \tilde{f}_k\|_{D_{\alpha+\varepsilon}} < 1/k$ and, therefore,

$$\lim_{k\to\infty} \|\partial f_k(\xi) - \partial\tilde{f}_k(\xi)\| = 0\,.$$

According to (2.13.13) and (2.13.14) this completes the proof.

(iii) Let V_ξ, W_ξ and u be as in the proof of Lemma 2.13.6 (i). Then it follows from (2.13.7) that, for some $\delta > 0$,

$$\operatorname{Re} u(z) < -\delta \quad \text{for all} \quad z \in D_{\alpha+\delta} \cap (V_\xi \setminus W_\xi)\,. \tag{2.13.15}$$

Consequently, we can define $\ln u \in \mathcal{O}(D_{\alpha+\delta} \cap (V_\xi \setminus \overline{W}_\xi))$. By Theorem 2.12.3 (iv) (cp. also the remark following Definition 2.12.1), the $\bar\partial$-equation can be solved in $D_{\alpha+\delta}$. By Theorem 2.10.3, this implies that every holomorphic Cousin problem over $D_{\alpha+\delta}$ has a solution. Hence $\ln u = v_1 - v_2$ for some $v_1 \in \mathcal{O}(V_\xi \cap D_{\alpha+\delta})$ and $v_2 \in \mathcal{O}(D_{\alpha+\delta} \setminus \overline{W}_\xi)$. Define $f := u\,e^{-v_1}$ in $D_{\alpha+\delta} \cap V_\xi$ and $f := e^{-v_2}$ in $D_{\alpha+\delta} \setminus \overline{W}_\xi$. Then f is holomorphic in $D_{\alpha+\delta}$, $f(\xi) = 0$, and it follows from (2.13.7) that $f(z) \neq 0$ for all $\xi \neq z \in \overline{D}_\alpha$. ■

2.13.7. Corollary. *Let X be a Stein manifold, and let z, ξ be different points in X. Then there exists $f \in \mathcal{O}(X)$ such $f(z) \neq f(\xi)$.*

Proof. By Theorem 2.13.5, there is a strictly plurisubharmonic exhausting C^2-function ϱ for X. Without loss of generality we can assume that $\varrho(z) \leqq \varrho(\xi)$. Then the required function $f \in \mathcal{O}(X)$ can be obtained from Lemma 2.13.6 (iii) and the approximation Theorem 2.12.3 (ii). ■

Proof of Theorem 2.13.5. That the condition is necessary follows from Theorem 2.13.4. Conversely, suppose that X admits a strictly plurisubharmonic exhausting C^2-function ϱ. Then X is countable at infinity and condition (ii) in Definition 2.13.3 follows from Lemma 2.13.6 (ii). Let $D_\alpha := \{z \in X : \varrho(z) < \alpha\}$, $\alpha \in \mathbb{R}$. It remains to prove that, for every $\alpha \in \mathbb{R}$, the set $\overline{D}_\alpha$ is $\mathcal{O}(X)$-convex. Let $\xi \in X \setminus \overline{D}_\alpha$. By Lemma 2.12.2 we can find a strictly plurisubharmonic exhausting C^2-function φ for X such that Crit (φ) is discrete and which is so close to ϱ that $D_\alpha \subset\subset G_{\varphi(\xi)}$, where $G_\beta := \{z \in X : \varphi(z) < \beta\}$ for $\beta \in \mathbb{R}$. After adding a small constant to φ, we can assume that, moreover, $d\varphi(z) \neq 0$ for $z \in \partial G_{\varphi(\xi)}$. Then, by Lemma 2.13.6 (i), there exists $f \in \mathcal{O}(X)$ such that $f(\xi) = 1$ but $|f| < 1$ on $\overline{D}_\alpha$. ■

2.13.8. Proposition. *Let X be a Stein manifold. Then every strictly pseudoconvex open set $D \subset\subset X$ is a Stein manifold.*

Proof. Let θ, ϱ be as in Definition 2.11.1 of strictly pseudoconvex open sets, and let ϱ_1 be a strictly plurisubharmonic exhausting C^2-function for X (Theorem 2.13.4). Choose $\varepsilon > 0$ so small that $\{z \in \theta : -\varepsilon < \varrho(z) < 0\} \subset\subset \theta$, and choose a real-valued C^∞-function χ on $(-\infty, 0)$ with the following properties: $\chi(t) = 0$ for $t < -\varepsilon$, $\chi(t) \to \infty$ for $t \to 0$, χ is strictly convex on $(-\varepsilon, 0)$. Then, by Theorem 1.4.12 (ii), $\chi \circ \varrho$ is strictly plurisubharmonic on $\{z \in \theta : -\varepsilon < \varrho(z) < 0\}$ and $\varrho_1 + \chi \circ \varrho$ is a strictly plurisubharmonic exhausting C^2-function for D. By Theorem 2.13.5, this implies that D is a Stein manifold. ■

Notes

The Oka-Hefer theorem 2.5.4 was obtained in Oka [8] and Hefer [1]. Solvability of Cousin problems in domains of holomorphy in $\mathbb{C}^n$ was first proved in 1937 by Oka [2]. Together with the solvability of the $\bar\partial$-equation in polydiscs obtained by Dolbeault and Grothendieck (see Cartan [3], Dolbeault [1, 2], Malgrange [1], Hörmander [1], Andreian Cazacu [1]) this gives solvability of the $\bar\partial$-equation in domains of holomorphy (Dolbeault [1, 2]). The identity of pseudoconvex open sets and domains of holomorphy in $\mathbb{C}^n$ was proved first for $n = 2$ in 1942 by Oka [4] and then in 1953/54 for the general case by Oka [6], Bremermann [1] and Norguet [1]. The Approximation Theorem 2.7.1 (for $\mathcal{O}(D)$-convex compact subsets of domains of holomorphy D) was obtained in 1935/36 by

WEIL [1] and OKA [7]. Observe that, in the pioneering works of OKA and WEIL, the integral representation formula for holomorphic functions in polynomial polyhedra found by BERGMAN [1] and WEIL [1] (see Chapter 4) served as the basic tool. OKA's results on solvability of Cousin problems as well as the Oka-Weil approximation theorem in domains of holomorphy in $\mathbb{C}^n$ were generalized to Stein manifolds in the seminars of H. CARTAN in 1950/51 (CARTAN [3]). In these seminars the more general theory of coherent analytic sheaves on Stein manifolds was developed (see also OKA [5], GUNNING/ROSSI [1], GRAUERT/REMMERT [1], ANDREIAN CAZACU [1] and HÖRMANDER [1]). Some special results of this theory will be given without proof in Chapter 4. The solution of the Levi problem for manifolds (Theorem 2.13.5) is due to GRAUERT [4]. The approximation theorem for pseudoconvex compact sets in Stein manifolds (Theorem 2.12.3 (iii)) was proved by DOCQUIER/GRAUERT [1]. Solvability of the $\bar{\partial}$-equation directly (that is, without use of the solution of the Levi problem) in pseudoconvex open sets and complex manifolds with strictly plurisubharmonic exhausting function for the first time was obtained by means of L_2-estimates in the $\bar{\partial}$ Neumann problem by HÖRMANDER [1, 2], KOHN [1] and MORREY [1]. This method made it possible to obtain the main results with appropriate L_2-estimates on the boundary in the strictly pseudoconvex case (KOHN, MORREY) and in the domain in the general case (HÖRMANDER). Solvability of the $\bar{\partial}$-equation with uniform estimates in C^2 strictly pseudoconvex open sets on Stein manifolds was obtained in 1970 by GRAUERT/LIEB [1], HENKIN [1], KERZMAN [1, 2], LIEB [2] and ØVRELID [1] by means of integral formulas constructed by HENKIN [4] and RAMIREZ [1]. Hölder estimates were proved by HENKIN/ROMANOV [1], KERZMAN [1, 2] and LIEB [2]. Observe the following result of KRANTZ [1]: If $D \subset\subset \mathbb{C}^n$ is a domain with smooth boundary such that the $\bar{\partial}$-equation can be solved with 1/2-Hölder estimates in D, then D is strictly pseudoconvex. SIBONY [1] constructed a pseudoconvex domain with smooth boundary without uniform estimates for the $\bar{\partial}$-equation. The Approximation Theorem 2.9.2 was proved by HENKIN [4] (other proofs were obtained by KERZMAN [2] and LIEB [1]). DIEDERICH/FORNAESS [1] constructed a pseudoconvex domain with smooth boundary, where the Approximation Theorem 2.9.2 is not valid.

The solution of $\bar{\partial}u = f$ (for (0, 1)-forms f) given by Theorem 2.6.1 (ii) is the unique solution defined by the condition

$$L^w_{\partial D} u = 0 . \qquad (*)$$

The solution of $\bar{\partial}u = f$ obtained by the method of KOHN [2] is defined by the condition

$$Su = 0 , \qquad (**)$$

where S is the orthogonal projection from $L^2(\partial D)$ onto the subspace spanned by the holomorphic functions on a neighbourhood of $\bar{D}$ (the Szegö projection). After the work of FEFFERMAN [1], where asymptotics for the Bergman and Szegö kernels are given, it was proved by FOLLAND/STEIN [1] and GREINER/STEIN [1] that the solution of $\bar{\partial}u = f$ satisfying (**) in strictly pseudoconvex C^2-domains admits the same estimates as the solution defined by (*). Moreover, for the ball, the solution of $\bar{\partial}u = f$ satisfying (*) coincides with the solution satisfying (**). In the case of an arbitrary strictly pseudoconvex domain with smooth boundary, the solution satisfying (*) gives a „parametrix" for the solution satisfying (**) and conversely. KERZMAN and STEIN [1] proved the formula $S = L^w_{\partial D}(I - A)^{-1}$, where $A = K^* - K$ is compact.

For pseudoconvex domains with C^∞-boundary, by means of the method of L_2-estimates, it was proved by KOHN [3] that the $\bar{\partial}$-equation can be solved with C^∞-estimates on the boundary. Using results of DIEDERICH/FORNAESS [2], for pseudoconvex domains with real-analytic boundary, KOHN [4] also obtained subelliptic L_2-estimates on the boundary for the $\bar{\partial}$-Neumann problem. The $\bar{\partial}$-equation in domains with piecewise smooth boundary will be considered in Chapter 4.

Exercises, remarks and problems

1 a) Prove that the Šilov boundary of a bounded convex domain with C^∞-boundary in $\mathbb{C}^n$ is the closure of the points of strict pseudoconvexity (the Šilov boundary of a domain

$D \subset\subset \mathbb{C}^n$ is by definition the smallest closed set $K \subseteq \partial D$ such that $\max_{\bar{D}} |f| \leq \max_K |f|$ for every continuous function f on $\bar{D}$ that is holomorphic in D). Hint. The integration in the Leray formula is essentially only over the points of strict pseudoconvexity.

b) Remark. This is true also for arbitrary pseudoconvex domains with smooth boundary (Basener [1], Debiard/Gaveau [1], Hakim/Sibony [1], Rossi [1]).

c) (Byčkov [1], Arenson [1]). Prove that the Šilov boundary of an arbitrary convex domain $D \subset\subset \mathbb{C}^n$ consists of all points which are not contained in some complex disc $\subseteq \partial D$.

d) **Open problem** (Vituškin [2]). Has the Šilov boundary of an arbitrary domain $\subset\subset \mathbb{C}^n$ positive Hausdorff measure of order n? Remark that Vituškin [2] constructed an example of a domain of holomorphy in $\mathbb{C}^2$, whose Šilov boundary is of topological dimension zero.

2. Prove that the Šilov boundary of D_n (see point 25 at p. 65) is equal to $\{Z : ZZ^* = I\}$.

3. Let $D \subset\subset X$ be a C^2 strictly pseudoconvex open set in a Stein manifold X, and let B be a holomorphic vector bundle over D which is continuous over $\bar{D}$ (this means, by definition, that there exists a covering $\{U_j\}$ of $\bar{D}$ by relatively open sets $U_j \subseteq \bar{D}$ such that B is defined by transition functions which are continuous on $\overline{U_i \cap U_j}$ and holomorphic in $U_i \cap U_j$). As in Subsection 2.11.2, then we can define the spaces $Z^0_{(0,q)}(\bar{D}, B)$ and $H^{1/2}_{(0,q-1)}(\bar{D}, B)$, $1 \leq q \leq n$. Prove that there exists a bounded linear operator T from $Z^0_{(0,q)}(\bar{D}, B)$ into $H^{1/2}_{(0,q-1)}(\bar{D}, B)$ such that $\bar{\partial} T f = f$ in D for every $f \in Z^0_{(0,q)}(\bar{D}, B)$. Hint. Use a modification of the proof of Theorem 2.12.3 (i) (in this proof it is not important that the bundle B admits a holomorphic continuation to X). For another proof see Leiterer [2], where a more general situation of sheaves over $\bar{D}$ is considered which are analytic over D and satisfy some coherence condition.

4. Let $D \subset\subset X$ be a C^2 strictly pseudoconvex open set in a Stein manifold X and let $A(\bar{D})$ be the algebra of continuous functions on $\bar{D}$ which are holomorphic in D. Let $f_1, \dots, f_N \in A(\bar{D})$ such that, for every $z \in \bar{D}$, $|f_1(z)| + \dots + |f_N(z)| \neq 0$. Prove that there are $g_1, \dots, g_N \in A(\bar{D})$ such that $f_1 g_1 + \dots + f_N g_N = 1$ on $\bar{D}$. Hint. Use Exercise 3. Remark. This is equivalent to the fact that the space of maximal ideals in $A(\bar{D})$ is equal to $\bar{D}$. By means of results of Kohn [3], Hakim/Sibony [2] proved that this is true also for arbitrary pseudoconvex domains $D \subset\subset \mathbb{C}^n$ with C^∞-boundary.

5. Let X be a Stein manifold, and let $L(N, \mathbb{C})$ be the space of complex $N \times N$ matrices.

a) Let $M: X \to L(N, \mathbb{C})$ be a holomorphic map such that rank $M(z)$ = constant for $z \in X$. Prove that there exists a holomorphic map $V: X \to L(N, \mathbb{C})$ such that $MVM = M$ and $VMV = V$ on X. Hint. Use Theorem 2.12.3 (iv).

b) Let $D \subset\subset X$ be a C^2 strictly pseudoconvex open set and let $M: \bar{D} \to L(N, \mathbb{C})$ be continuous on $\bar{D}$ and holomorphic in D such that rank $M(z)$ = constant for $z \in \bar{D}$. Prove that there exists a continuous map $V: \bar{D} \to L(N, \mathbb{C})$ which is holomorphic in D such that $VMV = V$ and $MVM = M$ on $\bar{D}$. Hint. Use Exercise 3.

6 (Øvrelid [2]). Let $D \subset\subset X$ be a C^2 strictly pseudoconvex open set in a Stein manifold X and let $A(\bar{D})$ be the algebra of continuous functions on $\bar{D}$ which are holomorphic in D. Let $f_1, \dots, f_N \in A(\bar{D})$ such that $|f_1(z)| + \dots + |f_N(z)| \neq 0$ for all $z \in \bar{D}$. Prove that then for every collection $g_1, \dots, g_N \in A(\bar{D})$ the following two conditions are equivalent:

(i) $g_1 f_1 + \dots + g_N f_N = 0$ on $\bar{D}$.

(ii) There exist $\varphi_{jk} \in A(\bar{D})$, $j, k = 1, \dots, N$, such that $\varphi_{jk} = -\varphi_{kj}$ and $g_k = \varphi_{k1} f_1 + \dots + \varphi_{kN} f_N$ on $\bar{D}$.

Hint. Find continuous functions φ_{jk} and use Exercise 3.

7. **Open problem.** Let $D \subset\subset \mathbb{C}^n$ be a pseudoconvex open set with smooth real-analytic boundary. Do there exist uniform estimates for the $\bar{\partial}$-equation? Remark. Range [1] proved that this is the case for convex D with smooth real-analytic boundary in $\mathbb{C}^2$. Sibony [1] constructed a pseudoconvex C^∞-domain $D \subset\subset \mathbb{C}^2$ without uniform estimates for the $\bar{\partial}$-equation.

8. Open problem. Does there exist a convex open set $D \subset\subset \mathbb{C}^n$ without uniform estimates for the $\bar\partial$-equation? In particular, this problem is open for the domain of all complex $N \times N$ matrices Z such that $ZZ^* - I$ is strictly negative-definite (cp. Exercise 25 in Chapter 1).

9. Open problem. Let $D \subset\subset \mathbb{C}^n$ be a pseudoconvex domain with C^∞-boundary. Let $k \in \{0, 1, 2, \ldots\}$, $0 < \alpha < 1$, and let f be a $\bar\partial$-closed $C^k_{(0,1)}$-form in D whose derivatives of order $\leqq k$ belong to $H^\alpha(\bar D)$. Does there exist a continuous function u on $\bar D$ such that $\bar\partial u = f$ in D? (For $k + \alpha \geqq N(n)$ see KOHN [3].)

10. We denote by $GL(N, \mathbb{C})$ the group of all invertible complex $N \times N$ matrices. Let X be a Stein manifold of complex dimension n, and let A be an $N \times N$ matrix of continuous (0, 1)-forms on X such that $\bar\partial A + A \wedge A = 0$ on X. (The product $A \wedge A$ is defined by the rules of matrix multiplication, where the product of elements in A is the exterior product.)

a) For every continuous $(0, q)$-form f in an open set $D \subset\subset X$, $0 \leqq q \leqq n$, we define

$$\bar\partial_A f := \bar\partial f + A \wedge f,$$

and we denote by $Z^0_{(0,q)}(\bar D, A)$ the Banach space of all continuous $(0, q)$-forms on $\bar D$ such that $\bar\partial_A f = 0$ in D. Prove that, for every point $\xi \in X$ and all $\varepsilon > 0$, there exist a neighbourhood V of ξ and bounded linear operators $R_q: Z^0_{(0,q)}(\bar V, A) \to C^0_{(0,q-1)}(\bar V)$, $1 \leqq q \leqq n$, such that $\bar\partial_A \circ R_q = \text{id}$ (:= identity map) and $\|R_q\| \leqq \varepsilon$. Hint. It follows from Theorems 2.2.1 and 2.2.2 that, for every $\delta > 0$, we can find a neighbourhood V of ξ such that there exist bounded linear operators $T_q: C^0_{(0,q)}(\bar V) \to C^0_{(0,q-1)}(\bar V)$, $1 \leqq q \leqq n$, such that $\|T_q\| \leqq \delta$ and $\bar\partial T_q f = f$ for all $f \in Z^0_{(0,q)}(\bar V)$. Define by setting $Sf := T_q(f + T_{q+1}A \wedge f)$ a bounded linear operator S from $Z^0_{(0,q)}(\bar V, A)$ into $C^0_{(0,q-1)}(\bar V)$. Then $\bar\partial_A \circ S = \text{id} + M$, where $Mf := T_{q+1}A \wedge f + A \wedge T_q(f + T_{q+1}A \wedge f)$. If δ is sufficiently small, then $\|M\| \leqq 1/2$ and $\|S\| \leqq \varepsilon/2$. Set $R := S \circ (\text{id} + M)^{-1}$.

b) Prove that for every point $\xi \in X$ there exist a neighbourhood V of ξ and a continuous map $U: V \to GL(N, \mathbb{C})$ such that

$$\bar\partial U = UA \quad \text{in} \quad V.$$

Hint. Use part a). (For another proof see MALGRANGE [1].)

c) Prove that for every continuous $(0,q)$-form f on X, $1 \leqq q \leqq n$, such that $\bar\partial f + A \wedge f = 0$ there exists a continuous $(0, q - 1)$-form u on X such that

$$\bar\partial u + A \wedge u = f \quad \text{on} \quad X.$$

Hint. By part b) we can find an open covering $\{V_j\}$ of X and continuous maps U_j: $V_j \to GL(N, \mathbb{C})$ such that $\bar\partial U_j = U_j A$ in V_j. Then the maps $U_i U_j^{-1}$ are holomorphic in $V_i \cap V_j$, and the system $\{U_j f_j\}$ defines a $\bar\partial$-closed $(0, q)$-form with values in the corresponding holomorphic vector bundle. Use Theorem 2.12.3 (iv).

11. Let X be a complex manifold of complex dimension n, and let $GL(N, \mathbb{C})$ be the group of invertible complex $N \times N$ matrices. Prove that the following two conditions are equivalent:

(i) For every $N \times N$ matrix A of continuous (0, 1)-forms on X such that $\bar\partial A + A \wedge A = 0$, there exists a continuous map $U: X \to GL(N, \mathbb{C})$ such that

$$\bar\partial U = UA \quad \text{on} \quad X.$$

(ii) Let $\{V_j\}$ be an open covering of X and let $U_j: V_j \to GL(N, \mathbb{C})$ be continuous maps such that $U_i U_j^{-1}$ is holomorphic in $V_i \cap V_j$. Then there exist holomorphic maps H_j: $V_j \to GL(N, \mathbb{C})$ such that $U_i U_j^{-1} = H_i H_j^{-1}$ in $V_i \cap V_j$.

Hint. Use Exercise 10b) to prove that (ii) $\Rightarrow$ (i). Set $A := U_j^{-1}\, \bar\partial U_j$ in V_j for the proof of the implication (i) $\Rightarrow$ (ii). Remark. For Stein manifolds X, condition (ii) is fulfilled. This is a profound theorem of GRAUERT [1, 2, 3] (see also CARTAN [2] and CORNALBA/GRIFFITHS [1]).

12. Open problem. Let $X \subseteqq \mathbb{C}^n$ be an open set such that the equivalent conditions (i) and (ii) in Exercise 11 are fulfilled. Is then X a domain of holomorphy? Remark. LEITERER [3] proved that the answer is affirmative if we suppose that, in addition, for every $\bar\partial$-closed $C^\infty_{(0,1)}$-form f on X, the equation $\bar\partial u = f$ has a continuous solution u on X.

13. Let $D \subset\subset \mathbb{C}^n$ be a C^2 strictly pseudoconvex open set, and let ϱ be a strictly plurisubharmonic C^2-function in some neighbourhood θ of $\bar D$ such that $D = \{z \in \theta : \varrho(z) < 0\}$. For $\xi \in \partial D$ we set $\hat H_\xi(\delta) := \{z \in \theta : |\hat F_\varrho(z, \xi)| < \delta^2\}$ (see (1.4.18) for the definition of $\hat F_\varrho$). Let f be an (n, n)-form in D such that

$$\int_{\hat H_\xi(\delta) \cap D} |f| = O(\delta^{2n}) \quad \text{for} \quad \delta \to 0\,,$$

uniformly in $\xi \in \partial D$.

a) (Hörmander [3], see also Henkin [3]). Prove that f defines a continuous linear functional on $\mathfrak{H}^1(D)$, where $\mathfrak{H}^1(D)$ is defined analogously as in Remark 8 in the Exercises, remarks and problems at the end of Chapter 1.

b) Prove that there is an $(n, n-1)$-form in D that is bounded on ∂D such that $\bar\partial u = f$. Hint. Use part a) and formulas for solving the $\bar\partial$-equation.

c) Remark. Henkin [3] and Varopoulos [1] proved the following theorem: *If f is a (p, q)-form in D which fulfils the condition*

$$\int_{\hat H_\xi(\delta) \cap D} (|f| + |\varrho|^{-1/2}\, \|f \wedge \bar\partial\varrho\|)\, d\sigma_{2n} = O(\delta^{2n}) \quad \text{for} \quad \delta \to 0\,,$$

uniformly in $\xi \in \partial D$ ($d\sigma_{2n}$ is the Lebesgue measure), then there is a solution of $\bar\partial u = f$, whose boundary values belong to each $L^s(\partial D)$, $s < \infty$. Moreover, Varopoulos [1] proved that this solution can be chosen to have boundary values in $BMO(\partial D)$; here $BMO(\partial D)$ is the space of all measurable functions f on ∂D with

$$\sup_{\xi \in \partial D,\, \delta > 0} \frac{1}{|\hat H_\xi(\delta) \cap \partial D|} \int_{\hat H_\xi(\delta) \cap \partial D} |f - f_{\xi,\delta}|\, d\sigma_{2n-1} < \infty\,,$$

where $d\sigma_{2n-1}$ is the volume measure on ∂D,

$$|\hat H_\xi(\delta) \cap \partial D| := \int_{\hat H_\xi(\delta) \cap \partial D} d\sigma_{2n-1}\,, \qquad f_{\xi,\delta} := \frac{1}{|\hat H_\xi(\delta) \cap \partial D|} \int_{\hat H_\xi(\delta) \cap \partial D} f\, d\sigma_{2n-1}\,.$$

Observe that $L^\infty(\partial D) \subsetneqq BMO(\partial D) \subsetneqq \bigcap_{s<\infty} L^s(\partial D)$. Varopoulos [1] proved also that, for $q < n$, a solution with bounded boundary values on ∂D need not exist.

14. Let $D \subset\subset \mathbb{C}^n$ be the unit ball, and let $f_1, \dots, f_N$ be bounded holomorphic functions in D such that $\inf_D (|f_1| + \dots + |f_N|) > 0$.

a) Let $A(\bar D)$ be the space of continuous functions on $\bar D$ that are holomorphic in D. For $\varphi \in L^\infty(\partial D)$ we write $\varphi \perp A(\bar D)$ if, for all $h \in A(\bar D)$, $\int_{\partial D} \varphi(\zeta)\, h(\zeta)\, \omega'(\bar\zeta) \wedge \omega(\zeta) = 0$. Prove that, for every $\varphi \in L^\infty(\partial D)$ with $\varphi \perp A(\bar D)$, there exist $\varphi_1, \dots, \varphi_N \in L^\infty(\partial D)$ such that $\varphi_1, \dots, \varphi_N \perp A(\bar D)$ and $\varphi = f_1\varphi_1 + \dots + f_N\varphi_N$. Hint. Use Exercise 13b) and an appropriate analogue of Carleson's construction in the proof of the Corona theorem (Carleson [3]).

b) **Open problem** (the Corona problem for the ball in $\mathbb{C}^n$, $n \geqq 2$). Do there exist bounded holomorphic functions $h_1, \dots, h_N$ in D such that $f_1h_1 + \dots + f_Nh_N = 1$ in D?

Remark. Henkin [3] and Varopoulos [1] (see also Amar [2]) proved that there exist holomorphic functions $h_j \in \bigcap_{p<\infty} \mathfrak{H}^p(D)$ such that $\sum f_jh_j = 1$ in D, where $\mathfrak{H}^p(D)$ is defined analogously as in Remark 8 in the Exercises, remarks and problems at the end of Chapter 1. Observe that the „hard" part in Carleson's proof of the Corona theorem (cp. Remark 8 in the Exercises, remarks and problems at the end of Chapter 1) works also in this situation. However, the corresponding $\bar\partial$-equation (the „soft" part) can be solved with boundary values in $BMO(\partial D)$ only. (Note that for the unit disc D, for every function u in D with $u|_{\partial D} \in BMO(\partial D)$, there exists a holomorphic function g in D such that $u + g$ is bounded. In fact, Fefferman/Stein [1] proved that such functions u can be written $u = f + \tilde h$, where $f|_{\partial D}, h|_{\partial D} \in L^\infty(\partial D)$ and $\tilde h$ is the Hilbert transform of h (that is $h + i\tilde h \in \mathcal{O}(D)$ and $\tilde h(0) = 0$). Set $g := i(h + i\tilde h)$.)

15 (LELONG [2], BREMERMANN [2]). Prove that, for every continuous plurisubharmonic function u in the pseudoconvex open set $D \subseteqq \mathbb{C}^n$, there exist holomorphic functions f_m in D such that

$$u(z) = \lim_{j\to\infty} \sup_{m \geqq j} \frac{1}{m} \ln |f_m(z)|, \qquad z \in D,$$

where the convergence is uniform on each compact set. Hint. Prove that the set D^* $:= \{(z, w): z \in D,\ w \in \mathbb{C}^1,\ |u(z)| < -\ln |w|\}$ is pseudoconvex; consequently (since the Levi problem is solved — Theorem 2.7.2), there is a holomorphic function $f(z, w)$ in D^* which cannot be continued holomorphically to a larger open set; take for $f_m(z)$ the coefficients of the power series $f(z, w) = \sum f_m(z)\, w^m$ with respect to w (see also RONKIN [1]).

16 (**Dolbeault's theorem**). Let X be a complex manifold of complex dimension n, let $\mathfrak{U} = \{U_j\}_{j\in I}$ be an open covering of X such that every U_j is a Stein manifold, and let B be a holomorphic vector bundle over X. For $0 \leqq q \leqq n$ we denote by $C^q(\mathfrak{U}, B)$ the group of all collections $\{f_{j_0\ldots j_q}\}_{(j_0,\ldots,j_q)\in I^{q+1}}$, where $f_{j_0\ldots j_q}$ is a holomorphic section of B over $U_{j_0} \cap \ldots \cap U_{j_q}$. By setting

$$(\delta^q f)_{j_0\ldots j_{q+1}} := \sum_{k=0}^{q+1} (-1)^k f_{j_0\ldots \hat{k} \ldots j_{q+1}}$$

we define a homomorphism $\delta^q\colon C^q(\mathfrak{U}, B) \to C^{q+1}(\mathfrak{U}, B)$. We denote by $\bar{\partial}^q$ the $\bar{\partial}$-operator considered as an operator from the space of all B-valued $C^\infty_{(0,q)}$-forms over X into the space of all B-valued $C^\infty_{(0,q+1)}$-forms over X. Prove the following generalization of Theorem 2.10.3: *For* $q = 1, \ldots, n$, *the factor groups* kernel δ^q/image δ^{q-1} *and* kernel $\bar{\partial}^q$/image $\bar{\partial}^q$ *are isomorphic.*

3. Theory of functions on strictly pseudoconvex sets with non-smooth boundary in $\mathbb{C}^n$

Summary. In this chapter we consider strictly pseudoconvex open sets $D \subset\subset \mathbb{C}^n$ (Definition 1.5.15) whose boundary need not be smooth. In Section 3.1 we construct a Koppelman-Leray formula for such open sets, where the Leray map from Theorem 2.5.5 is used. (Since the boundary of D need not be smooth, the boundary integrals in the Koppelman-Leray formula (1.12.1) will be replaced by volume integrals.) In Sections 3.2 and 3.3 we prove that the solution of $\bar{\partial} u = f$ given by this formula admits uniform estimates. (Hölder estimates are not obtained — however, see point 2 in the Exercises, remarks and problems at the end of this chapter.) Moreover, these estimates contain an estimation of u by the Hörmander diameter (Subsection 3.2.1) of the support of f. In Section 3.4 we prove a theorem on decomposition of singularities with uniform estimates for bounded holomorphic functions in D as well as for continuous functions on $\bar{D}$ which are holomorphic in D. In Section 3.5 we show that every continuous function on $\bar{D}$ that is holomorphic in D can be approximated uniformly on $\bar{D}$ by holomorphic functions in some neighbourhood of $\bar{D}$. Section 3.6 contains the following theorem: Let $X \subseteqq \mathbb{C}^n$ be a complex plane. Then every bounded holomorphic function f in $X \cap D$ admits a bounded holomorphic extension to D. If f is continuous on $\bar{D}$, then this extension can be chosen to be continuous on $\bar{D}$. In Chapter 4 this theorem will be generalized to the case that X is an arbitrary complex submanifold of some neighbourhood of $\bar{D}$ and D is a strictly pseudoconvex open set in a Stein manifold.

3.1. The Koppelman-Leray formula

In the theory developed in Chapter 2 for smooth strictly pseudoconvex open sets, the Koppelman-Leray formula (1.12.1) plays an important role. We would like to use this formula also in the present chapter, where the case of not necessarily smooth strictly pseudoconvex open sets is considered. Recall that the Leray map in Section 2.5 was already constructed for the general case of strictly pseudoconvex open sets with not necessarily smooth boundary. However, in view of the boundary integrals, we cannot immediately write the corresponding Koppelman-Leray formula. Therefore, in the present section, we give some modifications of this formula replacing the boundary integrals by volume integrals. The first step, which will be carried out in the following Subsection 3.1.1, consists in a modification of the support function $\Phi(z, \zeta)$.

Roughly speaking, we will replace the Levi polynomial $\hat{F}_\varrho(z, \zeta)$ by $\hat{F}_\varrho(z, \zeta) - 2\varrho(\zeta)$, and so we obtain a function $\tilde{\Phi}(z, \zeta)$ such that $\tilde{\Phi}(z, \zeta) = \Phi(z, \zeta)$ for $\zeta \in \partial D$ and (in distinction to $\Phi(z, \zeta)$) $\tilde{\Phi}(z, \zeta) \neq 0$ for all $z, \zeta \in D$. Using Stokes' formula, we can then replace all boundary integrals by volume integrals.

In the present chapter we also need information about the dependence of the Koppelman-Leray formula on „small“ changes of D. Therefore, together with D we also consider sets D_m which are „close" to D, and we will do all constructions „uniformly in m".

3.1.1. Construction of Φ, $\tilde{\Phi}$, w. Let $D \subset\subset \mathbb{C}^n$ be a strictly pseudoconvex open set, let $\theta \subset\subset \mathbb{C}^n$ be a neighbourhood of ∂D, and let ϱ be a strictly plurisubharmonic C^2-function in a neighbourhood of $\bar{\theta}$ such that

$$D \cap \theta = \{z \in \theta : \varrho(z) < 0\}\,. \tag{3.1.1}$$

Remark. We do not assume that $d\varrho(\zeta) \neq 0$ if $\zeta \in \partial D$, and we also do not assume that $D \neq \emptyset$.

Define $N(\varrho) := \{z \in \bar{\theta} : \varrho(z) = 0\}$ and suppose that $\varrho \neq 0$ on $\partial\theta$, that is,

$$N(\varrho) \subset\subset \theta\,. \tag{3.1.2}$$

By $x_j = x_j(\zeta)$ $(j = 1, \dots, 2n)$, we denote the real coordinates of $\zeta \in \mathbb{C}^n$ so that $\zeta_j = x_j + ix_{j+n}$ $(j = 1, \dots, n)$.

Since $N(\varrho) \subset\subset \theta$ and since ϱ is strictly plurisubharmonic and C^2, we can choose numbers $\varepsilon, \beta > 0$ and C^1-functions a_{jk} on $\bar{\theta}$ such that the following estimates hold

$$\operatorname{dist}\big(N(\varrho), \partial\theta\big) > 2\varepsilon\,, \tag{3.1.3}$$

$$\inf_{\zeta \in \bar{\theta}} \sum_{j,k=1}^{n} \frac{\partial^2 \varrho(\zeta)}{\partial \zeta_j\, \partial \bar{\zeta}_k}\, \xi_j \bar{\xi}_k > 3\,\beta\, |\xi|^2 \quad \text{for all} \quad 0 \neq \xi \in \mathbb{C}^n\,, \tag{3.1.4}$$

$$\sup_{\zeta \in \bar{\theta}} \left| \frac{\partial^2 \varrho(\zeta)}{\partial \zeta_j\, \partial \zeta_k} - a_{jk}(\zeta) \right| < \frac{\beta}{n^2}\,, \tag{3.1.5}$$

$$\left| \frac{\partial^2 \varrho(\zeta)}{\partial x_j\, \partial x_k} - \frac{\partial^2 \varrho(z)}{\partial x_j\, \partial x_k} \right| < \frac{\beta}{2n^2} \quad \text{for} \quad \zeta, z \in \bar{\theta} \quad \text{with} \quad |\zeta - z| \leqq 2\varepsilon\,. \tag{3.1.6}$$

We define

$$F(z, \zeta) := 2 \sum_{j=1}^{n} \frac{\partial \varrho(\zeta)}{\partial \zeta_j} (\zeta_j - z_j) - \sum_{j,k=1}^{n} a_{jk}(\zeta)\, (\zeta_j - z_j)\, (\zeta_k - z_k)\,. \tag{3.1.7}$$

Lemma A. *We have the estimate*

$$\operatorname{Re} F(z, \zeta) \geqq \varrho(\zeta) - \varrho(z) + \beta\, |\zeta - z|^2 \qquad (\zeta, z \in \bar{\theta}, |\zeta - z| \leqq 2\varepsilon)\,, \tag{3.1.8}$$

and there exist a neighbourhood $U \subset\subset \theta$ of $N(\varrho)$ and C^1-functions $\Phi(z, \zeta)$, $\tilde{\Phi}(z, \zeta)$, $M(z, \zeta)$, and $\tilde{M}(z, \zeta)$ for $\zeta \in U$ and $z \in U \cup D$ such that the following conditions are fulfilled:

(i) *$\Phi(z, \zeta)$ and $\tilde{\Phi}(z, \zeta)$ depend holomorphically on $z \in U \cup D$.*

(ii)

$$\Phi(z, \zeta) \neq 0 \quad \text{and} \quad \tilde{\Phi}(z, \zeta) \neq 0 \quad \text{for} \quad \zeta \in U, z \in D \cup U \quad \text{with} \quad |\zeta - z| \geqq \varepsilon; \tag{3.1.9}$$

$$M(z, \zeta) \neq 0 \quad \text{and} \quad \tilde{M}(z, \zeta) \neq 0 \quad \text{for} \quad \zeta \in U, z \in D \cup U; \tag{3.1.10}$$

$$\Phi(z, \zeta) = F(z, \zeta)\, M(z, \zeta) \quad \text{and} \quad \tilde{\Phi}(z, \zeta) = \big(F(z, \zeta) - 2\varrho(\zeta)\big)\, \tilde{M}(z, \zeta)$$
$$\text{for} \quad \zeta \in U, z \in D \cup U \quad \text{with} \quad |\zeta - z| \leqq \varepsilon\,. \tag{3.1.11}$$

(iii)

$$\tilde{\Phi}(z, \zeta) = \Phi(z, \zeta) \quad \text{for} \quad \zeta \in N(\varrho)\,, \qquad z \in U \cup D\,. \tag{3.1.12}$$

(iv) *Let* $\{\varrho_m\}_{m=1}^{\infty}$ *be a sequence of strictly plurisubharmonic* C^2*-functions in a neighbourhood of* $\bar{\theta}$ *tending, together with the first and second-order derivatives, uniformly on* $\bar{\theta}$ *to* ϱ. *Set*

$$F_m(z, \zeta) = 2 \sum_{j=1}^{n} \frac{\partial \varrho_m(\zeta)}{\partial \zeta_j} - \sum_{j,k=1}^{n} a_{jk}(\zeta)\,(\zeta_j - z_j)\,(\zeta_k - z_k)\,.$$

Then there exist an integer m_0 *and* C^1*-functions* $\Phi_m(z, \zeta)$, $\tilde{\Phi}_m(z, \zeta)$, $M_m(z, \zeta)$, *and* $\tilde{M}_m(z, \zeta)$ $(m \geqq m_0)$ *defined for* $\zeta \in U$ *and* $z \in D \cup U$ *such that the following statements hold:*

(1) *There are strictly pseudoconvex open sets* $D_m \subset\subset \mathbb{C}^n$ $(m \geqq m_0)$ *such that* $\partial D_m \subseteqq U$ *and* $D_m \cap \theta = \{z \in \theta : \varrho_m(z) < 0\}$.

(2) $N(\varrho_m) := \{z \in \bar{\theta} : \varrho_m(z) = 0\} \subseteqq U$.

(3) $D_m \cup U = D \cup U$.

(4) *Estimate* (3.1.8) *and conditions* (i), (ii), (iii) *are fulfilled if we replace* Φ *by* Φ_m, ϱ *by* ϱ_m, *etc.,*

(5) $\Phi_m(z, \zeta)$, $\tilde{\Phi}_m(z, \zeta)$, $M_m(z, \zeta)$, *and* $\tilde{M}_m(z, \zeta)$, *together with the first-order derivatives, uniformly tend in* $\zeta \in U$, $z \in U \cup D$ *to* $\Phi(z, \zeta)$, $\tilde{\Phi}(z, \zeta)$, $M(z, \zeta)$, *and* $\tilde{M}(z, \zeta)$, *respectively.*

Remark. The functions Φ and Φ_m agree with the functions Φ given by Theorem 2.4.3 for D and D_m, respectively. The construction of these functions given below coincides with the construction in the proof of Theorem 2.4.3. However, in distinction to Theorem 2.4.3, we give a parallel construction of $\tilde{\Phi}_m$ and, moreover, we have to look at the dependence on m of both constructions.

Proof of Lemma A. Estimate (3.1.8) follows from Lemma 2.4.2. In view of (3.1.3), we can choose a neighbourhood W of $N(\varrho)$ so small that

$$\operatorname{dist}(W, \partial\theta) > 2\varepsilon \tag{3.1.13}$$

and

$$\max_{\zeta \in W} |\varrho(\zeta)| < \varepsilon^2\beta/3\,. \tag{3.1.14}$$

Now we choose open sets U_0 and U such that $D \cup U_0$ is strictly pseudoconvex and

$$N(\varrho) \subset\subset U \subset\subset U_0 \subset\subset W\,. \tag{3.1.15}$$

Next we prove that

$$\begin{gathered}\operatorname{Re} F(z, \zeta) > \varepsilon^2\beta/3 \quad \text{and} \quad \operatorname{Re} F(z, \zeta) - 2\varrho(\zeta) > \varepsilon^2\beta/3 \\ \text{for} \quad \zeta \in \overline{W}, z \in \overline{W} \cup D \quad \text{with} \quad \varepsilon \leqq |\zeta - z| \leqq 2\varepsilon\,.\end{gathered} \tag{3.1.16}$$

To do this we fix points $\zeta \in \overline{W}$ and $z \in \overline{W} \cup D$ with $\varepsilon \leqq |\zeta - z| \leqq 2\varepsilon$. Then, by (3.1.13), $z \in \theta$ and it follows from (3.1.8) that $\operatorname{Re} F(z, \zeta) \geqq \varrho(\zeta) - \varrho(z) + \beta\varepsilon^2$. Together with (3.1.14) this implies

$$\operatorname{Re} F(z, \zeta) > -\varrho(z) + \tfrac{2}{3}\beta\varepsilon^2 \tag{3.1.17}$$

and

$$\operatorname{Re} F(z, \zeta) - 2\varrho(\zeta) > -\varrho(z) + \tfrac{2}{3}\beta\varepsilon^2\,. \tag{3.1.18}$$

Further, since $z \in \theta$ and $z \in \overline{W} \cup D$, at least one of the following relations holds: $z \in \overline{W}$ or $z \in \theta \cap D$. If $z \in \overline{W}$, it follows from (3.1.14) that $\varrho(z) < \varepsilon^2\beta/3$, and if $z \in \theta \cap D$, it follows from (3.1.1) that $\varrho(z) < 0$. In both cases $-\varrho(z) > -\frac{1}{3}\,\varepsilon^2\beta$. Together with (3.1.17) and (3.1.18) this implies (3.1.16).

Now we shall construct the functions Φ, $\tilde{\Phi}$, M, $\tilde{M}$. In view of (3.1.16), for $\zeta \in \overline{W}$, $z \in \overline{W} \cup D$ with $\varepsilon \leqq |\zeta - z| \leqq 2\varepsilon$, we can define $\ln F(z,\zeta)$ and $\ln\left(F(z,\zeta) - 2\varrho(\zeta)\right)$. Choose a C^∞-function χ on $\mathbb{C}^1$ such that $\chi(\eta) = 1$ if $|\eta| \leqq \varepsilon + \varepsilon/4$ and $\chi(\eta) = 0$ if $|\eta| \geqq 2\varepsilon - \varepsilon/4$. For $\zeta \in W$, $z \in W \cup D$ we define

$$f(z,\zeta) := \begin{cases} \bar{\partial}_z[\chi(\zeta - z) \ln \left(F(z,\zeta)\right)] & \text{if } \varepsilon \leqq |\zeta - z| \leqq 2\varepsilon\,, \\ 0 & \text{otherwise}\,, \end{cases} \tag{3.1.19}$$

and

$$\tilde{f}(z,\zeta) := \begin{cases} \bar{\partial}_z[\chi(\zeta - z) \ln \left(F(z,\zeta) - 2\varrho(\zeta)\right)] & \text{if } \varepsilon \leqq |\zeta - z| \leqq 2\varepsilon\,, \\ 0 & \text{otherwise}\,. \end{cases} \tag{3.1.20}$$

Since $F(z,\zeta)$ is holomorphic in z, $f(z,\zeta)$ and $\tilde{f}(z,\zeta)$ are $\bar{\partial}_z$-closed $C^\infty_{(0,1)}$-forms on $W \cup D$, and the maps $W \ni \zeta \to f(\cdot,\zeta)$ and $W \ni \zeta \to \tilde{f}(\cdot,\zeta)$, considered as maps with values in the Fréchet space $Z^\infty_{(0,1)}(W \cup D)$ (cp. the beginning of Section 2.4), are continuously differentiable. Since $D \cup U_0$ is strictly pseudoconvex and since $D \cup U_0 \subset\subset D \cup W$, from Lemma 2.4.1 we obtain a continuous linear operator

$$T\colon Z^\infty_{(0,1)}(W \cup D) \to C^\infty(U_0 \cup D)$$

such that $\bar{\partial} T\varphi = \varphi$ on $U_0 \cup D$ for all $\varphi \in Z^\infty_{(0,1)}(W \cup D)$. Now we define for all $\zeta \in U_0$ and $z \in U_0 \cup D$

$$\begin{aligned} u(z,\zeta) &:= \left(Tf(\cdot,\zeta)\right)(z)\,, & \tilde{u}(z,\zeta) &:= \left(T\tilde{f}(\cdot,\zeta)\right)(z)\,, \\ \tilde{M}(z,\zeta) &:= \exp\left[-u(z,\zeta)\right], & \tilde{M}(z,\zeta) &:= \exp\left[-\tilde{u}(z,\zeta)\right], \end{aligned} \tag{3.1.21}$$

$$\Phi(z,\zeta) := \begin{cases} F(z,\zeta)\, M(z,\zeta) & \text{if } |\zeta - z| \leqq \varepsilon\,, \\ \exp\left[\chi(\zeta - z) \ln F(z,\zeta) - u(z,\zeta)\right] & \text{if } |\zeta - z| \geqq \varepsilon\,, \end{cases} \tag{3.1.22}$$

$$\Phi(z,\zeta) := \begin{cases} \left(F(z,\zeta) - 2\varrho(\zeta)\right) \tilde{M}(z,\zeta) & \text{if } |\zeta - z| \leqq \varepsilon\,, \\ \exp\left[\chi(\zeta - z) \ln \left(F(z,\zeta) - 2\varrho(\zeta)\right) - \tilde{u}(z,\zeta)\right] & \text{if } |\zeta - z| \geqq \varepsilon\,. \end{cases} \tag{3.1.23}$$

It is easy to check that Φ and $\tilde{\Phi}$ fulfil conditions (i)—(iii).

Consider condition (iv). In view of the uniform convergence $\varrho_m \to \varrho$ together with the first- and second-order derivatives, we can choose m_0 so large that the estimates (3.1.4)—(3.1.6) remain valid if we replace, for $m \geqq m_0$, ϱ by ϱ_m. Then it follows from Lemma 2.4.2 that (3.1.8) also remains valid if we replace ϱ by ϱ_m and F by F_m. It is clear that for sufficiently large m the statements (1), (2), (3) in condition (iv) hold.

Finally, we choose m_0 so large that, for $m \geqq m_0$, (3.1.16) remains valid if we replace ϱ by ϱ_m and F by F_m. For $m \geqq m_0$ we can then define f_m, $\tilde{f}_m$, u_m, $\tilde{u}_m$, M_m, $\tilde{M}_m$, Φ_m, and $\tilde{\Phi}_m$ replacing ϱ by ϱ_m, F by F_m etc. in (3.1.19)—(3.1.23). It is easy to check that the functions so defined have the required properties. We only remark that for the proof of the uniform convergence $u_m \to u$ and $\tilde{u}_m \to \tilde{u}$ on $(U \cup D) \times U$ together with the first-order derivatives, we have to use that $U \subset\subset U_0$, because the linear operator T: $Z^\infty_{(0,1)}(W \cup D) \to C^\infty(U_0 \cup D)$ is only continuous with respect to the Fréchet topology of these spaces, which implies uniform convergence (with respect to z) on compact subsets only. ∎

Lemma B. *Let U be the neighbourhood from Lemma A, let $\Phi(z,\zeta)$ be the function from Lemma A, and let V_1, V_0 be neighbourhoods of $N(\varrho)$ such that $V_0 \cup D$ is strictly pseudoconvex and $V_1 \subset\subset V_0 \subset\subset U$. Then there exist continuous linear operators of Fréchet* *spaces* (cp. Section 2.5)

$$T_j\colon \mathcal{O}(U \cup D) \to \mathcal{O}((V_0 \cup D) \times (V_0 \cup D))$$

with the following properties:

(a) *The C^1-map $w = (w_1, \dots, w_n)$ defined by $w_j(z, \zeta) := T_j(\Phi(\cdot, \zeta))(z, \zeta)$ ($\zeta \in V_0$, $z \in V_0 \cup D$) is holomorphic in $z \in V_0 \cup D$ and*

$$\langle w(z, \zeta), \zeta - z\rangle = \Phi(z, \zeta) \qquad (\zeta \in V_0, z \in V_0 \cup D)\,.$$

(b) *If $\Phi_m(z, \zeta)$ are functions as in condition* (iv) *of Lemma A, then the C^1-map $w^m = (w_1^m, \dots, w_n^m)$ defined by*

$$w_j^m(z, \zeta) = T_j(\Phi_m(\cdot, \zeta))(z, \zeta) \qquad (\zeta \in V_0, z \in V_0 \cup D)$$

is holomorphic in $z \in V_0 \cup D$, $\langle w^m(z, \zeta), \zeta - z\rangle = \Phi_m(z, \zeta)$ ($\zeta \in V_0, z \in V_0 \cup D$), and $\lim\limits_{m\to\infty} w^m(z, \zeta) = w(z, \zeta)$ uniformly in $\zeta \in V_1$, $z \in V_1 \cup D$ together with the first-order derivatives.

Proof. The statement follows by using the proof of Theorem 2.5.5, the uniform convergence $\Phi_m \to \Phi$, together with the first-order derivatives, and the fact that $V_1 \subset\subset V_0$ and the linear operators T_j are continuous with respect to the topology of uniform convergence on compact sets. ∎

3.1.2. Definition of L_D and R_D. Let $D \subset\subset \mathbb{C}^n$, Φ, $\tilde{\Phi}$, w, $V_1 \subset\subset V_0 \subset\subset U$ as in Construction 3.1.1. Further, we choose a neighbourhood V_2 of $N(\varrho)$ such that $V_2 \subset\subset V_1$ and a C^∞-function χ on $\mathbb{C}^n$ such that

$$\chi = 0 \text{ on } \mathbb{C}^n \setminus V_1 \quad\text{and}\quad \chi = 1 \text{ on } V_2\,. \tag{3.1.24}$$

It follows from Construction 3.1.1, Lemma A (cp. (3.1.8)–(3.1.11)) that, for every fixed $z \in D$, there is a neighbourhood V_z of $N(\varrho)$ such that $\tilde{\Phi}(z, \zeta) \neq 0$ for all $\zeta \in (D \cap U) \cup V_z$. Together with (3.1.24) this implies that, for every fixed $z \in D$, $\chi(\zeta)\, w(z, \zeta)/\tilde{\Phi}(z, \zeta)$, $\zeta \in D \cup V_z$, is a C^1-map on $D \cup V_z$.

Consequently, for every fixed $z \in D$, the differential form

$$\omega_\zeta(\chi(\zeta)\, w(z, \zeta)/\tilde{\Phi}(z, \zeta)) := \bigwedge_{j=1}^{n} \mathrm{d}_\zeta(\chi(\zeta)\, w_j(z, \zeta)/\tilde{\Phi}(z, \zeta))$$

is continuous for $\zeta \in \bar{D}$. For every measurable bounded function f on D, we can therefore define

$$L_D f(z) := \frac{n!}{(2\pi i)^n} \int\limits_D f(\zeta)\, \omega_\zeta(\chi(\zeta)\, w(z, \zeta)/\tilde{\Phi}(z, \zeta)) \wedge \omega(\zeta)\,, \qquad z \in D\,. \tag{3.1.25}$$

Since $w(z, \zeta)$ and $\tilde{\Phi}(z, \zeta)$ are holomorphic in z, then $L_D f$ is holomorphic in D.

Further, setting

$$\tilde{\eta}_j(z, \zeta, \lambda) := (1 - \lambda)\frac{\chi(\zeta)\, w_j(z, \zeta)}{\tilde{\Phi}(z, \zeta)} + \lambda \frac{\bar{\zeta}_j - \bar{z}_j}{|\zeta - z|^2}, \qquad \tilde{\eta} := (\tilde{\eta}_1, \dots, \tilde{\eta}_n)\,,$$

$$\overline{\omega}(\tilde{\eta}(z, \zeta, \lambda)) := \bigwedge_{j=1}^{n} (\bar{\partial}_{z,\zeta} + \mathrm{d}_\lambda)\, \eta_j(z, \zeta, \lambda)\,,$$

we obtain a continuous differential form $\overline{\omega}(\tilde{\eta}(z, \zeta, \lambda))$ defined for $z \in D$, $\zeta \in \bar{D} \setminus z$, $0 \leq \lambda \leq 1$. The forms $\bar{\partial}_{(\zeta, z)} \dfrac{\bar{\zeta}_j - \bar{z}_j}{|\zeta - z|^2}$ have a singularity of order 2 at $\zeta = z$, and the forms $\bar{\partial}_{\zeta, z} \dfrac{w_j(z, \zeta)}{\tilde{\Phi}(z, \zeta)}$ are continuous at $\zeta = z$.

Hence the monomials in $\overline{\omega}(\tilde{\eta}(z, \zeta, \lambda))$ which are of degree 1 in λ have a singularity of order $\leqq 2n - 1$ at $\zeta = z$. Since only such monomials contribute to the integral (3.1.26) below, for every measurable bounded differential form f on D, we can therefore define the differential form

$$R_D f(z) := \frac{n!}{(2\pi i)^n} \int_{(\zeta,\lambda)\in D\times[0,1]} f(\zeta) \wedge \overline{\omega}(\tilde{\eta}(z, \zeta, \lambda)) \wedge \omega(\zeta)\,, \qquad z \in D\,. \tag{3.1.26}$$

Then it is easy to see that $R_D f$ is continuous in D, and $R_D f$ is a $(0, q - 1)$-form if f is a $(0, q)$-form ($R_D f = 0$ if f is a function).

3.1.3. Theorem (Koppelman-Leray formula). *We use the notations from Definition 3.1.2. Then*

(i) *for every continuous bounded function f on D such that $\bar{\partial} f$ is also continuous and bounded on D, we have*

$$f = L_D f + R_D \bar{\partial} f \quad in \quad D\,, \tag{3.1.27}$$

(ii) *for every continuous bounded $(0, q)$-form on D such that $\bar{\partial} f$ is also continuous and bounded in D, $1 \leqq q \leqq n$, we have*

$$f = \bar{\partial} R_D f + R_D \bar{\partial} f \quad in \quad D\,. \tag{3.1.28}$$

Proof. Let ϱ be the strictly plurisubharmonic function from Construction 3.1.1 which defines D. First we prove the theorem for the special case that $d\varrho(\zeta) \neq 0$ for $\zeta \in \partial D$ and f and $\bar{\partial} f$ are continuous on $\overline{D}$.

For $z \in D$, $\zeta \in \overline{D} \setminus z$, $0 \leqq \lambda \leqq 1$, consider the differential form

$$\overline{\omega}'(\tilde{\eta}(z, \zeta, \lambda)) := \sum_{j=1}^{n} (-1)^{j+1} \tilde{\eta}_j(z, \zeta, \lambda) \bigwedge_{k \neq j} (\bar{\partial}_{\zeta,z} + d_\lambda)\, \tilde{\eta}_k(z, \zeta, \lambda)\,.$$

This form is continuous for $z \in D$, $\zeta \in \overline{D} \setminus z$, $0 \leqq \lambda \leqq 1$. The monomials in $\overline{\omega}'(\tilde{\eta}(z, \zeta, \lambda))$ which are of degree 1 in λ have a singularity of order $\leqq 2n - 2$ at $\zeta = z$. Therefore, for every continuous bounded differential form f on D, the integral

$$\int_{(\zeta,\lambda)\in D\times[0,1]} f(\zeta) \wedge \overline{\omega}'(\tilde{\eta}(z, \zeta, \lambda)) \wedge \omega(\zeta)\,, \qquad z \in D\,,$$

converges and depends continuously differentiably on $z \in D$, where differentiation and integration can be interchanged. In particular, since $\dim_R (D \times [0, 1])$ is odd,

$$\bar{\partial}_z \int_{D\times[0,1]} f(\zeta) \wedge \overline{\omega}'(\tilde{\eta}(z, \zeta, \lambda)) \wedge \omega(\zeta) = - \int_{D\times[0,1]} \bar{\partial}_z[f(\zeta) \wedge \overline{\omega}'(\tilde{\eta}(z, \zeta, \lambda)) \wedge \omega(\zeta)]\,,$$
$$z \in D\,. \tag{3.1.29}$$

Now let f be a continuous $(0, q)$-form on $\overline{D}$ such that $\bar{\partial} f$ is also continuous on $\overline{D}$, $0 \leqq q \leqq n$. Since

$$(\bar{\partial}_{\zeta,z} + d_\lambda)\, \overline{\omega}'(\tilde{\eta}(z, \zeta, \lambda)) = n\overline{\omega}(\tilde{\eta}(z, \zeta, \lambda))\,,$$

we then obtain the relation

$$d_{\zeta,\lambda}[f(\zeta) \wedge \overline{\omega}'(\tilde{\eta}(z, \zeta, \lambda)) \wedge \omega(\zeta)] = \bar{\partial}_\zeta f(\zeta) \wedge \overline{\omega}'(\tilde{\eta}(z, \zeta, \lambda)) \wedge \omega(\zeta)$$
$$+ (-1)^q n f(\zeta) \wedge \overline{\omega}(\tilde{\eta}(z, \zeta, \lambda)) \wedge \omega(\zeta) - \bar{\partial}_z[f(\zeta) \wedge \overline{\omega}'(\tilde{\eta}(z, \zeta, \lambda)) \wedge \omega(\zeta)]\,.$$

Since the singularities at $\zeta = z$ of the monomials in $\overline{\omega}'(\tilde{\eta}(z, \zeta, \lambda))$ which are of degree 1 in λ are of order $\leqq 2n - 2$, we can apply Stokes' formula and, taking into account that $\partial(D \times [0, 1]) = \partial D \times [0, 1] - D \times 0 + D \times 1$ ($\dim_{\mathbb{R}} D$ is even!), we obtain

$$\int\limits_{D\times[0,1]} \bar{\partial}_\zeta f(\zeta) \wedge \overline{\omega}'(\tilde{\eta}(z, \zeta, \lambda)) \wedge \omega(\zeta) + (-1)^q \frac{(2\pi i)^n}{(n-1)!} R_D f(z)$$

$$- \int\limits_{D\times[0,1]} \bar{\partial}_z[f(\zeta) \wedge \overline{\omega}'(\tilde{\eta}(z, \zeta, \lambda)) \wedge \omega(\zeta)] = \int\limits_{\partial D\times[0,1]} f(\zeta) \wedge \overline{\omega}'(\tilde{\eta}(z, \zeta, \lambda)) \wedge \omega(\zeta)$$

$$- \int\limits_{D\times 0} f(\zeta) \wedge \overline{\omega}'(\tilde{\eta}(z, \zeta, \lambda)) \wedge \omega(\zeta) + \int\limits_{D\times 1} f(\zeta) \wedge \overline{\omega}'(\tilde{\eta}(z, \zeta, \lambda)) \wedge \omega(\zeta), \qquad z \in D. \tag{3.1.30}$$

If $\zeta \in \partial D$, then, by Construction 3.1.1, $\tilde{\Phi}(z, \zeta) = \Phi(z, \zeta) = \langle w(z, \zeta), \zeta - z\rangle$ and, therefore,

$$\tilde{\eta}(z, \zeta, \lambda) = (1 - \lambda) \frac{w(z, \zeta)}{\langle w(z, \zeta), \zeta - z\rangle} + \lambda \frac{\bar{\zeta} - \bar{z}}{|\zeta - z|^2}.$$

Consequently,

$$\int\limits_{\partial D\times[0,1]} f(\zeta) \wedge \overline{\omega}'(\tilde{\eta}(z, \zeta, \lambda)) \wedge \omega(\zeta) = \frac{(2\pi i)^n}{(n-1)!} R^w_{\partial D} f(z), \qquad z \in D, \tag{3.1.31}$$

where $R^w_{\partial D} f$ was introduced in Section 1.8. Taking into account that $\tilde{\eta}(z, \zeta, 1) = (\bar{\zeta} - \bar{z})/|\zeta - z|^2$ and using the same arguments as in the proof of Proposition 1.7.3, we obtain

$$\int\limits_{D\times 1} f(\zeta) \wedge \overline{\omega}'(\tilde{\eta}(z, \zeta, \lambda)) \wedge \omega(\zeta) = \frac{(2\pi i)^n}{(n-1)!} B_D f(z), \qquad z \in D, \tag{3.1.32}$$

where $B_D f$ was introduced in Section 1.8. Since $\tilde{\eta}(z, \zeta, 0) = \chi(\zeta)\, w(z, \zeta)/\tilde{\Phi}(z, \zeta)$ is holomorphic in z, we further obtain that

$$\int\limits_{D\times 0} f(\zeta) \wedge \overline{\omega}'(\tilde{\eta}(z, \zeta, \lambda)) \wedge \omega(\zeta) = \int\limits_{D} f(\zeta) \wedge \omega'_\zeta\left(\frac{\chi(\zeta)\, w(z, \zeta)}{\tilde{\Phi}(z, \zeta)}\right) \wedge \omega(\zeta), \qquad z = D. \tag{3.1.33}$$

Using the abbreviation $T_q := (-1)^q (R^w_{\partial D} + B_D)$, we obtain from (3.1.29)–(3.1.33) that, for every continuous $(0, q)$-form f on $\overline{D}$ such that $\bar{\partial} f$ is also continuous on $\overline{D}$, $0 \leqq q \leqq n$, the following relation holds:

$$R_D f(z) = T_q f(z) + (-1)^{q+1} \frac{(n-1)!}{(2\pi i)^n} \left[\bar{\partial}_z \int\limits_{D\times[0,1]} f(\zeta) \wedge \overline{\omega}'(\tilde{\eta}(z, \zeta, \lambda)) \wedge \omega(\zeta)\right.$$

$$\left. + \int\limits_{D} f(\zeta) \wedge \omega'_\zeta\left(\frac{\chi(\zeta)\, w(z, \zeta)}{\tilde{\Phi}(z, \zeta)}\right) \wedge \omega(\zeta) + \int\limits_{D\times[0,1]} \bar{\partial}_\zeta f(\zeta) \wedge \overline{\omega}'(\tilde{\eta}(z, \zeta, \lambda)) \wedge \omega(\zeta)\right],$$

$$z \in D. \tag{3.1.34}$$

Applying this to $\bar{\partial} f$, it follows that

$$R_D\, \bar{\partial} f(z) = T_{q+1}\, \bar{\partial} f(z)$$

$$+ (-1)^q \frac{(n-1)!}{(2\pi i)^n} \left[\bar{\partial}_z \int\limits_{D\times[0,1]} \bar{\partial}_\zeta f(\zeta) \wedge \overline{\omega}'(\tilde{\eta}(z, \zeta, \lambda)) \wedge \omega(\zeta)\right.$$

$$\left. + \int\limits_{D} \bar{\partial}_\zeta f(\zeta) \wedge \omega'_\zeta\left(\frac{\chi(\zeta)\, w(z,\zeta)}{\tilde{\Phi}(z, \zeta)}\right) \wedge \omega(\zeta)\right], \qquad z \in D, \tag{3.1.35}$$

for every continuous $(0, q)$-form f on $\bar{D}$ such that $\bar\partial f$ is also continuous on $\bar{D}$, $0 \leqq q \leqq n$. Further, by degree reasons for every such form,

$$\int_D f(\zeta) \wedge \omega'_\zeta\left(\frac{\chi(\zeta)\, w(z,\zeta)}{\tilde{\Phi}(z,\zeta)}\right) \wedge \omega(\zeta) = 0 \qquad (z \in D) \quad \text{if} \quad q \neq 1\,, \tag{3.1.36}$$

and, since $w(z, \zeta)$ and $\tilde{\Phi}(z, \zeta)$ are holomorphic in z,

$$\bar\partial_z \int_D f(\zeta) \wedge \omega'_\zeta\left(\frac{\chi(\zeta)\, w(z,\zeta)}{\tilde{\Phi}(z,\zeta)}\right) \wedge \omega(\zeta) = 0 \qquad (z \in D) \quad \text{if} \quad q = 1\,. \tag{3.1.37}$$

If follows from (3.1.34), (3.1.36), and (3.1.37) that, for every continuous $(0, q)$-form f on $\bar{D}$ such that $\bar\partial f$ is also continuous on $\bar{D}$, $0 \leqq q \leqq n$,

$$\begin{aligned} \bar\partial R_D f(z) &= \bar\partial T_q f(z) \\ &\quad + (-1)^{q+1} \frac{(n-1)!}{(2\pi i)^n} \bar\partial_z \int_{D\times[0,1]} \bar\partial_\zeta f(\zeta) \wedge \bar\omega'(\tilde{\eta}(z,\zeta,\lambda)) \wedge \omega(\zeta)\,, \\ &\quad z \in D\,. \end{aligned} \tag{3.1.38}$$

For $q \geqq 1$ it follows from (3.1.36) that the last integral on the right-hand side of (3.1.35) is zero. Therefore, for $1 \leqq q \leqq n$, from (3.1.35) and (3.1.38) we obtain the relation

$$R_D \bar\partial f + \bar\partial R_D f = \bar\partial T_q f + T_{q+1} \bar\partial f\,.$$

In view of the Koppelman-Leray formula (1.12.7), this implies (3.1.28).

Consider the case $q = 0$, that is, let f be a continuous function on $\bar{D}$ such that $\bar\partial f$ is also continuous on $\bar{D}$. Then, in view of the Koppelman-Leray formula (1.12.1),

$$f = L^w_{\partial D} f + T_1 \bar\partial f\,, \tag{3.1.39}$$

where $L^w_{\partial D} f$ was introduced in Section 1.8. Since, for $\zeta \in \partial D$, $\chi(\zeta) = 1$ and $\tilde{\Phi}(z, \zeta) = \Phi(z, \zeta) = \langle w(z, \zeta), \zeta - z\rangle$, from Proposition 1.7.3 we obtain

$$L^w_{\partial D} f(z) = \frac{(n-1)!}{(2\pi i)^n} \int_{\partial D} f(\zeta)\, \omega'_\zeta\left(\frac{\chi(\zeta)\, w(z,\zeta)}{\tilde{\Phi}(z,\zeta)}\right) \wedge \omega(\zeta)\,. \tag{3.1.40}$$

Since

$$d_\zeta \omega'_\zeta\left(\frac{\chi(\zeta)\, w(z,\zeta)}{\tilde{\Phi}(z,\zeta)}\right) = n\omega_\zeta\left(\frac{\chi(\zeta)\, w(z,\zeta)}{\tilde{\Phi}(z,\zeta)}\right),$$

it follows from Stokes' formula that

$$L^w_{\partial D} f(z) = \frac{(n-1)!}{(2\pi i)^n}\left[\int_D \bar\partial_\zeta f(\zeta) \wedge \omega'_\zeta\left(\frac{\chi(\zeta)\, w(z,\zeta)}{\tilde{\Phi}(z,\zeta)}\right) \wedge \omega(\zeta)\right] + L_D f(z)\,, \qquad z \in D\,. \tag{3.1.41}$$

Further, since $q = 0$, by degree reasons,

$$\int_{D\times[0,1]} \bar\partial_\zeta f(\zeta) \wedge \bar\omega'(\tilde{\eta}(z,\zeta,\lambda)) \wedge \omega(\zeta) = 0\,, \qquad z \in D\,. \tag{3.1.42}$$

From (3.1.35), (3.1.42), (3.1.41), and (3.1.39) follows (3.1.27). Now we pass to the general case, when not necessarily $d\varrho(z) \neq 0$ for $z \in \partial D$, and f is an arbitrary continuous bounded $(0, q)$-form in D such that $\bar\partial f$ is also continuous and bounded in D.

In view of Morse's lemma (see, for example, MILNOR [1], § 2, Lemma A), there are real-valued real-linear functions φ_m in $\mathbb{C}^m$ such that $|\varphi_m| < 1/m$ on $\bar{\theta}$ (θ is the same neighbourhood of $N(\varrho)$ as in Construction 3.1.1), and for each of the functions $\varrho + \varphi_m$ there is no more than a finite number of points $\zeta \in \bar{\theta}$ such that $\mathrm{d}(\varrho + \varphi_m)(\zeta) = 0$. Then for every m we can choose ε_m so that $1/m < \varepsilon_m < 2/m$ and $\mathrm{d}(\varrho + \varphi_m)(\zeta) \neq 0$ for all $\zeta \in \bar{\theta}$ with $\varrho(\zeta) + \varphi_m(\zeta) = -\varepsilon_m$. Define $\varrho_m(\zeta) = \varrho(\zeta) + \varphi_m(\zeta) + \varepsilon_m$. Then the functions ϱ_m together with the first and second-order derivatives tend to ϱ uniformly on $\bar{\theta}$. (The second-order derivatives of ϱ_m and ϱ even coincide.) Therefore, we have the situation which is described in Construction 3.1.1, Lemma A (iv) and Lemma B (b). In view of the special choice of ϱ_m, moreover:

$$\mathrm{d}\varrho_m(z) \neq 0 \quad \text{for} \quad z \in \partial D_m; \tag{3.1.43}$$

$$D_m \subset\subset D; \tag{3.1.44}$$

$$\left.\begin{array}{l}\text{for every compact set } K \subseteq D \text{ there is an integer } m_K \text{ such that} \\ K \subseteq D_m \text{ for } m \geqq m_K.\end{array}\right\} \tag{3.1.45}$$

Let L_{D_m} and R_{D_m} be the operators defined by (3.1.25) and (3.1.26) after replacing w by w^m and $\tilde{\Phi}$ by $\tilde{\Phi}_m$. Since $w^m \to w$ and $\tilde{\Phi}_m \to \tilde{\Phi}$ together with the first-order derivatives uniformly in ζ, $z \in V_1 \cup D$, and since $\tilde{\Phi}_m(z, \zeta) \neq 0$ ($\zeta \in U \cap \bar{D}$, $z \in D$) and $\chi(\zeta) = 0$ ($\zeta \in D \setminus V_1$, $z \in D$), it follows that for every fixed $z \in D$

$$\lim_{m\to\infty} \omega_\zeta \left(\frac{\chi(\zeta)\, w^m(z, \zeta)}{\tilde{\Phi}_m(z, \zeta)} \right) = \omega_\zeta \left(\frac{\chi(\zeta)\, w(z, \zeta)}{\tilde{\Phi}(z, \zeta)} \right) \tag{3.1.46}$$

uniformly in $\zeta \in \bar{D}$. In view of (3.1.45), this implies that for every continuous bounded function f on D and each $z \in D$

$$\lim_{m\to\infty} L_{D_m} f(z) = L_D f(z) \,. \tag{3.1.47}$$

Similarly as in the case of (3.1.46) we obtain that for every fixed $z \in D$

$$\begin{aligned} &\lim_{m\to\infty} |\zeta - z|^{2n-1}\, \bar{\omega} \left((1 - \lambda) \frac{\chi(\zeta)\, w^m(z, \zeta)}{\tilde{\Phi}_m(z, \zeta)} + \lambda \frac{\bar{\zeta} - \bar{z}}{|\zeta - z|^2} \right) \\ &= |\zeta - z|^{2n-1}\, \bar{\omega} \left((1 - \lambda) \frac{\chi(\zeta)\, w(z, \zeta)}{\tilde{\Phi}(z, \zeta)} + \lambda \frac{\bar{\zeta} - \bar{z}}{|\zeta - z|^2} \right) \end{aligned} \tag{3.1.48}$$

uniformly in $\zeta \in \bar{D}$ and $0 \leqq \lambda \leqq 1$. Since the singularity $1/|\zeta - z|^{2n-1}$ is integrable and independent of m, this together with (3.1.44) and (3.1.45) implies

$$\lim_{m\to\infty} R_{D_m} f(z) = R_D f(z) \tag{3.1.49}$$

for every continuous bounded differential form f on D and every fixed $z \in D$. Since for every D_m the theorem is already proved, from (3.1.47) and (3.1.49) part (i) of the theorem follows.

To prove (ii) we first observe that (3.1.48) not only holds for every fixed $z \in D$ but uniformly on every fixed compact subset of D. Consequently, (3.1.49) also holds uniformly on every compact subset of D. Since the theorem is already proved for every D_m, this implies

$$\lim_{m\to\infty} \bar{\partial} R_{D_m} f = f - \lim_{m\to\infty} R_{D_m} \bar{\partial} f = f - R_D\, \bar{\partial} f \tag{3.1.50}$$

uniformly on every compact subset of D and for every continuous bounded $(0, q)$-form f on D such that $\bar\partial f$ is also continuous and bounded on D. Taking into account that (3.1.49) and (3.1.50) hold uniformly on every compact subset of D, we conclude that $\bar\partial R_D f$ is continuous and (3.1.28) holds. ■

3.1.4. Corollary. *Using the notations introduced in Definition* 3.1.2, *we have:*

(i) *For every bounded holomorphic function f on D the following representation formula (Leray formula) holds:*

$$f = L_D f \quad in \quad D\,. \tag{3.1.51}$$

(ii) *If $1 \leqq q \leqq n$, then for every bounded continuous $(0, q)$-form f on D, such that $\bar\partial f = 0$ in D,*

$$u := R_D f \tag{3.1.52}$$

is a continuous $(0, q - 1)$-form on D which solves the equation $\bar\partial u = f$.

3.2. Uniform estimates for the $\bar\partial$-equation

In this section we prove that the solution of $\bar\partial u = f$ given by Corollary 3.1.4 (ii) admits uniform estimates. Moreover, we shall give more precise estimates by taking into account the *Hörmander diameter* of $\operatorname{supp} f$.

3.2.1. The Hörmander diameter. Let D, ϱ, θ be as in the beginning of Construction 3.1.1. For every $\xi \in \partial D$ and $\delta > 0$ we define (cp. also Subsection 1.5.13)

$$T_\xi := \left\{\zeta \in \mathbb{C}^n : \sum_{j=1}^{n} \frac{\partial\varrho(\xi)}{\partial\xi_j}(\zeta_j - \xi_j) = 0\right\}, \qquad E_\xi(\delta) := \{\zeta \in \mathbb{C}^n : |\zeta - \xi| < \delta\}\,,$$

$$\widetilde{H}_\xi(\delta) := E_\xi(\delta) \cap \{\zeta \in \mathbb{C}^n : \|d\varrho(\xi)\| \operatorname{dist}(\zeta, T_\xi) < \delta^2\}\,, \qquad H_\xi(\delta) := \widetilde{H}_\xi(\delta) \cap \overline{D}\,.$$

The set $H_\xi(\delta)$ will be called the *Hörmander ball* of radius δ centered at ξ. If $W \subseteqq D$ then the number

$$\operatorname{diam}_H W := \inf\{\delta > 0 : \text{There exists a } \xi \in \partial D \text{ with } W \subseteqq H_\xi(\delta)\}$$

will be called the *Hörmander diameter* of W.

Proposition. *Let $\Phi(z, \zeta)$ and $F(z, \zeta)$ be the functions introduced in Construction* 3.1.1. *Then there are constants $C < \infty$ and $\alpha > 0$ such that for all $\xi \in \partial D$ and $\delta > 0$*

$$\varrho(\zeta) < C\delta^2 \quad for \quad \zeta \in H_\xi(\delta) \cap \theta\,, \tag{3.2.1}$$

$$|F(z, \zeta) - F(z, \xi)| < C\delta^2 \quad for \quad z \in \overline{D}, \operatorname{dist}(z, H_\xi(\delta)) \leqq \delta \quad and$$
$$\zeta \in H_\xi(\delta) \cap \theta\,. \tag{3.2.2}$$

Proof. Since $\varrho(\xi) = 0$, it follows from Taylor's formula that for some $C_1 < \infty$ and all $\zeta \in \theta$

$$|\varrho(\zeta)| \leqq 2\left|\sum_{j=1}^{n} \frac{\partial\varrho(\xi)}{\partial\xi_j}(\zeta_j - \xi_j)\right| + C_1|\zeta - \xi|^2$$
$$\leqq C_1[\|d\varrho(\xi)\| \operatorname{dist}(\zeta, T_\xi) + |\zeta - \xi|^2]\,.$$

For $\zeta \in \theta \cap H_\xi(\delta)$ this implies (3.2.1) in view of the definition of $H_\xi(\delta)$. By definition of F, and since ϱ is a C^2-function, for some $C_2 < \infty$

$$|F(z,\zeta) - F(z,\xi)| \leqq C_2(|\zeta - \xi|\,|\zeta - z| + |\zeta - z|^2 + |\xi - z|^2)$$
$$+ 2\left|\sum_{j=1}^{n} \frac{\partial \varrho(\xi)}{\partial \xi_j}(\zeta_j - \xi_j)\right|, \qquad \zeta \in \theta\,, \qquad z \in \bar{D}\,.$$

If dist $(z, H_\xi(\delta)) \leqq \delta$ and therefore $|z - \xi| < 2\delta$, and if $\zeta \in H_\xi(\delta)$ and therefore $|\zeta - \xi| < \delta$, this implies that for some $C_3 < \infty$

$$|F(z,\zeta) - F(z,\xi)| \leqq C_3(\delta^2 + ||\mathrm{d}\varrho(\xi)||\,|\zeta - \xi|)\,.$$

This implies (3.2.2), because $|\zeta - \xi| \leqq$ dist (ζ, T_ξ) and $\zeta \in H_\xi(\delta)$ (cp. the definition of $H_\xi(\delta)$)

$$||\mathrm{d}\varrho(\xi)|| = \left(2\sum_{j=1}^{n}\left|\frac{\partial \varrho(\xi)}{\partial \xi_j}\right|^2\right)^{1/2}.$$

3.2.2. Theorem. *Let $D \subset\subset \mathbb{C}^n$ be a strictly pseudoconvex open set (with not necessarily smooth boundary), let χ, w, $\tilde{\Phi}$ be as in Definition* 3.1.2, *and let* $1 \leqq q \leqq n$. *Then* (for the definitions of the norm $||\cdot||_{0,D}$ and the Banach spaces $L^\infty_{(0,q)}(D)$ and $C^0_{(0,q-1)}(\bar{D})$ see Section 1.8):

(i) *For every $f \in L^\infty_{(0,q)}(D)$ and each point $z \in \bar{D}$, the integral*

$$R_D f(z) := \frac{n!}{(2\pi i)^n} \int\limits_{(\zeta,\lambda) \in \bar{D} \times [0,1]} f(\zeta) \wedge \bar{\omega}\left((1-\lambda)\frac{\chi(\zeta)\,w(z,\zeta)}{\tilde{\Phi}(z,\zeta)} + \lambda\frac{\bar{\zeta} - \bar{z}}{|\zeta - z|^2}\right) \wedge \omega(\zeta)$$

converges[1]), *and the so defined $(0, q-1)$-form $R_D f$ is continuous on $\bar{D}$.*

(ii) *There is a constant $C < \infty$ such that for every $f \in L^\infty_{(0,q)}(D)$*

$$||R_D f||_{0,D} \leqq C \operatorname{diam}_H (\operatorname{supp} f)\, ||f||_{0,D}\,, \tag{3.2.3}$$

and, moreover, if $\xi \in \partial D$, $\delta > 0$ such that $\operatorname{supp} f \subseteq H_\xi(\delta)$, *then for all $z \in \bar{D}$*

$$||R_D f(z)|| \leqq C\,[\delta/\text{dist}\,(z, H_\xi(\delta))]^{2n-1}\,\delta\,||f||_{0,D}\,. \tag{3.2.4}$$

(iii) *By parts* (i) *and* (ii) *the integral R_D defines a bounded linear operator from $L^\infty_{(0,q)}(D)$ into $C^0_{(0,q-1)}(\bar{D})$. This operator is compact*[2]).

3.2.3. Corollary. *Let $D \subset\subset \mathbb{C}^n$ be a strictly pseudoconvex open set (with not necessarily smooth boundary). Then there is a constant $C < \infty$ such that for every continuous bounded and $\bar{\partial}$-closed $(0, q)$-form f on D, the solution of $\bar{\partial}u = f$ in D, given by the formula $u := R_D f$* (cp. Corollary 3.1.4), *is continuous on $\bar{D}$ and admits the estimate*

$$||u||_{0,D} \leqq C \operatorname{diam}_H (\operatorname{supp} f)\, ||f||_{0,D}\,. \tag{3.2.5}$$

Moreover, if $\xi \in \partial D$, $\delta > 0$ such that $\operatorname{supp} f \subseteq H_\xi(\delta)$, *then for all $z \in \bar{D}$*

$$||u(z)|| \leqq C[\delta/\text{dist}\,(z, H_\xi(\delta))]^{2n-1}\,\delta\,||f||_{0,D}\,. \tag{3.2.5'}$$

Before proving Theorem 3.2.2, we give three lemmas which are also important for the subsequent sections of this chapter.

---

[1]) For $z \in D$ this integral was introduced in Definition 3.1.2, and in this case it is easy to see that it converges. Here we state the convergence also for $z \in \partial D$!

[2]) Recall that a linear operator between Banach spaces is called compact if every bounded set is mapped into a relatively compact set.

3.2.4. Lemma. *Let $D, \varrho, F, \Phi, \tilde{\Phi}$, and V_2 be as in Construction 3.1.1 and Definition 3.1.2. Then there exist real-valued quadratic polynomials $P(z, \zeta)$ and $Q(z, \zeta)$ in the real coordinates of ζ, whose coefficients are continuous functions in $z \in \overline{V}_2$, such that the following estimates hold:*

(i)

$$P(z, \zeta) = \operatorname{Im} F(z, \zeta) + o(|\zeta - z|^2) \quad \text{for} \quad \zeta, z \in V_2, \tag{3.2.6}$$

$$d_\zeta P(z, \zeta) = d_\zeta \operatorname{Im} F(z, \zeta) + O(|\zeta - z|) \quad \text{for} \quad \zeta, z \in V_2. \tag{3.2.7}$$

(ii)

$$Q(z, \zeta) = \varrho(\zeta) - \varrho(z) + o(|\zeta - z|^2) \quad \text{for} \quad \zeta, z \in V_2, \tag{3.2.8}$$

$$d_\zeta Q(z, \zeta) = d\varrho(\zeta) + O(|\zeta - z|) \quad \text{for} \quad \zeta, z \in V_2 \tag{3.2.9}$$

(iii)

$$\| d_\zeta F(z, \zeta) |_{\zeta = z} \wedge d\varrho(z) \| \geqq \frac{1}{\sqrt{n}} \| d\varrho(z) \|^2 \quad \text{for} \quad z \in V_2, \tag{3.2.10}$$

and for some constant $C < \infty$,

$$\| d_\zeta P(z, \zeta) \wedge d_\zeta Q(z, \zeta) \|$$

$$\geqq \frac{1}{\sqrt{n}} \| d\varrho(\zeta) \|^2 - C(\| d\varrho(\zeta) \| \, |\zeta - z| + |\zeta - z|^2) \quad \text{for} \quad \zeta, z \in V_2. \tag{3.2.10'}$$

(iv) *For some constant $\alpha > 0$,*

$$|\Phi(z, \zeta)| \geqq \alpha(|P(z, \zeta)| + |Q(z, \zeta)| + |\zeta - z|^2) \quad \text{for} \quad z \in V_2 \cap D, \zeta \in \partial D, \tag{3.2.11}$$

$$|\tilde{\Phi}(z, \zeta)| \geqq \alpha(|P(z, \zeta)| + |Q(z, \zeta)| + |\zeta - z|^2) \quad \text{for} \quad z, \zeta \in V_2 \cap D. \tag{3.2.12}$$

Proof. Let $x_j = x_j(\zeta)$ be the real coordinates of $\zeta \in \mathbb{C}^n$ so that $\zeta_j = x_j(\zeta) + i x_{j+n}(\zeta)$. Then

$$\operatorname{Im} F(z, \zeta) = \sum_{j=1}^{n} \frac{\partial \varrho(\zeta)}{\partial x_j} (x_{j+n}(\zeta) - x_{j+n}(z)) - \frac{\partial \varrho(\zeta)}{\partial x_{j+n}} (x_j(\zeta) - x_j(z))$$
$$+ \sum_{j,k=1}^{2n} u_{jk}(\zeta) (x_j(\zeta) - x_j(z)) (x_k(\zeta) - x_k(z)), \tag{3.2.13}$$

where u_{jk} are C^1-functions in a neighbourhood of $\overline{V}_2$. Define

$$P(z, \zeta) := \sum_{j=1}^{n} \left[\left\{ \frac{\partial \varrho(z)}{\partial x_j} + \sum_{s=1}^{2n} \frac{\partial^2 \varrho(z)}{\partial x_j \, \partial x_s} (x_s(\zeta) - x_s(z)) \right\} (x_{j+n}(\zeta) - x_{j+n}(z)) \right.$$
$$\left. - \left\{ \frac{\partial \varrho(z)}{\partial x_{j+n}} + \sum_{s=1}^{2n} \frac{\partial^2 \varrho(z)}{\partial x_{j+n} \, \partial x_s} (x_s(\zeta) - x_s(z)) \right\} (x_j(\zeta) - x_j(z)) \right]$$
$$+ \sum_{j,k=1}^{2n} u_{jk}(z) (x_j(\zeta) - x_j(z)) (x_k(\zeta) - x_k(z))$$

and

$$Q(z, \zeta) := \sum_{j=1}^{2n} \frac{\partial \varrho(z)}{\partial x_j} (x_j(\zeta) - x_j(z))$$
$$+ \frac{1}{2} \sum_{j,k=1}^{2n} \frac{\partial^2 \varrho(z)}{\partial x_j \, \partial x_k} (x_j(\zeta) - x_j(z)) (x_k(\zeta) - x_k(z)).$$

Then parts (i) and (ii) easily follow from Taylor's theorem.

We prove part (iii). By definition of $\|dF \wedge d\varrho\|$, we have

$$\|d_\zeta F(z,\zeta)\,|_{\zeta=z} \wedge d\varrho(z)\|^2$$

$$\geqq \sum_{1\leqq j\leqq k\leqq 2n} \left[\frac{\partial \operatorname{Im} F(z,\zeta)}{\partial x_j}\bigg|_{\zeta=z} \frac{\partial\varrho(z)}{\partial x_k} - \frac{\partial \operatorname{Im} F(z,\zeta)}{\partial x_k}\bigg|_{\zeta=z} \frac{\partial\varrho(z)}{\partial x_j}\right]^2.$$

In view of (3.2.13), for $j = 1, \dots, n$,

$$\frac{\partial \operatorname{Im} F(z,\zeta)}{\partial x_j}\bigg|_{\zeta=z} = \frac{\partial\varrho(z)}{\partial x_{j+n}} \quad \text{and} \quad \frac{\partial \operatorname{Im} F(z,\zeta)}{\partial x_{j+n}}\bigg|_{\zeta=z} = -\frac{\partial\varrho(z)}{\partial x_j}.$$

Taking the terms with $k = j + n$ only, this implies (3.2.10):

$$\|d_\zeta F(z,\zeta)|_{\zeta=z} \wedge d\varrho(z)\|^2 \geqq \sum_{j=1}^{n} \left[\left(\frac{\partial\varrho(z)}{\partial x_{j+n}}\right)^2 + \left(\frac{\partial\varrho(z)}{\partial x_j}\right)^2\right]^2$$

$$\geqq \frac{1}{n}\left[\sum_{j=1}^{2n} \left(\frac{\partial\varrho(z)}{\partial x_j}\right)^2\right]^2 = \frac{1}{n}\|d\varrho(z)\|^4.$$

Together with relations (3.2.7), (3.2.9), and $d\varrho(z) = d\varrho(\zeta) + O(|\zeta - z|)$, this implies (3.2.10′).

Finally observe that part (iv) follows from (3.2.6) and (3.2.8) and the estimates

$$|\Phi(z,\zeta)| \geqq \alpha(|\varrho(z)| + |\operatorname{Im} F(z,\zeta)| + |\zeta - z|^2) \qquad (z \in V_2 \cap D, \zeta \in \partial D),$$

$$|\tilde\Phi(z,\zeta)| \geqq \alpha(|\varrho(\zeta)| + |\varrho(z)| + |\operatorname{Im} F(z,\zeta)| + |\zeta - z|^2) \qquad (z, \zeta \in V_2 \cap D),$$

which hold in view of Construction 3.1.1, Lemma A, relations (3.1.8)–(3.1.11). ■

3.2.5. Lemma. *Let D, ϱ, $\tilde\Phi$, w, and V_2 be as in Construction* 3.1.1 *and Definition* 3.1.2. *Then there is a constant $C < \infty$ such that*

$$|w(z,\zeta)| \leqq C(\|d\varrho(\zeta)\| + |\zeta - z|) \quad \text{for} \quad \zeta, z \in V_2, \tag{3.2.14}$$

$$\left|\frac{\partial\tilde\Phi(z,\zeta)}{\partial\bar\zeta_j}\right| \leqq C\left(\left|\frac{\partial\varrho(\zeta)}{\partial\bar\zeta_j}\right| + |\zeta - z| + |\varrho(\zeta)|\right) \quad \text{for} \quad \zeta, z \in V_2; \qquad j = 1, \dots, n, \tag{3.2.15}$$

and for every differential form f

$$\|f(\zeta) \wedge \bar\partial_\zeta\tilde\Phi(z,\zeta)\| \leqq C[\|f(\zeta) \wedge \bar\partial\varrho(\zeta)\| + \|f(\zeta)\|\,(|\zeta - z| + |\varrho(\zeta)|)]$$
$$\text{for} \quad \zeta, z \in V_2. \tag{3.2.15′}$$

Proof. In view of Construction 3.1.1, Lemma A, condition (ii), and Lemma B, property (a)

$$\langle w(z,\zeta), \zeta - z\rangle = \Phi(z,\zeta) = F(z,\zeta)\,M(z,\zeta) \quad \text{for} \quad \zeta, z \in V_2 \quad \text{with} \quad |\zeta - z| \leqq \varepsilon.$$

Consequently, for $j = 1, \dots, n$ and $\zeta, z \in V_2$ with $|\zeta - z| \leqq \varepsilon$,

$$w_j(z,\zeta) + O(|\zeta - z|) = \frac{\partial F(z,\zeta)}{\partial\zeta_j} M(z,\zeta) + F(z,\zeta)\frac{\partial M(z,\zeta)}{\partial\zeta_j}.$$

Since $\partial F(z,\zeta)/\partial\zeta_j = 2\partial\varrho(\zeta)/\partial\zeta_j + O(|\zeta - z|)$, this implies (3.2.14). In view of Construction 3.1.1, Lemma A, condition (ii),

$$\tilde\Phi(z,\zeta) = (F(z,\zeta) - 2\varrho(\zeta))\,\tilde M(z,\zeta) \qquad (\zeta, z \in V_2; |\zeta - z| \leqq \varepsilon).$$

Since $\bar\partial_\zeta F(z,\zeta) = O(|\zeta - z|)$, this implies (3.2.15) and (3.2.15′). ■

3.2.6. Lemma. *Let $D, \varrho, \widetilde{\Phi}$, and V_2 be as in Construction* 3.1.1 *and Definition* 3.1.2. *Then there exists a constant $C < \infty$ such that for all $\xi \in \partial D, \delta > 0, z \in \bar{D} \cap V_2$, and $\varkappa = 1, 2$, the following estimates are fulfilled:*

$$\text{(i)} \qquad \int\limits_{\zeta \in H_\xi(\delta)} \frac{d\sigma_{2n}}{|\zeta - z|^{2n-\varkappa}} \leqq C\delta^\varkappa \min\{1, d^{2n-\varkappa}\},$$

$$\text{(ii)} \qquad \int\limits_{\zeta \in V_2 \cap H_\xi(\delta)} \frac{\|d\varrho(\zeta)\| \, d\sigma_{2n}}{|\widetilde{\Phi}(z, \zeta)| \, |\zeta - z|^{2n-1-\varkappa}} \leqq C\delta^\varkappa \min\{1, d^{2n+1-\varkappa}\},$$

$$\text{(iii)} \qquad \int\limits_{\zeta \in V_2 \cap H_\xi(\delta)} \frac{\|d\varrho(\zeta)\|^2 \, d\sigma_{2n}}{|\widetilde{\Phi}(z, \zeta)|^2 \, |\zeta - z|^{2n-2-\varkappa}} \leqq C\delta^\varkappa \min\{1, d^{2n+2-\varkappa}\},$$

where $d := \delta/\operatorname{dist}(z, H_\xi(\delta))$ and $d\sigma_{2n}$ is the Lebesgue measure in $\mathbb{C}^n$.

Proof. In this proof we denote all "large" constants by $C, C', \ldots$ and all "small" constants by $\alpha, \alpha', \ldots$. An expression of the form $a(x) \leqq Cb(x)$ must be read as follows: There exists a constant $C < \infty$ such that $a(x) \leqq Cb(x)$ for all x which are of interest in the corresponding situation.

(i). Since $H_\xi(\delta) \subseteqq E_\xi(\delta)$, it follows from Proposition 5 (i), Appendix 1, that

$$\int\limits_{H_\xi(\delta)} \frac{d\sigma_{2n}}{|\zeta - z|^{2n-\varkappa}} \leqq C\delta^\varkappa.$$

Moreover, it is clear that

$$\int\limits_{\zeta \in H_\xi(\delta)} \frac{d\sigma_{2n}}{|\zeta - z|^{2n-\varkappa}} \leqq \frac{1}{[\operatorname{dist}(z, H_\xi(\delta))]^{2n-\varkappa}} \int\limits_{H_\xi(\delta)} d\sigma_{2n} \leqq C \frac{\delta^{2n}}{[\operatorname{dist}(z, H_\xi(\delta))]^{2n-\varkappa}}.$$

(ii). Since for $\zeta \in H_\xi(\delta) \cap V_2$, $\|d\varrho(\zeta)\| \leqq d\varrho(\xi) + C\delta$, and, by (3.2.12), $|\widetilde{\Phi}(z, \zeta)| \geqq \alpha |\zeta - z|^2$, we have

$$\int\limits_{\zeta \in V_2 \cap H_\xi(\delta)} \frac{\|d\varrho(\zeta)\| \, d\sigma_{2n}}{|\widetilde{\Phi}(z, \zeta)| \, |\zeta - z|^{2n-1-\varkappa}} \leqq C \frac{\|d\varrho(\xi)\| + \delta}{[\operatorname{dist}(z, H_\xi(\delta))]^{2n+1-\varkappa}} \int\limits_{H_\xi(\delta)} d\sigma_{2n}.$$

Since

$$\int\limits_{H_\xi(\delta)} d\sigma_{2n} \leqq C \begin{cases} \delta^{2n} & \text{if } \|d\varrho(\xi)\| \leqq \delta, \\ \delta^{2n+2}/\|d\varrho(\xi)\|^2 & \text{if } \|d\varrho(\xi)\| \geqq \delta, \end{cases} \tag{3.2.16}$$

this implies that

$$\int\limits_{\zeta \in V_2 \cap H_\xi(\delta)} \frac{\|d\varrho(\zeta)\| \, d\sigma_{2n}}{|\widetilde{\Phi}(z, \zeta)| \, |\zeta - z|^{2n-1-\varkappa}} \leqq C \frac{\delta^{2n+1}}{[\operatorname{dist}(z, H_\xi(\delta))]^{2n+1-\varkappa}}.$$

This proves part (ii) for the case $d < 1$, that is, $\operatorname{dist}(z, H_\xi(\delta)) > \delta$. Now we consider the case

$$\operatorname{dist}(z, H_\xi(\delta)) \leqq \delta. \tag{3.2.17}$$

From (3.2.9) and (3.2.12) it follows

$$\int_{\zeta\in V_2\cap H_\xi(\delta)} \frac{\|d\varrho(\zeta)\|\, d\sigma_{2n}}{|\tilde{\Phi}(z,\zeta)|\, |\zeta - z|^{2n-1-\varkappa}}$$

$$\leqq C \int_{V_2\cap H_\xi(\delta)} \frac{\|d_\zeta Q(z,\zeta)\|\, d\sigma_{2n}}{(|Q(z,\zeta)| + |\zeta - z|^2)\, |\zeta - z|^{2n-1-\varkappa}} + C \int_{\zeta\in H_\xi(\delta)} \frac{d\sigma_{2n}}{|\zeta - z|^{2n-\varkappa}}.$$

The second integral on the right-hand side can be estimated in view of part (i). Since $|Q(z,\zeta)| \leqq C\,|\zeta - z|$, the first integral on the right-hand side can be estimated by the integral

$$\int_{V_2\cap H_\xi(\delta)} \frac{\|d_\zeta Q(z,\zeta)\|\, d\sigma_{2n}}{(|Q(z,\zeta)| + (|\zeta - z| + |Q(z,\zeta)|)^2)\, (|\zeta - z| + |Q(z,\zeta)|)^{2n-1-\varkappa}}.$$

In view of Proposition 4, Appenduix 1, this integral can further be estimated by

$$\int_{\substack{t\in R^{2n},\, |t|<R,\\ t_1\in Q(z,H_\xi(\delta))}} \frac{dt_1 \wedge \dots \wedge dt_{2n}}{(|t_1| + |t|^2)\, |t|^{2n-1-\varkappa}}, \tag{3.2.18}$$

where $R := \sup\limits_{\zeta\in H_\xi(\delta)\cap V_2} (|\zeta - z| + |Q(z,\zeta)|) \leqq C\delta$. In view of (3.2.17), it follows from (3.2.1) and (3.2.8) that for all $\zeta \in V_2 \cap H_\xi(\delta)$

$$|Q(z,\zeta) - Q(z,\xi)| \leqq |\varrho(\zeta)| + C\,|\zeta - \xi|^2 \leqq C\delta^2. \tag{3.2.19}$$

Consequently, (3.2.18) can be estimated by

$$\int_{\substack{t\in R^{2n},\, |t|<C\delta,\\ |t_1 - Q(z,\xi)|\leqq C\delta^2}} \frac{dt_1 \wedge \dots \wedge dt_{2n}}{(|t_1| + |t|^2)\, |t|^{2n-1-\varkappa}},$$

which is $\leqq C\delta^\varkappa$ as follows from Proposition 5 (ii), Appendix 1.

(iii). For dist $(z, H_\xi(\delta)) \geqq \delta$ we can use similar arguments as in the case of part (ii). We omit these arguments and consider the case (3.2.17). It follows from (3.2.10′) and (3.2.12) that

$$\int_{\zeta\in V_2\cap H_\xi(\delta)} \frac{\|d\varrho(\zeta)\|^2\, d\sigma_{2n}}{|\tilde{\Phi}(z,\zeta)|^2\, |\zeta - z|^{2n-2-\varkappa}}$$

$$\leqq C \int_{\zeta\in V_2\cap H_\xi(\delta)} \frac{\|d_\zeta P(z,\zeta) \wedge d_\zeta Q(z,\zeta)\|\, d\sigma_{2n}}{(|P(z,\zeta)| + |Q(z,\zeta)| + |\zeta - z|^2)^2\, |\zeta - z|^{2n-2-\varkappa}}$$

$$+ C \int_{V_2\cap GH_\xi(\delta)} \frac{\|d\varrho(\zeta)\|\, d\sigma_{2n}}{|\tilde{\Phi}(z,\zeta)|\, |\zeta - z|^{2n-1-\varkappa}} + C \int_{\zeta\in H_\xi(\delta)} \frac{d\sigma_{2n}}{|\zeta - z|^{2n-\varkappa}}.$$

The last two integrals can be estimated in view of parts (i) and (ii). Since

$$|P(z,\zeta)| + |Q(z,\zeta)| \leqq C\,|\zeta - z|,$$

the first integral on the right-hand side can be estimated by the integral

$$\int_{\zeta\in V_2\cap H_\xi(\delta)} \frac{\|d_\zeta P \wedge d_\zeta Q\|\, d\sigma_{2n}}{(|P| + |Q| + (|P| + |Q| + |\zeta - z|)^2)^2\, (|P| + |Q| + |\zeta - z|)^{2n-2-\varkappa}}.$$

In view of (3.2.19) and Proposition 4, Appendix 1, this integral can further be estimated by

$$\int\limits_{\substack{t\in R^{2n},\ |t|<C\delta,\\ |t_1-Q(z,\xi)|\leqq C\delta^2}} \frac{dt_1 \wedge \dots \wedge dt_{2n}}{(|t_1| + |t_2| + |t|^2)^2\, |t|^{2n-2-\varkappa}}.$$

The required estimate now follows from Proposition 5 (iii), Appendix 1. ■

3.2.7. Proof of Theorem 3.2.2. We shall use the notations C and α for "large" and "small" constants, respectively, as was explained in the beginning of the proof of Lemma 3.2.6.

I. Estimation of $R_D f(z)$. For $W \subseteqq \overline{D}$ and $f \in L^\infty_{(0,q)}(D)$ we write

$$I_W(z) := \left\| \int\limits_{W\times[0,1]} f(\zeta) \wedge \overline{\omega}\left[(1-\lambda)\frac{\chi(\zeta)\, w(z,\zeta)}{\widetilde{\Phi}(z,\zeta)} + \lambda \frac{\overline{\zeta}-\overline{z}}{|\zeta - z|^2}\right] \wedge \omega(\zeta) \right\|, \qquad z \in \overline{D}.$$

We want to prove that for all $\xi \in \partial D$, $\delta > 0$, $f \in L^\infty_{(0,q)}(D)$, and $z \in \overline{D}$

$$I_{H_\xi(\delta)}(z) \leqq C\, \|f\|_{0,D}\, \delta \min\{1, [\delta/\operatorname{dist}(z, H_\xi(\delta))]^{2n-1}\}. \tag{3.2.20}$$

Choose a neighbourhood V of ∂D such that $V \subset\subset V_2$. The monomials in the differential form in $I(z)$ which are of degree 1 in λ can be estimated by $1/|\zeta - z|^{2n-1}$ uniformly for $\zeta \in \overline{D}$ and $z \in \overline{D} \setminus V$ (cp. the arguments given in Definition 3.1.2). In view of Lemma 3.2.6 (i), this implies that

$$I_{H_\xi(\delta)}(z) \leqq C\, \|f\|_{0,D}\, \delta \min\{1, [\delta/\operatorname{dist}(z, H_\xi(\delta))]^{2n-1}\}, \quad z \in \overline{D} \setminus V. \tag{3.2.21}$$

By Construction 3.1.1, Lemma A, condition (ii), and estimate (3.1.8),

$$\inf_{\zeta\in D\setminus V_2,\, z\in\overline{D}} |\widetilde{\Phi}(z,\zeta)| > 0.$$

Further, since $V \subset\subset V_2$, $\inf\limits_{\zeta\in D\setminus V_2,\, z\in V} |\zeta - z| > 0$. Consequently,

$$\sup_{z\in\overline{D}\cap V} I_{H_\xi(\delta)\setminus V_2}(z) \leqq C\, \|f\|_{0,D} \int\limits_{E_\xi(\delta)\cap D} d\sigma_{2n}.$$

Since, clearly,

$$\int\limits_{E_\xi(\delta)\cap D} d\sigma_{2n} \leqq C\delta \min\{1, [\delta/\operatorname{dist}(z, H_\xi(\delta))]^{2n-1}\},$$

this together with (3.2.21) implies that for the proof of (3.2.20) it is sufficient to show that

$$I_{H_\xi(\delta)\cap V_2}(z) \leqq C\, \|f\|_{0,D}\, \delta \min\{1, [\delta/\operatorname{dist}(z, H_\xi(\delta))]^{2n-1}\}, \qquad z \in \overline{D} \cap V. \tag{3.2.22}$$

Taking into account that $\chi(\zeta) = 1$ for $\zeta \in V_2$, we obtain for $\zeta, z \in V_2 \cap \overline{D}$ with $\zeta \neq z$ (cp. Definition 1.7.6 for $\det_n$)

$$\overline{\omega}\left((1-\lambda)\frac{\chi(\zeta)\, w(z,\zeta)}{\widetilde{\Phi}(z,\zeta)} + \lambda\frac{\overline{\zeta}-\overline{z}}{|\zeta-z|^2}\right)$$
$$= \frac{1}{n!} \det_n\left((\overline{\partial}_{\zeta,z} + d_\lambda)\left[(1-\lambda)\frac{w(z,\zeta)}{\widetilde{\Phi}(z,\zeta)} + \lambda\frac{\overline{\zeta}-\overline{z}}{|\zeta-z|^2}\right]\right).$$

Since $\bar\partial_z w(z,\zeta) = 0$ and $\bar\partial_z \tilde\Phi(z,\zeta) = 0$, one has

$$(\bar\partial_{\zeta,z} + \mathrm{d}_\lambda)\left[(1-\lambda)\frac{w(z,\zeta)}{\tilde\Phi(z,\zeta)} + \lambda\frac{\bar\zeta - \bar z}{|\zeta - z|^2}\right]$$
$$= \left(\frac{\bar\zeta - \bar z}{|\zeta - z|^2} - \frac{w(z,\zeta)}{\tilde\Phi(z,\zeta)}\right)\mathrm{d}\lambda + (1-\lambda)\left(\frac{\bar\partial_\zeta w(z,\zeta)}{\tilde\Phi(z,\zeta)} - \frac{w(z,\zeta)\,\bar\partial_\zeta\tilde\Phi(z,\zeta)}{\tilde\Phi^2(z,\zeta)}\right)$$
$$+ \lambda\,\bar\partial_{\zeta,z}\frac{\bar\zeta - \bar z}{|\zeta - z|^2}.$$

Expanding the determinant (using the same arguments as in the proof of Lemma 2.2.1), we thus obtain for $\zeta, z \in V_2 \cap \bar D, \zeta \neq z$,

$$\bar\omega\left((1-\lambda)\frac{\chi(\zeta)\,w(z,\zeta)}{\tilde\Phi(z,\zeta)} + \lambda\frac{\bar\zeta - \bar z}{|\zeta - z|^2}\right)$$
$$= \sum_{s=0}^{n-2} p_s(\lambda)\,\mathrm{d}\lambda \wedge \det_{1,1,s,n-s-2}\left(\frac{\bar\zeta - \bar z}{|\zeta - z|^2}, \frac{w(z,\zeta)\,\bar\partial_\zeta\tilde\Phi(z,\zeta)}{\tilde\Phi^2(z,\zeta)}, \frac{\bar\partial_\zeta w(z,\zeta)}{\tilde\Phi(z,\zeta)}, \bar\partial_{\zeta,z}\frac{\bar\zeta - \bar z}{|\zeta - z|^2}\right)$$
$$+ \sum_{s=0}^{n-1} q_s(\lambda)\,\mathrm{d}\lambda \wedge \det_{1,s,n-s-1}\left(\frac{\bar\zeta - \bar z}{|\zeta - z|^2}, \frac{\bar\partial_\zeta w(z,\zeta)}{\tilde\Phi(z,\zeta)}, \bar\partial_{\zeta,z}\frac{\bar\zeta - \bar z}{|\zeta - z|^2}\right)$$
$$+ \sum_{s=0}^{n-1} r_s(\lambda)\,\mathrm{d}\lambda \wedge \det_{1,s,n-s-1}\left(\frac{w(z,\zeta)}{\tilde\Phi(z,\zeta)}, \frac{\bar\partial_\zeta w(z,\zeta)}{\tilde\Phi(z,\zeta)}, \bar\partial_{\zeta,z}\frac{\bar\zeta - \bar z}{|\zeta - z|^2}\right), \tag{3.2.23}$$

where p_s, q_s, r_s are some polynomials in λ. Observe that

$$\left\|\bar\partial_{\zeta,z}\frac{\bar\zeta - \bar z}{|\zeta - z|^2}\right\| \leqq C/|\zeta - z|^2 \quad \text{for} \quad \zeta, z \in V_2$$

and, by Construction 3.1.1, Lemma A, condition (ii), and estimate (3.1.8),

$$|\tilde\Phi(z,\zeta)| \geqq \alpha(|\zeta - z|^2 + |\varrho(\zeta)|) \quad \text{for} \quad \zeta, z \in V_2 \cap D\,.$$

Together with (3.2.14), (3.2.15) in Lemma 3.2.5, and (3.2.12) in Lemma 3.2.4 this implies that for $\zeta, z \in V_2 \cap \bar D$

$$\left\|\det_{1,1,s,n-s-2}\left(\frac{\bar\zeta - \bar z}{|\zeta - z|^2}, \frac{w(z,\zeta)\,\bar\partial_\zeta\tilde\Phi(z,\zeta)}{\tilde\Phi^2(z,\zeta)}, \frac{\bar\partial_\zeta w(z,\zeta)}{\tilde\Phi(z,\zeta)}, \bar\partial_{\zeta,z}\frac{\bar\zeta - \bar z}{|\zeta - z|^2}\right)\right\|$$
$$\leqq C\frac{(\|\mathrm{d}\varrho(\zeta)\| + |\zeta - z|)\,(\|\mathrm{d}\varrho(\zeta)\| + |\zeta - z| + |\varrho(\zeta)|)}{|\tilde\Phi(z,\zeta)|^{s+2}\,|\zeta - z|^{2n-2s-3}}$$
$$\leqq C'\frac{\|\mathrm{d}\varrho(\zeta)\|^2}{|\tilde\Phi(z,\zeta)|^2\,|\zeta - z|^{2n-3}} + C'\frac{\|\mathrm{d}\varrho(\zeta)\|}{|\tilde\Phi(z,\zeta)|\,|\zeta - z|^{2n-2}} + \frac{C'}{|\zeta - z|^{2n-1}},$$
$$\left\|\det_{1,s,n-s-1}\left(\frac{\bar\zeta - \bar z}{|\zeta - z|^2}, \frac{\bar\partial_\zeta w(z,\zeta)}{\tilde\Phi(z,\zeta)}, \bar\partial_{\zeta,z}\frac{\bar\zeta - \bar z}{|\zeta - z|^2}\right)\right\| \tag{3.2.23'}$$
$$\leqq \frac{C}{|\tilde\Phi(z,\zeta)|^s\,|\zeta - z|^{2n-2s-1}} \leqq \frac{C'}{|\zeta - z|^{2n-1}},$$
$$\left\|\det_{1,s,n-s-1}\left(\frac{w(z,\zeta)}{\tilde\Phi(z,\zeta)}, \frac{\bar\partial_\zeta w(z,\zeta)}{\tilde\Phi(z,\zeta)}, \bar\partial_{\zeta,z}\frac{\bar\zeta - \bar z}{|\zeta - z|^2}\right)\right\|$$
$$\leqq C\frac{\|\mathrm{d}\varrho(\zeta)\| + |\zeta - z|}{|\tilde\Phi(z,\zeta)|^{s+1}|\zeta - z|^{2n-2s-2}} \leqq C'\frac{\|\mathrm{d}\varrho(\zeta)\|}{|\tilde\Phi(z,\zeta)|\,|\zeta - z|^{2n-2}} + \frac{C'}{|\zeta - z|^{2n-1}}.$$

It follows from (3.2.23) and the last three estimates that for $z \in \bar{D} \cap V_2$

$$I_{H_\xi(\delta)\cap V_2}(z) \leqq C\,||f||_{0,D} \int\limits_{\zeta\in H_\xi(\delta)} \frac{d\sigma_{2n}}{|\zeta - z|^{2n-1}}$$

$$+ C\,||f||_{0,D} \int\limits_{\zeta\in V_2\cap H_\xi(\delta)} \frac{||d\varrho(\zeta)||\,d\sigma_{2n}}{|\tilde{\Phi}(z,\zeta)|\,|\zeta - z|^{2n-2}} + C\,||f||_{0,D} \int\limits_{\zeta\in V_2\cap H_\xi(\delta)} \frac{||d\varrho(\zeta)||^2\,d\sigma_{2n}}{|\tilde{\Phi}(z,\zeta)|^2\,|\zeta - z|^{2n-3}}.$$

Now estimate (3.2.22) (and therefore (3.2.20)) follows in view of Lemma 3.2.6.

II. End of the proof of Theorem 3.2.2. In view of (3.2.20), the integral $R_D f(z)$ converges for all $z \in \bar{D}$ and the estimates (3.2.3) and (3.2.4) hold. Therefore, it remains to prove that $R_D f$ is continuous on $\bar{D}$ and that the operator $R_D\colon L^\infty_{(0,q)}(D) \to C^0_{(0,q-1)}(\bar{D})$ is compact.

For every $\tau > 0$ we choose a real C^∞-function χ_τ in $\mathbb{C}^n$ such that $0 \leqq \chi_\tau(\zeta) \leqq 1$ for all $\zeta \in \mathbb{C}^n$, $\chi_\tau(\zeta) = 1$ for $|\zeta| \leqq \tau/2$, and $\chi_\tau(\zeta) = 0$ for $|\zeta| \geqq \tau$. Define for $f \in L^\infty_{(0,q)}(D)$, $\tau > 0$, and $z \in \bar{D}$

$$R^\tau_D f(z) := \int\limits_{(\zeta,\lambda)\in\bar{D}\times[0,1]} \chi_\tau(\zeta - z)\, f(\zeta)\, \bar{\omega}\left((1-\lambda)\frac{\chi(\zeta)\,w(z,\zeta)}{\tilde{\Phi}(z,\zeta)} + \frac{\bar{\zeta}-\bar{z}}{|\zeta - z|^2}\right) \wedge \omega(\zeta).$$

Then for every $\tau > 0$ the kernel of the integral operator $R^\tau_D - \dfrac{(2\pi i)^n}{n!} R_D$ is continuous for $\zeta, z \in \bar{D}$. Therefore, this operator is compact from $L^\infty_{(0,q)}(D)$ into $C^0_{(0,q-1)}(\bar{D})$. Consequently, the proof of the theorem will be completed if we show that

$$\lim_{\tau\to 0} \sup_{f\in L^\infty_{(0,q)}(D),\ ||f||_{0,D}=1} ||R^\tau_D f||_{0,D} = 0. \tag{3.2.24}$$

It follows from the definition of $H_\xi(\delta)$ that for all $\xi \in \partial D$ and $0 < \delta < 1$

$$\bar{D} \cap E_\xi(\delta) \subseteqq H_\xi(\sqrt{r\delta}), \tag{3.2.25}$$

where $r := \max\limits_{\xi\in\partial D} ||d\varrho(\xi)|| + 1$. For $\sigma > 0$ we set $U(\sigma) := \{z \in \bar{D}\colon \operatorname{dist}(z, \partial D) < \sigma\}$. Then for all $\tau, \sigma > 0$, $f \in L^\infty_{(0,q)}(D)$, and $z \in U(\sigma)$ there is some $\xi \in \partial D$ such that

$$\operatorname{supp} \chi_\tau(\zeta - z)\, f(\zeta) \subseteqq E_z(\tau) \subseteqq E_\xi(\tau + \sigma).$$

In view of (3.2.25) and (3.2.20) this implies that for every $f \in L^\infty_{(0,q)}(D)$ and all $\tau, \sigma > 0$ with $\tau + \sigma < 1$

$$\sup_{z\in U(\sigma)} ||R^\tau_D f(z)|| \leqq C\,||f||_{0,D} \sqrt{\tau + \sigma}. \tag{3.2.26}$$

On the other hand, for every fixed $\sigma > 0$, the monomials in the differential form

$$\bar{\omega}\left((1-\lambda)\frac{\chi(\zeta)\,w(z,\zeta)}{\tilde{\Phi}(z,\zeta)} + \lambda\frac{\bar{\zeta}-\bar{z}}{|\zeta - z|^2}\right),$$

which are of degree 1 in λ, can be estimated uniformly in $\zeta \in \bar{D}$, $z \in D \setminus U(\sigma)$ by $C/|\zeta - z|^{2n-1}$. Hence for every fixed $\sigma > 0$

$$\lim_{\tau\to 0} \sup_{\substack{f\in L^\infty_{(0,q)}(D),\\ ||f||_{0,D}=1}} \sup_{z\in D\setminus U(\sigma)} ||R^\tau_D f(z)|| = 0.$$

Together with (3.2.26) this implies (3.2.24). ■

3.3. Improvement of the estimates on the boundary

In the previous section we saw that the solution of $\bar\partial u = f$ given by Corollary 3.1.4 admits uniform estimates and can be continued continuously to the boundary (Corollary 3.2.3). In the present section we give a more precise estimate for the boundary values of this solution. These estimates will be used in the next section to prove a theorem on decomposition of singularities. For this purpose it is sufficient to consider (0, 1)-forms. In the present section we restrict ourselves to this case.

Notation. Throughout this section let $D, \varrho, \Phi, \tilde\Phi, w, \chi, V_2$ be as in Construction 3.1.1 and Definition 3.1.2. We define

$$\Phi^*(z,\zeta) := \Phi(\zeta, z)\,, \qquad w^*(z,\zeta) := -w(\zeta, z) \quad \text{for} \quad z \in V_2 \quad \text{and} \quad \zeta \in V_2 \cup D.$$

In view of Construction 3.1.1, Lemma A, condition (ii), and estimate (3.1.8),

$$\Phi^*(z,\zeta) \neq 0 \quad \text{and} \quad \tilde\Phi(z,\zeta) \neq 0 \quad \text{for} \quad z \in \partial D\,, \quad \zeta \in D\,. \tag{3.3.1}$$

Therefore, for every fixed point $z \in \partial D$ the differential form

$$\bar\omega\left((1-\lambda)\frac{\chi(\zeta)\,w(z,\zeta)}{\tilde\Phi(z,\zeta)} + \lambda\frac{w^*(z,\zeta)}{\Phi^*(z,\zeta)}\right)$$
$$:= \bigwedge_{j=1}^{n}(\bar\partial_\zeta + d_\lambda)\left((1-\lambda)\frac{\chi(\zeta)\,w_j(z,\zeta)}{\tilde\Phi(z,\zeta)} + \lambda\frac{w_j^*(z,\zeta)}{\Phi^*(z,\zeta)}\right)$$

is continuous for $(\zeta, \lambda) \in D \times [0, 1]$.

3.3.1. Lemma. (i) *For every $f \in L^\infty_{(0,1)}(D)$ and all $z \in \partial D$ the integral*

$$R_D^*f(z) := \frac{n!}{(2\pi i)^n}\int\limits_{(\zeta,\lambda)\in D\times[0,1]} f(\zeta)\wedge\bar\omega\left((1-\lambda)\frac{\chi(\zeta)\,w(z,\zeta)}{\tilde\Phi(z,\zeta)} + \lambda\frac{w^*(z,\zeta)}{\Phi^*(z,\zeta)}\right)\wedge\omega(\zeta) \tag{3.3.2}$$

converges.

(ii) *There exists a constant $C < \infty$ with the following property: If $\xi \in \partial D$, $\delta > 0$, and $f \in L^\infty_{(0,1)}(D)$ such that*

$$\operatorname{supp} f \subseteqq H_\xi(\delta)^{1)}\,, \tag{3.3.3}$$

$$\|f(\zeta)\| \leqq 1/\delta + \|d\varrho(\zeta)\|/\delta^2 \quad \textit{for} \quad \zeta \in H_\xi(\delta) \cap V_2\,, \tag{3.3.4}$$

$$\|f(\zeta)\wedge\bar\partial\varrho(\zeta)\| \leqq \|d\varrho(\zeta)\|/\delta \quad \textit{for} \quad \zeta \in H_\xi(\delta) \cap V_2\,, \tag{3.3.5}$$

then for all $z \in \partial D$

$$\|R_D^*f(z)\| \leqq C\min\{1, [\delta/\operatorname{dist}(z, H_\xi(\delta))]^{2n-1}\}\,. \tag{3.3.6}$$

Proof. We shall use the notations $C, C', \ldots$ and $\alpha, \alpha', \ldots$ for "large" and "small" constants, respectively, in the sense explained at the beginning of the proof of Lemma 3.2.6.

It is sufficient to show that for every form $f \in L^\infty_{(0,1)}(D)$ with the properties (3.3.3) to (3.3.5) (for some $\xi \in \partial D$ and $\delta > 0$) the integral (3.3.2) converges and satisfies estimate (3.3.6). In the general case the convergence of (3.3.2) then follows by choosing δ so large that $D \subseteqq H_\xi(\delta)$. For this purpose, let $\delta > 0$, $\xi \in \partial D$, and $f \in L^\infty_{(0,1)}(D)$ so that

[1]) $H_\xi(\delta)$ was defined in Subsection 3.2.1.

(3.3.3)—(3.3.5) hold. For every $W \subseteqq D$ and $z \in \partial D$ we shall write

$$I_W(z) := \int\limits_{(\zeta,\lambda)\in W\times[0,1]} \left| f(\zeta) \wedge \overline{\omega}\left((1-\lambda)\frac{\chi(\zeta)\, w(z,\zeta)}{\tilde{\Phi}(z,\zeta)} + \lambda \frac{w^*(z,\zeta)}{\Phi^*(z,\zeta)}\right) \wedge \omega(\zeta)\right|.$$

We have to prove that

$$I_D(z) \leqq C \min\{1, [\delta/\operatorname{dist}(z, H_\xi(\delta))]^{2n-1}\}, \qquad z \in \partial D. \tag{3.3.7}$$

At first consider $I_{D\setminus V_2}(z)$. Since $\Phi^*(z,\zeta) = \Phi(\zeta, z)$, it follows from Construction 3.1.1, Lemma A, condition (ii), and estimate (3.1.8) that

$$\inf_{z\in\partial D,\, \zeta\in D\setminus V_2} \min\{|\tilde{\Phi}(z,\zeta)|, |\Phi^*(z,\zeta)|\} > 0.$$

Hence, by (3.3.3),

$$I_{D\setminus V_2}(z) \leqq C\, ||f||_{0,D} \int\limits_{H_\xi(\delta)} d\sigma_{2n}, \qquad z \in \partial D.$$

Since by definition of $H_\xi(\delta)$

$$\int\limits_{H_\xi(\delta)} d\sigma_{2n} \leqq C \min\{1, \delta^{2n}, \delta^{2n+2}/||d\varrho(\xi)||^2\},$$

this, together with (3.3.4), implies that for all $z \in \partial D$

$$I_{D\setminus V_2}(z) \leqq C \min\{1, \delta^{2n-1}\} \leqq C' \min\{1, [\delta/\operatorname{dist}(z, H_\xi(\delta))]^{2n-1}\}.$$

To complete the proof of (3.3.7), therefore, it remains to estimate the integral $I_{D\cap V_2}(z)$. Using determinants (cp. Definition 1.7.6) and taking into account that $\chi = 1$ on V_2, we obtain for $\zeta \in V_2 \cap D$ and $z \in \partial D$

$$\overline{\omega}\left((1-\lambda)\frac{\chi(\zeta)\, w(z,\zeta)}{\tilde{\Phi}(z,\zeta)} + \lambda\frac{w^*(z,\zeta)}{\Phi^*(z,\zeta)}\right)$$

$$= \frac{1}{n!}\det_n\left((d_\lambda + \bar\partial_\zeta)\left[(1-\lambda)\frac{w(z,\zeta)}{\tilde{\Phi}(z,\zeta)} + \lambda\frac{w^*(z,\zeta)}{\Phi^*(z,\zeta)}\right]\right).$$

Since $\bar\partial_\zeta w^*(z,\zeta) = 0$ and $\bar\partial_\zeta \Phi^*(z,\zeta) = 0$, we have for $\zeta \in V_2 \cap D$ and $z \in \partial D$

$$(d_\lambda + \bar\partial_\zeta)\left[(1-\lambda)\frac{w(z,\zeta)}{\tilde{\Phi}(z,\zeta)} + \lambda\frac{w^*(z,\zeta)}{\Phi^*(z,\zeta)}\right]$$

$$= \left(\frac{w^*(z,\zeta)}{\Phi^*(z,\zeta)} - \frac{w(z,\zeta)}{\tilde{\Phi}(z,\zeta)}\right) d\lambda + (1-\lambda)\left(\frac{\bar\partial_\zeta w(z,\zeta)}{\tilde{\Phi}(z,\zeta)} - \frac{w(z,\zeta)\, \bar\partial_\zeta\tilde{\Phi}(z,\zeta)}{\tilde{\Phi}^2(z,\zeta)}\right).$$

Expanding the determinant (using the same arguments as in the proof of Lemma 2.2.1), we thus obtain for $z \in \partial D$ and $\zeta \in V_2 \cap D$

$$\overline{\omega}\left((1-\lambda)\frac{w(z,\zeta)}{\tilde{\Phi}(z,\zeta)} + \lambda\frac{w^*(z,\zeta)}{\Phi^*(z,\zeta)}\right) = p(\lambda)\, d\lambda \wedge \det_{1,n-1}\left(\frac{w(z,\zeta)}{\tilde{\Phi}(z,\zeta)}, \frac{\bar\partial_\zeta w(z,\zeta)}{\tilde{\Phi}(z,\zeta)}\right)$$

$$+ q(\lambda)\, d\lambda \wedge \det_{1,1,n-2}\left(\frac{w^*(z,\zeta)}{\Phi^*(z,\zeta)}, \frac{w(z,\zeta)\, \bar\partial_\zeta\tilde{\Phi}(z,\zeta)}{\tilde{\Phi}^2(z,\zeta)}, \frac{\bar\partial_\zeta w(z,\zeta)}{\tilde{\Phi}(z,\zeta)}\right), \tag{3.3.8}$$

where $p(\lambda)$, $q(\lambda)$ are some polynomials in λ. From (3.2.14), (3.2.12), (3.3.3), and (3.3.4) it follows that for $z \in \partial D$ and $\zeta \in V_2 \cap D$

$$\left\| f(\zeta) \wedge \det_{1,n-1}\left(\frac{w(z,\zeta)}{\tilde{\Phi}(z,\zeta)}, \frac{\bar{\partial}_\zeta w(z,\zeta)}{\tilde{\Phi}(z,\zeta)}\right)\right\| \leqq C\left(\frac{1}{\delta} + \frac{\|d\varrho(\zeta)\|}{\delta^2}\right)\left(\frac{\|d\varrho(\zeta)\|}{|\tilde{\Phi}(z,\zeta)|^n} + \frac{|\zeta - z|}{|\tilde{\Phi}(z,\zeta)|^n}\right)$$

$$\leqq \frac{C'}{\delta}\frac{\|d\varrho(\zeta)\|}{|\tilde{\Phi}(z,\zeta)|\,|\zeta - z|^{2n-2}} + \frac{C'}{\delta}\frac{1}{|\zeta - z|^{2n-1}}$$

$$+ \frac{C'}{\delta^2}\frac{\|d\varrho(\zeta)\|^2}{|\tilde{\Phi}(z,\zeta)|^2\,|\zeta - z|^{2n-4}} + \frac{C'}{\delta^2}\frac{\|d\varrho(\zeta)\|}{|\tilde{\Phi}(z,\zeta)|\,|\zeta - z|^{2n-3}}. \tag{3.3.9}$$

Since $w^*(z, z) = -w(z, z)$ and therefore, $w^*(z, \zeta) = -w(z, \zeta) + O(|\zeta - z|)$, we have for $z \in \partial D$, $\zeta \in V_2 \cap D$

$$\left\| f(\zeta) \wedge \det_{1,1,n-2}\left(\frac{w^*(z,\zeta)}{\Phi^*(z,\zeta)}, \frac{w(z,\zeta)\,\bar{\partial}_\zeta\tilde{\Phi}(z,\zeta)}{\tilde{\Phi}^2(z,\zeta)}, \frac{\bar{\partial}_\zeta w(z,\zeta)}{\tilde{\Phi}(z,\zeta)}\right)\right\|$$

$$\leqq C\frac{|\zeta - z|\,|w(z,\zeta)|\,\|f(\zeta) \wedge \bar{\partial}_\zeta\tilde{\Phi}(z,\zeta)\|}{|\Phi^*(z,\zeta)|\,|\tilde{\Phi}(z,\zeta)|^n}. \tag{3.3.10}$$

Since $|\varrho(\zeta)| \leqq C\,|\zeta - z|$ for $z \in \partial D$ and in view of (3.2.14) and (3.2.15′) we obtain for $z \in \partial D, \zeta \in V_2 \cap D$

$$|w(z,\zeta)|\,\|f(\zeta) \wedge \bar{\partial}_\zeta\tilde{\Phi}(z,\zeta)\|$$

$$\leqq C(\|d\varrho(\zeta)\| + |\zeta - z|)\,(\|f(\zeta) \wedge \bar{\partial}\varrho(\zeta)\| + \|f(\zeta)\|\,|\zeta - z|)\,.$$

In view of (3.3.3)–(3.3.5) we thus obtain for $z \in \partial D$, $\zeta \in V_2 \cap D$

$$|w(z,\zeta)|\,\|f(\zeta) \wedge \bar{\partial}_\zeta\tilde{\Phi}(z,\zeta)\|$$

$$\leqq C(\|d\varrho(\zeta)\| + |\zeta - z|)\left(\frac{\|d\varrho(\zeta)\|}{\delta} + \frac{|\zeta - z|}{\delta} + \frac{|\zeta - z|\,\|d\varrho(\zeta)\|}{\delta^2}\right)$$

$$\leqq \frac{C'}{\delta}(\|d\varrho(\zeta)\|^2 + |\zeta - z|\,\|d\varrho(\zeta)\| + |\zeta - z|^2)$$

$$+ \frac{C'}{\delta^2}(|\zeta - z|\,\|d\varrho(\zeta)\|^2 + |\zeta - z|^2\|d\varrho(\zeta)\|)\,.$$

Together with (3.3.10), (3.2.11), and (3.2.12) this implies that for $z \in \partial D$, $\zeta \in V_2 \cap D$

$$\left\| f(\zeta) \wedge \det_{1,1,n-2}\left(\frac{w^*(z,\zeta)}{\Phi^*(z,\zeta)}, \frac{w(z,\zeta)\,\bar{\partial}_\zeta\tilde{\Phi}(z,\zeta)}{\tilde{\Phi}^2(z,\zeta)}, \frac{\bar{\partial}_\zeta w(z,\zeta)}{\tilde{\Phi}(z,\zeta)}\right)\right\|$$

$$\leqq \frac{C}{\delta}\left[\frac{\|d\varrho(\zeta)\|^2}{|\tilde{\Phi}(z,\zeta)|^2\,|\zeta - z|^{2n-3}} + \frac{\|d\varrho(\zeta)\|}{|\tilde{\Phi}(z,\zeta)|\,|\zeta - z|^{2n-2}} + \frac{1}{|\zeta - z|^{2n-1}}\right]$$

$$+ \frac{C}{\delta^2}\left[\frac{\|d\varrho(\zeta)\|^2}{|\tilde{\Phi}(z,\zeta)|^2\,|\zeta - z|^{2n-4}} + \frac{\|d\varrho(\zeta)\|}{|\tilde{\Phi}(z,\zeta)|\,|\zeta - z|^{2n-3}}\right]. \tag{3.3.11}$$

It follows from (3.3.3), (3.3.8), (3.3.9), and (3.3.11) that for $z \in \partial D$

$$I_{D\cap V_z}(z) \leqq \frac{C}{\delta} \int\limits_{\zeta \in H_\xi(\delta)} \frac{\mathrm{d}\sigma_{2n}}{|\zeta - z|^{2n-1}} + \frac{C}{\delta} \int\limits_{\zeta \in H_\xi(\delta)\cap V_z} \frac{||\mathrm{d}\varrho(\zeta)||\, \mathrm{d}\sigma_{2n}}{|\tilde{\Phi}(z, \zeta)|\, |\zeta - z|^{2n-2}}$$
$$+ \frac{C}{\delta} \int\limits_{\zeta \in H_\xi(\delta)\cap V_z} \frac{||\mathrm{d}\varrho(\zeta)||^2\, \mathrm{d}\sigma_{2n}}{|\tilde{\Phi}(z, \zeta)|^2\, |\zeta - z|^{2n-3}} + \frac{C}{\delta^2} \int\limits_{\zeta \in H_\xi(\delta)\cap V_z} \frac{||\mathrm{d}\varrho(\zeta)||\, \mathrm{d}\sigma_{2n}}{|\tilde{\Phi}(z, \zeta)|\, |\zeta - z|^{2n-3}}$$
$$+ \frac{C}{\delta^2} \int\limits_{\zeta \in H_\xi(\delta)\cap V_z} \frac{||\mathrm{d}\varrho(\zeta)||^2\, \mathrm{d}\sigma_{2n}}{|\tilde{\Phi}(z, \zeta)|^2\, |\zeta - z|^{2n-4}}.$$

The required estimate now follows from Lemma 3.2.6. ■

Notation. Let $\Delta := \{\lambda = (\lambda_1, \lambda_2) : 0 \leqq \lambda_1, \lambda_2 \leqq 1, \lambda_1 + \lambda_2 = 1\}$ endowed with an arbitrary orientation. In view of estimates (3.2.11) and (3.2.12) in Lemma 3.2.4, for every fixed $z \in \partial D$ the differential form

$$\overline{\omega}\left(\lambda_1 \frac{\chi(\zeta)\, w(z, \zeta)}{\tilde{\Phi}(z, \zeta)} + \lambda_2 \frac{w^*(z, \zeta)}{\Phi^*(z, \zeta)} + (1 - \lambda_1 - \lambda_2) \frac{\overline{\zeta} - \overline{z}}{|\zeta - z|^2}\right)$$
$$:= \bigwedge_{j=1}^{n} (\overline{\partial}_\zeta + \mathrm{d}_\lambda) \left(\lambda_1 \frac{\chi(\zeta)\, w_j(z, \zeta)}{\tilde{\Phi}(z, \zeta)} + \lambda_2 \frac{w_j^*(z, \zeta)}{\Phi^*(z, \zeta)} + (1 - \lambda_1 - \lambda_2) \frac{\overline{\zeta}_j - \overline{z}_j}{|\zeta - z|^2}\right)$$

is continuous for $(\zeta, \lambda) \in (\overline{D} \setminus z) \times \Delta$.

3.3.2. Lemma. *For every $z \in \partial D$ there exists a constant $C < \infty$ such that for every bounded 1-form f on D and all $\delta > 0$*

$$I(\delta) := \Bigg| \int\limits_{\zeta \in D,\, |\zeta - z| = \delta,\, \lambda \in \Delta} f(\zeta) \wedge \overline{\omega}\left(\lambda_1 \frac{\tilde{w}(z, \zeta)}{\tilde{\Phi}(z, \zeta)} + \lambda_2 \frac{w^*(z, \zeta)}{\Phi^*(z, \zeta)} + (1 - \lambda_1 - \lambda_2) \frac{\overline{\zeta} - \overline{z}}{|\zeta - z|^2}\right) \wedge \omega(\zeta) \Bigg|$$
$$\leqq C\, ||f||_{0,D}\, \delta\, |\ln \delta|\,. \tag{3.3.12}$$

Proof. We shall use the notations $C, C', \ldots$ and $\alpha, \alpha', \ldots$ for "large" and "small" constants, respectively, in the sense explained at the beginning of the proof of Lemma 3.2.6.

Since $\overline{\partial}_\zeta w^*(z, \zeta) = 0$ and $\overline{\partial}_\zeta \Phi^*(z, \zeta) = 0$, and since $\chi(\zeta) = 1$ for $|\zeta - z| \leqq \delta$ if $z \in \partial D$ and δ is small enough, we have for all sufficiently small $\delta > 0$ (for the definition of the determinant cp. Definition 1.7.6)

$$\overline{\omega}\left(\lambda_1 \frac{\chi w}{\tilde{\Phi}} + \lambda_2 \frac{w^*}{\Phi^*} + (1 - \lambda_1 - \lambda_2) \frac{\overline{\zeta} - \overline{z}}{|\zeta - z|^2}\right)$$
$$= \frac{1}{n!} \det{}_n \left[\left(\frac{w}{\tilde{\Phi}} - \frac{\overline{\zeta} - \overline{z}}{|\zeta - z|^2}\right) \mathrm{d}\lambda_1 + \left(\frac{w^*}{\Phi^*} - \frac{\overline{\zeta} - \overline{z}}{|\zeta - z|^2}\right) \mathrm{d}\lambda_2 \right.$$
$$\left. + \lambda_1 \left(\frac{\overline{\partial}_\zeta w}{\tilde{\Phi}} - \frac{w\, \overline{\partial}_\zeta \tilde{\Phi}}{\tilde{\Phi}^2}\right) + (1 - \lambda_1 - \lambda_2)\, \overline{\partial}_\zeta \frac{\overline{\zeta} - \overline{z}}{|\zeta - z|^2}\right]$$
$$= \sum_{s=0}^{n-2} p_s(\lambda)\, \mathrm{d}\lambda_1 \wedge \mathrm{d}\lambda_2 \wedge \det{}_{1,1,s,n-2-s}\left(\frac{w}{\tilde{\Phi}}, \frac{w^*}{\Phi^*}, \frac{\overline{\partial}_\zeta w}{\tilde{\Phi}}, \overline{\partial}_\zeta \frac{\overline{\zeta} - \overline{z}}{|\zeta - z|^2}\right)$$
$$+ \sum_{s=0}^{n-2} q_s(\lambda)\, \mathrm{d}\lambda_1 \wedge \mathrm{d}\lambda_2 \wedge \det{}_{1,1,s,n-2-s}\left(\frac{w}{\tilde{\Phi}}, \frac{\overline{\zeta} - \overline{z}}{|\zeta - z|^2}, \frac{\overline{\partial}_\zeta w}{\tilde{\Phi}}, \overline{\partial}_\zeta \frac{\overline{\zeta} - \overline{z}}{|\zeta - z|^2}\right)$$

$$+ \sum_{s=0}^{n-2} r_s(\lambda)\, d\lambda_1 \wedge d\lambda_2 \wedge \det_{1,1,s,n-2-s}\left(\frac{w^*}{\Phi^*}, \frac{\bar\zeta - \bar z}{|\zeta - z|^2}, \frac{\bar\partial_\zeta w}{\tilde\Phi}, \bar\partial_\zeta \frac{\bar\zeta - \bar z}{|\zeta - z|^2}\right)$$

$$+ \sum_{s=0}^{n-3} t_s(\lambda)\, d\lambda_1 \wedge d\lambda_2 \wedge \det_{1,1,1,s,n-3-s}\left(\frac{w^*}{\Phi^*}, \frac{\bar\zeta - \bar z}{|\zeta - z|^2}, \frac{w\, \bar\partial_\zeta \tilde\Phi}{\tilde\Phi^2}, \frac{\bar\partial_\zeta w}{\tilde\Phi}, \bar\partial_\zeta \frac{\bar\zeta - \bar z}{|\zeta - z|^2}\right), \tag{3.3.13}$$

where p_s, q_s, r_s, t_s are some polynomials. Since $|\varrho(\zeta)| \leqq C\,|\zeta - z|$ for $z \in \partial D$,

$$\left\| \bar\partial_\zeta \frac{\bar\zeta - \bar z}{|\zeta - z|^2} \right\| \leqq C/|\zeta - z|^2,$$

and $w^*(z, \zeta) = -w(z, \zeta) + O(|\zeta - z|)$, and in view of estimates (3.2.14), (3.2.15) in Lemma 3.2.5, we have

$$\left\| \det_{1,1,s,n-2-s}\left(\frac{w}{\tilde\Phi}, \frac{w^*}{\Phi^*}, \frac{\bar\partial_\zeta w}{\tilde\Phi}, \bar\partial_\zeta \frac{\bar\zeta - \bar z}{|\zeta - z|^2}\right)\right\| \leqq C \frac{\|d\varrho(z)\| + |\zeta - z|}{|\Phi^*|\, |\tilde\Phi|^{s+1}\, |\zeta - z|^{2n-2s-5}}, \tag{3.3.14}$$

$$\left\| \det_{1,1,s,n-2-s}\left(\frac{w}{\tilde\Phi}, \frac{\bar\zeta - \bar z}{|\zeta - z|^2}, \frac{\bar\partial_\zeta w}{\tilde\Phi}, \bar\partial_\zeta \frac{\bar\zeta - \bar z}{|\zeta - z|^2}\right)\right\| \leqq C \frac{\|d\varrho(z)\| + |\zeta - z|}{|\tilde\Phi|^{s+1}\, |\zeta - z|^{2n-2s-3}}, \tag{3.3.15}$$

$$\left\| \det_{1,1,s,n-2-s}\left(\frac{w^*}{\Phi^*}, \frac{\bar\zeta - \bar z}{|\zeta - z|^2}, \frac{\bar\partial_\zeta w}{\tilde\Phi}, \bar\partial_\zeta \frac{\bar\zeta - \bar z}{|\zeta - z|^2}\right)\right\| \leqq C \frac{\|d\varrho(z)\| + |\zeta - z|}{|\Phi^*|\, |\tilde\Phi|^{s} |\zeta - z|^{2n-2s-3}}, \tag{3.3.16}$$

$$\left\| \det_{1,1,1,s,n-3-s}\left(\frac{w^*}{\Phi^*}, \frac{\bar\zeta - \bar z}{|\zeta - z|^2}, \frac{w\, \bar\partial_\zeta \tilde\Phi}{\tilde\Phi^2}, \frac{\bar\partial_\zeta w}{\tilde\Phi}, \bar\partial_\zeta \frac{\bar\zeta - \bar z}{|\zeta - z|^2}\right)\right\|$$

$$\leqq C \frac{(\|d\varrho(z)\| + |\zeta - z|)^2}{|\Phi^*|\, |\tilde\Phi|^{s+2}\, |\zeta - z|^{2n-2s-6}}. \tag{3.3.17}$$

If $d\varrho(z) = 0$, then it follows from estimates (3.3.13)–(3.3.17) and estimates (3.2.11), (3.2.12) in Lemma 3.2.4 that

$$I(\delta) \leqq C\, \|f\|_{0,D} \int\limits_{|\zeta - z| = \delta} \frac{d\sigma_{2n-1}^\delta}{|\zeta - z|^{2n-2}} \leqq C'\delta\, \|f\|_{0,D},$$

where $d\sigma_{2n-1}^\delta$ is the volume form of the sphere $|\zeta - z| = \delta$. If $d\varrho(z) \neq 0$, then these estimates give

$$I(\delta) \leqq C\, \|f\|_{0,D} \int\limits_{|\zeta - z| = \delta} \frac{d\sigma_{2n-1}^\delta}{|\tilde\Phi|\, |\zeta - z|^{2n-3}} + C\, \|f\|_{0,D} \int\limits_{|\zeta - z| = \delta} \frac{d\sigma_{2n-1}^\delta}{|\Phi^*|\, |\tilde\Phi|^2\, |\zeta - z|^{2n-6}}. \tag{3.3.18}$$

Further, it follows from Construction 3.1.1, Lemma A, condition (ii), and estimate (3.1.8) that

$$|\tilde\Phi(z, \zeta)| \geqq \alpha(|\operatorname{Im} F(z, \zeta)| + |\varrho(\zeta)| + |\zeta - z|^2) \quad \text{for} \quad \zeta, z \in V_2 \cap D, \tag{3.3.19}$$

and

$$|\Phi^*(z, \zeta)| \geqq \alpha(|\operatorname{Im} F(\zeta, z)| + |\varrho(\zeta)| + |\zeta - z|^2) \text{ for } z \in \partial D,\ \zeta \in V_2 \cap D. \tag{3.3.20}$$

Since $F(\zeta, z) + F(z, \zeta) = O(|\zeta - z|^2)$, (3.3.20) implies that

$$|\Phi^*(z, \zeta)| \geqq \alpha\big(|\operatorname{Im} F(z, \zeta)| + |\varrho(\zeta)| + |\zeta - z|^2\big) \quad \text{for} \quad z \in \partial D\,, \quad \zeta \in V_2 \cap D\,. \tag{3.3.21}$$

Finally in view of Lemma 3.2.4 (iii), for some $\delta_0 > 0$ we can find real coordinates $t_1(\zeta), \dots, t_{2n}(\zeta)$ in the neighbourhood $|\zeta - z| < \delta_0$ of z such that $t_1(\zeta) = \varrho(\zeta)$ and $t_2(\zeta) = \operatorname{Im} F(z, \zeta)$. Now it follows from (3.3.18), (3.3.19), and (3.3.21) that for $\delta_0 > \delta > 0$

$$I(\delta) \leqq \frac{C\,\|f\|_{0,D}}{\delta^{2n-3}} \int\limits_{|\zeta - z| = \delta} \frac{d\sigma_{2n-1}^{\delta}}{|t_1(\zeta)| + \delta^2} + \frac{C\,\|f\|_{0,D}}{\delta^{2n-6}} \int\limits_{|\zeta - z| = \delta} \frac{d\sigma_{2n-1}^{\delta}}{(|t_1(\zeta)| + |t_2(\zeta)| + \delta^2)^3}\,.$$

In view of Proposition 6, Appendix 1, this implies the required estimate. ■

3.3.3. Lemma. *Let f be a continuous bounded $(0, 1)$-form on D such that $\bar\partial f = 0$ in D. Then for all $z \in \partial D$*

$$R_D^* f(z) = R_D f(z)\,,$$

where R_D^* is defined in Lemma 3.3.1 *and R_D is defined in Theorem* 3.2.2 (i).

Proof. Fix $z \in \partial D$. We would like to apply Stokes' formula to the differential form

$$f(\zeta) \wedge \bar\omega\left(\lambda_1 \frac{\chi(\zeta)\, w(z, \zeta)}{\tilde\Phi(z, \zeta)} + \lambda_2 \frac{w^*(z, \zeta)}{\Phi^*(z, \zeta)} + (1 - \lambda_1 - \lambda_2) \frac{\bar\zeta - \bar z}{|\zeta - z|^2}\right) \wedge \omega(\zeta) \tag{3.3.22}$$

on $\bar D \times \Delta$. However, this form has a singularity at $\zeta = z$. Therefore, we introduce open sets D_δ as follows: Let $K \subseteqq \partial D$ be the set of points $\zeta \in \partial D$ with $d\varrho(\zeta) = 0$. In view of Theorem 1.4.21, K is locally contained in a smooth real n-dimensional submanifold. Therefore, for every $\tau > 0$ we can find a neighbourhood $U_K(\tau)$ of K with piecewise smooth boundary $\partial U_K(\tau)$ such that

$$\sigma_{2n-1}\big(\partial U_K(\tau)\big) \leqq C\tau^{n-1}\,, \tag{3.3.23}$$

where $\sigma_{2n-1}\big(\partial U_K(\tau)\big)$ denotes the $(2n - 1)$-dimensional Euclidean volume of $\partial U_K(\tau)$. Further, for $\delta > 0$ let $E_z(\delta)$ be the open Euclidean ball of radius δ centered at z. Since for every fixed $\delta > 0$ the form (3.3.22) is bounded on $\big(\bar D \setminus E_z(\delta)\big) \times \Delta$, and in view of (3.3.23), for every $\delta > 0$ we can find $\tau_\delta > 0$ so small that

$$\int\limits_{[D \cap (\partial U_K(\tau_\delta) \setminus E_z(\delta))] \times \Delta} \left| f \wedge \bar\omega\left(\lambda_1 \frac{\chi w}{\tilde\Phi} + \lambda_2 \frac{w^*}{\Phi^*} + (1 - \lambda_1 - \lambda_2) \frac{\bar\zeta - \bar z}{|\zeta - z|^2}\right) \wedge \omega(\zeta)\right| \leqq \delta\,. \tag{3.3.24}$$

Define $D_\delta := D \setminus \big(E_z(\delta) \cup U_K(\tau_\delta)\big)$. Since $\bar\partial f = 0$, the form (3.3.22) is closed on $D_\delta \times \Delta$. Consequently, it follows from Stokes' formula and estimates (3.3.24) and (3.3.12) that

$$\lim_{\delta \to 0} \Bigg[\int\limits_{(\zeta, \lambda_1) \in D_\delta \times [0,1]} f \wedge \bar\omega\left(\lambda_1 \frac{\chi w}{\tilde\Phi} + (1 - \lambda_1) \frac{\bar\zeta - \bar z}{|\zeta - z|^2}\right) \wedge \omega(\zeta)$$

$$- \int\limits_{(\zeta, \lambda_2) \in D_\delta \times [0,1]} f \wedge \bar\omega\left(\lambda_2 \frac{w^*}{\Phi^*} + (1 - \lambda_2) \frac{\bar\zeta - \bar z}{|\zeta - z|^2}\right) \wedge \omega(\zeta)$$

$$+ \int\limits_{(\zeta, \lambda_2) \in D_\delta \times [0,1]} f \wedge \bar\omega\left((1 - \lambda_2) \frac{\chi w}{\tilde\Phi} + \lambda_2 \frac{w^*}{\Phi^*}\right) \wedge \omega(\zeta)\Bigg] = 0\,. \tag{3.3.25}$$

Since $\langle w^*(z,\zeta), \zeta - z\rangle = \langle -w(\zeta, z), \zeta - z\rangle = \Phi(\zeta, z) = \Phi^*(z, \zeta)$ and, therefore,

$$\left\langle \lambda_2 \frac{w^*}{\Phi^*} + (1-\lambda_2)\frac{\bar\zeta - \bar z}{|\zeta - z|^2}, \zeta - z\right\rangle = 1 \quad \text{for} \quad \zeta \in D\,, \qquad 0 \leqq \lambda_2 \leqq 1\,,$$

we obtain that

$$\bar\omega\left(\lambda_2 \frac{w^*}{\Phi^*} + (1-\lambda_2)\frac{\bar\zeta - \bar z}{|\zeta - z|^2}\right) = \bigwedge_{j=1}^{n} (\mathrm{d}_{\lambda_2} + \bar\partial_\zeta)\left(\lambda_2 \frac{w_j^*}{\Phi^*} + (1-\lambda_2)\frac{\bar\zeta_j - \bar z_j}{|\zeta - z|^2}\right) = 0\,.$$

Consequently, the second integral in (3.3.25) is zero. Hence (3.3.25) implies the required equality $R_D^* f(z) = R_D f(z)$. ■

3.3.4. Theorem. *Let $D \subset\subset \mathbb{C}^n$ be a strictly pseudoconvex open set (with not necessarily smooth boundary), and let ϱ be a strictly plurisubharmonic C^2-function in some neighbourhood θ of ∂D such that $D \cap \theta = \{z \in \theta : \varrho(z) < 0\}$. Then there exists a constant $C < \infty$ with the following property: Let $\xi \in \partial D$, $\delta > 0$, and let f be a continuous, bounded, and $\bar\partial$-closed $(0, 1)$-form on D such that the following estimates are fulfilled* (for the definition of $H_\xi(\delta)$ see Subsection 3.2.1):

$$\operatorname{supp} f \subseteqq H_\xi(\delta)\,, \tag{3.3.26}$$

$$\|f(\zeta)\| \leqq \frac{1}{\delta} + \frac{\|\mathrm{d}\varrho(\zeta)\|}{\delta^2} \quad \textit{for} \quad \zeta \in H_\xi(\delta)\,, \tag{3.3.27}$$

$$\|f(\zeta) \wedge \bar\partial\varrho(\zeta)\| \leqq \frac{\|\mathrm{d}\varrho(\zeta)\|}{\delta} \quad \textit{for} \quad \zeta \in H_\xi(\delta)\,. \tag{3.3.28}$$

Then the solution of $\bar\partial u = f$ given by the formula $u := R_D f$ is continuous on $\bar D$ (cp. Corollary 3.1.4) *and admits the estimate*

$$|u(z)| \leqq C \min\{1, [\delta/\operatorname{dist}(z, H_\xi(\delta))]^{2n-1}\}\,, \qquad z \in \partial D\,. \tag{3.3.29}$$

Proof. Since by Lemma 3.3.3, for $z \in \partial D$, $R_D f(z) = R_D^* f(z)$, the statement follows from Lemma 3.3.1. ■

Remark. Estimate (3.3.29) was proved for $z \in \partial D$ only. It is not clear whether it also holds for $z \in D$.

3.4. Decomposition of singularities

First we observe the following proposition, which is a simple consequence of the Leray formula proved in Theorem 3.1.3 (i) and the estimates given for the operator R_D in Theorem 3.2.2 (i):

3.4.1. Proposition. *Let $D \subset\subset \mathbb{C}^n$ be a strictly pseudoconvex open set (with not necessarily smooth boundary), and let $U_j \subseteqq \mathbb{C}^n$ $(j = 1, \ldots, N)$ be open sets such that $\bar D \subseteqq \bigcup_{j=1}^{N} U_j$. Then every bounded holomorphic function f on D admits a decomposition*

$$f = \sum_{j=1}^{N} f_j\,, \tag{3.4.1}$$

where every f_j is bounded and holomorphic in some neighbourhood of $\bar D \setminus (\partial D \cap U_j)$. If, moreover, f admits a continuous continuation to $\bar D$, then the functions f_j also admit continuous continuations to $\bar D$.

Proof. Choose C^∞-functions χ_j on $\mathbb{C}^n$ such that $\sum \chi_j = 1$ on $\bar{D}$ and $\chi_j = 0$ on $\mathbb{C}^n \setminus U_j$, and define $f_j := L_D \chi_j f$, where L_D is the operator defined in Definition 3.1.2. Since $\chi_j = 0$ on $\mathbb{C}^n \setminus U_j$, then f_j is holomorphic in some neighbourhood of $\bar{D} \setminus (\partial D \cap U_j)$. Since, in view of the Leray formula stated in Theorem 3.1.3 (i), $f = L_D f$, (3.4.1) follows.

If f is continuous on $\bar{D}$, then all f_j are continuous on $\bar{D}$ since, by Theorem 3.1.3 (i),

$$f_j = \chi_j f - R_D f\, \bar{\partial} \chi_j ,$$

and since, by Theorem 3.2.2 (i), the functions $R_D f\, \bar{\partial}\chi_j$ are continuous on $\bar{D}$. ■

The aim of the present section is the proof of a more precise result.

Notation. Throughout this section $D \subset\subset \mathbb{C}^n$ is a strictly pseudoconvex open set and ϱ is a strictly plurisubharmonic C^2-function in a neighbourhood θ of ∂D such that

$$D \cap \theta = \{z \in \theta : \varrho(z) < 0\} .$$

If $\xi_1, \dots, \xi_N \in \partial D$ and $\delta > 0$, then by $d(\xi_1, \dots, \xi_N; \delta)$ we denote the order of the system of sets $\{H_{\xi_j}(\delta)\}_{j=1}^N$[1]), that is, the number d such that at least one point belongs to d of the sets $H_{\xi_j}(\delta)$, but no point belongs to $d + 1$ of these sets.

3.4.2. Theorem. *There exists a constant $C < \infty$ with the following property: If $\xi_1, \dots, \xi_N \in \partial D$ and $\delta > 0$, then every bounded holomorphic function f on D admits a decomposition $f = \sum_{j=0}^{N} f_j$ such that the following conditions are fulfilled:*

(i) *For $j = 1, \dots, N$ the function f_j is holomorphic in some neighbourhood of $\bar{D} \setminus (\partial D \cap H_{\xi_j}(\delta))$ and admits the estimate* (for the definition of $\|\cdot\|_{0,D}$ see Section 1.8)

$$\|f_j\|_{0,D} \leqq C d(\xi_1, \dots, \xi_N; \delta)\, \|f\|_{0,D} . \tag{3.4.2}$$

(ii) *f_0 is holomorphic in some neighbourhood of*

$$D \cup [\partial D \cap \bigcup_{j=1}^{N} H_{\xi_j}(\delta/2)] . \tag{3.4.3}$$

(iii) *If, moreover, f admits a continuous continuation to $\bar{D}$, then the functions $f_0, \dots, f_N$ also are continuous on $\bar{D}$.*

For the proof of this theorem we use the following

3.4.3. Lemma. *There exists a constant $C < \infty$ with the following property: For all $\xi \in \partial D$ and $\delta > 0$ there is a C^∞-function χ on $\mathbb{C}^n$ such that* (for the definition of $\tilde{H}_\xi(\delta)$ see Subsection 3.2.1)

$$\left.\begin{array}{lll} 0 \leqq \chi \leqq 1 & \text{on} & \mathbb{C}^n , \\ \chi = 1 & \text{on} & \tilde{H}_\xi(\delta) , \\ \operatorname{supp} \chi \subset\subset \tilde{H}_\xi(2\delta) ; & & \end{array}\right\} \tag{3.4.4}$$

$$\|d\chi(\zeta)\| \leqq C(1/\delta + \|d\varrho(\zeta)\|/\delta^2) \quad \textit{for} \quad \zeta \in \tilde{H}_\xi(2\delta) \cap \theta , \tag{3.4.5}$$

$$\|\bar{\partial}\chi(\zeta) \wedge \bar{\partial}\varrho(\zeta)\| \leqq C \|d\varrho(\zeta)\|/\delta \quad \textit{for} \quad \zeta \in \tilde{H}_\xi(2\delta) \cap \theta . \tag{3.4.6}$$

Proof. If $\|d\varrho(\xi)\| \leqq 2\delta$, then $\tilde{H}_\xi(2\delta) = E_\xi(2\delta)$, and it is enough to choose χ so that $\chi = 1$ on $E_\xi(\delta)$ $(\supseteq \tilde{H}_\xi(\delta))$, $\chi = 0$ on $\mathbb{C}^n \setminus E_\xi(2\delta)$ and for some $C_1 < \infty$, $\|d\chi(\zeta)\| \leqq C_1/\delta$

[1]) $H_\xi(\delta)$ was defined in Subsection 3.2.1.

for $\zeta \in \mathbb{C}^n$. Therefore, we can assume that

$$||d\varrho(\xi)|| > 2\delta . \tag{3.4.7}$$

Now we choose C^∞-functions $\varphi(\lambda)$ on $\mathbb{C}^1$ and $\psi(\eta)$ on $\mathbb{C}^{n-1}$ such that

$$0 \leqq \varphi \leqq 1 \quad \text{on} \quad \mathbb{C}^1 , \qquad \varphi(\lambda) = \begin{cases} 1 & \text{for} \quad |\lambda| < \delta^2/||d\varrho(\xi)|| , \\ 0 & \text{for} \quad |\lambda| > 2\delta^2/||d\varrho(\xi)|| , \end{cases} \tag{3.4.8}$$

$$0 \leqq \psi \leqq 1 \quad \text{on} \quad \mathbb{C}^n , \qquad \psi(\eta) = \begin{cases} 1 & \text{for} \quad |\eta| < \delta , \\ 0 & \text{for} \quad |\eta| > \delta 3/2 , \end{cases} \tag{3.4.9}$$

and for some constant $C_3 < \infty$,

$$||d\varphi(\lambda)|| \leqq C_3 \, ||d\varrho(\xi)||/\delta^2 , \tag{3.4.10}$$

$$||d\psi(\lambda)|| \leqq C_3/\delta . \tag{3.4.11}$$

Further, let $\vartheta^{(1)}, \dots , \vartheta^{(n-1)}$ be an orthonormal basis of the linear space $T_\xi - \xi := \{\zeta - \xi : \zeta \in T_\xi\}$. Define

$$\vartheta^{(n)} = \frac{2}{||d\varrho(\xi)||} \{\partial\varrho(\xi)/\partial\bar{\xi}_j\}_{j=1}^n .$$

Since ϱ is real and, therefore,

$$||d\varrho(\xi)|| = 2\Big(\sum_{j=1}^n |\partial\varrho(\xi)/\partial\bar{\xi}_j|^2\Big)^{1/2} ,$$

then $\vartheta^{(1)}, \dots , \vartheta^{(n)}$ is an orthonormal basis in $\mathbb{C}^n$. Let $\beta_j = \beta_j(\zeta), \zeta \in \mathbb{C}^n$, be the complex coordinates in $\mathbb{C}^n$ such that $\zeta = \sum_{j=1}^n \beta_j(\zeta) \, \vartheta^{(j)}, \zeta \in \mathbb{C}^n$. Then

$$\beta_n(\zeta) = \frac{2}{||d\varrho(\xi)||} \sum_{j=1}^n \frac{\partial\varrho(\xi)}{\partial\xi_j} \zeta_j , \qquad \zeta \in \mathbb{C}^n . \tag{3.4.12}$$

Define $\chi(\zeta) = \varphi\big(\beta_n(\zeta - \xi)\big) \, \psi\big(\beta_1(\zeta - \xi), \dots , \beta_{n-1}(\zeta - \xi)\big)$, and check the conditions (3.4.4)—(3.4.6).

Let $\zeta \in \widetilde{H}_\xi(\delta)$. Then $\big(\sum_{j=1}^{n-1} |\beta_j(\zeta - \xi)|^2\big)^{1/2} \leqq |\zeta - \xi| \leqq \delta$ and

$$|\beta_n(\zeta - \xi)| = \operatorname{dist}(\zeta, T_\xi) \leqq \delta^2/||d\varrho(\xi)|| .$$

Consequently, by (3.4.8) and (3.4.9), $\chi(\zeta) = 1$.

Let $\zeta \in \mathbb{C}^n$ and $\chi(\zeta) \neq 0$. Then by (3.4.8), (3.4.9), and (3.4.7), $\big(\sum_{j=1}^{n-1} |\beta_j(\zeta - \xi)|^2\big)^{1/2} < \delta 3/2$ and $|\beta_n(\zeta - \xi)| \leqq 2\delta^2/||d\varrho(\xi)|| < \delta$. It follows that

$$|\zeta - \xi| = \Big(\sum_{j=1}^n |\beta_j(\zeta - \xi)|^2\Big)^{1/2} < (9\delta^2/4 + \delta^2)^{1/2} < 2\delta$$

and

$$||d\varrho(\xi)|| \operatorname{dist}(\zeta, T_\xi) = ||d\varrho(\xi)|| \, |\beta_n(\zeta - \xi)| < 2\delta^2 = (\sqrt{2}\delta)^2 ,$$

that is,

$$\zeta \in \widetilde{H}_\xi(\sqrt{2}\delta) \subset\subset \widetilde{H}_\xi(2\delta) .$$

It remains to prove estimates (3.4.5) and (3.4.6). Since $\vartheta^{(1)}, \dots , \vartheta^{(n)}$ is an orthonormal basis in $\mathbb{C}^n$, we have $|\partial\beta_j(\zeta)/\partial\zeta_k| \leqq 1$ for $\zeta \in \mathbb{C}^n$ and $j, k = 1, \dots , n$. Together with (3.4.8)—(3.4.11) this implies that for some constant $C_4 < \infty$

$$||d\chi(\zeta)|| \leqq C_4(1/\delta + ||d\varrho(\xi)||/\delta^2) .$$

Taking into account that for $\zeta \in \widetilde{H}_\xi(2\delta) \cap \theta$

$$\|\mathrm{d}\varrho(\xi)\| \leqq \|\mathrm{d}\varrho(\zeta)\| + |\zeta - \xi| \leqq \|\mathrm{d}\varrho(\zeta)\| + 2\delta\,,$$

we thus obtain (3.4.5).

From (3.4.12) it follows that

$$\begin{aligned}\bar\partial_\zeta\varphi(\beta_n(\zeta-\xi)) &= \frac{\partial\varphi(\lambda)}{\partial\bar\lambda}\bigg|_{\lambda=\beta_n(\zeta-\xi)} \mathrm{d}_\zeta\overline{\beta_n(\zeta-\xi)}\\ &= \frac{\partial\varphi(\lambda)}{\partial\bar\lambda}\bigg|_{\lambda=\beta_n(\zeta-\xi)} \frac{2}{\|\mathrm{d}\varrho(\xi)\|}\sum_{j=1}^{n}\frac{\partial\varrho(\xi)}{\partial\bar\xi_j}\,\mathrm{d}\bar\zeta_j\\ &= \frac{\partial\varphi(\lambda)}{\partial\bar\lambda}\bigg|_{\lambda=\beta_n(\zeta-\xi)} \frac{2}{\|\mathrm{d}\varrho(\xi)\|}\left(\bar\partial\varrho(\zeta) + \mathrm{O}(|\zeta-\xi|)\right).\end{aligned}$$

Consequently, for $\zeta \in \theta$,

$$\begin{aligned}\bar\partial\chi(\zeta) \wedge \bar\partial\varrho(\zeta) &= \varphi(\beta_n(\zeta-\xi))\,\bar\partial_\zeta\psi(\beta_1(\zeta-\xi), \dots, \beta_{n-1}(\zeta-\xi)) \wedge \bar\partial\varrho(\zeta)\\ &+ \frac{\partial\varphi(\lambda)}{\partial\bar\lambda}\bigg|_{\lambda=\beta_n(\zeta-\xi)} \frac{\psi(\beta_1(\zeta-\xi), \dots, \beta_{n-1}(\zeta-\xi))}{\|\mathrm{d}\varrho(\xi)\|}\,\mathrm{O}(|\zeta-\xi|) \wedge \bar\partial\varrho(\zeta)\,.\end{aligned}$$

In view of (3.4.8)—(3.4.11), we can therefore conclude that for some constant $C_5 < \infty$ and all $\zeta \in \theta$

$$\|\bar\partial\chi(\zeta) \wedge \bar\partial\varrho(\zeta)\| \leqq C_5(\|\mathrm{d}\varrho(\zeta)\|/\delta + |\zeta-\xi|\,\|\mathrm{d}\varrho(\zeta)\|/\delta^2)\,.$$

This implies (3.4.6), because $|\zeta - \xi| < 2\delta$ for $\zeta \in \widetilde{H}_\xi(2\delta)$. ∎

Proof of Theorem 3.4.2. In view of Lemma 3.4.3 there are C^∞-functions $\tilde\chi_j$ ($j = 1, \dots, N$) on $\mathbb{C}^n$ such that $0 \leqq \tilde\chi_j \leqq 1$ on $\mathbb{C}^n$, $\tilde\chi_j = 1$ on $\widetilde{H}_{\xi_j}(\delta/2)$, supp $\tilde\chi_j \subset\subset \widetilde{H}_{\xi_j}(\delta)$, and for some constant $C_1 < \infty$

$$\|\bar\partial\tilde\chi_j(\zeta)\| \leqq C_1(1/\delta + \|\mathrm{d}\varrho(\zeta)\|/\delta^2) \quad \text{for} \quad \zeta \in \theta\,,$$

$$\|\bar\partial\tilde\chi_j(\zeta) \wedge \bar\partial\varrho(\zeta)\| \leqq C_1\,\|\mathrm{d}\varrho(\zeta)\|/\delta \quad \text{for} \quad \zeta \in \theta\,.$$

Define $\tilde\chi_0 := (1 - \tilde\chi_1) \dots (1 - \tilde\chi_N)$ and $\chi_j = \tilde\chi_j/(\tilde\chi_0 + \dots + \tilde\chi_N)$ for $j = 0, 1, \dots, N$. Then

$$\begin{aligned}&\sum_{j=0}^{N}\chi_j = 1 \quad \text{on} \quad \mathbb{C}^n\,,\\ &\chi_0 = 0 \quad \text{on} \quad \bigcup_{j=1}^{N} H_{\xi_j}(\delta/2)\,,\\ &\operatorname{supp}\chi_j \subset\subset \widetilde{H}_{\xi_j}(\delta) \quad \text{for} \quad j = 1, \dots, N\,,\end{aligned} \tag{3.4.13}$$

and for some constant $C_2 < \infty$

$$\|\bar\partial\chi_j(\zeta)\| \leqq C_2 d(\xi_1, \dots, \xi_N; \delta)\,(1/\delta + \|\mathrm{d}\varrho(\zeta)\|/\delta^2)$$
$$\text{for} \quad \zeta \in \theta \quad \text{and} \quad j = 1, \dots, N\,, \tag{3.4.14}$$

$$\|\bar\partial\chi_j(\zeta) \wedge \bar\partial\varrho(\zeta)\| \leqq C_2 d(\xi_1, \dots, \xi_N; \delta)\frac{\|\mathrm{d}\varrho(\zeta)\|}{\delta}$$
$$\text{for} \quad \zeta \in \theta \quad \text{and} \quad j = 1, \dots, N\,. \tag{3.4.15}$$

Define $f_j = L_D\chi_j f$ for $j = 0, 1, \dots, N$, where L_D was defined in Definition 3.1.2. Since the kernel of the integral L_D depends holomorphically on z, it follows from (3.4.13) that f_j is holomorphic in some neighbourhood of $\overline{D} \setminus (\partial D \cap H_{\xi_j}(\delta))$. Since $\sum_{j=0}^{N} \chi_j = 1$, it follows from the Leray formula stated in Theorem **3.1.3** (i) that $f = \sum_{j=0}^{N} f_j$.

Now we prove estimate (3.4.2). In view of Corollary 3.2.3, $R_D f \, \bar\partial\chi_j$ is continuous on $\overline{D}$ and, in view of Theorem 3.3.4, it follows from estimates (3.4.14) and (3.4.15) that for some $C_3 < \infty$

$$\sup_{z \in \partial D} |R_D f \, \bar\partial\chi_j| \leqq C_3 d(\xi_1, \dots, \xi_N; \delta) \, \|f\|_{0,D} \quad \text{for} \quad j = 1, \dots, N .$$

Taking into account that, by Theorem **3.1.3** (i),

$$f_j = \chi_j f - R_D f \, \bar\partial\chi_j , \tag{3.4.16}$$

we thus obtain that $|f_j(z)| \leqq (1 + 2C_3 d(\xi_1, \dots, \xi_N; \delta)) \, \|f\|_{0,D}$ for all $z \in D$ which are sufficiently close to ∂D (f_j need not be continuous on $\overline{D}$). By the maximum principle for holomorphic functions this implies (3.4.2).

Finally, we consider the case that f is continuous on $\overline{D}$. Since, by Corollary 3.2.3, for all $j = 0, \dots, N$ the function $R_D f \, \bar\partial\chi_j$ is continuous on $\overline{D}$, and since (3.4.16) holds for $j = 0, \dots, N$, we then conclude that the functions $f_0, \dots, f_N$ are continuous on $\overline{D}$, too. ■

For functions which are continuous on $\overline{D}$, now we shall give a more precise result, which will be used in the next section to prove an approximation theorem for holomorphic functions in D which are continuous on $\overline{D}$. First we introduce some notations.

Notation. Let $D, \varrho, F, \tilde\Phi, \Phi, w, \chi$, and $N(\varrho) \subset\subset V_2 \subset\subset \theta$ be as in Construction 3.1.1 and Definition 3.1.2. Choose a C^∞-function ϑ on $\mathbb{C}^n$ such that $\vartheta = 1$ in some neighbourhood of $\overline{D}$ and $\vartheta = 0$ in some neighbourhood of $\mathbb{C}^n \setminus (D \cup V_2)$.

By Construction 3.1.1, Lemma A, condition (ii) and estimate (3.1.8),

$$\Phi(z, \zeta) \neq 0 \quad \text{for} \quad z \in D \quad \text{and} \quad \zeta \in V_2 \setminus D .$$

Consequently, for every C^1-function φ on $\mathbb{C}^n$ and all $z \in D$, we can define

$$L_{\partial D}\varphi(z) := -\frac{(n-1)!}{(2\pi i)^n} \int\limits_{V_2 \setminus D} \bar\partial(\vartheta(\zeta)\,\varphi(\zeta)) \wedge \frac{\omega'_\zeta(w(z,\zeta)) \wedge \omega(\zeta)}{\Phi^n(z,\zeta)} . \tag{3.4.16'}$$

Remark that, by Stokes' formula, in the case of smooth ∂D, $L_{\partial D}\varphi = L^w_{\partial D}\varphi$, where $L^w_{\partial D}$ is the boundary integral introduced in Subsection 1.8.6. Further it follows from Stokes' formula that the definition of $L_{\partial D}\varphi$ is independent of the choice of ϑ (also for the general case when ∂D is not necessarily smooth).

3.4.4. Lemma. *For every C^1-function φ on $\mathbb{C}^n$ we have*

$$L_{\partial D}\varphi(z) = L_D\varphi(z) + \frac{(n-1)!}{(2\pi i)^n} \int\limits_D \bar\partial\varphi(\zeta) \wedge \frac{\omega'_\zeta(\chi(\zeta)\, w(z,\zeta)) \wedge \omega(\zeta)}{\tilde\Phi^n(z,\zeta)} \quad \textit{for} \quad z \in D , \tag{3.4.17}$$

where L_D is the operator introduced in Definition 3.1.2.

Proof. Consider the differential form γ on $\mathbb{C}^n$ defined by

$$\gamma(\zeta) := \begin{cases} \varphi(\zeta) \dfrac{\omega'_\zeta(\chi(\zeta)\, w(z,\zeta)) \wedge \omega(\zeta)}{\tilde\Phi^n(z,\zeta)} & \text{if} \quad \zeta \in D , \\[2ex] \vartheta(\zeta)\,\varphi(\zeta) \, \dfrac{\omega'_\zeta(w(z,\zeta)) \wedge \omega(\zeta)}{\Phi^n \quad ,\zeta)} & \text{if} \quad \zeta \in V_2 \setminus D . \end{cases}$$

Since $\chi(\zeta) = 1$ for $\zeta \in V_2$ and $\tilde{\Phi}(z, \zeta) = \Phi(z, \zeta)$ for $\zeta \in \partial D$, the form γ is continuous on $\mathbb{C}^n$. Taking into account that

$$\mathrm{d}_\zeta \frac{\omega'_\zeta(w(z,\zeta)) \wedge \omega(\zeta)}{\Phi^n(z,\zeta)} = 0 \quad \text{for} \quad \zeta \in V_2$$

(cp. Proposition 1.7.1) and

$$\mathrm{d}_\zeta \frac{\omega'_\zeta(\chi(\zeta)\, w(z,\zeta)) \wedge \omega(\zeta)}{\tilde{\Phi}^n(z,\zeta)} = n\omega_\zeta\left(\frac{\chi(\zeta)\, w(z,\zeta)}{\tilde{\Phi}(z,\zeta)}\right) \wedge \omega(\zeta) \quad \text{for} \quad \zeta \in D \cup V_2$$

(cp. Proposition 1.7.3), we see that $\mathrm{d}\gamma$ is continuous on $\mathbb{C}^n \setminus \partial D$ and admits continuous continuations from $\mathbb{C}^n \setminus \overline{D}$ to ∂D and from D to ∂D, where

$$\mathrm{d}\gamma(\zeta) = \bar{\partial}(\vartheta(\zeta)\, \varphi(\zeta)) \wedge \frac{\omega'_\zeta(w(z,\zeta)) \wedge \omega(\zeta)}{\Phi^n(z,\zeta)} \quad \text{for} \quad \zeta \in \mathbb{C}^n \setminus D$$

and

$$\mathrm{d}\gamma(\zeta) = \bar{\partial}\varphi(\zeta) \wedge \frac{\omega'_\zeta(\chi(\zeta)\, w(z,\zeta)) \wedge \omega(\zeta)}{\tilde{\Phi}^n(z,\zeta)} + n\varphi(\zeta)\, \omega_\zeta\left(\frac{\chi(\zeta)\, w(z,\zeta)}{\tilde{\Phi}(z,\zeta)}\right) \wedge \omega(\zeta)$$

for $\zeta \in D$.

If ∂D is smooth, then (3.4.17) immediately follows from Stokes' formula. In the general case we can similarly proceed as in the proof of Lemma 3.3.3 using the fact that the set $K := \{\zeta \in \partial D : \mathrm{d}\varrho(\zeta) = 0\}$ is locally contained in a smooth real n-dimensional submanifold (Theorem 1.4.21). ■

3.4.5. Lemma. *There is a constant $C < \infty$ with the following property: If $\xi \in \partial D$, $\delta > 0$, and $f \in L^\infty_{(0,1)}(D)$ such that*

$$\|f(\zeta)\| \leqq 1/\delta + \|\mathrm{d}\varrho(\zeta)\|/\delta^2 \quad \text{for} \quad \zeta \in H_\xi(\delta) \cap V_2\,, \tag{3.4.18}$$

then for all $z \in \overline{D}$

$$\int\limits_{H_\xi(\delta)} \left| f(\zeta) \wedge \frac{\omega'_\zeta(\chi(\zeta)\, w(z,\zeta)) \wedge \omega(\zeta)}{\tilde{\Phi}^n(z,\zeta)} \right| \leqq C \min\{1, [\delta/\operatorname{dist}(z, H_\xi(\delta))]^{2n-1}\}\,. \tag{3.4.19}$$

Proof. In view of Construction 3.1.1, Lemma A, condition (ii), and estimate (3.1.8)

$$\inf_{z\in\overline{D},\, \zeta\in D\setminus V_2} |\tilde{\Phi}(z,\zeta)| > 0 \quad \text{and} \quad \inf_{z\in D\setminus V_2,\, \zeta\in D} |\tilde{\Phi}(z,\zeta)| > 0\,.$$

Further, $\int\limits_{H_\xi(\delta)} \mathrm{d}\sigma_{2n} \leqq C_0 \min\{\delta^{2n}, \delta^{2n+2}/\|\mathrm{d}\varrho(\xi)\|^2\}$ for some $C_0 < \infty$ and, by (3.4.18), there is a constant $C_1 < \infty$ such that

$$\|f(\zeta)\| \leqq C_1(1/\delta + \|\mathrm{d}\varrho(\xi)\|/\delta^2) \quad \text{for} \quad \zeta \in H_\xi(\delta)\,. \tag{3.4.20}$$

This implies that for some constant $C_2 < \infty$

$$\sup_{z\in\overline{D}} \int\limits_{H_\xi(\delta)\setminus V_2} \left| f(\zeta) \wedge \frac{\omega'_\zeta(\chi(\zeta)\, w(z,\zeta)) \wedge \omega(\zeta)}{\tilde{\Phi}^n(z,\zeta)} \right|$$
$$\leqq C_2(1/\delta + \|\mathrm{d}\varrho(\xi)\|/\delta^2) \min\{\delta^{2n}, \delta^{2n+2}/\|\mathrm{d}\varrho(\xi)\|^2\} \leqq 2C_2\delta^{2n-1}$$

and

$$\sup_{z\in D\setminus V_2} \int\limits_{H_\xi(\delta)} \left| f(\zeta) \frac{\omega'_\zeta(\chi(\zeta)\, w(z,\zeta)) \wedge \omega(\zeta)}{\tilde{\Phi}^n(z,\zeta)} \right| \leqq 2C_2\delta^{2n-1}\,.$$

Therefore it remains to prove that for some $C_3 < \infty$

$$\sup_{z \in V_2} \int_{H_\xi(\delta) \cap V_2} \left| f(\zeta) \wedge \frac{\omega'_\zeta(\chi(\zeta)\, w(z,\zeta)) \wedge \omega(\zeta)}{\tilde{\Phi}^n(z,\zeta)} \right| \leqq C_3 \min \{1, [\delta/\operatorname{dist}(z, H_\xi(\delta))]^{2n-1}\}.$$

This follows from Lemma 3.2.6 because, by (3.4.18), (3.2.12), and (3.2.14), for some $C_4 < \infty$

$$\left\| f(\zeta) \frac{\omega'_\zeta(\chi(\zeta)\, w(z,\zeta)) \wedge \omega(\zeta)}{\tilde{\Phi}^n(z,\zeta)} \right\| \leqq \frac{C_4}{\delta} \frac{\|d\varrho(\zeta)\|}{|\tilde{\Phi}(z,\zeta)|\, |\zeta - z|^{2n-2}} + \frac{C_4}{\delta} \frac{1}{|\zeta - z|^{2n-1}}$$
$$+ \frac{C_4}{\delta^2} \frac{\|d\varrho(\zeta)\|^2}{|\tilde{\Phi}(z,\zeta)|^2\, |\zeta - z|^{2n-4}} + \frac{C_4}{\delta^2} \frac{\|d\varrho(\zeta)\|}{|\tilde{\Phi}(z,\zeta)|\, |\zeta - z|^{2n-3}}. \blacksquare$$

3.4.6. Corollary. *For every $f \in L^\infty_{(0,1)}(D)$ the function*

$$\int_D f(\zeta) \wedge \frac{\omega'_\zeta(\chi(\zeta)\, w(z,\zeta)) \wedge \omega(\zeta)}{\tilde{\Phi}^n(z,\zeta)}, \qquad z \in D, \tag{3.4.21}$$

is holomorphic in D and continuous on $\bar{D}$.

Proof. The holomorphicity follows from the fact that $\tilde{\Phi}(z,\zeta)$ and $w(z,\zeta)$ holomorphically depend on $z \in D$ and $\tilde{\Phi}(z,\zeta) \neq 0$ for $\zeta, z \in D$. Further it is clear that for all $\xi \in \partial D$ and $\delta > 0$ the function

$$\int_{D \setminus H_\xi(\delta)} f(\zeta) \wedge \frac{\omega'_\zeta(\chi(\zeta)\, w(z,\zeta)) \wedge \omega(\zeta)}{\tilde{\Phi}^n(z,\zeta)}$$

depends continuously on $z \in H_\xi(\delta)$, and it follows from Lemma 3.4.5 that for all $\xi \in \partial D$

$$\lim_{\delta \to 0} \sup_{z \in D} \left| \int_{H_\xi(\delta)} f(\zeta) \wedge \frac{\omega'_\zeta(\chi(\zeta)\, w(z,\zeta)) \wedge \omega(\zeta)}{\tilde{\Phi}^n(z,\zeta)} \right| = 0.$$

This implies the continuity of (3.4.21) on $\bar{D}$. $\blacksquare$

The right-hand side in (3.4.17) is defined not only for C^1-functions φ on $\mathbb{C}^n$ but, by Corollary 3.4.6, for all bounded continuous functions φ on D such that $\bar\partial\varphi \in L^\infty_{(0,1)}(D)$. We use this to extend the definition of the operator $L_{\partial D}$:

Definition. For every bounded continuous function φ on D, such that $\bar\partial\varphi \in L^\infty_{(0,1)}(D)$ we define

$$L_{\partial D}\varphi(z) = L_D\varphi(z) + \frac{(n-1)!}{(2\pi i)^n} \int_D \bar\partial\varphi(\zeta) \wedge \frac{\omega'_\zeta(\chi(\zeta)\, w(z,\zeta)) \wedge \omega(\zeta)}{\tilde{\Phi}^n(z,\zeta)}, \qquad z \in D. \tag{3.4.22}$$

3.4.7. Proposition. (i) *For every bounded continuous function φ on $\bar{D}$, such that $\bar\partial\varphi \in L^\infty_{(0,1)}(D)$, the function $L_{\partial D}\varphi$ is continuous on $\bar{D}$ and holomorphic in D.*

(ii) *There exists a constant $C < \infty$ with the following property: If $\xi \in \partial D$, $\delta > 0$, and φ is a bounded continuous function on $\bar{D}$ such that $\bar\partial\varphi \in L^\infty_{(0,1)}(D)$ and the following* *conditions are fulfilled*

$$\operatorname{supp} \bar\partial\varphi \subseteqq H_\xi(\delta) ,$$
$$\|\bar\partial\varphi(\zeta)\| \leqq 1/\delta + \|d\varrho(\zeta)\|/\delta^2 \quad \textit{for} \quad \zeta \in H_\xi(\delta) \cap V_2 ,$$
$$\|\bar\partial\varphi(\zeta) \wedge \bar\partial\varrho(\zeta)\| \leqq \|d\varrho(\zeta)\|/\delta \quad \textit{for} \quad \zeta \in H_\xi(\delta) \cap V_2 ,$$

then for all $z \in \partial D$

$$|L_{\partial D}\varphi(z)| \leqq |\varphi(z)| + C \min \{1, [\delta/\operatorname{dist}(z, H_\xi(\delta))]^{2n-1}\} .$$

Proof. In view of the Leray formula stated in Theorem 3.1.3 (i), $L_D\varphi = \varphi - R_D\,\bar\partial\varphi$ on D. By Theorem 3.3.4 this implies that $L_D\varphi$ is continuous on $\bar D$ and admits the estimate

$$|L_D\varphi(z)| \leqq |\varphi(z)| + C \min \{1, [\delta/\operatorname{dist}(z, H_\xi(\delta))]^{2n-1}\} , \qquad z \in \partial D .$$

Together with Lemma 3.4.5 and Corollary 3.4.6 this proves the proposition. ■

3.4.8. Theorem. *Let $D \subset\subset \mathbb{C}^n$ be a strictly pseudoconvex open set. Then there exists a constant $C < \infty$ with the following property: If $\xi_1, \dots, \xi_N \in \partial D$, $\delta > 0$, and if f is a continuous function on $\bar D$ that is holomorphic in D, then f admits a decomposition*

$$f = f_0 + L_{\partial D}\,\varphi + \sum_{j=1}^{N} f_j , \tag{3.4.23}$$

where the following conditions are fulfilled:

(i) *For $j = 1, \dots, N$ the function f_j is continuous on $\bar D$ and holomorphic in some neighbourhood of $\bar D \setminus (\partial D \cap H_{\xi_j}(\delta))$. Further, for $j = 1, \dots, N$ and all $z \in \partial D$ we have the estimate*

$$|f_j(z)| \leqq C d(\xi_1, \dots, \xi_N; \delta)\, \omega(\delta, f) \min \{1, [\delta/\operatorname{dist}(z, H_{\xi_j}(\delta))]^{2n-1}\} , \tag{3.4.24}$$

where $\omega(\delta, f) := \sup\limits_{\zeta,\eta\in \bar D;\ |\zeta-\eta|\leqq\delta} |f(\zeta) - f(\eta)|.$

(ii) *φ is a C^∞-function on $\mathbb{C}^n$ and $\|\varphi\|_{0,\mathbb{C}^n} \leqq \|f\|_{0,D}$.*

(iii) *f_0 is continuous on $\bar D$ and holomorphic in some neighbourhood of*

$$D \cup \Big(\partial D \cap \bigcup_{j=1}^{N} H_{\xi_j}(\delta/2)\Big) . \tag{3.4.25}$$

Proof. Let χ_j $(j = 0, 1, \dots, N)$ be the same functions as in the proof of Theorem 3.4.2. Define $f_0 := L_{\partial D}\chi_0 f$, $f_j := L_{\partial D}\chi_j(f - f(\xi_j))$ for $j = 1, \dots, N$, $\varphi := \sum_{j=1}^{N} \chi_j f(\xi_j)$. Since $\bar\partial f = 0$, it follows from the Leray formula stated in Theorem 3.1.3 (i) and the definition of $L_{\partial D}$ that $L_{\partial D}f = L_D f = f$. Together with $\sum_{j=0}^{N} \chi_j = 1$ this gives (3.4.23).

Now we prove condition (i). In view of Proposition 3.4.7 (i), the functions $f_1, \dots, f_N$ are continuous on $\bar D$. Since $\operatorname{supp} \chi_j \subset\subset \tilde H_{\xi_j}(\delta)$, it follows that f_j is holomorphic in some neighbourhood of $\bar D \setminus (\partial D \cap H_{\xi_j}(\delta))$. Since $\operatorname{supp} \bar\partial\chi_j(f - f(\xi_j)) \subseteqq H_{\xi_j}(\delta)$, and since by (3.4.14) and (3.4.15), for $\zeta \in H_{\xi_j}(\delta)$,

$$\|\bar\partial\chi_j(\zeta)\,(f(\zeta) - f(\xi_j))\| \leqq C d(\xi_1, \dots, \xi_N; \delta)\,(1/\delta + \|d\varrho(\zeta)\|/\delta^2)\,\omega(\delta, f)$$

and

$$\|\bar\partial\chi_j(\zeta)\,(f(\zeta) - f(\xi_j)) \wedge \bar\partial\varrho(\zeta)\| \leqq C d(\xi_1, \dots, \xi_N; \delta)\,\frac{\|d\varrho(\zeta)\|}{\delta}\,\omega(\delta, f) ,$$

it follows from Proposition 3.4.7 (ii) that for some $C' < \infty$ and all $z \in \partial D$

$$|f_j(z)| \leqq C'[|\chi_j(z)\,(f(z) - f(\xi_j))| + \min\{1, [\delta/\operatorname{dist}(z, H_{\xi_j}(\delta))]^{2n-1}\}]\, d(\xi_1, \dots, \xi_N; \delta)\, \omega(\delta, f)\,.$$

This implies (3.4.24), because $\chi_j(z) = 0$ for $z \notin H_{\xi_j}(\delta)$ and, therefore, for some $C'' < \infty$ and all $z \in \partial D$,

$$|\chi_j(z)\,(f(z) - f(\xi_j))| \leqq C'' \min\{1, [\delta/\operatorname{dist}(z, H_{\xi_j}(\delta))]^{2n-1}\}\,.$$

Condition (ii) immediately follows from the definition of φ.

Finally we prove condition (iii). By Proposition 3.4.7 (i) the function f_0 is continuous on $\overline{D}$. Since $\chi_0 = 0$ on $\bigcup_{j=1}^{N} \widetilde{H}_{\xi_j}(\delta/2)$, f_0 is holomorphic in some neighbourhood of the set (3.4.25). ■

3.5. Uniform approximation

Notation. Throughout this section $D \subset\subset \mathbb{C}^n$ is a strictly pseudoconvex, open set (not necessarily $D \neq \emptyset$), and ϱ is a strictly plurisubharmonic C^2-function in a neighbourhood θ of ∂D such that

$$D \cap \theta = \{z \in \theta : \varrho(z) < 0\} \quad \text{and} \quad N(\varrho) := \{z \in \theta : \varrho(z) = 0\} \subset\subset \theta\,.$$

Set $K(\varrho) := D \cup N(\varrho)$.

In the present section we prove the following

3.5.1. Theorem. *Every continuous function f on $K(\varrho)$ that is holomorphic in D can be approximated uniformly on $K(\varrho)$ by functions which are holomorphic in some neighbourhood of $K(\varrho)$.*

Remark. This theorem contains two special cases. If $D = \emptyset$, then $K(\varrho) = N(\varrho)$ is the zero set of a non-negative strictly plurisubharmonic function defined in a neighbourhood of $K(\varrho)$. In this case the theorem asserts that every continuous function on $K(\varrho)$ can be approximated uniformly on $K(\varrho)$ by functions which are holomorphic in a neighbourhood of $K(\varrho)$. If $D \neq \emptyset$ and $N(\varrho) = \partial D$, then $K(\varrho) = \overline{D}$ is the closure of a strictly pseudoconvex open set $D \subset\subset \mathbb{C}^n$. Recall (Theorem 1.5.19) that for every strictly pseudoconvex open set D, the function ϱ can be chosen so that $N(\varrho) = \partial D$.

Before starting the proof of Theorem 3.5.1 we give two lemmas.

Notation. Further on in this section we use the notations introduced in Section 3.4.

3.5.2. Lemma. *For sufficiently small $\delta > 0$ and every C^∞-function φ on $\mathbb{C}^n$ we define $U_\delta(\overline{D}) := D \cup \{z \in \mathbb{C}^n : \operatorname{dist}(z, \overline{D}) < \delta\}$ and for all $z \in U_\delta(\overline{D})$*

$$L_\delta\varphi(z) := -\frac{(n-1)!}{(2\pi i)^n} \int_{V_2 \setminus U_\delta(\overline{D})} \bar\partial(\vartheta(\zeta)\,\varphi(\zeta)) \wedge \frac{\omega'_\zeta(w(z,\zeta)) \wedge \omega(\zeta)}{\Phi^n(z,\zeta)}\,.$$

Then $L_\delta\varphi$ is holomorphic in $U_\delta(\overline{D})$ and

$$\lim_{\delta \to 0} \|L_\delta\varphi - L_{\partial D}\varphi\|_{0,D} = 0\,.$$

Proof. The conclusion that $L_\delta\varphi$ is holomorphic in $U_\delta(\overline{D})$ follows from the fact that $w(z,\zeta)$ and $\Phi(z,\zeta)$ are holomorphic in z and $\Phi(z,\zeta) \neq 0$ for $z \in U_\delta(\overline{D})$ and $\zeta \in V_2 \setminus U_\delta(\overline{D})$ (cp. Construction 3.1.1, Lemma A, conditions (i), (ii), and estimate (3.1.8)). Further, from (3.4.16′) it follows that for some $C_1 < \infty$

$$|L_\delta \varphi(z) - L_{\partial D}\varphi(z)| \leqq C_1 \int\limits_{U_\delta(\bar{D})\setminus D} \frac{|w(z,\zeta)|\, d\sigma_{2n}}{|\Phi(z,\zeta)|^n}, \qquad z \in D,$$

where $d\sigma_{2n}$ is the Lebesgue measure in $\mathbb{C}^n$. From Construction 3.1.1, condition (ii), and estimate (3.1.8) one obtains that for some $\alpha > 0$

$$|\Phi(z,\zeta)| \geqq \alpha(|\varrho(\zeta) - \varrho(z)| + |\zeta - z|^2) \quad \text{for} \quad z \in D, \quad \zeta \in U_\delta(\bar{D}) \setminus D.$$

Together with Lemma 3.2.4 (ii) and estimate (3.2.14) in Lemma 3.2.5 this implies that for some $C_2 < \infty$

$$|L_\delta\varphi(z) - L_{\partial D}\varphi(z)|$$
$$\leqq C_2 \int\limits_{U_\delta(\bar{D})\setminus D} \frac{\|d_\zeta Q(z,\zeta)\|\, d\sigma_{2n}}{(|Q(z,\zeta)| + |\zeta - z|^2)\, |\zeta - z|^{2n-2}} + C_2 \int\limits_{U_\delta(\bar{D})\setminus D} \frac{d\sigma_{2n}}{|\zeta - z|^{2n-1}},$$

where $d\sigma_{2n}$ is the Lebesgue measure in $\mathbb{C}^n$. The second integral on the right-hand side tends to zero for $\delta \to 0$. Therefore, we only have to consider the first integral. Since, by (3.2.8), $Q(z,\zeta) = O(|\zeta - z|)$, for some $C_3 < \infty$ we have

$$\int\limits_{U_\delta(\bar{D})\setminus D} \frac{\|d_\zeta Q(z,\zeta)\|\, d\sigma_{2n}}{(|Q(z,\zeta)| + |\zeta - z|^2)\, |\zeta - z|^{2n-2}}$$
$$\leqq C_3 \int\limits_{U_\delta(\bar{D})\setminus D} \frac{\|d_\zeta Q(z,\zeta)\|\, d\sigma_{2n}}{(|Q(z,\zeta)| + (|Q(z,\zeta)| + |\zeta - z|)^2)\, (|Q(z,\zeta)| + |\zeta - z|)^{2n-2}}.$$

In view of Proposition 4, Appendix 1, we thus obtain for some $C_4 < \infty$,

$$\int\limits_{U_\delta(\bar{D})\setminus D} \frac{\|d_\zeta Q(z,\zeta)\|\, d\sigma_{2n}}{(|Q(z,\zeta)| + |\zeta - z|^2)\, |\zeta - z|^{2n-2}} \leqq C_4 \int\limits_{t \in T_z(U_\delta(\bar{D})\setminus D)} \frac{dt_1 \wedge \ldots \wedge dt_{2n}}{(|t_1| + |t|^2)\, |t|^{2n-2}},$$
$$z \in D,$$

where, if $x_j(\zeta)$ are the real coordinates of $\zeta \in \mathbb{C}^n$,

$$T_z(U_\delta(\bar{D}) \setminus D) := \bigcup_{j=1}^{2n} \{(|Q(z,\zeta)|, |x_1(\zeta - z)|, \ldots \overset{\wedge}{\underset{j}{}} \ldots, |x_{2n}(\zeta - z)|) : \zeta \in U_\delta(\bar{D}) \setminus D\}.$$

In view of Proposition 5 (ii), Appendix 1, the integral on the right-hand side converges for every $\delta > 0$. Since the integrand does not depend on δ and since the measure of $T_z(U_\delta(\bar{D}) \setminus D)$ tends to zero for $\delta \to 0$, uniformly in z, this completes the proof. ■

Notation. Set $\varrho_m(\zeta) := \varrho(\zeta) - 1/m$ for $m = 1, 2, \ldots$, and let m_0, $N(\varrho_m)$, $\tilde{\Phi}_m$, Φ_m, w^m, D_m be as in Construction 3.1.1, Lemma A, condition (iv), and Lemma B, property (b). Let V_2 and χ be as in Definition 3.1.2, suppose that $D_m \subseteqq D \cup V_2$ and $N(\varrho_m) \subseteqq V_2$ for $m \geqq m_0$, and let L_{D_m}, R_{D_m} $(m \geqq m_0)$ be the operators introduced in Definition 3.1.2. Further on, we need the estimates given in Lemmas 3.2.4 and 3.2.5 "uniformly in m".

Therefore, we prove

3.5.3. Lemma

$$|\Phi_m(z,\zeta)| \geqq \alpha(|P(z,\zeta)| + |Q(z,\zeta)| + |\zeta - z|^2) \quad \text{for} \quad z \in V_2 \cap D_m,$$
$$\zeta \in \partial D_m\,, \tag{3.5.1}$$
$$|\tilde{\Phi}_m(z,\zeta)| \geqq \alpha(|P(z,\zeta)| + |Q(z,\zeta)| + |\zeta - z|^2) \quad \text{for} \quad z,\zeta \in V_2 \cap D_m\,, \tag{3.5.2}$$
$$|w^m(z,\zeta)| \leqq C(||d\varrho(\zeta)|| + |\zeta - z|) \quad \text{for} \quad \zeta, z \in V_2\,, \tag{3.5.3}$$
$$\left|\frac{\partial\tilde{\Phi}_m(z,\zeta)}{\partial\bar{\zeta}_j}\right| \leqq C\left(\left|\frac{\partial\varrho(\zeta)}{\partial\bar{\zeta}_j}\right| + |\zeta - z| + |\varrho_m(\zeta)|\right) \quad \text{for} \quad \zeta, z \in V_2; \quad j = 1, \dots, n\,. \tag{3.5.4}$$

Proof. Let $\varepsilon > 0$, M, M_m, $\tilde{M}$, $\tilde{M}_m$ be as in Construction 3.1.1.

(3.5.1). It follows from Construction 3.1.1, Lemma A that $\Phi_m(z,\zeta) = F(z,\zeta)\,M_m(z,\zeta)$ for $\zeta, z \in V_2$; $|\zeta - z| \leqq \varepsilon$; $m \geqq m_0$ (remark that $d\varrho_m = d\varrho$ and therefore $F_m = F$). In view of the uniform convergence $M_m \to M$ it follows that for some $\alpha_1 > 0$,

$$|\Phi_m(z,\zeta)| \geqq \alpha_1\,|F(z,\zeta)| \quad \text{for} \quad \zeta, z \in V_2; \qquad |\zeta - z| \leqq \varepsilon; \qquad m \geqq m_0\,.$$

If $z \in V_2 \cap D_m$, $\zeta \in \partial D_m$, then $\varrho(z) < 1/m$, $\varrho(\zeta) = 1/m$ and therefore $\varrho(\zeta) - \varrho(z) = |\varrho(\zeta) - \varrho(z)|$. Taking into account (3.1.8), thus we obtain that for some $\alpha_2 > 0$,

$$|\Phi_m(z,\zeta)| \geqq \alpha_2(|\varrho(\zeta) - \varrho(z)| + |\mathrm{Im}\,F(z,\zeta)| + |\zeta - z|^2)$$

for $z \in V_2 \cap D_m, \zeta \in \partial D_m, |\zeta - z| \leqq \varepsilon, m \geqq m_0$. Together with (3.2.6) and (3.2.8) this implies that for some $\alpha_3 > 0$, $|\Phi_m(z,\zeta)| \geqq \alpha_3(|P(z,\zeta)| + |Q(z,\zeta)| + |\zeta - z|^2)$ for $z \in V_2 \cap D_m$; $\zeta \in \partial D_m$; $|\zeta - z| \leqq \varepsilon$; $m \geqq m_0$. Since $\Phi_m(z,\zeta) \neq 0$ if $\zeta, z \in V_2$ and $|\zeta - z| \geqq \varepsilon$, and in view of the uniform convergence $\Phi_m \to \Phi$, the last estimate holds for $|\zeta - z| \geqq \varepsilon$, too, that is, (3.5.1) is proved.

(3.5.2). From Construction 3.1.1, Lemma A it follows that

$$\tilde{\Phi}_m(z,\zeta) = \big(F(z,\zeta) - 2\varrho_m(\zeta)\big)\,\tilde{M}_m(z,\zeta) \quad \text{for} \quad \zeta, z \in V_2; \qquad |\zeta - z| \leqq \varepsilon;$$
$$m \geqq m_0\,. \tag{3.5.5}$$

In view of the uniform convergence $\tilde{M}_m \to \tilde{M}$ it follows that for some $\alpha_1 > 0$, $|\tilde{\Phi}_m(z,\zeta)| \geqq \alpha_1\,|F(z,\zeta) - 2\varrho_m(\zeta)|$ for ζ, $z \in V_2$; $|\zeta - z| \leqq \varepsilon$; $m \geqq m_0$. Since by (3.1.8), for all $\zeta, z \in V_2$ with $|\zeta - z| \leqq \varepsilon$,

$$\mathrm{Re}\,F(z,\zeta) \geqq \varrho(\zeta) - \varrho(z) + \beta|\zeta - z|^2 = \varrho_m(\zeta) - \varrho_m(z) + \beta\,|\zeta - z|^2\,,$$

and since $\varrho_m < 0$ in D_m, thus we obtain that for some $\alpha_2 > 0$

$$|\tilde{\Phi}_m(z,\zeta)| \geqq \alpha_2(|\mathrm{Im}\,F(z,\zeta)| + |\varrho(\zeta) - \varrho(z)| + |\zeta - z|^2) \quad \text{for}$$
$$\zeta, z \in V_2 \cap D_m; \qquad |\zeta - z| \leqq \varepsilon; \qquad m \geqq m_0\,.$$

Together with (3.2.6) and (3.2.8) this implies that (3.5.2) holds for $|\zeta - z| \leqq \varepsilon$. Taking into account that $\tilde{\Phi}_m(z,\zeta) \neq 0$ for $\zeta, z \in V_2$ with $|\zeta - z| \geqq \varepsilon$ and that $\tilde{\Phi}_m \to \tilde{\Phi}$, we conclude the proof of (3.5.2).

(3.5.3). Since $\langle w^m(z,\zeta), \zeta - z\rangle = F(z,\zeta)\,M_m(z,\zeta)$ for $\zeta, z \in V_2$ with $|\zeta - z| \leqq \varepsilon$, we have

$$w_j^m(z,\zeta) = \frac{\partial F(z,\zeta)}{\partial \zeta_j}\,M_m(z,\zeta) + F(z,\zeta)\,\frac{\partial M_m(z,\zeta)}{\partial \zeta_j} + \mathrm{O}(|\zeta - z|)$$

for $\zeta, z \in V_2$, $1 \leqq j \leqq n$, $m \geqq m_0$, where $\mathrm{O}(|\zeta - z|)$ is uniform in m, in view of the uniform convergence $w^m \to w$ together with the first-order derivatives. Taking into account the uniform convergence $M_m \to M$ together with the first-order derivatives, and taking into account that $F(z,\zeta) = O(|\zeta - z|)$ and $\partial_\zeta F(z,\zeta) = 2\partial\varrho(\zeta) + O(|\zeta - z|)$, we conclude that (3.5.3) holds.

(3.5.4). We obtain (3.5.4) from (3.5.5) if we take into account the relations $F(z,\zeta) = O(|\zeta - z|)$, $\bar\partial_\zeta F(z,\zeta) = O(|\zeta - z|)$, and the uniform convergence $\widetilde{M}_m \to \widetilde{M}$ together with the first-order derivatives. ■

3.5.4. Lemma. *For every C^∞-function f in $\mathbb{C}^n$ that is holomorphic in some neighbourhood of $\bar D$,*

$$\lim_{m\to\infty} \|f - L_{D_m} f\|_{0,K(\varrho)} = 0 .$$

Proof. In view of the Leray formula stated in Theorem 3.1.3 (i), $f = L_{D_m} f + R_{D_m} \bar\partial f$ in D_m. Therefore, we have to prove that

$$\lim_{m\to\infty} \|R_{D_m} \bar\partial f\|_{0,K(\varrho)} = 0 .$$

By hypothesis, $\bar\partial f = 0$ in some neighbourhood $U_{\bar D}$ of $\bar D$. Consequently, for all $z \in K(\varrho)$

$$R_{D_m} \bar\partial f(z) = \frac{n!}{(2\pi i)^n} \int\limits_{(D_m \setminus U_{\bar D}) \times [0,1]} \bar\partial f(\zeta) \wedge \bar\omega\left((1-\lambda)\frac{\chi(\zeta)\, w^m(z,\zeta)}{\widetilde{\Phi}_m(z,\zeta)} + \lambda \frac{\bar\zeta - \bar z}{|\zeta - z|^2}\right) \wedge \omega(\zeta) . \tag{3.5.6}$$

As in the proof of Theorem 3.2.2, we now obtain that for $z \in K(\varrho)$ and $\zeta \in D_m \setminus U_{\bar D}$ relation (3.2.23) remains valid if we replace w by w_m and $\widetilde{\Phi}$ by $\widetilde{\Phi}_m$. Further, from (3.5.2) and (3.2.8) it follows that for some $\alpha_1 > 0$

$$|\widetilde{\Phi}_m(z,\zeta)| \geqq \alpha_1(|\varrho(\zeta)| + |\zeta - z|^2) \quad \text{for} \quad z \in K(\varrho), \quad \zeta \in D_m \setminus U_{\bar D}, \quad m \geqq m_0.$$

Using (3.5.3) and (3.5.4), we can therefore proceed further as in the proof of Theorem 3.2.2, and we conclude that, for $z \in K(\varrho)$, $\zeta \in D_m \setminus U_{\bar D}$, $m \geqq m_0$, estimates (3.2.23') remain valid if we replace w by w^m and $\widetilde{\Phi}$ by $\widetilde{\Phi}_m$. Thus we obtain that for some $C_1 < \infty$ and all $z \in K(\varrho)$, $m \geqq m_0$,

$$|R_{D_m} \bar\partial f(z)| \leqq C_1 \|\bar\partial f\|_{0,D_m} \left(\int\limits_{D_m \setminus U_{\bar D}} \frac{d\sigma_{2n}}{|\zeta - z|^{2n-1}} + \int\limits_{D_m \setminus U_{\bar D}} \frac{\|d\varrho(\zeta)\|\, d\sigma_{2n}}{|\widetilde{\Phi}_m(z,\zeta)|\, |\zeta - z|^{2n-2}} + \int\limits_{D_m \setminus U_{\bar D}} \frac{\|d\varrho(\zeta)\|^2\, d\sigma_{2n}}{|\widetilde{\Phi}_m(z,\zeta)|^2\, |\zeta - z|^{2n-3}} \right),$$

where $d\sigma_{2n}$ is the Lebesgue measure in $\mathbb{C}^n$. In view of (3.5.2) in Lemma 3.5.3, (3.2.9) and (3.2.10') in Lemma 3.2.4, and in view of the relations $Q(z,\zeta) = O(|\zeta - z|)$ and $P(z,\zeta) = O(|\zeta - z|)$ (which follow from (3.2.6) and (3.2.8)), this implies that for some $C_2 < \infty$ and all $z \in K(\varrho)$, $m \geqq m_0$,

$$|R_{D_m} \bar\partial f(z)| \leqq C_2 \|\bar\partial f\|_{0,D_m} \left(\int\limits_{D_m \setminus U_{\bar D}} \frac{d\sigma_{2n}}{|\zeta - z|^{2n-1}} + \int\limits_{D_m \setminus U_{\bar D}} \frac{\|d_\zeta Q\|\, d\sigma_{2n}}{(|Q| + (|\zeta - z| + |Q|)^2)\,(|\zeta - z| + |Q|)^{2n-2}} + \int\limits_{D_m \setminus U_{\bar D}} \frac{\|d_\zeta Q \wedge d_\zeta P\|\, d\sigma_{2n}}{(|Q| + |P| + (|\zeta - z| + |Q| + |P|)^2)^2\,(|\zeta - z| + |P| + |Q|)^{2n-3}} \right).$$

In view of Proposition 4, Appendix 1, it follows that for some $C_3 < \infty$ and all $z \in K(\varrho)$, $m \geqq m_0$,

$$|R_{D_m} \bar\partial f(z)| \leqq C_3 \|\bar\partial f\|_{0,D_m} \left(\int\limits_{t \in T_1(D_m \setminus U_{\bar D}, z)} \frac{dt_1 \wedge \ldots \wedge dt_{2n}}{|t|^{2n-1}} \right.$$

$$\left. + \int\limits_{t \in T_2(D_m \setminus U_{\bar D}, z)} \frac{dt_1 \wedge \ldots \wedge dt_{2n}}{(|t_1| + |t|^2)\, |t|^{2n-2}} + \int\limits_{t \in T_3(D_m \setminus U_{\bar D}, z)} \frac{dt_1 \wedge \ldots \wedge dt_{2n}}{(|t_1| + |t_2| + |t|^2)^2\, |t|^{2n-3}} \right),$$

where, if $x_j(\zeta)$ are the real coordinates of $\zeta \in \mathbb{C}^n$,

$$T_1(D_m \setminus U_{\bar D}, z) := \{(|x_1(\zeta - z)|, \ldots, |x_{2n}(\zeta - z)|) : \zeta \in D_m \setminus U_{\bar D}\},$$

$$T_2(D_m \setminus U_{\bar D}, z) := \bigcup_{j=1}^{2n} \{(|Q(z,\zeta)|, |x_1(\zeta - z)|, \ldots \underset{j}{\wedge} \ldots, |x_{2n}(\zeta - z)|) : \zeta \in D_m \setminus U_{\bar D}\},$$

$$T_3(D_m \setminus U_{\bar D}, z) := \bigcup_{1 \leqq j < k \leqq 2n} \{(|Q(z,\zeta)|, |P(z,\zeta)|, |x_1(\zeta - z)|, \ldots \underset{j,k}{\wedge} \ldots, |x_{2n}(\zeta - z)|) : \zeta \in D_m \setminus U_{\bar D}\}.$$

In view of Proposition 5 (i)—(iii), Appendix 1, the integrals on the right-hand side of this estimate converge for all $m \geqq m_0$ and $z \in K(\varrho)$. Since the integrands are independent of m, and since the measures of the sets $T_j(D_m \setminus U_{\bar D}, z)$ tend uniformly in $z \in K(\varrho)$ to zero if $m \to \infty$,[1]) this completes the proof of Lemma 3.5.4. ■

3.5.5. Proof of Theorem 3.5.1. For $n = 1$ this theorem follows from VITUŠKIN's theorem (VITUŠKIN [1]). Therefore, we can assume that $n \geqq 2$. Further, by Lemma 3.5.4 we only have to prove the following proposition: Every continuous function f on $\bar D$ that is holomorphic in D, can be approximated uniformly on $\bar D$ by holomorphic functions in some neighbourhood of $\bar D$.

In fact, combining this proposition and the Weierstrass approximation theorem, we obtain that every continuous function on $K(\varrho)$ that is holomorphic in D can be approximated uniformly on $K(\varrho)$ by functions which are C^∞ on $\mathbb{C}^n$ and holomorphic in some neighbourhood of $\bar D$, and by Lemma 3.5.4 for such functions the theorem is already proved.

And so, we consider a continuous function f on $\bar D$ that is holomorphic in D. Set

$$\operatorname{Crit}(\varrho) := \{\xi \in \partial D : d\varrho(\xi) = 0\}.$$

Since, by Theorem 1.4.21, Crit (ϱ) is locally contained in a smooth real n-dimensional manifold, there is a constant $C_1 < \infty$ depending on ϱ only such that: For every $\delta > 0$ there exists a finite set $S_\delta \subseteqq \operatorname{Crit}(\varrho)$ with the following properties:

(1) $\operatorname{Crit}(\varrho) \subseteqq \bigcup_{\xi \in S_\delta} \tilde H_\xi(\delta/2)$ [2]) (3.5.7)

(2) For every point $z \in \mathbb{C}^n$, there is a decomposition $S_\delta = \bigcup_{k=0}^{\infty} S^z_{\delta,k}$, where $S^z_{\delta,k} \cap S^z_{\delta,j} = \emptyset$ if $k \neq j$ ($S^z_{\delta,k} = \emptyset$ for k large enough) such that for $k = 0, 1, 2, \ldots$

$$\operatorname{card}(S^z_{\delta,k}) \leqq \min\{C_1, C_1 k^{n-1}\}, \tag{3.5.8}$$

[1]) This follows from the fact that the measure of $D_m \setminus U_{\bar D}$ tends to zero, which follows from the fact that the set $K(\varrho) \setminus U_{\bar D}$ is locally contained in a smooth real n-dimensional submanifold (Theorem 1.4.21).

[2]) Recall that, for $\xi \in \operatorname{Crit}(\varrho)$, $\tilde H_\xi(\delta) = E_\xi(\delta)$ is the Euclidean ball of radius δ centered at ξ (cp. Subsection 3.2.1).

where card $(S^z_{\delta,k})$ denotes the number of points in $S^z_{\delta,k}$, and for $k = 1, 2, \ldots$ and $\xi \in S^z_{\delta,k}$ we have the estimate

$$\operatorname{dist}\left(z, \tilde{H}_\xi(\delta)\right) \geqq k\delta\,. \tag{3.5.9}$$

(3) $$d(S_\delta, \delta) \leqq C_1\,, \tag{3.5.10}$$

where $d(S_\delta, \delta)$ is the order of the system of sets $\{\tilde{H}_\xi(\delta)\}_{\xi \in S_\delta}$, that is, at least one point belongs to $d(S_\delta, \delta)$ of the sets $\tilde{H}_\xi(\delta)$ $(\xi \in S_\delta)$, but no point belongs to $d(S_\delta, \delta) + 1$ of these sets.

In view of (3.5.7) and (3.5.10) and Theorem 3.4.8, one obtains the following proposition: There is a constant $C_2 < \infty$ such that for every $\delta > 0$ the function f admits a decomposition $f = f_0^\delta + L_{\partial D}\varphi^\delta + \sum_{\xi \in S_\delta} f_\xi^\delta$ with the following properties:

(i) For every $\xi \in S_\delta$, the function f_ξ^δ is continuous on $\bar{D}$ and holomorphic in some neighbourhood of $\bar{D} \setminus (\partial D \cap H_\xi(\delta))$, and for every $z \in \partial D$

$$|f_\xi^\delta(z)| \leqq C_1 C_2 \omega(\delta, f) \min\{1, [\delta/\operatorname{dist}(z, H_\xi(\delta))]^{2n-1}\}\,. \tag{3.5.11}$$

(ii) φ^δ is a C^∞-function on $\mathbb{C}^n$.

(iii) f_0^δ is continuous on $\bar{D}$ and holomorphic in some neighbourhood of $D \cup \operatorname{Crit}(\varrho)$.

We already know that the functions $L_{\partial D}\varphi^\delta$ and f_0^δ can be approximated uniformly on $\bar{D}$. For $L_{\partial D}\varphi^\delta$ this follows from Lemma 3.5.2. For f_0^δ we can use the following argument: Let U be a neighbourhood of Crit (ϱ) such that f_0^δ is holomorphic in $D \cup U$. Then we can choose a strictly pseudoconvex open set D' with C^2-boundary such that $\bar{D} \subseteqq \bar{D}' \subseteqq \bar{D} \cup U$, and the assertion follows from Theorem 2.9.2.

To complete the proof it is therefore sufficient to show that

$$\lim_{\delta \to 0} \max_{z \in D} \Big|\sum_{\xi \in S_\delta} f_\xi^\delta(z)\Big| = 0\,. \tag{3.5.12}$$

From (3.5.11), (3.5.8), and (3.5.9) it follows that $\sum_{\xi \in S^z_{\delta,0}} |f_\xi^\delta(z)| \leqq C_1^2 C_2 \omega(\delta, f)$ for $z \in \partial D$, and for $k = 1, 2, \ldots,$ $\sum_{\xi \in S^z_{\delta,k}} |f_\xi^\delta(z)| \leqq C_1^2 C_2 \omega(\delta, f)\, 1/k^n$ for $z \in \partial D$. Consequently,

$$\Big|\sum_{\xi \in S_\delta} f_\xi^\delta(z)\Big| \leqq C_1^2 C_2 \omega(\delta, f) \left(1 + \sum_{k=1}^{\infty} \frac{1}{k^n}\right) \quad \text{for} \quad z \in \partial D\,.$$

This implies (3.5.12), because $n \geqq 2$, $\omega(\delta, f) \to 0$ for $\delta \to 0$, and the functions $\sum_{\xi \in S_\delta} f_\xi^\delta(z)$ are holomorphic (maximum principle!). ■

3.6. Bounded extension of holomorphic functions from complex planes

In this section we prove that every bounded holomorphic function f defined on the intersection of a strictly pseudoconvex open set D with a complex plane X admits a bounded holomorphic continuation to D. If, moreover, f is continuous on $\bar{D} \cap X$, then this continuation can be chosen to be continuous on $\bar{D}$. In Section 4.11 we shall generalize this result to the case that X is an arbitrary closed complex submanifold in some neighbourhood of $\bar{D}$.

3.6.1. Notation and preliminaries. Let $D \subset\subset \mathbb{C}^n$ be a strictly pseudoconvex open set (with not necessarily smooth boundary). Let ϱ, F, $\tilde{\Phi}$, Φ, w, χ, and $N(\varrho) \subseteqq V_2$ be as in Construction 3.1.1 and Definition 3.1.2. Set $\varrho_m(\zeta) := \varrho(\zeta) - 1/m$ for $m = 1, 2, \ldots$, and let m_0, $N(\varrho_m)$, $\tilde{\Phi}_m$, Φ_m, w^m, D_m be as in Construction 3.1.1, Lemma A, condition (iv) and Lemma B, property (b). Suppose that $D_m \subseteqq D \cup V_2$ and $N(\varrho_m) \subseteqq V_2$ for $m \geqq m_0$. Set

$$X := \{z \in \mathbb{C}^n : z_n = 0\} .$$

For $\zeta = (\zeta_1, \ldots, \zeta_n) \in \mathbb{C}^n$ we write $\zeta' := (\zeta_1, \ldots, \zeta_{n-1})$,

$$\bar{\partial}_{\zeta'} := \sum_{j=1}^{n-1} \frac{\partial}{\partial \bar{\zeta}_j} \, d\bar{\zeta}_j , \quad \partial_{\zeta'} := \sum_{j=1}^{n-1} \frac{\partial}{\partial \zeta_j} d\zeta_j ,$$

$$d_{\zeta'} := \bar{\partial}_{\zeta'} + \partial_{\zeta'} , \qquad \omega_{\zeta'}(\zeta) = d\zeta_1 \wedge \ldots \wedge d\zeta_{n-1} .$$

Sometimes we shall write $\varrho_\infty := \varrho$, $\Phi_\infty := \Phi$ etc. For $m_0 \leqq m \leqq \infty$ we introduce the notation $(w^m(z, \zeta))' := (w_1^m(z, \zeta), \ldots, w_{n-1}^m(z, \zeta))$ and the differential form

$$\bar{\omega}_{\zeta'}\left(\frac{\chi(\zeta)\,(w^m(z, \zeta))'}{\tilde{\Phi}_m(z, \zeta)}\right) := \bigwedge_{j=1}^{n-1} \bar{\partial}_{\zeta'} \frac{\chi(\zeta)\, w_j^m(z, \zeta)}{\tilde{\Phi}_m(z, \zeta)} . \tag{3.6.1}$$

From Construction 3.1.1, Lemma A, condition (ii), and estimate (3.1.8) it follows that for all $m_0 \leqq m \leqq \infty$ there exists a neighbourhood $U_{\partial D_m \setminus X}$ of $\partial D_m \setminus X$ such that

$$\tilde{\Phi}_m(z, \zeta) \neq 0 \quad \text{for} \quad \zeta \in X \cap \bar{D}_m , \qquad z \in D_m \cup U_{\partial D_m \setminus X} .$$

Consequently, for every fixed $z \in D_m \cup U_{\partial D_m \setminus X}$, $m_0 \leqq m \leqq \infty$, the differential form (3.6.1) is continuous on $X \cap \bar{D}_m$. For every bounded holomorphic function f on $X \cap D_m$ and $z \in D_m \cup U_{\partial D_m \setminus X}$, therefore we can define

$$E_m f(z) := \frac{(n-1)!}{(2\pi i)^{n-1}} \int_{X \cap D_m} f(\zeta)\, \bar{\omega}_{\zeta'}\left(\frac{\chi(\zeta)\,(w^m(z, \zeta))'}{\tilde{\Phi}_m(z, \zeta)}\right) \wedge \omega_{\zeta'}(\zeta) .$$

3.6.2. Lemma. *For every bounded, holomorphic function f on $X \cap D_m$, $E_m f$ is holomorphic on $D_m \cup U_{\partial D_m \setminus X}$ and*

$$f(z) = E_m f(z) \quad \textit{for} \quad z \in X \cap D_m \qquad (m_0 \leqq m \leqq \infty) .$$

Proof. That $E_m f$ is holomorphic follows from the fact that $\tilde{\Phi}_m(z, \zeta)$ and $w^m(z, \zeta)$ are holomorphic in z. The equality $f = Ef|_{X \cap D_m}$ follows from the Leray formula stated in Theorem 3.1.3 (i) if we take into account that Construction 3.1.1 applied to $X \cap D_m$ gives the restriction of $\tilde{\Phi}_m$ and $(w^m)'$ to X. ■

3.6.3. Further notation and preliminaries. For $z \in V_2$, $\zeta \in V_2 \cap D_m$, and $m_0 \leqq m \leqq \infty$, we define

$$\Phi_m^*(z, \zeta) := \Phi_m(\zeta, z) , \qquad {}^*w^m(z, \zeta) := -w^m(\zeta, z) ,$$

$$({}^*w^m(z, \zeta))' := ({}^*w_1^m(z, \zeta), \ldots, {}^*w_{n-1}^m(z, \zeta)) .$$

In view of (3.5.1) and (3.5.2) in Lemma 3.5.3, $\Phi_m^*(z, \zeta) \neq 0$ and $\tilde{\Phi}_m(z, \zeta) \neq 0$ for $z \in \partial D_m \setminus X$, $\zeta \in X \cap \bar{D}_m$, $m_0 \leqq m \leqq \infty$. Consequently, for every fixed $z \in \partial D_m \setminus X$, $m_0 \leqq m \leqq \infty$, the differential form (for $\det_{1,n-1}$ cp. Definition 1.7.6)

$$\det_{1,n-1}\left(\frac{{}^*w^m(z, \zeta)}{\Phi_m^*(z, \zeta)}, \bar{\partial}_{\zeta'} \frac{\chi(\zeta)\, w^m(z, \zeta)}{\tilde{\Phi}_m(z, \zeta)}\right)$$

is continuous on $\bar{D}_m \cap X$.

3.6.4. Lemma. *For every bounded holomorphic function f on $X \cap D_m$ and all $z \in \partial D_m \setminus X$, $m_0 \leqq m \leqq \infty$, we have*

$$E_m f(z) = z_n \frac{(-1)^n}{(2\pi i)^{n-1}} \int\limits_{\zeta \in X \cap D_m} f(\zeta) \det_{1,n-1}\left(\frac{{}^*w^m(z,\zeta)}{\Phi_m^*(z,\zeta)}, \bar\partial_{\zeta'} \frac{\chi(\zeta)\, w^m(z,\zeta)}{\tilde\Phi_m(z,\zeta)}\right) \wedge \omega_{\zeta'}(\zeta) .$$

Proof. In this proof we omit the index m. Fix $z \in \partial D \setminus X$. Then

$$(-1)^n \det_{1,n-1}\left(\frac{{}^*w}{\Phi^*}, \bar\partial_{\zeta'} \frac{\chi w}{\tilde\Phi}\right)$$
$$= -\frac{{}^*w_n}{\Phi^*} \det_{n-1}\left(\bar\partial_{\zeta'} \frac{\chi w'}{\tilde\Phi}\right) + (n-1)\, \bar\partial_{\zeta'} \frac{\chi w_n}{\tilde\Phi} \wedge \det_{1,n-2}\left(\frac{{}^*w'}{\Phi^*}, \bar\partial_{\zeta'} \frac{\chi w'}{\tilde\Phi}\right).$$

Since

$$\det_{n-1}\left(\bar\partial_{\zeta'} \frac{\chi w'}{\tilde\Phi}\right) = (n-1)!\, \bar\omega_{\zeta'}\left(\frac{\chi w'}{\tilde\Phi}\right)$$

and since, by Construction 3.1.1, Lemma B, for $\zeta \in X \cap D$,

$$-z_n \frac{{}^*w_n}{\Phi^*} = 1 - \sum_{j=1}^{n-1} (\zeta_j - z_j) \frac{{}^*w_j}{\Phi^*},$$

$$z_n \bar\partial_{\zeta'} \frac{\chi w_n}{\tilde\Phi} = \bar\partial_{\zeta'}\left[\sum_{j=1}^{n-1} (\zeta_j - z_j) \frac{\chi w_j}{\tilde\Phi} - \frac{\chi\Phi}{\tilde\Phi}\right] = \sum_{j=1}^{n-1} (\zeta_j - z_j)\, \bar\partial_{\zeta'} \frac{\chi w_j}{\tilde\Phi} - \bar\partial_{\zeta'} \frac{\chi\Phi}{\tilde\Phi},$$

it follows that for $\zeta \in X \cap D$,

$$z_n(-1)^n \det_{1,n-1}\left(\frac{{}^*w}{\Phi^*}, \bar\partial_{\zeta'} \frac{\chi w}{\tilde\Phi}\right)$$
$$= \left(1 - \sum_{j=1}^{n-1} (\zeta_j - z_j) \frac{{}^*w_j}{\Phi^*}\right)(n-1)!\, \bar\omega_{\zeta'}\left(\frac{\chi w'}{\tilde\Phi}\right)$$
$$+ (n-1) \sum_{j=1}^{n-1} (\zeta_j - z_j)\, \bar\partial_{\zeta'} \frac{\chi w_j}{\tilde\Phi} \wedge \det_{1,n-2}\left(\frac{{}^*w'}{\Phi^*}, \bar\partial_{\zeta'} \frac{\chi w'}{\tilde\Phi}\right)$$
$$- (n-1)\, \bar\partial_{\zeta'} \frac{\chi\Phi}{\tilde\Phi} \wedge \det_{1,n-2}\left(\frac{{}^*w'}{\Phi^*}, \bar\partial_{\zeta'} \frac{\chi w'}{\tilde\Phi}\right). \tag{3.6.2}$$

A computation gives that for $j = 1, \dots, n-1$,

$$\bar\partial_{\zeta'} \frac{\chi w_j}{\tilde\Phi} \wedge \det_{1,n-2}\left(\frac{{}^*w'}{\Phi^*}, \bar\partial_{\zeta'} \frac{\chi w'}{\tilde\Phi}\right) = \frac{{}^*w_j}{\Phi^*} (n-2)!\, \bar\omega_{\zeta'}\left(\frac{\chi w'}{\tilde\Phi}\right).$$

Together with (3.6.2) this implies that for $\zeta \in X \cap D$,

$$z_n(-1)^n \det_{1,n-1}\left(\frac{{}^*w}{\Phi^*}, \bar\partial_{\zeta'} \frac{\chi w}{\tilde\Phi}\right)$$
$$= (n-1)!\, \bar\omega_{\zeta'}\left(\frac{\chi w'}{\tilde\Phi}\right) - (n-1)\, \bar\partial_{\zeta'} \frac{\chi\Phi}{\tilde\Phi} \wedge \det_{1,n-2}\left(\frac{{}^*w'}{\Phi^*}, \bar\partial_{\zeta'} \frac{\chi w'}{\tilde\Phi}\right).$$

To complete the proof therefore it is sufficient to prove that

$$I(z) := \int\limits_{X \cap D} f\, \bar\partial_{\zeta'} \frac{\chi\Phi}{\tilde\Phi} \wedge \det_{1,n-2}\left(\frac{{}^*w'}{\Phi^*}, \bar\partial_{\zeta'} \frac{\chi w'}{\tilde\Phi}\right) \wedge \omega_{\zeta'}(\zeta) = 0 .$$

To do this we would like to apply Stokes' formula. Since $\partial D \cap X$ is not necessarily smooth, we replace $X \cap D$ by open sets $D_\delta \subseteqq X \cap D$ with piecewise smooth boundary: Let $S'(\varrho) \subseteqq \partial D \cap X$ be the set of points $\zeta \in \partial D \cap X$ with $d_{\zeta'}\varrho(\zeta) = 0$. In view of Theorem 1.4.21, $S'(\varrho)$ is locally contained in a smooth real $(n-1)$-dimensional submanifold of X. Therefore, we can find neighbourhoods $U(\delta) \subseteqq X$ of $S'(\varrho)$, $\delta > 0$, such that for some constant $C < \infty$ and all $\delta > 0$

$$\operatorname{dist}(\zeta, \partial D \cap X) \leqq C\delta \quad \text{for} \quad \zeta \in U(\delta)\,, \tag{3.6.3}$$

$$\sigma_{2n-2}(U(\delta)) \leqq C\delta^{n-1} \leqq C\delta\,, \tag{3.6.4}$$

$$\sigma_{2n-3}(\partial U(\delta)) \leqq C\delta^{n-2} \leqq C\,, \tag{3.6.5}$$

where σ_{2n-2} and σ_{2n-3} denote the $(2n-2)$-dimensional and the $(2n-3)$-dimensional Euclidean volume, respectively. Set $D_\delta := (D \cap X) \setminus U(\delta)$. Then by (3.6.4)

$$I(z) = O(\delta) + \int_{D_\delta} f \bar\partial_{\zeta'} \frac{\chi\Phi}{\tilde\Phi} \wedge \det{}_{1,n-2}\left(\frac{{}^*w'}{\Phi^*}, \bar\partial_{\zeta'} \frac{\chi w'}{\tilde\Phi}\right) \wedge \omega_{\zeta'}(\zeta)\,.$$

Since $f(\zeta)$, ${}^*w(z,\zeta)$, and $\Phi^*(z,\zeta)$ are holomorphic in ζ, we have

$$d_{\zeta'}\left[f \det{}_{1,n-2}\left(\frac{{}^*w'}{\Phi^*}, \bar\partial_{\zeta'} \frac{\chi w'}{\tilde\Phi}\right) \wedge \omega_{\zeta'}(\zeta)\right] = 0\,. \tag{3.6.6}$$

Therefore, it follows from Stokes' formula that

$$I(z) = O(\delta) + \int_{\partial D_\delta} f \frac{\chi\Phi}{\tilde\Phi} \wedge \det{}_{1,n-2}\left(\frac{{}^*w'}{\Phi^*}, \bar\partial_{\zeta'} \frac{\chi w'}{\tilde\Phi}\right) \wedge \omega_{\zeta'}(\zeta)\,.$$

Since $\chi(\zeta)\,\Phi(z,\zeta)/\tilde\Phi(z,\zeta) = 1$ for $\zeta \in \partial D$, $\zeta \neq z$, and in view of (3.6.3) and (3.6.5), this implies that

$$I(z) = O(\delta) + \int_{\partial D_\delta} f \det{}_{1,n-2}\left(\frac{{}^*w'}{\Phi^*}, \bar\partial_{\zeta'} \frac{\chi w'}{\tilde\Phi}\right) \wedge \omega_{\zeta'}(\zeta)\,.$$

Therefore by (3.6.6) and Stokes' formula we can conclude that $I(z) = O(\delta)$, that is, $I(z) = 0$. ∎

For $\delta > 0$ and $z \in \partial D_m$ we denote by $H_z^m(\delta)$ the Hörmander ball with respect to ϱ_m, D_m introduced in Section 3.2.1.

3.6.5. Lemma. *There exists a number $\delta > 0$ such that for all $z \in \partial D_m \setminus X$ and $\zeta \in H_z^m\left(\delta \left|\frac{\partial\varrho(z)}{\partial z_n} z_n\right|^{1/2}\right) \cap V_2 \cap X$, $m_0 \leqq m \leqq \infty$, the following estimates hold:*

$$\|d_{z'}\varrho(z)\|\,|\zeta' - z'| \geqq \left|\frac{\partial\varrho(z)}{\partial z_n} z_n\right|, \tag{3.6.7}$$

$$\|d_{\zeta'}P(z,\zeta) \wedge d_{\zeta'}Q(z,\zeta)\| \geqq \frac{1}{\sqrt{2n}}\|d_{z'}\varrho(z)\|^2\,, \tag{3.6.8}$$

where P, Q are the polynomials from Lemma 3.2.4.

Proof. By definition of the Hörmander ball and by (3.2.2), for $z \in \partial D_m \setminus X$, $\zeta \in H_z^m\left(\delta \left|\frac{\partial\varrho(z)}{\partial z_n} z_n\right|^{1/2}\right) \cap V_2$, $m_0 \leqq m \leqq \infty$, we have the inequalities

$$|\zeta - z|^2 \leqq \delta^2 \left|\frac{\partial\varrho(z)}{\partial z_n} z_n\right| \tag{3.6.9}$$

and

$$|F(z,\zeta)| \leqq C\delta^2 \left|\frac{\partial\varrho(z)}{\partial z_n} z_n\right|.$$

Since

$$2\sum_{j=1}^{n} \frac{\partial\varrho(z)}{\partial z_j}(\zeta_j - z_j) = F(z,\zeta) + O(|\zeta - z|^2),$$

this implies that for some $C' < \infty$ and all $z \in \partial D_m \setminus X$, $\zeta \in H_z^m\left(\delta\left|\frac{\partial\varrho(z)}{\partial z_n} z_n\right|^{1/2}\right) \cap V_2 \cap X$

$$\left|-\frac{\partial\varrho(z)}{\partial z_n} z_n + \sum_{j=1}^{n-1} \frac{\partial\varrho(z)}{\partial z_j}(\zeta_j - z_j)\right| \leqq C'\delta^2 \left|\frac{\partial\varrho(z)}{\partial z_n} z_n\right|.$$

If $\delta^2 < 1/2C'$, we obtain therefore that for all $z \in \partial D_m \setminus X$, $\zeta \in H_z^m\left(\delta\left|\frac{\partial\varrho(z)}{\partial z_n} z_n\right|^{1/2}\right) \cap V_2 \cap X$, $m_0 \leqq m \leqq \infty$,

$$\left|\sum_{j=1}^{n-1} \frac{\partial\varrho(z)}{\partial z_j}(\zeta_j - z_j)\right| \geqq \frac{1}{2}\left|\frac{\partial\varrho(z)}{\partial z_n} z_n\right|.$$

This implies (3.6.7), because of the Schwarz inequality and the relation

$$||d_{z'}\varrho(z)|| = 2\left(\sum_{j=1}^{n-1}\left|\frac{\partial\varrho(z)}{\partial z_j}\right|^2\right)^{1/2}.$$

From (3.6.7) and (3.6.9) we obtain that for all $z \in \partial D_m \setminus X, \zeta \in H_z^m\left(\delta\left|\frac{\partial\varrho(z)}{\partial z_n} z_n\right|^{1/2}\right) \cap V_2 \cap X, m_0 \leqq m \leqq \infty$,

$$|\zeta - z| \leqq \delta^2 \, ||d_{z'}\varrho(z)|| . \tag{3.6.10}$$

In the same way as in the proof of (3.2.10') in Lemma 3.2.4, one can prove that for some $C'' < \infty$ and all $\zeta, z \in V_2$

$$||d_{\zeta'}P(z,\zeta) \wedge d_{\zeta'}Q(z,\zeta)|| \geqq \frac{1}{\sqrt{n}} ||d_{z'}\varrho(z)||^2 - C''(||d_{z'}\varrho(z)|| \, |\zeta - z| + |\zeta - z|^2).$$

Together with (3.6.10) this implies (3.6.8) if $\delta > 0$ is sufficiently small. ■

3.6.6. Lemma. *There is a constant $C < \infty$ such that for all $z \in \partial D_m \setminus X$ and $m_0 \leqq m \leqq \infty$ the following estimates hold:*

(i)

$$\int_{\zeta \in X \cap D_m} \frac{d\sigma_{2n-2}}{|\zeta - z|^{2n-1}} \leqq C/|z_n|,$$

(ii)

$$\int_{\zeta \in X \cap D_m \cap V_2} \frac{||d_{\zeta'}\varrho(\zeta)|| \, d\sigma_{2n-2}}{|\tilde{\Phi}_m(z,\zeta)\, \Phi_m^*(z,\zeta)| \, |\zeta - z|^{2n-4}} \leqq C/|z_n|,$$

(iii)

$$\int_{\zeta \in X \cap D_m \cap V_2} \frac{||d_{\zeta'}\varrho(\zeta)||^2 \, d\sigma_{2n-2}}{|\tilde{\Phi}_m(z,\zeta)|^2 \, |\Phi_m^*(z,\zeta)| \, |\zeta - z|^{2n-5}} \leqq C/|z_n|,$$

(iv)

$$\int_{\zeta \in X \cap D_m \cap V_2} \frac{|\partial\varrho(\zeta)/\partial\zeta_n| \, d\sigma_{2n-2}}{|\tilde{\Phi}_m(z,\zeta)\, \Phi_m^*(z,\zeta)| \, |\zeta - z|^{2n-4}} \leqq C/|z_n|,$$

(v)

$$\int\limits_{\zeta\in X\cap D_m\cap V_2} \frac{\|\mathrm{d}_{\zeta'}\varrho(\zeta)\|\,|\partial\varrho(\zeta)/\partial\zeta_n|\,\mathrm{d}\sigma_{2n-2}}{|\tilde{\Phi}_m(z,\zeta)|^2\,|\Phi_m^*(z,\zeta)|\,|\zeta - z|^{2n-5}} \leqq C/|z_n|\,,$$

where $\mathrm{d}\sigma_{2n-2}$ *is the Euclidean volume form on* X.

Proof. In this proof we use the notations $C, C', \ldots$ and $\alpha, \alpha', \ldots$ for "large" and "small" constants, respectively, as explained at the beginning of the proof of Lemma 3.2.6.

(i)

$$\int\limits_{X\cap D_m} \frac{\mathrm{d}\sigma_{2n-2}}{|\zeta - z|^{2n-1}} \leqq \int\limits_{X\cap D_m} \frac{\mathrm{d}\sigma_{2n-2}}{(|z_n|^2 + |\zeta' - z'|^2)^2\,|\zeta' - z'|^{2n-5}}$$

$$\leqq \int\limits_{\substack{t\in R^{2n-2}\\ |t|<\mathrm{diam}(D_m)}} \frac{\mathrm{d}t_1\wedge\ldots\wedge\mathrm{d}t_{2n-2}}{(|z_n|^2 + |t|^2)^2\,|t|^{2n-5}}\,,$$

where diam (D_m) is the Euclidean diameter of D_m. In view of Proposition 7, Appendix 1, this implies the required estimate.

(ii). In view of (3.2.11), (3.2.12), (3.5.1), (3.5.2), and (3.2.9) the integral in (ii) can be estimated by the integral in (i) and the following integral:

$$\int\limits_{X\cap D_m\cap V_2} \frac{\|\mathrm{d}_{\zeta'}Q(z,\zeta)\|\,\mathrm{d}\sigma_{2n-2}}{(|Q(z,\zeta)| + |\zeta - z|^2)^2\,|\zeta - z|^{2n-4}}\,.$$

Since by (3.2.8), $|Q(z,\zeta)| \leqq C\,|\zeta - z|$, and, consequently, for $\zeta\in X\cap D_m\cap V_2$, $z\in\partial D_m\setminus X$, $m_0\leqq m\leqq\infty$,

$$(|Q(z,\zeta)| + |\zeta - z|^2)^2\,|\zeta - z|^{2n-4}$$

$$\geqq \alpha\big(|z_n|^2 + |Q(z,\zeta)| + (|Q(z,\zeta)| + |\zeta' - z'|)^2\big)^2\,(|Q(z,\zeta)| + |\zeta' - z'|)^{2n-4}\,,$$

we can apply Proposition 4, and obtain that the last integral can be estimated by

$$\int\limits_{t\in R^{2n-2},\ |t|<R} \frac{\mathrm{d}t_1\wedge\ldots\wedge\mathrm{d}t_{2n-2}}{(|z_n|^2 + |t_1| + |t|^2)^2\,|t|^{2n-4}}\,,$$

where $R<\infty$ is so large that $|Q(z,\zeta)| + |\zeta' - z'| < R$ for all $z\in\partial D_m$, $\zeta\in X\cap D_m$, $m_0\leqq m\leqq\infty$. The required estimate now follows from Proposition 7, Appendix 1.

(iii). In the same way as in the proof of (3.2.10') we can show that for $\zeta, z\in V_2$

$$\|\mathrm{d}_{\zeta'}P(z,\zeta)\wedge\mathrm{d}_{\zeta'}Q(z,\zeta)\| \geqq \frac{1}{\sqrt{n}}\|\mathrm{d}_{\zeta'}\varrho(\zeta)\|^2 - C(\|\mathrm{d}_{\zeta'}\varrho(\zeta)\|\,|\zeta - z| + |\zeta - z|^2)\,.$$

Together with (3.2.11), (3.2.12), (3.5.1), (3.5.2) this implies that the integral (iii) can be estimated by the integrals in (i), (ii), and the following integral:

$$\int\limits_{X\cap D_m\cap V_2} \frac{\|\mathrm{d}_{\zeta'}P(z,\zeta)\wedge\mathrm{d}_{\zeta'}Q(z,\zeta)\|\,d\sigma_{2n-2}}{(|P(z,\zeta)| + |Q(z,\zeta)| + |\zeta - z|^2)^3\,|\zeta - z|^{2n-5}}\,.$$

Since by (3.2.6) and (3.2.8), $|P(z,\zeta)| + |Q(z,\zeta)| \leqq C\,|\zeta - z|$ and, consequently, for all $z\in\partial D_m\setminus X$, $\zeta\in X\cap D_m\cap V_2$, $m_0\leqq m\leqq\infty$,

$$(|P| + |Q| + |\zeta - z|^2)^3\,|\zeta - z|^{2n-5}$$

$$\geqq \big(|z_n|^2 + |P| + |Q| + (|P| + |Q| + |\zeta' - z'|)^2\big)^3\,(|P| + |Q| + |\zeta' - z'|)^{2n-5}\,,$$

we can apply Proposition 4, Appendix 1, and obtain that the last integral can be estimated by

$$\int\limits_{t\in R^{2n-2},\,|t|<R} \frac{dt_1 \wedge \dots \wedge dt_{2n-2}}{(|z_n|^2 + |t_1| + |t_2| + |t|^2)^3\, |t|^{2n-5}},$$

where $R < \infty$ is so large that $|P(z, \zeta)| + |Q(z, \zeta)| + |\zeta - z| < R$ for all $z \in \partial D_m$, $\zeta \in X \cap D_m \cap V_2$, $m_0 \leqq m \leqq \infty$. The required estimate now follows from Proposition 7, Appendix 1.

(iv). First we remark that, in view of (3.2.11), (3.2.12), (3.5.1), (3.5.2), and the estimate $|\partial\varrho(\zeta)/\partial\zeta_n| \leqq |\partial\varrho(z)/\partial z_n| + C\,|\zeta - z|$, the integral in part (iv) can be estimated by the integral in part (i) and the integral which is obtained from the integral in part (iv) after replacing $|\partial\varrho(\zeta)/\partial\zeta_n|$ by $|\partial\varrho(z)/\partial z_n|$, that is, we only have to prove that for all $z \in \partial D_m \setminus X$, $m_0 \leqq m \leqq \infty$,

$$\int\limits_{\zeta\in X\cap D_m\cap V_2} \frac{d\sigma_{2n-2}}{|\tilde{\Phi}_m(z, \zeta)\, \Phi_m^*(z, \zeta)|\, |\zeta - z|^{2n-4}} \leqq C \left|\frac{\partial\varrho(z)}{\partial z_n} z_n\right|^{-1}. \tag{3.6.11}$$

Let $\delta > 0$ be the number from Lemma 3.6.5. For simplicity we write

$$H_z^m := H_z^m\left(\delta \left|\frac{\partial\varrho(z)}{\partial z_n} z_n\right|^{1/2}\right).$$

In view of Construction 3.1.1, Lemma A, we have $|\tilde{\Phi}_m(z, \zeta)|, |\Phi_m^*(z, \zeta)| \geqq \alpha\, |\varrho_m(\zeta)|$ for $z \in \partial D_m, \zeta \in D_m \cap V_2$, $m_0 \leqq m \leqq \infty$. Together with (3.2.1) (it is clear from the proof of (3.2.1) that (3.2.1) holds uniformly in m), this implies that

$$|\tilde{\Phi}_m(z, \zeta)|, |\Phi_m^*(z, \zeta)| \geqq \alpha\delta^2 \left|\frac{\partial\varrho(z)}{\partial z_n} z_n\right|$$
$$\text{for} \quad z \in \partial D_m, \qquad \zeta \in (V_2 \cap D_m) \setminus H_z^m, \qquad m_0 \leqq m \leqq \infty. \tag{3.6.12}$$

Together with (3.2.11), (3.2.12), (3.5.1), (3.5.2) this implies that for all $z \in \partial D_m \setminus X$, $m_0 \leqq m \leqq \infty$,

$$\int\limits_{\zeta\in(X\cap V_2\cap D_m)\setminus H_z^m} \frac{d\sigma_{2n-2}}{|\tilde{\Phi}_m(z, \zeta)\, \Phi_m^*(z, \zeta)|\, |\zeta - z|^{2n-4}}$$
$$\leqq C \int\limits_{\zeta\in X\cap D_m} \frac{d\sigma_{2n-2}}{\left(\left|\frac{\partial\varrho(z)}{\partial z_n} z_n\right| + |\zeta' - z'|^2\right)^2 |\zeta' - z'|^{2n-4}}.$$

In view of Proposition 7, Appendix 1, the integral on the right-hand side can be estimated by $\left|\frac{\partial\varrho(z)}{\partial z_n} z_n\right|^{-1}$. Therefore to prove (3.6.11) it is sufficient to prove that for all $z \in \partial D_m \setminus X$, $m_0 \leqq m \leqq \infty$,

$$I_m(z) := \int\limits_{\zeta\in H_z^m\cap V_2\cap X} \frac{d\sigma_{2n-2}}{|\tilde{\Phi}_m(z, \zeta)\, \Phi_m^*(z, \zeta)|\, |\zeta - z|^{2n-4}} \leqq C \left|\frac{\partial\varrho(z)}{\partial z_n} z_n\right|^{-1}. \tag{3.6.13}$$

It follows from estimate (3.6.8) in Lemma 3.6.5 and the estimates (3.2.11), (3.2.12), (3.5.1), (3.5.2) that for all $z \in \partial D_m \setminus X$, $m_0 \leqq m \leqq \infty$,

$$I_m(z) \leqq C \,||d_{z'}\varrho(z)||^{-2} \int\limits_{\zeta \in H_z^m \cap V_z \cap X} \frac{||\mathrm{d}_{\zeta'} P(z,\zeta) \wedge \mathrm{d}_{\zeta'} Q(z,\zeta)|| \, \mathrm{d}\sigma_{2n-2}}{(|P(z,\zeta)| + |Q(z,\zeta)| + |\zeta - z|^2)^2 \, |\zeta - z|^{2n-4}}$$

$$\leqq C \,||\mathrm{d}_{z'}\varrho(z)||^{-2} \int\limits_{\zeta \in H_z^m \cap V_z \cap X} \frac{||\mathrm{d}_{\zeta'} P(z,\zeta) \wedge \mathrm{d}_{\zeta'} Q(z,\zeta)|| \, \mathrm{d}\sigma_{2n-2}}{(|P(z,\zeta)| + |Q(z,\zeta)| + |\zeta - z|^2)^{3/2} |\zeta - z|^{2n-3}}.$$

In view of estimate (3.6.7) in Lemma 3.6.5, this implies that for all $z \in \partial D_m \setminus X$, $m_0 \leqq m \leqq \infty$,

$$I_m(z) \leqq C \left| \frac{\partial \varrho(z)}{\partial z_n} z_n \right|^{-2} \int\limits_{\zeta \in H_z^m \cap V_z \cap X} \frac{||\mathrm{d}_{\zeta'} P(z,\zeta) \wedge \mathrm{d}_{\zeta'} Q(z,\zeta)|| \, \mathrm{d}\sigma_{2n-2}}{(|P(z,\zeta)| + |Q(z,\zeta)| + |\zeta - z|^2)^{3/2} |\zeta - z|^{2n-5}}.$$

We now distinguish the cases $n = 2$ and $n \geqq 3$. If $n = 2$, then we use the fact that by definition of H_z^m for $\zeta \in H_z^m$

$$|\zeta - z| \leqq \delta \left| \frac{\partial \varrho(z)}{\partial z_n} z_n \right|^{1/2}.$$

For $n \geqq 3$ we use the inequality $|P(z,\zeta)| + |Q(z,\zeta)| + |\zeta' - z'| \leqq C\,|\zeta - z|$, which follows from (3.2.6) and (3.2.8). Then we obtain that for all $z \in \partial D_m \setminus X$, $m_0 \leqq m \leqq \infty$,

$$I_m(z) \leqq C \left| \frac{\partial \varrho(z)}{\partial z_2} z_2 \right|^{-3/2} \int\limits_{H_z^m \cap V_z \cap X} \frac{||\mathrm{d}_{\zeta'} P(z,\zeta) \wedge \mathrm{d}_{\zeta'} Q(z,\zeta)|| \, \mathrm{d}\sigma_{2n-2}}{(|P(z,\zeta)| + |Q(z,\zeta)|)^{3/2}} \quad \text{if} \quad n = 2\,,$$

and

$$I_m(z) \leqq C \left| \frac{\partial \varrho(z)}{\partial z_n} z_n \right|^{-2} \int\limits_{H_z^m \cap V_z \cap X} \frac{||\mathrm{d}_{\zeta'} P \wedge \mathrm{d}_{\zeta'} Q|| \, \mathrm{d}\sigma_{2n-2}}{(|P| + |Q|)^{3/2} \, (|P| + |Q| + |\zeta' - z'|)^{2n-5}} \quad \text{if} \quad n \geqq 3\,.$$

In view of Proposition 4, Appendix 1, this implies that for all $z \in \partial D_m \setminus X$, $m_0 \leqq m \leqq \infty$,

$$I_m(z) \leqq C \left| \frac{\partial \varrho(z)}{\partial z_2} z_2 \right|^{-3/2} \int\limits_{t \in \mathbb{R}^2,\, |t| < r_m(z)} \frac{\mathrm{d}t_1 \wedge \mathrm{d}t_2}{(|t_1| + |t_2|)^{3/2}} \quad \text{if} \quad n = 2\,,$$

and

$$I_m(z) \leqq C \left| \frac{\partial \varrho(z)}{\partial z_n} z_n \right|^{-2} \int\limits_{\substack{t \in \mathbb{R}^{2n-2},\, |t| < R_m(z), \\ |t_1| + |t_2| < r_m(z)}} \frac{\mathrm{d}t_1 \wedge \ldots \wedge \mathrm{d}t_{2n-2}}{(|t_1| + |t_2|)^{3/2} \, |t|^{2n-5}} \quad \text{if} \quad n \geqq 3\,,$$

where $r_m(z) := \sup\limits_{\zeta \in H_z^m} \{|P(z,\zeta)| + |Q(z,\zeta)|\}$ and $R_m(z) := \sup\limits_{\zeta \in H_z^m} \{|P(z,\zeta)| + |Q(z,\zeta)| + |\zeta' - z'|\}$. In view of (3.2.1), (3.2.2), and the definition of H_z^m, for all $z \in \partial D_m \setminus X$, $\zeta \in H_z^m$, $m_0 \leqq m \leqq \infty$, we have the estimates

$$|\zeta - z| \leqq \delta \left| \frac{\partial \varrho(z)}{\partial z_n} z_n \right|^{1/2}$$

and

$$|\varrho_m(\zeta)|,\ |F(z,\zeta)| \leqq C\delta^2 \left| \frac{\partial \varrho(z)}{\partial z_n} z_n \right|.$$

Further, by (3.2.6) and (3.2.8), for all $z \in \partial D_m \setminus X$, $\zeta \in H_z^m$, $m_0 \leqq m \leqq \infty$, $|P(z,\zeta)| \leqq |F(z,\zeta)| + C\,|\zeta - z|^2$ and $|Q(z,\zeta)| \leqq |\varrho(\zeta) - \varrho(z)| + C\,|\zeta - z|^2 = |\varrho_m(\zeta)| + C\,|\zeta - z|^2$. This implies that for all $z \in \partial D_m \setminus X$, $m_0 \leqq m \leqq \infty$,

$$r_m(z) \leqq C \left| \frac{\partial \varrho(z)}{\partial z_n} z_n \right| \quad \text{and} \quad R_m(z) \leqq C \left| \frac{\partial \varrho(z)}{\partial z_n} z_n \right|^{1/2}.$$

Now estimate (3.6.13) follows from Proposition 8, Appendix 1.

(v). Let $\delta > 0$ be the number from Lemma 3.6.5. For short, we write

$$H_z^m := H_z^m \left(\delta \left| \frac{\partial \varrho(z)}{\partial z_n} z_n \right|^{1/2} \right).$$

First we prove that

$$\int\limits_{\zeta \in (X \cap D_m \cap V_z) \setminus H_z^m} \frac{\|d_{\zeta'} \varrho(\zeta)\|\, |\partial \varrho(\zeta)/\partial \zeta_n|\, d\sigma_{2n-2}}{|\widetilde{\Phi}_m(z,\zeta)|^2\, |\Phi_m^*(z,\zeta)|\, |\zeta - z|^{2n-5}} \leqq C/|z_n|\,. \tag{3.6.14}$$

Observe that $|\partial \varrho(\zeta)/\partial \zeta_n| \leqq |\partial \varrho(z)/\partial z_n| + C\,|\zeta - z|$ and, by (3.2.9), $\|d_{\zeta'} \varrho(\zeta)\| \leqq \|d_{\zeta'} Q(z,\zeta)\| + C\,|\zeta - z|$. Together with (3.2.11), (3.2.12), (3.5.1), (3.5.2) this implies that the integral on the left-hand side of (3.6.14) can be estimated by the integrals in parts (i), (ii), (iv) of the lemma and the integral which is obtained from the integral in (3.6.14) after replacing $\|d_{\zeta'} \varrho(\zeta)\|\, |\partial \varrho(\zeta)/\partial \zeta_n|$ by $\|d_{\zeta'} Q(z,\zeta)\|\, |\partial \varrho(z)/\partial z_n|$, that is, we only have to prove that for all $z \in \partial D_m \setminus X$, $m_0 \leqq m \leqq \infty$,

$$\int\limits_{\zeta \in (X \cap D_m \cap V_z) \setminus H_z^m} \frac{\|d_{\zeta'} Q(z,\zeta)\|\, d\sigma_{2n-2}}{|\widetilde{\Phi}_m(z,\zeta)|^2\, |\Phi_m^*(z,\zeta)|\, |\zeta - z|^{2n-5}} \leqq C \left| \frac{\partial \varrho(z)}{\partial z_n} z_n \right|^{-1}. \tag{3.6.15}$$

From (3.6.12) and (3.2.11), (3.2.12), (3.5.1), (3.5.2) it follows that for all $z \in \partial D_m \setminus X$, $m_0 \leqq m \leqq \infty$,

$$\int\limits_{\zeta \in (X \cap D_m \cap V_z) \setminus H_z^m} \frac{\|d_{\zeta'} Q(z,\zeta)\|\, d\sigma_{2n-2}}{|\widetilde{\Phi}_m(z,\zeta)|^2\, |\Phi_m^*(z,\zeta)|\, |\zeta - z|^{2n-5}}$$

$$\leqq C \int\limits_{\zeta \in X \cap D_m \cap V_z} \frac{\|d_{\zeta'} Q(z,\zeta)\|\, d\sigma_{2n-2}}{\left(\left| \frac{\partial \varrho(z)}{\partial z_n} z_n \right| + |Q(z,\zeta)| + |\zeta - z|^2 \right)^3 |\zeta - z|^{2n-5}}.$$

Since by (3.2.8), $|Q(z,\zeta)| \leqq C\,|\zeta - z|$, in the integral on the right-hand side, $|\zeta - z|$ can be replaced by $|\zeta' - z'| + |Q(z,\zeta)|$. Therefore, we can apply Propositions 4 and 7, Appendix 1, and obtain (3.6.15). It remains to prove that for all $z \in \partial D_m \setminus X$, $m_0 \leqq m \leqq \infty$,

$$\int\limits_{\zeta \in H_z^m \cap X \cap V_z} \frac{\|d_{\zeta'} \varrho(\zeta)\|\, |\partial \varrho(\zeta)/d\zeta_n|\, \partial\sigma_{2n-2}}{|\widetilde{\Phi}_m(z,\zeta)|^2\, |\Phi_m^*(z,\zeta)|\, |\zeta - z|^{2n-5}} \leqq C/|z_n|\,. \tag{3.6.16}$$

Since $d\varrho(\zeta) = d\varrho(z) + O(|\zeta - z|)$, the integral on the left-hand side in (3.6.16) can be estimated by the integrals in parts (i), (ii), (iv) of the lemma and the integral which is obtained from the integral in (3.6.16) after replacing $\|d_{\zeta'} \varrho(\zeta)\|\, |\partial \varrho(\zeta)/\partial \zeta_n|$ by $\|d_{z'} \varrho(z)\|\, |\partial \varrho(z)/\partial z_n|$, that is, we only have to prove that for all $z \in \partial D_m \setminus X$,

$m_0 \leqq m \leqq \infty$,

$$J_m(z) := \int\limits_{\zeta \in H_z^m \cap X \cap V_2} \frac{\|d_{z'}\varrho(z)\|^2 \, d\sigma_{2n-2}}{|\tilde{\Phi}_m(z,\zeta)|^2 \, |\Phi_m^*(z,\zeta)| \, |\zeta - z|^{2n-5}} \leqq C \frac{\|d_{z'}\varrho(z)\|}{|(\partial\varrho(z)/\partial z_n) \, z_n|}. \tag{3.6.17}$$

In view of (3.2.11), (3.2.12), (3.5.1), (3.5.2), and (3.6.8), for all $z \in \partial D_m \setminus X$, $m_0 \leqq m \leqq \infty$,

$$J_m(z) \leqq C \int\limits_{\zeta \in H_z^m \cap X \cap V_2} \frac{\|d_{\zeta'}P(z;\zeta) \wedge d_{\zeta'}Q(z,\zeta)\| \, d\sigma_{2n-2}}{(|P(z,\zeta)| + |Q(z,\zeta)| + |\zeta - z|^2)^3 \, |\zeta - z|^{2n-5}}.$$

Together with (3.6.7) this implies that for all $z \in D_m \setminus X$, $m_0 \leqq m \leqq \infty$,

$$J_m(z) \leqq C \int\limits_{\zeta \in H_z^m \cap X \cap V_2} \frac{\|d_{\zeta'}P \wedge d_{\zeta'}Q\| \, d\sigma_{2n-2}}{\left(\left|\frac{\partial\varrho(z)}{\partial z_n} z_n\right|^2 \Big/ \|d_{z'}\varrho(z)\|^2 + |P| + |Q| + |\zeta - z|^2\right)^3 |\zeta - z|^{2n-5}}.$$

Since by (3.2.6), and (3.2.8) $|P(z,\zeta)| + |Q(z,\zeta)| \leqq C \, |\zeta - z|$, in the last integral, we can replace $|\zeta - z|$ by $|P(z,\zeta)| + |Q(z,\zeta)| + |\zeta' - z'|$. Therefore, we can apply Propositions 4 and 7, Appendix 1, and we obtain (3.6.17). ■

3.6.7. Lemma. *There is a constant $C < \infty$ such that for all $m_0 \leqq m \leqq \infty$, and every bounded holomorphic function f on $X \cap D_m$ the following estimate holds:*

$$|E_m f(z)| \leqq C \, \|f\|_{0, D_m \cap X} \quad \text{for all} \quad z \in \bar{D}_m \setminus (\partial D_m \cap X). \tag{3.6.18}$$

Proof. Since, by Lemma 3.6.2, $E_m f$ is holomorphic in $D_m \cup U_{\partial D_m \setminus X}$, it is sufficient to consider the case $z \in \partial D_m \setminus X$, because then the general case follows by the maximum principle for holomorphic functions. It follows from Lemma 3.6.4 that for some constant $C_1 < \infty$ and all $z \in \partial D_m \setminus X$, $m_0 \leqq m \leqq \infty$,

$$|E_m f(z)| \leqq C_1 \, |z_n| \, \|f\|_{0, D_m \cap X} \int\limits_{X \cap D_m} \left\| \det_{1,n-1}\left(\frac{{}^*w^m}{\Phi_m^*}, \bar{\partial}_{\zeta'} \frac{\chi w^m}{\tilde{\Phi}_m}\right) \wedge \omega_{\zeta'}(\zeta) \right\| d\sigma_{2n-2}.$$

Therefore, since $\chi = 1$ on V_2 and since, by Construction 3.1.1, for some $\alpha > 0$ and all $z \in \partial D_m$, $\zeta \in D_m \setminus V_2$, $m_0 \leqq m \leqq \infty$,

$$|\tilde{\Phi}_m(z,\zeta)|, \, |\Phi_m^*(z,\zeta)| \geqq \alpha,$$

it is, sufficient to estimate for all $z \in \partial D_m \setminus X$, $m_0 \leqq m \leqq \infty$, the integral over $X \cap D_m \cap V_2$ of

$$\left\| \det_{1,n-1}\left(\frac{{}^*w^m}{\Phi_m^*}, \bar{\partial}_{\zeta'} \frac{\chi w^m}{\tilde{\Phi}_m}\right) \right\| \leqq C_2 \left\| \det_{1,n-1}\left(\frac{{}^*w^m}{\Phi_m^*}, \frac{\bar{\partial}_{\zeta'} w^m}{\tilde{\Phi}_m}\right) \right\|$$
$$+ C_3 \left\| \det_{1,1,n-2}\left(\frac{{}^*w^m}{\Phi_m^*}, \frac{w^m \bar{\partial}_{\zeta'} \tilde{\Phi}^m}{\tilde{\Phi}_m^2}, \frac{\bar{\partial}_{\zeta'} w^m}{\tilde{\Phi}_m}\right) \right\|. \tag{3.6.19}$$

Since ${}^*w^m(z,z) = -w^m(z,z)$ and since $w^m \to w$ uniformly together with the first-order derivatives, there is a constant $C_4 < \infty$ such that for all $\zeta, z \in V_2$, $m_0 \leqq m \leqq \infty$,

$$|{}^*w(z,\zeta) + w^m(z,\zeta)| \leqq C_4 \, |\zeta - z|.$$

Further it is clear that for some $C_5 < \infty$ and all $z \in \partial D_m$, $\zeta \in V_2$, $|\varrho_m(\zeta)| \leqq C_5 \, |\zeta - z|$. We recall also that for some $\alpha > 0$ and all $z \in \partial D_m$, $\zeta \in V_2 \cap D_m$, $m_0 \leqq m \leqq \infty$, (cp. (3.2.11), (3.2.12), (3.5.1), (3.5.2)) $|\tilde{\Phi}_m(z,\zeta)|, |\Phi^*_m(z,\zeta)| \geqq \alpha\,|\zeta - z|^2$. Together with (3.2.14), (3.2.15), (3.5.3), (3.5.4) the last three estimates imply that for some $C_6 < \infty$ and all $z \in \partial D_m \setminus X$, $\zeta \in X \cap D_m \cap V_2$, $m_0 \leqq m \leqq \infty$,

$$\left\| \det_{1,n-1}\left(\frac{{}^*w^m(z,\zeta)}{\Phi^*_m(z,\zeta)}, \frac{\bar{\partial}_{\zeta'} w^m(z,\zeta)}{\tilde{\Phi}_m(z,\zeta)}\right)\right\| \leqq \frac{C_6}{|\zeta - z|^{2n-1}} + \frac{C_6\,\|d\varrho(\zeta)\|}{|\tilde{\Phi}_m(z,\zeta)\,\Phi^*_m(z,\zeta)|\,|\zeta - z|^{2n-4}} \tag{3.6.20}$$

and

$$\left\| \det_{1,1,n-2}\left(\frac{{}^*w^m(z,\zeta)}{\Phi^*_m(z,\zeta)}, \frac{w^m(z,\zeta)\,\bar{\partial}_{\zeta'}\tilde{\Phi}_m(z,\zeta)}{\tilde{\Phi}^2_m(z,\zeta)}, \frac{\bar{\partial}_{\zeta'} w^m(z,\zeta)}{\tilde{\Phi}_m(z,\zeta)}\right)\right\|$$
$$\leqq \frac{C_6}{|\zeta - z|^{2n-1}} + \frac{C_6\,\|d\varrho(\zeta)\|}{|\tilde{\Phi}_m(z,\zeta)\,\Phi^*_m(z,\zeta)|\,|\zeta - z|^{2n-4}} + \frac{C_6\,\|d\varrho(\zeta)\|\,\|d_{\zeta'}\varrho(\zeta)\|}{|\tilde{\Phi}_m(z,\zeta)|^2\,|\Phi^*_m(z,\zeta)|\,|\zeta - z|^{2n-5}}. \tag{3.6.21}$$

By taking into account that $\|d\varrho(\zeta)\| \leqq \|d_{\zeta'}\varrho(\zeta)\| + 2\,|\partial\varrho(\zeta)/\partial\zeta_n|$, now the required estimate (3.6.18) follows from (3.6.19)–(3.6.21) and Lemma 3.6.6. ■

3.6.8. Theorem. *Let $D \subset\subset \mathbb{C}^n$ be a strictly pseudoconvex open set (with not necessarily smooth boundary), and let $X = \{z \in \mathbb{C}^n : z_{k+1} = \dots = z_n = 0\}$, $1 \leqq k \leqq n-1$. Then*

(i) *For every bounded holomorphic function f on $D \cap X$ there exists a bounded holomorphic function F on D such that*

$$F = f \quad \text{on} \quad X \cap D.$$

(ii) *For every continuous function f on $X \cap \bar{D}$ that is holomorphic in $X \cap D$ there exists a continuous function F on $\bar{D}$ that is holomorphic in D such that*

$$F = f \quad \text{on} \quad X \cap \bar{D}.$$

Proof. First we consider the case that $k = n - 1$. By setting $F = E_\infty f$, then part (i) immediately follows from Lemmas 3.6.2 and 3.6.7.

Now we prove part (ii) for $k = n - 1$. In view of the Approximation Theorem 3.5.1, we can find holomorphic functions f_j $(j = 1, 2, \dots)$ in some neighbourhoods of $X \cap \bar{D}$ such that

$$\sum_{j=1}^{\infty} \|f_j\|_{0,D\cap X} < \infty \quad \text{and} \quad f = \sum_{j=1}^{\infty} f_j \quad \text{on} \quad X \cap \bar{D}.$$

For every j choose $m_j \geqq m_0$ so large that f_j is holomorphic in a neighbourhood of $X \cap \overline{D_{m_j}}$. Then by Lemmas 3.6.2 and 3.6.7, $F := \sum_{j=1}^{\infty} E_{m_j} f_j$ gives the required extension.

In the general case, when $1 \leqq k \leqq n-1$, the theorem follows by applying k times the special case proved above. ■

Notes

Without the factor $\operatorname{diam}_H(\operatorname{supp} f)$ in estimate (3.2.3) and without estimate (3.2.4), Theorem 3.2.2 was proved in 1974 by Henkin (unpublished). In its formulation which is given here, Theorem 3.2.2 was obtained in 1980 by Henkin/Leiterer [2]. In its proof the constructions given by Henkin [2, 3] and Skoda [1, 2] for solving the $\bar{\partial}_b$-equations play an important role. For the case of smooth ∂D, see the Notes at the end of Chapter 2. For smooth ∂D, Theorem 3.4.2 was obtained in 1969 by Henkin [4], where instead of estimate (3.4.2) the estimate $\|f_j\|_{0,D} \leqq C \ln (C + C/\delta) \|f\|_{0,D}$ was obtained. In its formulation given here, Theorem 3.4.2 as well as Theorem 3.4.8 was proved in Henkin/Leiterer [2]. These theorems form a natural generalization of Vituškin's [1] theorem on decomposition of singularities for one complex variable to the case of several complex variables. The Approximation Theorem 3.5.1 was announced by Henkin in Čirka/Henkin [1]. Its proof was published in Henkin/Leiterer [2]. For the case that $K(\varrho) = \overline{D}$, where D is a strictly pseudoconvex open set with C^2-boundary, this theorem was proved in 1969 by Henkin [4] and, independently, by Kerzman [1] and Lieb [1]. For the case that $K(\varrho)$ is a compact subset of a totally real C^1-submanifold X (cp. Corollary 1.4.20), Theorem 3.5.1 was proved in 1971 by Harvey/Wells [1] and, for $X \in C^\infty$, in 1967 by Hörmander/Wermer [1] and Nirenberg/Wells [1]. For non-negative ϱ the proof of Theorem 3.5.1 is essentially simpler and was published by Henkin in Čirka/Henkin [1] (1975) (see also Sakai [1] (1979)). For the Extension Theorem 3.6.8 see the Notes at the end of Chapter 4. To obtain the estimates in this chapter, we essentially used an idea of Range/Siu [1] which consists in replacing $F(z, \zeta)$ and $\varrho(\zeta) - \varrho(z)$ by polynomials (Lemma 3.2.4) and applying Proposition 4, Appendix 1.

Exercises, remarks and problems

1. Let $X \subseteqq \mathbb{C}^n$ be a complex plane of complex dimension $\leqq n - 1$, and let $D \subset\subset X$ be a strictly pseudoconvex open set (with not necessarily smooth boundary). Prove that there exists a strictly pseudoconvex open set $\widetilde{D} \subset\subset \mathbb{C}^n$ with C^2-boundary such that $X \cap \widetilde{D} = D$.

2. Conjecture (with an idea how to prove). Let $D \subset\subset \mathbb{C}^n$ be a strictly pseudoconvex open set (with not necessarily smooth boundary). Then, for every continuous $(0, q)$-form f on $\overline{D}$ such that $\bar\partial f = 0$ in D, $1 \leqq q \leqq n$, there exists a 1/2-Hölder continuous $(0, q - 1)$-form on $\overline{D}$ such that $\bar\partial u = f$ in D. We propose to prove this in the following way: By point 1 the statement follows from the corresponding assertion for the case of smooth ∂D if an appropriate generalization of the Extension Theorem 3.6.8 to $\bar\partial$-closed $(0, q)$-forms is proved.

3. Open problem. Let $D \subset\subset \mathbb{C}^n$ be a strictly pseudoconvex open set (with not necessarily smooth boundary), and let $X \subseteqq \mathbb{C}^n$ be a complex plane of complex dimension $\leqq n - 1$. By Lemmas 3.6.2 and 3.6.7 the extension of bounded holomorphic functions stated in Theorem 3.6.8 (i) can be given by a bounded linear operator from the Banach space of bounded holomorphic functions in $X \cap D$ into the Banach space of bounded holomorphic functions in D.

Problem. Does there exist a bounded linear operator $E: A(X \cap \overline{D}) \to A(\overline{D})$ such that $Ef = f$ on $X \cap \overline{D}$ for all $f \in A(X \cap \overline{D})$? Here $A(X \cap \overline{D})$ $(A(\overline{D}))$ is the Banach space of continuous functions on $X \cap \overline{D}$ $(\overline{D})$ which are holomorphic in $X \cap D$ (D).

Remark. For the case of smooth ∂D, such an operator E exists (Henkin [5]).

4. Let $D \subset\subset \mathbb{C}^n$ be a C^2 strictly convex open set, $0 \in D$, and let ϱ be a strictly convex C^2-function in a neighbourhood θ of $\overline{D}$ such that $D = \{z \in \theta: \varrho(z) < 0\}$. Denote by $N(D)$ the class of holomorphic functions f in D satisfying the condition

$$\sup_{r<1} \int_{\partial D} |\ln |f(r\zeta)|| \, d\sigma_{2n-1} < \infty ,$$

where $d\sigma_{2n-1}$ is the Euclidean volume form on ∂D. Let $f \in N(D)$ and let M_j, k_j, $d\sigma_{2n-2}$ be as in point 22 in the Exercises, remarks and problems at the end of Chapter 1. Prove that

$$\sum_j k_j \int_{M_j} |\varrho| \, d\sigma_{2n-2} < \infty .$$

Remark. Henkin [2, 3] and Skoda [1, 2] proved that also the converse is true: If M is a complex submanifold of D with $\int_M |\varrho| \, d\sigma_{2n-2} < \infty$, then there exists an $f \in N(D)$ such that $M = \{z \in D : f(z) = 0\}$.

5. Let $D \subset\subset \mathbb{C}^2$ be the unit ball. Prove that there exists a complex submanifold M, $\dim_{\mathbb{C}} M = 1$, of D with $\int_M (1 - |\zeta|)^\alpha \, d\sigma_2 < \infty$ ($d\sigma_2$ is the Euclidean volume form on M) for all

$\alpha > 0$ and such that the following condition is fulfilled: If f is a bounded holomorphic function in D such that $f = 0$ on M, then $f = 0$ in D.

Remark. BERNDTSSON [1] proved the following theorem: If $D \subset\subset \mathbb{C}^2$ is a C^2 strictly pseudoconvex open set, and M is a complex submanifold of D such that $\int_M d\sigma_2 < \infty$, then there exists a bounded holomorphic function f in D such that $M = \{z \in D : f(z) = 0\}$.

6. (HENKIN/ČIRKA [1], VAL'SKIJ [1]). Let g be a continuous function on $\bar{D}$ that is holomorphic and $\neq 0$ in D, and let $K := \{z \in \partial D : g(z) = 0\}$. Then, for every continuous function f on K, there exists a continuous function F on $\bar{D}$ that is holomorphic in D such that $F = f$ on K.

7. Let $D \subset\subset \mathbb{C}^n$ be a C^2 strictly pseudoconvex open set, and let $M \subseteq \partial D$ be a real C^1-submanifold such that, for every $\xi \in M$, the real tangent plane of M at ξ is contained in the complex tangent plane of ∂D at ξ (cp. Subsection 1.5.13).

a) (HENKIN/TUMANOV [1]). Prove that then M is totally real (cp. Definition 1.4.16).

b) (HENKIN/TUMANOV [1]). Construct a C^1-function $\tilde{g}$ in a neighbourhood of $\bar{D}$ such that $\tilde{g} \neq 0$ in D, $M = \{z \in \partial D : \tilde{g}(z) = 0\}$, $d\tilde{g}(z)' \neq 0$ for $z \in M$ but $\bar{\partial}\tilde{g}(z) = 0$ for $z \in M$.

c) (RUDIN [1] and HENKIN/TUMANOV [1] (for the case that M is of class C^2)). Prove that, for every continuous function f on M, there exists a continuous function F on $\bar{D}$ that is holomorphic in D such that $F = f$ on M. Hint. Use an appropriate solution of $\bar{\partial}u = \bar{\partial}(1/\tilde{g})$, where $\tilde{g}$ is the function from part b), and prove that $g := \tilde{g}/(1 - u\tilde{g})$ has the properties in point 6.

For further similar results see BURNS/STOUT [1], CHAUMAT/CHOLLET [1], NAGEL [1], TUMANOV [2], HENRIKSEN [1].

8. Let $D \subset\subset \mathbb{C}^n$ be a C^2 strictly pseudoconvex open set, and let $M \subseteq \partial D$ be a real C^1-submanifold with the following property: For every continuous function f on M there exists a continuous function F on $\bar{D}$ that is holomorphic in D such that $F = f$ on M.

Open problem. Does it follow that, for every $\xi \in M$, the real tangent plane of M at ξ is contained in the complex tangent plane of ∂D at ξ (cp. Subsection 1.5.13)?

Remark. HENKIN/TUMANOV [1] (see also NAGEL/RUDIN [1]) proved that the answer is affirmative if M is of class $C^{1+\varepsilon}$, $\varepsilon > 0$.

9. Let B be a holomorphic vector bundle over a complex manifold X, let $D \subset\subset X$ be a strictly pseudoconvex open set (with not necessarily smooth boundary), and let $1 \leqq q \leqq n$. Prove that then the following two statements hold:

(i) There exists a bounded linear operator S from $Z^0_{(0,q)}(\bar{D}, B)$ into $C^0_{(0,q-1)}(\bar{D}, B)$ such that $\bar{\partial} \circ S = \mathrm{id} + K$, where K is compact.

(ii) The image of $\bar{\partial}$ as an operator between $C^0_{(0,q-1)}(\bar{D}, B)$ and $Z^0_{(0,q)}(\bar{D}, B)$ is a closed and finitely codimensional subspace of $Z^0_{(0,q)}(\bar{D}, B)$.

Hint. Use Lemma 1.5.23 (ii), Corollary 3.2.3, Theorem 3.2.2 (iii) and the scheme of the proof of Theorem 2.11.3.

10. Prove that under the hypotheses of Theorem 2.12.3, for every $\alpha \in \mathbb{R}$ (without the hypothesis $d\varrho(z) \neq 0$, $z \in \partial D$), the following two statements hold:

(i)′ For every $1 \leqq q \leqq n$, there exists a bounded linear operator $T: Z^0_{(0,q)}(\bar{D}_\alpha, B) \to C^0_{(0,q-1)}(\bar{D}_\alpha, B)$ such that $\bar{\partial} \circ T = \mathrm{id}$.

(ii)′ Every continuous section $f: \bar{D}_\alpha \to B$ that is holomorphic in D_α can be approximated uniformly on $\bar{D}_\alpha$ by holomorphic sections of B over X.

Hint. Use Exercise 9 and the scheme of the proof of Theorem 2.12.3.

11. (The uniqueness theorem of PINČUK [1]). Let $D \subset\subset \mathbb{C}^n$ be an open set with C^1-boundary and let $M \subseteq \partial D$ be an n-dimensional totally real C^1-submanifold of ∂D. Let f be a continuous function on $\bar{D}$ that is holomorphic in D such that $f = 0$ on M. Prove that $f \equiv 0$ on $\bar{D}$.

Hint. Choose a bordered real C^1-submanifold $N \subseteq \bar{D}$ such that: 1. $\partial D \cap \partial N \subseteq M$, 2. $\partial N \cap D$ is also a smooth totally real submanifold and "sufficiently close" to $\partial D \cap \partial N$; then we only have to prove that, for every C^∞-function φ with compact support in D,

$$\int_{z \in \partial N \cap D} f(z)\, \varphi(z)\, dz_1 \wedge \ldots \wedge dz_n = 0;$$

in view of the Approximation Theorems 2.7.1 and 3.5.1 (cp. the remark following Theorem 3.5.1 and Corollary 1.4.20) and since $\partial N \cap D$ is "sufficiently close" to $\partial N \cap \partial D$, φ can be approximated on $\partial N \cap D$ uniformly by holomorphic functions $\tilde{\varphi}$ in a neighbourhood of N; since $f = 0$ on $\partial D \cap \partial N$, it follows from Stokes' formula that

$$\int_{\partial N \cap D} f(z)\, \tilde{\varphi}(z)\, dz_1 \wedge \ldots \wedge dz = \int_N d(f(z)\, \tilde{\varphi}(z)\, dz_1 \wedge \ldots \wedge dz_n) = 0\,.$$

12. (The uniqueness theorem of TUMANOV.) Let $D \subset\subset \mathbb{C}^n$ be a strictly pseudoconvex domain with real-analytic boundary. Denote by $A^k(\overline{D})$, $k = 0, 1, 2, \ldots$, the space of all holomorphic functions in D such that the derivatives of order k are continuous on $\overline{D}$. Let $A^{1/2}(\overline{D}) =: \mathcal{O}(D) \cap H^{1/2}(\overline{D})$ (cp. Section 1.8 for the definition of $H^{1/2}(\overline{D})$), and let $f \in A^{1/2}(\overline{D})$ (cp. point 10) such that $\operatorname{Im} f = 0$ on a set $E \subseteq \partial D$ of positive $(2n-1)$-dimensional Hausdorff measure.

a) Prove that $f = \text{const}$ in D. Remark. TUMANOV [1] proved this for $f \in A^2(D)$ and SIBONY [3] for $f \in A^1(\overline{D})$.

b) Prove that this is not true for $f \in A(\overline{D})$. Hint: Use the recent results of ALEKSANDROV [1] on the existence of inner functions in strictly pseudoconvex domains.

13 (NAGEL/RUDIN [1]). Let $D \subset\subset \mathbb{C}^n$ be a strictly pseudoconvex C^2-domain, and let Γ be a real C^2-curve on ∂D which is transversal to the complex tangent plane of ∂D at each point in Γ. Prove that every bounded holomorphic function in D has non-tangential boundary values almost everywhere on Γ.

14. Open problem. Let $D \subset\subset \mathbb{C}^n$ be a polydisc, let $X \subseteq \mathbb{C}^n$ be a complex plane of complex dimension $\leqq n-1$, and let f be a bounded holomorphic function on $X \cap D$. Does there exist a bounded holomorphic function F in D such that $F = f$ on $X \cap D$? Remark. POLJAKOV [1] proved this for $n = 2$.

4. Global integral formulas on Stein manifolds and applications

Summary. In this chapter we generalize the integral formulas presented in Chapters 1–3 to Stein manifolds (Sections 4.3, 4.5, 4.8, and 4.10). Moreover, in Sections 4.4 and 4.6 a formula will be constructed where for different pieces of the boundary different Leray maps are used. (For space-saving reasons we did not separately prove this formula for the case of $\mathbb{C}^n$.) This formula will be called the *Leray-Norguet formula* (for the case of functions) and the *Koppelman-Leray-Norguet formula* (for the general case of $(0, q)$-forms). The Koppelman-Leray-Norguet formula is especially useful for so-called real non-degenerate strictly pseudoconvex polyhedra (Section 4.9). This class of polyhedra contains the real non-degenerate analytic polyhedra as well as the real transversal intersections of C^2 strictly pseudoconvex open sets. For such polyhedra, by means of the Koppelman-Leray-Norguet formula, a formula for solving the $\bar{\partial}$-equation is obtained. (Under certain additional conditions the solution given by this formula admits uniform estimates — cp. the Notes at the end of this chapter. Unfortunately we could not include these estimates in this book.) For real non-degenerate analytic polyhedra, the Leray-Norguet formula implies a generalization of the Weil formula (mentioned in the preface to this book and in the Notes at the end of Chapter 2) to Stein manifolds (Section 4.7). In Section 4.10 we generalize the results from Sections 3.1 and 3.2 to the case of strictly pseudoconvex open sets (with not necessarily smooth boundary) in Stein manifolds. (The corresponding generalizations of Sections 3.3–3.5 can be obtained similarly and are left to the reader.) In Section 4.11, by means of the formulas and estimates presented in Section 4.10, we generalize the Extension Theorem 3.6.8 to the case of the intersection of a strictly pseudoconvex open set D (with not necessarily smooth boundary) in a Stein manifold with an arbitrary closed complex submanifold of some neighbourhood of $\bar{D}$. In Section 4.12 we show how to extend the integral formulas presented in this book to $(0, q)$-forms with values in holomorphic vector bundles.

Now let us explain the idea of the construction of the integral formulas on Stein manifolds presented in this chapter. Let X be a Stein manifold of complex dimension n, and let $D \subset\subset X$ be an open set with smooth boundary. Consider, for example, the Martinelli-Bochner formula (1.9.2) for holomorphic functions. Then the problem is, by what can the map $\zeta - z$ be replaced? First consider the case that there exists a holomorphic map $u(z, \zeta)$ from $X \times X$ into $\mathbb{C}^n$ such that the following two conditions are fulfilled:

(1) $u(z, \zeta) \neq 0$ if $\zeta \neq z$;

(2) for every fixed point $z \in X$ the map $u(z, \cdot)$ is biholomorphic in some neighbourhood of z, where $u(z, z) = 0$.

Then u can be used instead of $\zeta - z$, and, as in the case of (1.9.2), one can prove that, for every holomorphic function f in some neighbourhood of $\bar{D}$,

$$f(z) = \frac{(n-1)!}{(2\pi i)^n} \int_{\partial D} f(\zeta) \frac{\omega'_\zeta(\overline{u(z, \zeta)}) \wedge \omega_\zeta(u(z, \zeta))}{|u(z, \zeta)|^{2n}}, \qquad z \in D .$$

However, such a map u need not exist. Moreover, if X is not parallelizable, then it is even impossible to satisfy only condition (2) (this follows from results of Schneider [1], for more details cp. also Henkin/Leiterer [1]).

In order to avoid this difficulty, in 1974 A. Dynin proposed to use a holomorphic map $s(z, \zeta)$ with values in the complex tangent bundle $T(X)$ of X (cp. Subsection 4.2.3) such that $s(z, \zeta) \in T_z(X)$ for all $(z, \zeta) \in X \times X$ and such that the following conditions are fulfilled: (1) $s(z, \zeta) \neq 0$ for $z \neq \zeta$; (2) for every fixed $z \in X$, $s(z, \cdot)$ is biholomorphic in some neighbourhood of z, where $s(z, z) = 0$. It is easy to find a fibre preserving C^∞-map σ from the complex tangent bundle $T(X)$ onto the complex cotangent bundle $T^*(X)$ of X such that $||a|| := \langle \sigma a, a\rangle^{1/2}$, $a \in T(X)$ defines a norm in the fibres of $T(X)$, where $\langle b, a\rangle$ is the value of the covector $b \in T^*_z(X)$ at $a \in T_z(X)$ (cp. Subsections 4.2.3 and 4.3.1). Further, it turns out (cp. Subsection 4.3.1) that the form $\omega'_\zeta(\sigma s(z, \zeta)) \wedge \omega_\zeta(s(z, \zeta))$ has an invariant meaning (whereas the forms $\omega'_\zeta(\sigma s(z, \zeta))$ and $\omega_\zeta(s(z, \zeta))$ can not be defined independently of the choice of local coordinates). Now one can prove that, for every holomorphic function in some neighbourhood of $\bar{D}$,

$$f(z) = \frac{(n-1)!}{(2\pi i)^{2n}} \int_{\partial D} f(\zeta) \frac{\omega'_\zeta(\sigma s(z, \zeta)) \wedge \omega_\zeta(s(z, \zeta))}{\langle \sigma s(z, \zeta), s(z, \zeta)\rangle^n}, \qquad z \in D. \qquad (*)$$

Unfortunately, this elegant approach of Dynin does not work in general, because such a map $s(z, \zeta)$ also need not exist (This follows also from the results of Schneider [1]. Observe that the obstructions for the existence of such a map $s(z, \zeta)$ are purely topological — for more details see Henkin/Leiterer [1] and Leiterer [4]). The construction presented in this chapter was found by Henkin/Leiterer [1] and consists in a combination of Dynin's idea with an idea of Bishop (cp. Theorem 7 in Bishop [1] and point 10 in the Exercises, remarks and problems at the end of this chapter). We construct a map $s(z, \zeta)$ satisfying condition (2) (whereas condition (1) can be violated), and, by means of Cartan's Theorem B from the theory of coherent analytic sheaves, we find a holomorphic function $\varphi(z, \zeta)$ on $X \times X$ and an integer $\varkappa \geqq 0$ such that $\varphi(z, z) = 1$ for all $z \in X$ and $\varphi^\varkappa(z, \zeta)/\langle \sigma s(z, \zeta), s(z, \zeta)\rangle$ is smooth for all $z \neq \zeta$. (This will be done in Section 4.2.) Then, for every integer $\nu \geqq 2n\varkappa$, one can prove that, for every holomorphic function f in some neighbourhood of $\bar{D}$,

$$f(z) = \frac{(n-1)!}{(2\pi i)^{2n}} \int_{\partial D} f(\zeta) \frac{\varphi^\nu(z, \zeta) \wedge \omega'_\zeta(\sigma s(z, \zeta)) \wedge \omega_\zeta(s(z, \zeta))}{\langle \sigma s(z, \zeta), s(z, \zeta)\rangle^n}, \qquad z \in D.$$

The function $\varphi^\nu(z, \zeta)$ removes the singularities of the kernel $\dfrac{\omega'_\zeta(\sigma s(z, \zeta)) \wedge \omega_\zeta(s(z, \zeta))}{\langle \sigma s(z, \zeta), s(z, \zeta)\rangle^n}$. Remark that, if such singularities do not exist, then, since $\varphi(z, \zeta)$ is holomorphic in ζ and $\varphi(z, z) = 1$, it follows from (*) that this factor does not change the value of the integral.

4.1. Coherent analytic sheaves

We know from Sections 2.10, 2.12, and 2.13 that every holomorphic Cousin problem in a holomorphic vector bundle over a Stein manifold can be solved. In this chapter we meet Cousin data which satisfy some additional conditions and we are interested in solutions which fulfil the same conditions. Consider an example. Let X be a Stein manifold, let $\{U_j\}$ be an open covering of X, and let $h_{ij} \in \mathcal{O}(U_i \cap U_j)$ be Cousin data. Then we know that $h_{ij} = h_i - h_j$ in $U_i \cap U_j$ for some system of functions $h_j \in \mathcal{O}(U_j)$. Now suppose that, moreover, for some closed complex submanifold $Y \subseteqq X$, $h_{ij} = 0$ in $Y \cap U_i \cap U_j$. Can we choose then the solution $\{h_j\}$ of the Cousin problem so that also $h_j = 0$ in $Y \cap U_j$? The answer is affirmative. This follows from Cartan's Theorem B from the theory of coherent analytic sheaves.

In the present section we give the formulations of some special results of this theory, which will be used in the subsequent sections. We omit the proofs. The proofs can be found, for example, in Cartan [3], Andreian Cazacu [1], Grauert/Remmert [2], Gunning/Rossi [1] and Hörmander [1]. Remark that the methods developed in the present book do not give any new aspects for the proofs of these results.

In order to avoid the abstract language of general sheaves, we consider subsheaves of sheaves of germs of holomorphic sections in holomorphic vector bundles only:

4.1.1. Definition. Let X be a complex manifold, and let B be a holomorphic vector bundle over X. We denote by ${}_X\mathcal{O}^B$ the map which attaches to every open set $U \subseteq X$ the space ${}_X\mathcal{O}^B(U)$ of holomorphic sections of B over U. We write also $\mathcal{O}(U, B) := {}_X\mathcal{O}^B(U)$. ${}_X\mathcal{O}^B$ is called the *sheaf of germs of holomorphic sections of* B. For the product bundle $X \times \mathbb{C}^N$ we write ${}_X\mathcal{O}^N := {}_X\mathcal{O}^{X \times \mathbb{C}^N}$ and $\mathcal{O}^N(U) := {}_X\mathcal{O}^{X \times \mathbb{C}^N}(U)$.

A map $\mathcal{F}$ which attaches to every open set $U \subseteq X$ a subset $\mathcal{F}(U)$ of $\mathcal{O}(U, B)$ is called an *analytic subsheaf of* ${}_X\mathcal{O}^B$ if the following conditions are fulfilled:

(i) For every open set $U \subseteq X$, $\mathcal{F}(U)$ is an $\mathcal{O}(U)$-module, that is, if $f, g \in \mathcal{F}(U)$ and $\alpha, \beta \in \mathcal{O}(U)$, then $\alpha f + \beta g \in \mathcal{F}(U)$.

(ii) If $V \subseteq U \subseteq X$ are open sets and $f \in \mathcal{F}(U)$, then the $f \mid V \in \mathcal{F}(V)$.

(iii) If $U_i \subseteq X$ are open sets, $f \in \mathcal{O}(\bigcup U_i. B)$ with $f \in \mathcal{F}(U_i)$, then $f \in \mathcal{F}(\bigcup U_i)$.

An analytic subsheaf $\mathcal{F}$ of ${}_X\mathcal{O}^B$ is called *coherent* if, for every point $z \in X$, there exist a neighbourhood U_z of z and finitely many $f_1, \dots, f_N \in \mathcal{F}(U_z)$ such that the following condition is fulfilled: If $\zeta \in U_z$, V_ζ is a neighbourhood of ζ, and $f \in \mathcal{F}(V_\zeta)$, then there exist a neighbourhood $W_\zeta \subseteq V_\zeta \cap U_z$ of ζ and $\psi_1, \dots, \psi_N \in \mathcal{O}(W_\zeta)$ such that $f = \psi_1 f_1 + \dots + \psi_N f_N$ over W.

4.1.2. Theorem (The Oka theorem). *Let X be a complex manifold and let A be an $r \times s$ matrix whose entries are holomorphic functions on X. Denote by $\mathcal{K}_A$ the analytic subsheaf of ${}_X\mathcal{O}^s$ which is defined by $\mathcal{K}_A(U) := \{f \in \mathcal{O}^s(U) : Af = 0 \text{ in } U\}$, $U \subseteq X$ open. Then $\mathcal{K}_A$ is coherent.*

4.1.3. Theorem. *Let X be a Stein manifold, let B be a holomorphic vector bundle over X, and let $\mathcal{F}$ be a coherent analytic subsheaf of ${}_X\mathcal{O}^B$.*

(i) **(Cartan's Theorem A)** *If $f \in \mathcal{F}(U)$, where $U \subseteq X$ is open, then for every $z \in U$ we can find finitely many $f_1, \dots, f_N \in \mathcal{F}(X)$ as well as holomorphic functions $\psi_1, \dots, \psi_N$ in some neighbourhood $V \subseteq U$ of z such that $f = \psi_1 f_1 + \dots + \psi_N f_N$ over V.*

(ii) **(Cartan's Theorem B)** *For every open covering $\{U_j\}_{j \in J}$ and every system $f_{ij} \in \mathcal{F}(U_i \cap U_j)$, $i, j \in J$, such that $f_{ij} + f_{jk} = f_{ik}$ in $U_i \cap U_j \cap U_k$ for all $i, j, k \in J$, there exists a system $f_j \in \mathcal{F}(U_j)$, $j \in J$, such that $f_{ij} = f_i - f_j$ in $U_i \cap U_j$ for all $i, j \in J$.*

4.1.4. Theorem. *Let X be a complex manifold, let B be a holomorphic vector bundle over X, and let $\mathcal{F}$ be a coherent analytic subsheaf of ${}_X\mathcal{O}^B$. Then, for every open set $U \subseteq X$, $\mathcal{F}(U)$ is a closed subspace of the Fréchet space $\mathcal{O}(U, B)$ endowed with the topology of uniform convergence on compact subsets.*

Examples of coherent analytic sheaves

4.1.5. Let Y be a closed complex submanifold of the complex manifold X. Define by

$$\mathcal{F}_Y(U) := \{f \in \mathcal{O}(U) : f(z) = 0 \text{ for } z \in U \cap Y\}, \qquad U \subseteq X \text{ open}, \tag{4.1.1}$$

an analytic subsheaf $\mathcal{F}_Y$ of ${}_X\mathcal{O}^1$. $\mathcal{F}_Y$ is coherent. The simple proof of this fact is left to the reader. Hint. Use appropriate local holomorphic coordinates (Corollary 1.1.19 (i)) and local power series expansion of holomorphic functions.

4.1.6. Let X be a complex manifold and $f = (f_1, \dots, f_N) \in \mathcal{O}^N(X)$. If $U \subseteq X$ is open, then we denote by $\mathcal{F}_f(U)$ the space of all functions $g \in \mathcal{O}(U)$ such that the following condition is fulfilled: For every point $z \in U$, there are a neighbourhood $V \subseteq U$ of z and holomorphic functions $g_j \in \mathcal{O}(V)$ such that $g = g_1 f_1 + \dots + g_N f_N$ in V. It is clear from the definition that the so defined analytic subsheaf $\mathcal{F}_f$ of ${}_X\mathcal{O}^1$ is coherent. $\mathcal{F}_f$ is called *the sheaf which is generated by* f.

4.1.7. Let X be a complex manifold, let B be a holomorphic vector bundle over X, and let $f\colon X \to B$ be a holomorphic section. Since B is locally holomorphically trivial, the section f can be locally represented by a vector of holomorphic functions, where two different representations are connected by an invertible holomorphic matrix. Therefore, as in Example 4.1.6, the section f generates a coherent analytic subsheaf $\mathcal{F}_f$ of ${}_X\mathcal{O}^1$.

4.1.8. Corollary. *Let X be a Stein manifold and let Y be a closed complex submanifold of X. Then, for every holomorphic function f on Y, there exists a holomorphic function F on X such that $F = f$ on Y.*

Proof. It is trivial that f admits local holomorphic extensions, that is, we can find an open covering $\{U_j\}$ of X and functions $F_j \in \mathcal{O}(U_j)$ such that $F_j = f$ on $Y \cap U_j$. Then $F_i - F_j \in \mathcal{F}_Y(U_i \cap U_j)$ (Example 4.1.5) and from Theorem 4.1.3 (ii) we obtain $H_j \in \mathcal{F}_Y(U_j)$ such that $F_i - F_j = H_i - H_j$ on $U_i \cap U_j$. Setting $F := F_j - H_j$ on U_j, we obtain the required extension F. ■

4.2. The section $s(z, \zeta)$ and the function $\varphi(z, \zeta)$

In this section we construct the map $s(z, \zeta)$ and the function $\varphi(z, \zeta)$ mentioned in the summary at the beginning of this chapter. We need the following

4.2.1. Lemma. *Let X be a Stein manifold, let B, B' be holomorphic vector bundles over X, and let $A\colon B \to B'$ be an injective*[1]) *holomorphic homomorphism of vector bundles. Then there exists a holomorphic homomorphism $A^{(-1)}\colon B' \to B$ of vector bundles such that* $A^{(-1)}A = \mathrm{id}$ (id := identity map).

First we prove the following local result:

4.2.2. Lemma. *Let $L(N, M, \mathbb{C})$ be the space of complex $N \times M$ matrices* (N = row index), *and let $GL(N, \mathbb{C})$ be the group of invertible complex $N \times N$ matrices. If U is a neighbourhood of $0 \in \mathbb{C}^n$ and $A\colon U \to L(N, M, \mathbb{C})$ is a holomorphic map such that* rank $A(0) = M$, *then there exist a neighbourhood $V \subseteq U$ of 0 and holomorphic maps*

$$A^{(-1)}\colon V \to L(M, N, \mathbb{C}), \qquad T\colon V \to GL(N, \mathbb{C})$$

such that, for all $z \in V$,

(i) $A^{(-1)}(z)\, A(z) = \mathrm{id}_M$ (id_M := unit matrix of order M),

(ii) $T(z)\, A(z) = A(0)$.

Proof. Since rank $A(0) = M$, there exists a complex $M \times N$ matrix B_0 such that $B_0 A(0) = \mathrm{id}_M$. Choose a neighbourhood V of 0 so small that, for all $z \in V$, the matrix $B_0 A(z)$ is invertible. Set $A^{(-1)}(z) := (B_0 A(z))^{-1} B_0$ and $T(z) := \mathrm{id}_N + (A(0) - A(z))\, A^{(-1)}(z)$. Then (i) and (ii) are fulfilled and, after shrinking V, $T(z)$ is invertible for all $z \in V$. ■

Proof of Lemma 4.2.1. From Lemma 4.2.2 (i) we obtain an open covering $\{U_j\}$ of X and holomorphic homomorphisms $A_j^{(-1)}\colon B'|_{U_j} \to B|_{U_j}$ such that $A_j^{(-1)}A = \mathrm{id}$ over U_j. For every open set $U \subseteq X$, we denote by $\mathcal{F}(U)$ the space of all holomorphic homomorphisms $F\colon B'|_U \to B|_U$ such that $FA = 0$ over U. It follows from Lemma 4.2.2 (ii) that there is a holomorphic vector bundle over X such that $\mathcal{F}$ is the sheaf of germs of holomorphic sections of this bundle (Definition 4.1.1). Clearly, $A_j^{(-1)} - A_i^{(-1)} \in \mathcal{F}(U_i \cap U_j)$.

[1]) Injectivity means that, for every $z \in X$, the kernel of the induced linear map between the fibres of B and B' over z is trivial.

Consequently, by Theorem 4.1.3 (ii) or by Theorems 2.13.4, 2.12.3 (iv) and 2.10.3, we can find $H_j \in \mathcal{F}(U_j)$ such that $A_j^{(-1)} - A_i^{(-1)} = H_i - H_j$ on $U_i \cap U_j$. Setting $A^{(-1)} := A_j^{(-1)} + H_j$, we complete the proof. ■

4.2.3. The complex tangent and cotangent bundles. Let X be a complex manifold of complex dimension n, and let $\{(U_j, h_j)\}$ be a holomorphic atlas of X. Suppose $U_j \subset\subset X$. Set $G_{ij}(z) := J_{h_i \circ h_j^{-1}}(z)$, $z \in U_i \cap U_j$, where $J_{h_i \circ h_j^{-1}}(z)$ is the Jacobi matrix of $h_i \circ h_j^{-1}$ at z (Definition 1.1.15). Then $G_{ij}G_{jk} = G_{ik}$ on $U_i \cap U_j \cap U_k$ (Corollary 1.1.17). The holomorphic vector bundle over X which is defined by the transition functions $\{G_{ij}\}$ will be denoted by $T(X)$ and is called the *complex tangent bundle of* X. The holomorphic vector bundle which is defined by the transition functions $\{(G_{ij}^t)^{-1}\}$, where G_{ij}^t is the transpose of G_{ij}, will be denoted by $T^*(X)$ and is called the *complex cotangent bundle of* X. The fibres of $T(X)$ and $T^*(X)$ over $z \in X$ will be denoted by $T_z(X)$ and $T_z^*(X)$.

4.2.4. Lemma. *Let X be a Stein manifold and let $T(X)$ be the complex tangent bundle of X* (Subsection 4.2.3). *Suppose, in addition, that X is a relatively compact open subset of some larger Stein manifold* (however, cp. the remark following this lemma). *Then there exist a holomorphic map $s: X \times X \to T(X)$ and a holomorphic function φ on $X \times X$ such that the following conditions are fulfilled*:

(i) *$s(z, \zeta) \in T_z(X)$ for all $z, \zeta \in X$ (that is, $s(z, \zeta)$ is a section in the pull-back of the bundle $T(X)$ with respect to the map $X \times X \ni (z, \zeta) \to z \in X$).*

(ii) *For every fixed $z \in X$, $s(z, z) = 0$ and the map $s(z, \zeta)$ from X into $T_z(X)$ is biholomorphic in some neighbourhood of $\zeta = z$.*

(iii) *$\varphi(z, z) = 1$ for all $z \in X$.*

(iv) *If $\mathcal{F}_s$ is the analytic subsheaf of ${}_{X \times X}\mathcal{O}^1$ generated by s* (Example 4.1.7), *then $\varphi \in \mathcal{F}_s((X \times X) \setminus \{(z, z) : z \in X\})$.*

(v) *There is an integer $\varkappa \geqq 0$ such that, for every norm $\|\cdot\|$*[1]) *in $T(X)$, the function $\varphi^\varkappa \|s\|^{-2}$ is C^2 on $(X \times X) \setminus \{(z, z) : z \in X\}$. (It is sufficient to choose $\varkappa \geqq 9$.)*

Remark. Lemma 4.2.4 is valid also without the hypothesis that X is a relatively compact open subset of some larger Stein manifold. However, then the proof is more complicated. In fact, in the following proof we use a holomorphic map $f: X \to \mathbb{C}^N$ (for some large but finite N) such that, for all $z \in X$, the Jacobi matrix (with respect to local coordinates) of f has maximal rank. In the special case that X is a relatively compact open subset of some larger Stein manifold, the existence of such a map f follows immediately from the definition of Stein manifolds. In the general case such f also exists, but a proof is necessary; remark that, moreover, f can be chosen to be biholomorphic from X onto some closed complex submanifold of $\mathbb{C}^N$ (see, for example, Gunning/Rossi [1] and Hörmander [1]). We do not need this generalization of Lemma 4.2.4, because we shall use the section $s(z, \zeta)$ and the function $\varphi(z, \zeta)$ from Lemma 4.2.4 in some neighbourhood of $\overline{D} \times \overline{D}$ only, where D is a relatively compact open subset of a Stein manifold.

Proof of Lemma 4.2.4. Let $f: X \to \mathbb{C}^N$ be as in the preceding remark. Denote by F the holomorphic homomorphism from $T(X)$ into the product bundle $X \times \mathbb{C}^N$ which is locally defined by the Jacobi matrix of f. Then F is injective, and we obtain from Lemma 4.2.1 a holomorphic homomorphism of vector bundles $F^{(-1)}: X \times \mathbb{C}^N \to T(X)$ such that $F^{(-1)}F = \mathrm{id}$ on X. Define $s(z, \zeta) := F^{(-1)}(z, f(\zeta) - f(z))$ for $z, \zeta \in X$. Then it is

---

1) A norm in $T(X)$ is a C^∞-map $\|\cdot\|: T(X) \to [0, +\infty)$ such that $\|\cdot\|$ is a norm in each fibre $T_z(X)$, $z \in X$. By means of local trivializations of $T(X)$ and a partition of unity, it is easy to construct such a norm (cp. Subsection 2.11.2).

clear that condition (i) is fulfilled, and it follows from the Taylor formula (1.1.6) that (ii) is satisfied.

Now we construct the function φ. Set $\Delta := \{(z, z) : z \in X\}$ and

$$Y := \{(z, \zeta) \in X \times X : s(z, \zeta) = 0\} \setminus \Delta'.$$

Then, by condition (ii), Y is a closed subset of $X \times X$, and the open sets $U_\Delta := (X \times X) \setminus Y$, $U_Y := (X \times X) \setminus \Delta$ form an open covering of $X \times X$. Since $\mathcal{F}_s$ is coherent and the constant function with value 1 belongs to $\mathcal{F}_s(U_\Delta \cap U_Y)$, from Theorem 4.1.3 (i) we obtain functions $\varphi_Y \in \mathcal{F}_s(U_Y)$ and $\varphi_\Delta \in \mathcal{F}_s(U_\Delta)$ such that $1 = \varphi_Y - \varphi_\Delta$ in $U_\Delta \cap U_Y$. Setting $\varphi := \varphi_Y$ in U_Y and $\varphi := 1 + \varphi_\Delta$ in U_Δ, we obtain $\varphi \in \mathcal{O}(X \times X)$ satisfying conditions (iii) and (iv).

It remains to prove condition (v). Consider some point $z \in (X \times X) \setminus \Delta$ and choose a neighbourhood $U \subseteq (X \times X) \setminus \Delta$ of z so small that $T(X)$ is holomorphically trivial over U. Then the map s can be represented by a vector $(s_1, \dots, s_n)$ ($n = \dim_{\mathbb{C}} X$) of holomorphic functions in U. After shrinking U, we can assume that $\varphi = \varphi_1 s_1 + \dots + \varphi_n s_n$ for some $\varphi_j \in \mathcal{O}(U)$ (this follows from the definition of $\mathcal{F}_s$). After shrinking U, this implies that there is a constant $C < \infty$ such that $|\varphi(z, \zeta)| \leqq C \,||s(z, \zeta)||$ for all $(z, \zeta) \in U$. Hence $\varphi^9 \,||s||^{-2}$ is a C^2-function in U. ■

4.3. The Martinelli-Bochner formula and the Leray formula

In this section we generalize the Martinelli-Bochner formula (1.9.1) and the Leray formula (1.10.1) to Stein manifolds.

4.3.1. Notation and preliminaries. X is a Stein manifold, and n is the complex dimension of X. If $D \subset\subset X$ is an open set, then the boundary ∂D of D will be called *piecewise* C^1 if there exist open sets $V_1, \dots, V_N \subseteq X$ and C^1-functions $\varrho_k : V_k \to \mathbb{R}$, $k = 1, \dots, N$, such that the following conditions are fulfilled:

(1) $\partial D \subseteq V_1 \cup \dots \cup V_N$.

(2) A point $z \in V_1 \cup \dots \cup V_N$ belongs to D if and only if, for every $1 \leqq k \leqq N$ with $z \in V_k$, $\varrho_k(z) < 0$.

(3) For every collection of indices $1 \leqq k_1 < \dots < k_l \leqq N$, we have $d\varrho_{k_1}(z) \wedge \dots \wedge d\varrho_{k_l}(z) \neq 0$ for all $z \in V_{k_1} \cap \dots \cap V_{k_l}$.

Further on we suppose that $D \subset\subset X$ is an open set with piecewise C^1-boundary.

We choose the following orientation in X: If $z_1, \dots, z_n$ are local holomorphic coordinates in X and x_j are the corresponding real coordinates such that $z_j = x_j + i x_{j+n}$ then the form $dx_1 \wedge \dots \wedge dx_{2n}$ defines the orientation of X (it is easy to see that in this way, in fact, an orientation of X can be defined).

Further let $\{(U_j, h_j)\}$ be a fixed holomorphic atlas of X such that $U_j \subset\subset X$ for all j. We use the corresponding notations introduced in Subsection 4.2.3. If $U \subseteq X$ is an open set, then we denote by $T(U)$ and $T^*(U)$ the restrictions of $T(X)$ and $T^*(X)$ to U. Further, we fix holomorphic trivializations $H_j : T(U_j) \to U_j \times \mathbb{C}^n$ and $H_j^* : T^*(U_j) \to U_j \times \mathbb{C}^n$ such that

$$(z, (G_{ij}(z))\, \xi) = H_i \circ H_j^{-1}(z, \xi)$$

and

$$(z, ((G_{ij}^t)^{-1}(z))\, \xi) = H_i^* \circ (H_j^*)^{-1}(z, \xi) \quad \text{for} \quad z \in U_i \cap U_j \quad \text{and} \quad \xi \in \mathbb{C}^n.$$

If $z \in U_j$ and $a \in T_z(X)$ $\big(a \in T_z^*(X)\big)$, then the vector $a_j \in \mathbb{C}^n$ with $H_j(a) = (z, a_j)$ $\big(H_j^*(a), = (z, a_j)\big)$ is called the *expression of a with respect to* (U_j, h_j). If $z \in U_i \cap U_j$, $a \in T_z(X)$,

$b \in T_z^*(X)$ and a_i, a_j, b_i, b_j are the expressions of a, b with respect to (U_i, h_i) and (U_j, h_j), respectively, then

$$a_i = G_{ij}(z)\, a_j \quad \text{and} \quad b_i = (G_{ij}^t)^{-1}(z)\, b_j\,. \tag{4.3.1}$$

Therefore, the following definition is correct: If $z \in X$, $a \in T_z(X)$ and $b \in T_z^*(X)$, then we choose an arbitrary j such that $z \in U_j$ and define

$$\langle b, a\rangle := \langle b_j, a_j\rangle\,, \tag{4.3.2}$$

where b_j and a_j are the expressions of b and a with respect to (U_j, h_j), and $\langle b_j, a_j\rangle$ is defined as in Section 1.7.

Now let $U \subseteq X$ be an open set, let Y be a real C^1-manifold, and let $a\colon U \times Y \to T(X)$ and $b\colon U \times Y \to T^*(X)$ be C^1-maps such that $a(z, y) \in T_z(X)$ and $b(z, y) \in T_z^*(X)$ for all $(z, y) \in U \times Y$. Let $a_j\colon (U \cap U_j) \times Y \to \mathbb{C}^n$ and $b_j\colon (U \cap U_j) \times Y \to \mathbb{C}^n$ be the expressions of a and b with respect to (U_j, h_j). Then it follows from (4.3.1), (1.7.5) and Proposition 1.7.4 that

$$\omega_y'\big(b_i(z, y)\big) \wedge \omega_y\big(a_i(z, y)\big) = \omega_y'\big(b_j(z, y)\big) \wedge \omega_y\big(a_j(z, y)\big) \quad \text{for} \quad z \in U \cap U_i \cap U_j \quad \text{and} \quad y \in Y\,. \tag{4.3.3}$$

Therefore, by setting

$$\omega_y'\big(b(z, y)\big) \wedge \omega_y\big(a(z, y)\big) := \omega_y'\big(b_j(z, y)\big) \wedge \omega_y\big(a_j(z, y)\big) \quad \text{for} \quad z \in U \cap U_j \quad \text{and} \quad y \in Y\,, \tag{4.3.4}$$

a continuous differential form $\omega_y'\big(b(z, y)\big) \wedge \omega_y\big(a(z, y)\big)$ on $U \times Y$ is correctly defined.

We are interested in integral formulas for functions on $\bar{D}$. Therefore, without loss of generality, we can assume that X is a relatively compact open subset of some larger Stein manifold and apply Lemma 4.2.4 (cp. also the remark following Lemma 4.2.4). Let $s(z, \zeta)$, $\varphi(z, \zeta)$ and $\varkappa$ be as in Lemma 4.2.4. Further, we fix a fibre preserving C^∞-map $\sigma\colon T(X) \to T^*(X)$ such that the following conditions are fulfilled: $\langle \sigma a, a\rangle \geqq 0$ for all $a \in T(X)$ and the map

$$\|a\|_\sigma := (\langle \sigma a, a\rangle)^{1/2}\,, \qquad a \in T(X), \tag{4.3.5}$$

defines a norm in each fibre of $T(X)$. Such a map σ can be defined as follows: If $z \in U_j$, $a \in T_z(X)$ and a_j is the expression of a with respect to (U_j, h_j), then we denote by $\sigma_j a$ the vector in $T_z^*(X)$ whose expression with respect to (U_j, h_j) is $\bar{a}_j$. Choose a C^∞-partition of unity χ_j subordinate to $\{U_j\}$ and define

$$\sigma a := \sum_j \chi_j(z)\, \sigma_j a \quad \text{for} \quad a \in T_z(X) \quad \text{and} \quad z \in X\,. \tag{4.3.6}$$

We write

$$\bar{s}(z, \zeta) := \sigma s(z, \zeta)\,. \tag{4.3.7}$$

($\bar{s}(z, \zeta)$ will be used instead of the map $\bar{\zeta} - \bar{z}$ in $\mathbb{C}^n$.)

4.3.2. Leray sections, $B_{\partial D}$ and $L_{\partial D}$ for functions, and B_D, $R_{\partial D}$ for 1-forms. We use the notations from Subsection 4.3.1. By Lemma 4.2.4 (v), $\varphi^\varkappa/\|s\|^2$ is a C^2-function for all $z, \zeta \in X$ with $z \neq \zeta$. Therefore, for every integer $\nu \geqq n\varkappa$ and every fixed $z \in X$, the differential form (cp. (4.3.4))

$$\frac{\varphi^\nu(z, \zeta)\; \omega_\zeta'\big(\bar{s}(z, \zeta)\big) \wedge \omega_\zeta\big(s(z, \zeta)\big)}{\|s(z, \zeta)\|_\sigma^{2n}}$$

is C^1 for $\zeta \in X \setminus z$, and the singularity at $\zeta = z$ is of order $2n - 1$. Therefore, for every integer $\nu \geqq n\varkappa$ and every bounded 1-form f on D, we can define

$$\big(B_D(\varphi^\nu, \bar{s}, s) f\big)(z) := \frac{(n-1)!}{(2\pi i)^n} \int\limits_{\zeta \in D} f(\zeta) \wedge \frac{\varphi^\nu(z,\zeta)\, \omega_\zeta'\big(\bar{s}(z,\zeta)\big) \wedge \omega_\zeta\big(s(z,\zeta)\big)}{\|s(z,\zeta)\|_\sigma^{2n}}, \quad z \in D. \tag{4.3.8}$$

Further, for every integer $\nu \geqq n\varkappa$ and every bounded measurable function f on ∂D, we define

$$\big(B_{\partial D}(\varphi^\nu, \bar{s}, s) f\big)(z) := \frac{(n-1)!}{(2\pi i)^n} \int\limits_{\zeta \in \partial D} f(\zeta) \frac{\varphi^\nu(z,\zeta)\, \omega_\zeta'\big(\bar{s}(z,\zeta)\big) \wedge \omega_\zeta\big(s(z,\zeta)\big)}{\|s(z,\zeta)\|_\sigma^{2n}}, \quad z \in D. \tag{4.3.9}$$

A *Leray section* for (D, s, φ) is by definition a couple $(s^*, \varkappa^*)$, where $\varkappa^* \geqq 0$ is an integer and $s^*(z,\zeta)$ is a C^1-map with values in $T^*(X)$ defined for ζ in some neighbourhood of ∂D and $z \in D$ such that the following conditions are fulfilled:

(i)* $s^*(z,\zeta) \in T_z^*(X)$ for all $z \in D$ and ζ in some neighbourhood of ∂D.

(ii)* $\langle s^*(z,\zeta), s(z,\zeta)\rangle \neq 0$ for $z \in D$, $\zeta \in \partial D$ with $\varphi(z,\zeta) \neq 0$, and the function

$$\frac{\varphi^{\varkappa^*}(z,\zeta)}{\langle s^*(z,\zeta),\, s(z,\zeta)\rangle}$$

is C^1 in some neighbourhood $\subseteqq D \times X$ of $D \times \partial D$. (This means, for every $K \subset\subset D$, there is a neighbourhood V_K of ∂D such that this function is C^1 for all $(z,\zeta) \in K \times V_K$.)

Example. $(\bar{s}, \varkappa)$ is a Leray section for (D, s, φ).

If $(s^*, \varkappa^*)$ is a Leray section for (D, s, φ), then, for every integer $\nu \geqq n\varkappa^*$ and every fixed $z \in D$, the differential form (cp. (4.3.4))

$$\frac{\varphi^\nu(z,\zeta)\, \omega_\zeta'\big(s^*(z,\zeta)\big) \wedge \omega\big(s(z,\zeta)\big)}{\langle s^*(z,\zeta), s(z,\zeta)\rangle^n}$$

is continuous for ζ in some neighbourhood of ∂D. Therefore, for every bounded measurable function f on ∂D, we can define

$$\big(L_{\partial D}(\varphi^\nu, s^*, s) f\big)(z) := \frac{(n-1)!}{(2\pi i)^n} \int\limits_{\zeta \in \partial D} f(\zeta) \frac{\varphi^\nu(z,\zeta)\, \omega_\zeta'\big(s^*(z,\zeta)\big) \wedge \omega_\zeta\big(s(z,\zeta)\big)}{\langle s^*(z,\zeta), s(z,\zeta)\rangle^n}, \quad z \in D. \tag{4.3.10}$$

If $(s^*, \varkappa^*)$ is a Leray section for (D, s, φ), then we set

$$t^*_{(s^*, \bar{s}, s)}(z,\zeta,\lambda) := (1-\lambda) \frac{s^*(z,\zeta)}{\langle s^*(z,\zeta), s(z,\zeta)\rangle} + \lambda \frac{\bar{s}(z,\zeta)}{\|s(z,\zeta)\|_\sigma^2}.$$

By condition (v) in Lemma 4.2.4 and condition (ii) in the definition of a Leray section, then, for every fixed $z \in D$, the map

$$\varphi^{\max(\varkappa, \varkappa^*)}(z,\zeta)\, t^*_{(s^*, \bar{s}, s)}(z,\zeta,\lambda)$$

is C^1 for $0 \leqq \lambda \leqq 1$ and ζ in some neighbourhood of ∂D. Therefore, by Proposition 1.7.3, for every integer $\nu \geqq \max(n\varkappa, n\varkappa^*)$ and every fixed $z \in D$, the differential form (cp. (4.3.4))

$$\varphi^\nu(z,\zeta)\, \omega_{\zeta,\lambda}'\big(t^*_{(s^*, \bar{s}, s)}(z,\zeta,\lambda)\big) \wedge \omega_\zeta\big(s(z,\zeta)\big)$$

is continuous for $0 \leqq \lambda \leqq 1$ and ζ in some neighbourhood of ∂D.

If $(s^*, \varkappa^*)$ is a Leray section for (D, s, φ), then, for every integer $\nu \geqq \max(n\varkappa, n\varkappa^*)$ and every bounded 1-form f on ∂D, we define

$$\big(R_{\partial D}(\varphi^\nu, s^*, \bar{s}, s) f\big)(z)$$
$$:= \frac{(n-1)!}{(2\pi i)^n} \int\limits_{(\zeta,\lambda)\in\partial D\times[0,1]} f(\zeta) \wedge \varphi^\nu(z,\zeta)\, \omega'_{\zeta,\lambda}\big(t^*(z,\zeta,\lambda)\big) \wedge \omega_\zeta\big(s(z,\zeta)\big), \quad z \in D, \tag{4.3.11}$$

where $t^* = t^*_{(s^*,\bar{s},s)}$.

4.3.3. Theorem (Martinelli-Bochner formula). *We use the notations from Subsection 4.3.1. Let $\nu \geqq 2\varkappa n$ be an integer. Then, for every continuous function f on $\bar{D}$ such that $\bar\partial f$ is also continuous on $\bar{D}$, we have*

$$f = B_{\partial D} f - B_D \bar\partial f \quad \text{in} \quad D, \tag{4.3.12}$$

where $B_{\partial D} = B_{\partial D}(\varphi^\nu, \bar{s}, s)$ and $B_D = B_D(\varphi^\nu, \bar{s}, s)$ (Subsection 4.3.2).

Proof. First we prove the formula for the constant function $f = 1$. Fix $z \in D$. Choose $U_{j_0}(\{(U_j, h_j)\}$ is the holomorphic atlas from Subsection 4.3.1) so that $z \in U_{j_0}$. Let $v(\zeta)$ be the expression of $\bar{s}(z,\zeta)$ and let $u(\zeta)$ be the expression of $s(z,\zeta)$ with respect to (U_{j_0}, h_{j_0}) (cp. Subsection 4.3.1). Then

$$\omega'_\zeta\big(\bar{s}(z,\zeta)\big) \wedge \omega_\zeta\big(s(z,\zeta)\big) = \omega'_\zeta\big(v(\zeta)\big) \wedge \omega_\zeta\big(u(\zeta)\big),$$

and we have to prove that $I = (2\pi i)^n/(n-1)!$, where

$$I := \int\limits_{\zeta\in\partial D} \varphi^\nu(z,\zeta) \frac{\omega'_\zeta\big(v(\zeta)\big) \wedge \omega_\zeta\big(u(\zeta)\big)}{\langle v(\zeta), u(\zeta)\rangle^n}.$$

For ζ in a neighbourhood of ∂D and $0 \leqq \lambda \leqq 1$, we define

$$\eta(\zeta,\lambda) := \varphi^\varkappa(z,\zeta)\left[(1-\lambda)\frac{\langle v(\zeta), u(\zeta)\rangle\, \overline{u(\zeta)}}{|u(\zeta)|^2} + \lambda v(\zeta)\right]$$

and

$$\Omega(\zeta,\lambda) := \varphi^\nu(z,\zeta) \frac{\omega'_{\zeta,\lambda}\big(\eta(\zeta;\lambda)\big) \wedge \omega_\zeta\big(u(\zeta)\big)}{\langle \eta(\zeta,\lambda), u(\zeta)\rangle^n}.$$

Since, for some $\alpha > 0$, $|u(\zeta)| \geqq \alpha\, \|s(z,\zeta)\|_\sigma$, the map $\eta(\zeta,\lambda)$ is C^1 in a neighbourhood of ∂D. Since $\langle \eta(\zeta,\lambda), u(\zeta)\rangle = \varphi^\varkappa(z,\zeta)\, \langle v(\zeta), u(\zeta)\rangle = \varphi^\varkappa(z,\zeta)\, \|s(z,\zeta)\|^2$ and since $\nu \geqq 2n\varkappa$, if follows that $\Omega(\zeta,\lambda)$ is continuous for ζ in some neighbourhood of ∂D and $0 \leqq \lambda \leqq 1$. Since $\varphi(z,\zeta)$ is holomorphic, it follows from Proposition 1.7.1 that $d_{\zeta,\lambda}\Omega(\zeta,\lambda) = 0$. Therefore, Stokes' formula gives

$$\int\limits_{\partial D\times 1} \Omega(\zeta,\lambda) = \int\limits_{\partial D\times 0} \Omega(\zeta,\lambda). \tag{4.3.13}$$

Taking into account the relations $\eta(\zeta, 0) = \varphi^\varkappa(z,\zeta)\, (\langle v(\zeta), u(\zeta)\rangle/|u(\zeta)|^2)\, \overline{u(\zeta)}$ and $\eta(\zeta,1) = \varphi^\varkappa(z,\zeta)\, v(\zeta)$, we obtain from Proposition 1.7.3 that

$$\Omega(\zeta,\lambda)|_{\lambda=0} = \varphi^\nu(z,\zeta) \frac{\omega'_\zeta\big(\overline{u(\zeta)}\big) \wedge \omega_\zeta\big(u(\zeta)\big)}{|u(\zeta)|^{2n}}$$

and

$$\Omega(\zeta,\lambda)|_{\lambda=1} = \varphi^\nu(z,\zeta) \frac{\omega'_\zeta\big(v(\zeta)\big) \wedge \omega_\zeta\big(u(\zeta)\big)}{\langle v(\zeta), u(\zeta)\rangle^n}.$$

Hence (4.3.13) can be written

$$I = \int_{\zeta\in\partial D} \varphi^\nu(z,\zeta)\,\frac{\omega'_\zeta\left(\overline{u(\zeta)}\right) \wedge \omega_\zeta\left(u(\zeta)\right)}{|u(\zeta)|^{2n}}. \tag{4.3.14}$$

By condition (ii) in Lemma 4.2.4, we can choose a neighbourhood $W \subset\subset D$ of z with smooth boundary ∂W such that u is biholomorphic in some neighbourhood of $\overline{W}$. Since φ is holomorphic and by Proposition 1.7.1, the form under the sign of integration on the right-hand side of (4.3.14) is closed for all $\zeta \in X$ with $u(\zeta) \neq 0$. Since $\nu \geqq n\varkappa$ and, therefore, this form is C^1 in $X \setminus z$, we conclude that this form is closed in $X \setminus z$. Hence, by Stokes' formula, it follows from (4.3.14) that

$$I = \int_{\zeta\in\partial W} \varphi^\nu(z,\zeta)\,\frac{\omega'_\zeta\left(\overline{u(\zeta)}\right) \wedge \omega_\zeta\left(u(\zeta)\right)}{|u(\zeta)|^{2n}}. \tag{4.3.15}$$

Since u is biholomorphic in a neighbourhood of $\overline{W}$, we obtain

$$I = \int_{\xi\in\partial(u(W))} \varphi^\nu\left(z, u^{-1}(\xi)\right)\frac{\omega'(\bar\xi) \wedge \omega(\xi)}{|\xi|^{2n}}.$$

Consequently, since the function $u(\overline{W}) \ni \xi \to \varphi\left(z, u^{-1}(\xi)\right)$ is holomorphic, and since $\varphi\left(z, u^{-1}(0)\right) = \varphi(z,z) = 1$ (condition (iii) in Lemma 4.2.4), it follows from the Martinelli-Bochner formula (1.9.2) in $\mathbb{C}^n$ that $I = (2\pi i)^n/(n-1)!$.

Now we consider the general case of a continuous function f on $\overline{D}$ such that $\bar\partial f$ is also continuous on $\overline{D}$. Fix $z \in D$. Since φ is holomorphic, then it follows from Proposition 1.7.1 that, for every $\zeta \in \overline{D}$ such that $s(z,\zeta) \neq 0$, we have the relation

$$\bar\partial_\zeta\left[f(\zeta)\,\frac{\varphi^\nu(z,\zeta)\,\omega'_\zeta\left(\bar s(z,\zeta)\right) \wedge \omega_\zeta\left(s(z,\zeta)\right)}{\|s(z,\zeta)\|_\sigma^{2n}}\right] = \bar\partial f(\zeta) \wedge \frac{\varphi^\nu(z,\zeta)\,\omega'_\zeta\left(\bar s(z,\zeta)\right) \wedge \omega_\zeta\left(s(z,\zeta)\right)}{\|s(z,\zeta)\|_\sigma^{2n}}. \tag{4.3.16}$$

Since $\nu \geqq n\varkappa$ and, therefore, the form $\varphi^\nu\omega'_\zeta(\bar s) \wedge \omega_\zeta(s)/\|s\|^{2n}$ is C^1 for $\zeta \in X \setminus z$, (4.3.16) is valid for all $\zeta \in D \setminus z$. Fix some local coordinates in a neighbourhood of z and denote by E_ε the open ball of radius ε centered at z with respect to these coordinates. Then, for all sufficiently small $\varepsilon > 0$, it follows from (4.3.16) and Stokes' formula that

$$(B_{D\setminus E_\varepsilon}\,\bar\partial f)(z) = (B_{\partial D} f)(z) - (B_{\partial E_\varepsilon} f)(z). \tag{4.3.17}$$

Since, for constant functions, the theorem is already proved, we obtain

$$(B_{\partial E_\varepsilon} f)(z) = f(z) + \left(B_{\partial E_\varepsilon}(f - f(z))\right)(z).$$

Since the form $\omega'_\zeta\left(\bar s(z,\zeta)\right) \wedge \omega_\zeta\left(s(z,\zeta)\right)/\|s(z,\zeta)\|^{2n}$ has a singularity of order $2n-1$ at $\zeta = z$, this implies that $(B_{\partial E_\varepsilon} f)(z)$ tends to $f(z)$ for $\varepsilon \to 0$. Hence (4.3.17) tends to (4.3.12) when $\varepsilon \to 0$. ■

4.3.4. Theorem (Leray formula). *We use the notations from Subsection* 4.3.1. *Let* $(s^*, \varkappa^*)$ *be a Leray section for* (D, s, φ) (Subsection 4.3.2), *and let* $\nu \geqq \max(2n\varkappa, n\varkappa^*)$ *be an integer. Then, for every continuous function f on $\overline{D}$ such that $\bar\partial f$ is also continuous on $\overline{D}$, we have*

$$f = L_{\partial D} f - R_{\partial D}\,\bar\partial f - B_D\,\bar\partial f \quad \textit{in} \quad D, \tag{4.3.18}$$

where $L_{\partial D} = L_{\partial D}(\varphi^\nu, s^*, s)$, $R_{\partial D} = R_{\partial D}(\varphi^\nu, s^*, \bar{s}, s)$ *and* $B_D = B_D(\varphi^\nu, \bar{s}, s)$ (cp. Subsection 4.3.2).

To prove this theorem we need the following lemma.

4.3.5. Lemma. *Under the hypotheses of Theorem 4.3.4, for every fixed $z \in D$, we have*

$$d_{\zeta,\lambda}[f(\zeta)\, \varphi^\nu(z,\zeta)\, \omega'_{\zeta,\lambda}(t^*(z,\zeta,\lambda)) \wedge \omega_\zeta(s(z,\zeta))]$$
$$= \bar{\partial} f(\zeta) \wedge \varphi^\nu(z,\zeta)\, \omega'_{\zeta,\lambda}(t^*(z,\zeta,\lambda)) \wedge \omega_\zeta(s(z,\zeta))$$

for $0 \leqq \lambda \leqq 1$ and ζ in some neighbourhood of ∂D, where $t^* = t^*_{(s^*,\bar{s},s)}$ (cp. Subsection 4.3.2).

Proof. Since the form in brackets contains a factor of bidegree $(n, 0)$, we only have to prove that $(\bar{\partial}_\zeta + d_\lambda)\, \Omega(\zeta, \lambda) = 0$, where

$$\Omega(\zeta,\lambda) := \varphi^\nu(z,\zeta)\, \omega'_{\zeta,\lambda}(t^*(z,\zeta,\lambda)) \wedge \omega_\zeta(s(z,\zeta)) .$$

Choose U_{j_0} ($\{(U_j, h_j)\}$ is the holomorphic atlas from Subsection 4.3.1) so that $z \in U_{j_0}$. Let $\mu := \max(\varkappa, \varkappa^*)$, let $v(\zeta,\lambda)$ be the expression of $\varphi^\mu(z,\zeta)\, t^*(z,\zeta,\lambda)$, and let $u(\zeta)$ be the expression of $s(z,\zeta)$ with respect to (U_{j_0}, h_{j_0}) (cp. Subsection 4.3.1). Then by Proposition 1.7.3

$$\Omega(\zeta,\lambda) = \varphi^{\nu-n\mu}(z,\zeta) \wedge \omega'(v(\zeta,\lambda)) \wedge \omega(u(\zeta)) .$$

Since $\omega(u(\zeta))$ is of bidegree $(n, 0)$, we have

$$\Omega(\zeta,\lambda) = \varphi^{\nu-n\mu}(z,\zeta) \sum_{k=1}^{n} (-1)^{k+1}\, v_k(\zeta,\lambda) \bigwedge_{l\neq k} (\bar{\partial}_\zeta + d_\lambda)\, v_l(\zeta,\lambda) \wedge \omega(u(\zeta)) .$$

Taking into account that $\varphi^{\nu-n\mu}$ is holomorphic, we thus obtain

$$(\bar{\partial}_\zeta + d_\lambda)\, \Omega(\zeta,\lambda) = n\varphi^{\nu-n\mu}(z,\zeta) \bigwedge_{k=1}^{n} (\bar{\partial}_\zeta + d_\lambda)\, v_k(\zeta,\lambda) \wedge \omega(u(\zeta)) . \tag{4.3.19}$$

Since

$$\sum_{k=1}^{n} v_k(\zeta,\lambda)\, u_k(\zeta) = \varphi^\mu(z,\zeta)\, \langle t^*(z,\zeta,\lambda), s(z,\zeta)\rangle = \varphi^\mu(z,\zeta) ,$$

and since $(\bar{\partial}_\zeta + d_\lambda)\, \varphi^\mu(z,\zeta) = (\bar{\partial}_\zeta + d_\lambda)\, u_k(\zeta) = 0$, we obtain

$$\sum_{k=1}^{n} u_k(\zeta)\, (\bar{\partial}_\zeta + d_\lambda)\, v_k(\zeta,\lambda) = 0 .$$

This implies that

$$\bigwedge_{k=1}^{n} (\bar{\partial}_\zeta + d_\lambda)\, v_k(\zeta,\lambda) = 0 , \tag{4.3.20}$$

because the form on the left-hand side of (4.3.20) is continuous and the set $\{\zeta \in X: u(\zeta) \neq 0\}$ is dense in X. From (4.3.20) and (4.3.19) it follows that $(\bar{\partial}_\zeta + d_\lambda)\, \Omega(\zeta,\lambda) = 0$. ■

Proof of Theorem 4.3.4. By the Martinelli-Bochner formula (4.3.12), it is sufficient to prove that

$$R_{\partial D}\, \bar{\partial} f = L_{\partial D} f - B_{\partial D} f \quad \text{in} \quad D .$$

This follows from Stokes' formula and Lemma 4.3.5 if we take into account that by Proposition 1.7.3

$$\omega'_{\zeta,\lambda}(t^*(z,\zeta,\lambda)) \wedge \omega_\zeta(s(z,\zeta))|_{\lambda=0} = \frac{\omega'_\zeta(s^*(z,\zeta)) \wedge \omega_\zeta(s(z,\zeta))}{\langle s^*(z,\zeta), s(z,\zeta)\rangle^n}$$

and

$$\omega'_{\zeta,\lambda}(t^*(z,\zeta,\lambda)) \wedge \omega_\zeta(s(z,\zeta))|_{\lambda=1} = \frac{\omega'_\zeta(\bar{s}(z,\zeta)) \wedge \omega_\zeta(s(z,\zeta))}{\|s(z,\zeta)\|_\sigma^{2n}},$$

where $t^* = t^*_{(s^*,\bar{s},s)}$ (cp. Subsection 4.3.2).

Remark. For $s^* = \bar{s}$ we have $R_{\partial D} = 0$, $L_{\partial D} = B_{\partial D}$ and the Leray formula (4.3.18) is the Martinelli-Bochner formula (4.3.12).

4.4. The Leray-Norguet formula

For many purposes it is appropriate to use different Leray maps for different pieces of the boundary. This is the case, for example, for real non-degenerate analytic polyhedra (Section 4.7) as well as in the more general situation of real non-degenerate strictly pseudoconvex polyhedra (Section 4.9). In the present section we prove a corresponding generalization of the Leray formula (4.3.18), which will be called the Leray-Norguet formula.

4.4.1. Notation and preliminaries. We use the notations from Subsection 4.3.1. Set $S_k := \{z \in \partial D \cap V_k : \varrho_k(z) = 0\}$ for $k = 1, \dots, N$, where V_k and ϱ_k are as in the definition of a piecewise C^1-boundary at the beginning of Subsection 4.3.1. For every ordered collection $K = (k_1, \dots, k_l)$ of integers $1 \leqq k_1, \dots, k_l \leqq N$, we define

$$S_K := \begin{cases} S_{k_1} \cap \dots \cap S_{k_l} & \text{if the integers } k_1, \dots, k_l \text{ are different in pairs}, \\ \emptyset & \text{otherwise}. \end{cases}$$

We choose the orientation on S_K so that

$$\partial D = \sum_{k=1}^{N} S_k \tag{4.4.1}$$

and

$$\partial S_K = \sum_{j=1}^{N} S_{Kj}, \tag{4.4.2}$$

where ∂D and ∂S_K carry the orientations induced from the orientations of D and S_K, respectively, and $Kj := (k_1, \dots, k_l, j)$ if $K = (k_1, \dots, k_l)$. Then the orientation of S_K is skew-symmetric in the components of K.

We denote by Δ the subset of all points $(\lambda_0, \lambda_1, \dots, \lambda_N) \in \mathbb{R}^{N+1}$ such that $\lambda_j \geqq 0$ for $j = 0, 1, \dots, N$ and $\sum_{j=0}^{N} \lambda_j = 1$. We orient Δ by the form $d\lambda_1 \wedge \dots \wedge d\lambda_N$. For every strictly increasing collection $I = (i_1, \dots, i_l)$ of integers $0 \leqq i_1 < \dots < i_l \leqq N$, we set

$$\Delta_I := \{\lambda \in \Delta : \sum_{r=1}^{l} \lambda_{i_r} = 1\}$$

and choose the orientation of Δ_I so that

$$\partial \Delta_I = \sum_{r=1}^{l} (-1)^{r-1} \Delta_{i_1,\dots,\hat{i}_r,\dots,i_l}, \tag{4.4.3}$$

where $\hat{i}_r$ means that i_r is omitted. It is easy to see that then, for every strictly increasing collection $K = (k_1, \dots, k_l)$ of integers $1 \leqq k_1 < \dots < k_l \leqq N$, Δ_{0K} carries the orientation defined by the form $d\lambda_{k_1} \wedge \dots \wedge d\lambda_{k_l}$ ($0K := (0, k_1, \dots, k_l)$). This implies that, for $k = 0, 1, \dots, N$, the one-point set Δ_k is oriented by $+1$.

For every collection $I = (i_1, \dots, i_l)$ of integers we write $|I| := l$.

Lemma.

$$\partial(\sum_K (-1)^{|K|} S_K \times \Delta_{0K}) = \sum_K S_K \times \Delta_K - \partial D \times \Delta_0 , \tag{4.4.4}$$

where the summation is over all strictly increasing collections $K = (k_1, \dots, k_l)$ *of integers* $1 \leqq k_1 < \dots < k_l \leqq N$.

Proof. Since $\dim_R S_K = 2n - |K|$, we obtain that the left-hand side of (4.4.4) is equal to

$$\sum_K (-1)^{|K|} (\partial S_K \times \Delta_{0K} + (-1)^{2n-|K|} S_K \times \partial \Delta_{0K}) .$$

By (4.4.2) and (4.4.3) this implies that the left-hand side of (4.4.4) is equal to

$$\sum_K (-1)^{|K|} \sum_{j=1}^N S_{Kj} \times \Delta_{0K} + \sum_K S_K \times \Delta_K + \sum_K S_K \times \sum_{r=1}^{|K|} (-1)^r \Delta_{0, k_1, \dots, \hat{k}_r, \dots, k_{|K|}}$$

if $K = (k_1, \dots, k_{|K|})$. Therefore, since $\sum_{|K|=1} S_K \times \Delta_0 = \partial D \times \Delta_0$, it remains to prove that

$$\sum_K (-1)^{|K|} \sum_{j=1}^N S_{Kj} \times \Delta_{0K} + \sum_{|K| \geqq 2} S_K \times \sum_{r=1}^{|K|} (-1)^r \Delta_{0, k_1, \dots, \hat{k}_r, \dots, k_{|K|}} = 0 . \tag{4.4.5}$$

For $l = 1, \dots, N$, we denote by P_l the set of all strictly increasing collections $K = (k_1, \dots, k_l)$ of integers $1 \leqq k_1 < \dots < k_l \leqq N$. If $K = (k_1, \dots, k_l) \in P_l$, then we denote by K^c the collection in P_{N-l} with $K \cup K^c = \{1, \dots, N\}$. If $K' = (k'_1, \dots, k'_{l+1}) \in P_{l+1}$ and $j \in K'$, that is, $j = k'_r$ for some $1 \leqq r \leqq l + 1$, then we write $\nu(K', j) := r$ and $L(K', j) := (k'_1, \dots, \hat{k}'_r, \dots, k'_{l+1})$. Since $S_{Kj} = \emptyset$ for $j \in K$, now (4.4.5) can be written

$$\sum_{l=1}^{N-1} (\sum_{\substack{K \in P_l \\ j \in K^c}} (-1)^l S_{Kj} \times \Delta_{0K} + \sum_{\substack{K' \in P_{l+1} \\ i \in K'}} (-1)^{\nu(K', i)} S_{K'} \times \Delta_{0L(K', i)}) = 0 . \tag{4.4.6}$$

For every couple (K, j), where $K \in P_l$ and $j \in K^c$, there exists one and only one couple (K', i), where $K' \in P_{l+1}$ and $i \in K'$, such that $L(K', i) = K$ and $i = j$. Moreover each such couple (K', i) can be obtained in this way. Therefore, it is sufficient to prove that for all $1 \leqq l \leqq N - 1$, $K' \in P_{l+1}$ and $j \in K'$,

$$(-1)^l S_{L(K', j)j} \times \Delta_{0L(K', j)} + (-1)^{\nu(K', j)} S_{K'} \times \Delta_{0L(K', j)} = 0 ,$$

which follows from the relation $S_{L(K', j)j} = (-1)^{\nu(K', j) - l - 1} S_{K'}$. ■

4.4.2. Leray-Norguet sections, L_{S_K} for functions and R_{S_K} for 1-forms. We use the notations from Subsections 4.3.1 and 4.4.1. A *Leray-Norguet section* for (D, s, φ) is by definition a collection $(s_1, \dots, s_N, \varkappa^*)$, where $\varkappa^* \geqq 0$ is an integer and, for every $1 \leqq k \leqq N$, $s_k^*(z, \zeta)$ is a C^1-map with values in $T^*(X)$ (cp. Subsection 4.2.3) defined for ζ in some neighbourhood of S_k and $z \in D$ such that the following conditions are fulfilled:

(i)* $s_k^*(z, \zeta) \in T_z^*(X)$ for all $z \in D$ and ζ in some neighbourhood of S_k.

(ii)* $\langle s_k^*(z, \zeta), s(z, \zeta)\rangle \neq 0$ if $z \in D, \zeta \in S_k$ and $\varphi(z, \zeta) \neq 0$, and the function

$$\frac{\varphi^{\varkappa^*}(z, \zeta)}{\langle s^*(z, \zeta), s(z, \zeta)\rangle}$$

is C^1 for (z, ζ) in some neighbourhood $\subseteqq D \times X$ of $D \times S_k$.

Let $(s_1, \ldots, s_N, \varkappa^*)$ be a Leray-Norguet section for (D, s, φ), and let $K = (k_1, \ldots, k_l)$ be a strictly increasing collection of integers $1 \leqq k_1 < \ldots < k_l \leqq N$. By condition (ii) and Proposition 1.7.3, then, for every fixed $z \in D$ and every integer $\nu \geqq n\varkappa^*$, the differential form (cp. (4.3.4))

$$\varphi^\nu(z, \zeta)\, \omega'_{\zeta,\lambda}\left(\sum_{k\in K} \lambda_k \frac{s_k^*(z, \zeta)}{\langle s_k^*(z, \zeta), s(z, \zeta)\rangle}\right) \wedge \omega_\zeta(s(z, \zeta))$$

is continuous for $\lambda \in \Delta_K$ and ζ in some neighbourhood of S_K. If f is a bounded measurable function on S_K and $\nu \geqq n\varkappa^*$ is an integer, then we define

$$\begin{aligned}&\left(L_{S_K}(\varphi^\nu, s_1^*, \ldots, s_N^*, s) f\right)(z)\\ &:= \frac{(n-1)!}{(2\pi i)^n} \int\limits_{(\zeta,\lambda)\in S_K \times \Delta_K} f(\zeta)\, \varphi^\nu(z, \zeta)\, \omega'_{\zeta,\lambda}\left(\sum_{k\in K} \lambda_k \frac{s_k^*(z, \zeta)}{\langle s_k^*(z, \zeta), s(z, \zeta)\rangle}\right) \wedge \omega_\zeta(s(z, \zeta))\,,\\ &\qquad z \in D\,. \qquad (4.4.7)\end{aligned}$$

For all (z, ζ, λ) with $\varphi(z, \zeta) \neq 0$ in some neighbourhood $\subseteqq D \times X \times \Delta_{0K}$ of $D \times S_K \times \Delta_{0K}$, we set

$$t^*_{(s_1^*,\ldots,s_N^*,\bar{s},s)}(z, \zeta, \lambda) := \lambda_0 \frac{\bar{s}(z, \zeta)}{\|s(z, \zeta)\|_\sigma^2} + \sum_{k\in K} \lambda_k \frac{s_k^*(z, \zeta)}{\langle s_k^*(z, \zeta), s(z, \zeta)\rangle}\,. \qquad (4.4.8)$$

Then, for every integer $\nu \geqq \max(n\varkappa, n\varkappa^*)$ and every fixed $z \in D$, the differential form

$$\varphi^\nu(z, \zeta)\, \omega'_{\zeta,\lambda}\left(t^*_{(s_1^*,\ldots,s_N^*,\bar{s},s)}(z, \zeta, \lambda)\right) \wedge \omega_\zeta(s(z, \zeta))$$

is continuous for $\lambda \in \Delta_{0K}$ and ζ in some neighbourhood of S_K. If f is a bounded 1-form on S_K and $\nu \geqq \max(n\varkappa^*, n\varkappa)$, then we define

$$\begin{aligned}&\left(R_{S_K}(\varphi^\nu, s_1^*, \ldots, s_N^*, \bar{s}, s) f\right)(z)\\ &:= (-1)^K \frac{(n-1)!}{(2\pi i)^n} \int\limits_{(\zeta,\lambda)\in S_K \times \Delta_{0K}} f(\zeta) \wedge \varphi^\nu(z,\zeta)\, \omega'_{\zeta,\lambda}(t^*(z, \zeta, \lambda)) \wedge \omega_\zeta(s(z,\zeta))\,, \quad z \in D\,,\\ &\qquad (4.4.9)\end{aligned}$$

where $t^* = t^*_{(s_1^*,\ldots,s_N^*,\bar{s},s)}$.

Remark

$$L_{S_K} = 0 \quad \text{if} \quad |K| \geqq n+1 \quad \text{and} \quad R_{S_K} = 0 \quad \text{if} \quad |K| \geqq n\,. \qquad (4.4.10)$$

This follows from the fact that $\dim_{\mathbb{R}} S_K = 2n - |K|$ and the form under the sign of integration in (4.4.7) and (4.4.9), respectively, is of degree $\geqq n$ and $\geqq n+1$, respectively, in ζ.

Remark. For $N = 1$, Definitions (4.4.7) and (4.4.9) agree with (4.3.10) and (4.3.11), respectively.

4.4.3. Theorem (Leray-Norguet formula). *We use the notations from Subsection* 4.4.1. *Let* $(s_1^*, \ldots, s_N^*, \varkappa^*)$ *be a Leray-Norguet section for* (D, s, φ) (cp. Subsection 4.4.2), *and let* $\nu \geqq \max(2n\varkappa, n\varkappa^*)$ *be an integer. Then, for every continuous function* f *on* $\overline{D}$ *such* *that $\bar\partial f$ is also continuous on $\bar D$, we have*

$$f = \sum_{|K|\leqq n} L_{S_K} f - \sum_{|K|\leqq n-1} R_{S_K} \bar\partial f - B_D \bar\partial f \quad in \quad D\,, \tag{4.4.11}$$

where the summations are over all strictly increasing collections $K = (k_1, \dots, k_l)$ of integers $1 \leqq k_1 < \dots < k_l \leqq N$ with $l \leqq n$ and $l \leqq n-1$, respectively, and $L_{S_K} = L_{S_K}(\varphi^\nu, s_1^*, \dots, s_N^*, s)$, $R_{S_K} = R_{S_K}(\varphi^\nu, s_1^*, \dots, s_N^*, \bar s, s)$ (cp. Subsection 4.4.2) *and $B_D = B_D(\varphi^\nu, \bar s, s)$* (cp. Subsection 4.3.2).

Remark. If $(s^*, \varkappa^*)$ is a Leray section for (D, s, φ), then by setting $s_k^* := s^*$ we obtain a Leray-Norguet section $(s_1^*, \dots, s_N^*, \varkappa^*)$ for (D, s, φ). Then $L_{S_K} = 0$, $R_{S_K} = 0$ if $|K| \geqq 2$ and $L_{S_1} + \dots + L_{S_N} = L_{\partial D}$, $R_{S_1} + \dots + R_{S_N} = R_{\partial D}$, that is, in this case, the Leray-Norguet formula (4.4.11) is the Leray formula (4.3.18).

To prove Theorem 4.4.3 we need the following lemma.

4.4.4. Lemma. *Under the hypotheses of Theorem 4.4.3, for every fixed $z \in D$ and every strictly increasing collection $K = (k_1, \dots, k_l)$ of integers $1 \leqq k_1 < \dots < k_l \leqq N$, we have*

$$d_{\zeta,\lambda}[f(\zeta)\, \varphi^\nu(z,\zeta)\, \omega'_{\zeta,\lambda}(t^*(z,\zeta,\lambda)) \wedge \omega_\zeta(s(z,\zeta))] = \bar\partial f(\zeta) \wedge \varphi^\nu(z,\zeta)\, \omega'_{\zeta,\lambda}(t^*(z,\zeta,\lambda))$$
$$\wedge\, \omega_\zeta(s(z,\zeta))$$

for all $\lambda \in \Delta_{0K}$ and ζ in some neighbourhood of S_K, where $t^* = t^*_{(s_1^*,\dots,s_N^*,\bar s, s)}$ (cp. Subsection 4.4.2).

Proof. Repetition of that of Lemma 4.3.5. ■

Proof of Theorem 4.4.3. According to the Martinelli-Bochner formula (4.3.12), it is sufficient to prove that

$$\sum_{|K|\leqq n-1} R_{S_K} \bar\partial f = \sum_{|K|\leqq n} L_{S_K} f - B_{\partial D} f\,.$$

This follows from relations (4.4.4), (4.4.10) and Lemma 4.4.4 if we take into account that Δ_0 carries the orientation $+1$ and, by Proposition 1.7.3,

$$\omega'_{\zeta,\lambda}(t^*(z,\zeta,\lambda)) \wedge \omega_\zeta(s(z,\zeta))\big|_{\partial D\times\Delta_0} = \frac{\omega'_\zeta(\bar s(z,\zeta)) \wedge \omega_\zeta(s(z,\zeta))}{\|s(z,\zeta)\|_\sigma^{2n}}\,. \ \blacksquare$$

4.4.5. Corollary. *If under the hypotheses of Theorem 4.4.3, for every fixed $z \in D$, the sections $s_1^*(z,\zeta), \dots, s_N^*(z,\zeta)$ are holomorphic for ζ in some neighbourhood of $S_1, \dots, S_N$, respectively, then $L_{S_K} f = 0$ if $|K| \neq n$ and, therefore,*

$$f = \sum_{|K|=n} L_{S_K} f - \sum_{|K|\leqq n-1} R_{S_K} \bar\partial f - B_D \bar\partial f \quad in \quad D\,. \tag{4.4.12}$$

In particular, if f is holomorphic in D, then

$$f = \sum_{|K|=n} L_{S_K} f \quad in \quad D\,. \tag{4.4.13}$$

Proof. If $s_1^*(z,\zeta), \dots, s_N^*(z,\zeta)$ are holomorphic in ζ, then for every collection $K = (k_1, \dots, k_l)$ of strictly increasing integers $1 \leqq k_1 < \dots < k_l \leqq N$, the form

$$f(\zeta) \wedge \varphi^\nu(z,\zeta)\, \omega'_{\zeta,\lambda}\left(\sum_{k\in K} \lambda_k \frac{s_k^*(z,\zeta)}{\langle s_k^*(z,\zeta), s(z,\zeta)\rangle}\right) \wedge \omega_\zeta(s(z,\zeta))$$

is of degree n in ζ. Since $\dim_{\mathbb R} S_K = 2n - |K|$, this implies that $L_{S_K} f = 0$ if $|K| \neq n$. ■

4.5. The Koppelman formula and the Koppelman-Leray formula

In this section we generalize the Koppelman formula (1.11.1) and the Koppelman-Leray formula (1.12.1) to Stein manifolds.

4.5.1. The operators $B_{\partial D}$, B_D, $L_{\partial D}$ and $R_{\partial D}$ for differential forms of arbitrary degree. We use the notations from Subsection 4.3.1. First we consider the following general situation. Let $V, W \subseteq X$ be open sets, let Λ be some real C^1-manifold, and let $a^*: V \times W \times \Lambda \to T^*(X)$ be a C^1-map such that $a^*(z, \zeta, \lambda) \in T_z^*(X)$ for all $z \in V$, $\zeta \in W$, $\lambda \in \Lambda$. For $z \in U_j \cap V$, $\zeta \in W$, $\lambda \in \Lambda$ ($\{(U_j, h_j)\}$ is the holomorphic atlas from Subsection 4.3.1) we denote by $a_j^*(z, \zeta, \lambda)$ the expression of $a^*(z, \zeta, \lambda)$ with respect to (U_j, h_j). For $(z, \zeta, \lambda) \in (U_j \cap V) \times W \times \Lambda$, we define (for $\det_{1,n-1}$ see Definition 1.7.6)

$$\overline{\omega}'_{z,\zeta,\lambda}(a_j^*(z,\zeta,\lambda)) := \frac{1}{(n-1)!} \det_{1,n-1}(a_j^*(z, \zeta, \lambda), (\bar{\partial}_{z,\zeta} + d_\lambda)\, a_j^*(z, \zeta, \lambda)) .$$

For $(z, \zeta) \in U_j \times X$, we denote by $s_j(z, \zeta)$ the expression of $s(z, \zeta)$ with respect to (U_j, h_j). Since, for every holomorphic function $\psi(z)$, $\bar{\partial}_z(\psi(z)\, a_j^*(z, \zeta, \lambda)) = \psi(z)\, \bar{\partial}_z a_j^*(z,\zeta,\lambda)$, then it follows from Proposition 1.7.4 and relation (4.3.1) that

$$\overline{\omega}'_{z,\zeta,\lambda}(a_j^*(z, \zeta, \lambda)) \wedge \omega_\zeta(s_j(z, \zeta)) = \overline{\omega}'_{z,\zeta,\lambda}(a_i^*(z, \zeta, \lambda)) \wedge \omega_\zeta(s_i(z, \zeta))$$

for all $(z, \zeta, \lambda) \in (U_j \cap U_i \cap V) \times W \times \Lambda$. Therefore, by setting

$$\overline{\omega}'_{z,\zeta,\lambda}(a^*(z, \zeta, \lambda)) \wedge \omega_\zeta(s(z, \zeta)) := \overline{\omega}'_{z,\zeta,\lambda}(a_j^*(z, \zeta, \lambda)) \wedge \omega_\zeta(s_j(z, \zeta)) \quad \text{for}$$
$$(z, \zeta, \lambda) \in (U_j \cap V) \times W \times \Lambda , \tag{4.5.1}$$

a continuous differential form $\overline{\omega}'_{z,\zeta,\lambda}(a^*(z, \zeta, \lambda)) \wedge \omega_\zeta(s(z, \zeta))$ on $V \times W \times \Lambda$ is correctly defined. In the same way as in the case of Proposition 1.7.3 it can be proved that, for every C^1-function $\psi(z, \zeta, \lambda)$ on $V \times W \times \Lambda$,

$$\overline{\omega}'_{z,\zeta,\lambda}(\psi(z, \zeta, \lambda)\, a^*(z, \zeta, \lambda)) \wedge \omega_\zeta(s(z, \zeta))$$
$$= \psi^n(z, \zeta, \lambda)\, \overline{\omega}'_{z,\lambda,\zeta}(a^*(z, \zeta, \lambda)) \wedge \omega_\zeta(s(z, \zeta)) . \tag{4.5.2}$$

Now we define the operators B_D and $B_{\partial D}$. By condition (v) in Lemma 4.2.4, for every integer $\nu \geqq n\varkappa$, the differential form (cp. 4.5.1)

$$\frac{\varphi^\nu(z, \zeta)\, \overline{\omega}'_{z,\zeta}(\bar{s}(z, \zeta)) \wedge \omega_\zeta(s(z, \zeta))}{\|s(z, \zeta)\|_\sigma^{2n}}$$

is C^1 on $(X \times X) \setminus \{(z, z) : z \in X\}$ and the singularity at $\zeta = z$ is of order $2n - 1$. Therefore, for every integer $\nu \geqq n\varkappa$ and every bounded differential form f on D, we can define

$$(B_D(\varphi^\nu, \bar{s}, s) f)\,(z)$$
$$:= \frac{(n-1)!}{(2\pi i)^n} \int_{\zeta \in D} f(\zeta) \wedge \frac{\varphi^\nu(z, \zeta)\, \overline{\omega}'_{z,\zeta}(\bar{s}(z, \zeta)) \wedge \omega_\zeta(s(z, \zeta))}{\|s(z, \zeta)\|_\sigma^{2n}}, \quad z \in D . \tag{4.5.3}$$

Further, for every integer $\nu \geqq n\varkappa$ and every bounded differential form f on ∂D, we define

$$(B_{\partial D}(\varphi^\nu, \bar{s}, s) f)\,(z)$$
$$:= \frac{(n-1)!}{(2\pi i)^n} \int_{\zeta \in \partial D} f(\zeta) \wedge \frac{\varphi^\nu(z, \zeta)\, \overline{\omega}'_{z,\zeta}(\bar{s}(z, \zeta)) \wedge \omega_\zeta(s(z, \zeta))}{\|s(z, \zeta)\|_\sigma^{2n}}, \quad z \in D . \tag{4.5.4}$$

Now we define the operator $L_{\partial D}$. Let $(s^*, \varkappa^*)$ be a Leray section for (D, s, φ) (Subsection 4.3.2). By condition (ii) in the definition of a Leray section, for every integer $\nu \geqq n\varkappa^*$, the differential form (cp. (4.5.1))

$$\frac{\varphi^\nu(z, \zeta)\, \overline{\omega}'_{z,\zeta}\big(s^*(z, \zeta)\big) \wedge \omega_\zeta\big(s(z, \zeta)\big)}{\langle s^*(z, \zeta), s(z, \zeta)\rangle^n}$$

is continuous in some neighbourhood $\subseteqq D \times X$ of $D \times \partial D$. Therefore, for every integer $\nu \geqq n\varkappa^*$ and every bounded differential form f on ∂D, we can define

$$\begin{aligned}&\big(L_{\partial D}(\varphi^\nu, s^*, s)\, f\big)\,(z)\\ &:= \frac{(n-1)!}{(2\pi i)^n} \int\limits_{\zeta \in \partial D} f(\zeta) \wedge \frac{\varphi^\nu(z, \zeta)\, \overline{\omega}'_{z,\zeta}\big(s^*(z, \zeta)\big) \wedge \omega_\zeta\big(s(z, \zeta)\big)}{\langle s^*(z, \zeta), s(z, \zeta)\rangle^n}\,, \qquad z \in D\,.\end{aligned} \tag{4.5.5}$$

Finally we define $R_{\partial D}$. Let $(s^*, \varkappa^*)$ be a Leray section for (D, s, φ). For all $0 \leqq \lambda \leqq 1$, $z \in D$ and ζ in some neighbourhood of ∂D such that $\langle s^*(z, \zeta), s(z, \zeta)\rangle \neq 0$, we define

$$t^*_{(s^*, \bar{s}, s)}(z, \zeta, \lambda) := (1 - \lambda)\, \frac{s^*(z, \zeta)}{\langle s^*(z, \zeta), s(z, \zeta)\rangle} + \lambda\, \frac{\bar{s}(z, \zeta)}{\|s(z, \zeta)\|_\sigma^2}\,. \tag{4.5.6}$$

By condition (v) in Lemma 4.2.4 and condition (ii)* in the definition of a Leray section (Subsection 4.3.2), the map

$$\varphi^{\max(\varkappa, \varkappa^*)}(z, \zeta)\, t^*_{(s^*, \bar{s}, s)}(z, \zeta, \lambda) \tag{4.5.7}$$

is C^1 for all (z, ζ, λ) in some neighbourhood $\subseteqq D \times X \times [0, 1]$ of $D \times \partial D \times [0, 1]$. By (4.5.2) this implies that, for every integer $\nu \geqq \max(n\varkappa, n\varkappa^*)$, the differential form (cp. (4.5.1))

$$\varphi^\nu(z, \zeta)\, \overline{\omega}'_{z,\zeta,\lambda}\big(t^*_{(s^*, \bar{s}, s)}(z, \zeta, \lambda)\big) \wedge \omega_\zeta\big(s(z, \zeta)\big)$$

is continuous in some neighbourhood $\subseteqq D \times X \times [0, 1]$ of $D \times \partial D \times [0, 1]$. Therefore, for every integer $\nu \geqq \max(n\varkappa, n\varkappa^*)$ and every bounded differential form f on ∂D we can define

$$\begin{aligned}&\big(R_{\partial D}(\varphi_\nu, s^*, \bar{s}, s)\, f\big)\,(z)\\ &:= \frac{(n-1)!}{(2\pi i)^n} \int\limits_{(\zeta, \lambda) \in \partial D \times [0,1]} f(\zeta) \wedge \varphi^\nu(z, \zeta)\, \overline{\omega}'_{z,\zeta,\lambda}\big(t^*(z, \zeta, \lambda)\big) \wedge \omega_\zeta\big(s(z, \zeta)\big)\,, \qquad z \in D\,,\end{aligned} \tag{4.5.8}$$

where $t^* = t^*_{(s^*, \bar{s}, s)}$.

Remark. In the case of functions or 1-forms respectively, the definitions of B_D, $B_{\partial D}$, $L_{\partial D}$ and $R_{\partial D}$ given in the present subsection agree with the corresponding definitions in Subsection 4.3.2.

4.5.2. Theorem (Koppelman formula). *We use the notations from Subsection* 4.3.1. *Let $\nu \geqq 2n\varkappa$ be an integer, and let f be a continuous $(0, q)$-form on $\overline{D}$ such that $\bar\partial f$ is also continuous on $\overline{D}$, $0 \leqq q \leqq n$. Then the forms $B_{\partial D} f$, $B_D\, \bar\partial f$, $B_D f$ and $\bar\partial B_D f$ are continuous in D, and we have*

$$(-1)^q f = B_{\partial D} f - B_D\, \bar\partial f + \bar\partial B_D f \quad in \quad D\,, \tag{4.5.9}$$

where $B_{\partial D} = B_{\partial D}(\varphi^\nu, \bar{s}, s)$ and $B_D = B_D(\varphi^\nu, \bar{s}, s)$ (Subsection 4.5.1).

Proof. If $q = 0$, then, by degree reasons $B_D f = 0$ and (4.5.9) is the Martinelli-Bochner formula (4.3.12). Therefore, we only have to prove the theorem for $1 \leqq q \leqq n$.

It is easy to see that $B_{\partial D}f$, $B_D\,\bar\partial f$ and B_Df are continuous. Therefore, it remains to prove that in the sense of distributions $\bar\partial B_D f = (-1)^q f - B_{\partial D}f + B_D\,\bar\partial f$ in D. That is, we have to prove that, for every C^∞-form v with compact support in D,

$$(-1)^q\int_D (B_Df)\wedge\bar\partial v = (-1)^q\int_D f\wedge v - \int_D (B_{\partial D}f)\wedge v + \int_D (B_D\,\bar\partial f)\wedge v\,. \quad (4.5.10)$$

Since f, $B_{\partial D}f$ and $B_D\,\bar\partial f$ are of bidegree $(0, q)$ and B_Df is of bidegree $(0, q-1)$, we can assume that v is of bidegree $(n, n-q)$. Using the notation

$$\theta(z,\zeta) := \frac{(n-1)!}{(2\pi i)^n}\,\frac{\varphi^\nu(z,\zeta)\,\overline{\omega}'_{z,\zeta}(\bar s(z,\zeta))\wedge\omega_\zeta(s(z,\zeta))}{\|s(z,\zeta)\|_\sigma^{2n}},$$

(4.5.10) can be written

$$(-1)^q\int_{(z,\zeta)\in D\times D} f(\zeta)\wedge\theta(z,\zeta)\wedge\bar\partial v(z) = (-1)^q\int_{z\in D} f(z)\wedge v(z)$$
$$-\int_{(z,\zeta)\in D\times\partial D} f(\zeta)\wedge\theta(z,\zeta)\wedge v(z) + \int_{(z,\zeta)\in D\times D}\bar\partial f(\zeta)\wedge\theta(z,\zeta)\wedge v(z)\,. \quad (4.5.11)$$

It follows from Proposition 1.7.1 that all monomials in $d_{z,\zeta}\theta(z,\zeta)$ contain at least one of the differentials $dz_1, \dots, dz_n$ (when $z_1, \dots, z_n$ are local holomorphic coordinates). Since $v(z)$ is of bidegree $(n, n-q)$, we obtain

$$d_{z,\zeta}(\theta(z,\zeta)\wedge v(z)) = -\theta(z,\zeta)\wedge\bar\partial v(z) \quad\text{for}\quad \zeta\neq z\,. \quad (4.5.12)$$

Moreover, since all monomials in $\theta(z,\zeta)$ contain the factor $d\zeta_1\wedge\dots\wedge d\zeta_n$ (when $\zeta_1, \dots, \zeta_n$ are local holomorphic coordinates), we get $df(\zeta)\wedge\theta(z,\zeta) = \bar\partial f(\zeta)\wedge\theta(z,\zeta)$ for $\zeta\neq z$. Hence

$$d_{z,\zeta}(f(\zeta)\wedge\theta(z,\zeta)\wedge v(z))$$
$$= \bar\partial f(\zeta)\wedge\theta(z,\zeta)\wedge v(z) - (-1)^q f(\zeta)\wedge\theta(z,\zeta)\wedge\bar\partial v(z) \quad\text{for}\quad \zeta\neq z\,. \quad (4.5.13)$$

By condition (ii) in Lemma 4.2.4 we can find a neighbourhood $U_\Delta \subseteqq X\times X$ of the diagonal $\Delta := \{(z,z) : z\in X\}$ such that for every fixed $z\in X$, the map $s(z,\zeta)$ is biholomorphic for all $\zeta\in X$ with $(z,\zeta)\in U_\Delta$. Consider the open sets

$$U_\varepsilon := \{(z,\zeta)\in U_\Delta\times U_\Delta : \|s(z,\zeta)\|_\sigma < \varepsilon\}\,,\qquad \varepsilon > 0\,.$$

Since $D\subset\subset X$, for sufficiently small $\varepsilon > 0$, $\partial U_\varepsilon\cap(D\times D)$ is smooth. Since $\operatorname{supp} v\subset\subset D$, for sufficiently small $\varepsilon > 0$, we have

$$\partial(D\times D\setminus U_\varepsilon)\cap((\operatorname{supp} v)\times X) = (D\times\partial D\cup\partial U_\varepsilon)\cap((\operatorname{supp} v)\times X)\,.$$

Taking into account also the orientations, this can be written

$$\partial(D\times D\setminus U_\varepsilon)\cap((\operatorname{supp} v)\times X) = (D\times\partial D - \partial U_\varepsilon)\cap((\operatorname{supp} v)\times X)\,.$$

Therefore, it follows from Stokes' formula and (4.5.13) that

$$\int_{(z,\zeta)\in D\times\partial D} f(\zeta)\wedge\theta(z,\zeta)\wedge v(z) - \int_{(z,\zeta)\in\partial U_\varepsilon} f(\zeta)\wedge\theta(z,\zeta)\wedge v(z)$$
$$= \int_{(z,\zeta)\in D\times D\setminus U_\varepsilon}\bar\partial f(\zeta)\wedge\theta(z,\zeta)\wedge v(z) - (-1)^q\int_{(z,\zeta)\in D\times D\setminus U_\varepsilon} f(\zeta)\wedge\theta(z,\zeta)\wedge\bar\partial v(z)\,.$$

It is clear that the integrals on the right-hand side tend to the corresponding integrals in (4.5.11) when $\varepsilon\to 0$. Therefore, it remains to prove that

$$\lim_{\varepsilon\to 0}\int_{(z,\zeta)\in\partial U_\varepsilon} f(\zeta)\wedge\theta(z,\zeta)\wedge v(z) = (-1)^q\int_{z\in D} f(z)\wedge v(z)\,. \quad (4.5.14)$$

Let $\{(U_j, h_j)\}$ be the holomorphic atlas of X from Subsection 4.3.1, and let $\{\chi_j\}$ be a C^∞-partition of unity subordinate to $\{U_j\}$. We write $v_j := \chi_j v$ and choose open sets $V_j \subset\subset U_j$ such that $\operatorname{supp} \chi_j \subseteqq V_j$. Then it is sufficient to prove that, for all j,

$$\lim_{\varepsilon\to 0} \int_{(z,\zeta)\in\partial U_\varepsilon} f(\zeta) \wedge \theta(z,\zeta) \wedge v_j(z) = (-1)^q \int_{z\in V_j} f(z) \wedge v_j(z) . \tag{4.5.15}$$

From now on let j be fixed. If ε is sufficiently small, then, for all $(z,\zeta) \in U_\varepsilon$ with $z \in V_j$, we have $\zeta \in U_j$. Therefore, (4.5.15) can be written

$$\lim_{\varepsilon\to 0} \int_{(z,\zeta)\in(V_j\times U_j)\cap\partial U_\varepsilon} f(\zeta) \wedge \theta(z,\zeta) \wedge v_j(z) = (-1)^q \int_{z\in V_j} f(z) \wedge v_j(z) . \tag{4.5.16}$$

To prove (4.5.16) we need some preparations. Let $u(z,\zeta)$ be the expression of $s(z,\zeta)$ with respect to (U_j, h_j) (cp. Subsection 4.3.1), and let $w(z,\zeta)$ be the expression of $\bar{s}(z,\zeta)$ with respect to (U_j, h_j). By condition (ii) in Lemma 4.2.4, we can find $\varepsilon_0 > 0$ such that the map T: $(U_j \times U_j) \cap U_{\varepsilon_0} \to \mathbb{C}^n \times \mathbb{C}^n$ defined by $T(z,\zeta) := (h_j(z), u(z,\zeta))$ is biholomorphic from $(U_j \times U_j) \cap U_{\varepsilon_0}$ onto some open set $W_{\varepsilon_0} \subseteqq \mathbb{C}^n \times \mathbb{C}^n$. Let T^{-1}: $W_{\varepsilon_0} \to (U_j \times U_j) \cap U_{\varepsilon_0}$ be the inverse of T. Let $\alpha(\eta, \xi)$: $W_{\varepsilon_0} \to (U_j$ be the holomorphic map which is defined by $\alpha(h_j(z), u(z,\zeta)) = \zeta$ for all $(z,\zeta) \in (U_j \times U_j) \cap U_{\varepsilon_0}$. Then

$$T^{-1}(\eta,\xi) = (h_j^{-1}(\eta), \alpha(\eta,\xi)) \quad \text{for all} \quad (\eta,\xi) \in W_{\varepsilon_0} . \tag{4.5.17}$$

Further, it is clear that

$$u(T^{-1}(\eta,\xi)) = \xi \quad \text{for all} \quad (\eta,\xi) \in W_{\varepsilon_0} \tag{4.5.18}$$

and

$$T^{-1}(\eta, 0) = (h_j^{-1}(\eta), h_j^{-1}(\eta)) \quad \text{for all} \quad \eta \in h_j(U_j) . \tag{4.5.19}$$

We set $Z_\varepsilon := T((V_j \times U_j) \cap \partial U_\varepsilon)$ for $0 < \varepsilon \leqq \varepsilon_0$. Since $V_j \subset\subset U_j$, we can find $0 < \varepsilon_1 \leqq \varepsilon_0$ such that

$$Z_\varepsilon \subset\subset W_{\varepsilon_0} \quad \text{for} \quad 0 < \varepsilon \leqq \varepsilon_1 , \tag{4.5.20}$$

and, for some constant $C < \infty$, $|u(z,\zeta)| \leqq C\,||s(z,\zeta)||_\sigma \leqq C\varepsilon$ for $0 < \varepsilon \leqq \varepsilon_1$ and $(z,\zeta) \in (V_j \times U_j) \cap \partial U_\varepsilon$, that is,

$$|\xi| \leqq C\varepsilon \quad \text{for} \quad 0 < \varepsilon \leqq \varepsilon_1 \quad \text{and} \quad (\eta,\xi) \in Z_\varepsilon . \tag{4.5.21}$$

From (4.5.19) we obtain $w(T^{-1}(\eta, 0)) = 0$ for all $\eta \in h_j(U_j)$. Since $w \circ T^{-1}$ is C^∞ on W_{ε_0}, together with (4.5.20) and (4.5.21) this implies that there is a constant $C < \infty$ such that

$$|w(T^{-1}(\eta,\xi))| \leqq C\varepsilon \quad \text{and} \quad ||d_\eta w(T^{-1}(\eta,\xi))|| \leqq C\varepsilon$$

for all $0 < \varepsilon \leqq \varepsilon_1$ and $(\eta,\xi) \in Z_\varepsilon$. Therefore, we can find a constant $C < \infty$ such that

$$||\overline{\omega}'_{\eta,\xi}(w(T^{-1}(\eta,\xi))) - \overline{\omega}'_\xi(w(T^{-1}(\eta,\xi)))|| \leqq C\varepsilon^2$$
$$\text{for all} \quad 0 < \varepsilon \leqq \varepsilon_1 \quad \text{and} \quad (\eta,\xi) \in Z_\varepsilon , \tag{4.5.22}$$

where

$$\overline{\omega}'_{\eta,\xi}(w) := \frac{1}{(n-1)!} \det{}_{1,n-1}\,(w, (\bar\partial_\eta + \bar\partial_\xi)\, w)$$

and

$$\overline{\omega}'_\xi(w) := \frac{1}{(n-1)!} \det{}_{1,n-1}\,(w, \bar\partial_\xi w) \qquad \text{(cp. Definition 1.7.6).}$$

Let $(h_j^{-1})^* f$ and $\alpha^* f$ be the pull-backs of f defined by h_j^{-1} and α, respectively, and let

$$(h_j^{-1})^* f(\eta) = \sum_{1\leqq i_1<\ldots<i_q\leqq n} f_{i_1\ldots i_q}(\eta)\, d\bar\eta_{i_1} \wedge \ldots \wedge d\bar\eta_{i_q} \quad \text{for} \quad \eta \in h_j(U_j) . \tag{4.5.23}$$

If $h_{j1}, \dots, h_{jn}$ are the components of h_j, then it follows that

$$f(\zeta) = \sum_{1 \leq i_1 < \dots < i_q \leq n} f_{i_1 \dots i_q}(h_j(\zeta)) \, \mathrm{d}\overline{h}_{ji_1}(\zeta) \dots \mathrm{d}\overline{h}_{ji_q}(\zeta) \quad \text{for} \quad \zeta \in U_j , \tag{4.5.24}$$

and

$$\alpha^* f(\eta, \xi) = \sum_{1 \leq i_1 < \dots < i_q \leq n} f_{i_1 \dots i_q}(h_j(\alpha(\eta, \xi))) \, \mathrm{d}_{\eta,\xi}\overline{h}_{ji_1}(\alpha(\eta, \xi)) \wedge \dots \wedge \mathrm{d}_{\eta,\xi}\overline{h}_{ji_q}(\alpha(\eta, \xi))$$

$$\text{for} \quad (\eta, \xi) \in W_{\varepsilon_0} . \tag{4.5.25}$$

Define

$$\alpha^*_\eta f(\xi, \eta) := \sum_{1 \leq i_1 < \dots < i_q \leq n} f_{i_1 \dots i_q}(h_j(\alpha(\eta, \xi))) \, \mathrm{d}_\eta\overline{h}_{ji_1}(\alpha(\eta, \xi)) \wedge \dots \wedge \mathrm{d}_\eta\overline{h}_{ji_q}(\alpha(\eta, \xi))$$

$$\text{for} \quad (\eta, \xi) \in W_{\varepsilon_0} . \tag{4.5.26}$$

From (4.5.19), (4.5.20) and (4.5.21) we obtain a constant C such that

$$|h_j(\alpha(\eta, \xi)) - \eta| \leq C\varepsilon \quad \text{and} \quad \|\mathrm{d}_\eta h_j(\alpha(\eta, \xi)) - \mathrm{d}\eta\| \leq C\varepsilon$$

for all $0 < \varepsilon \leq \varepsilon_1$ and $(\eta, \xi) \in Z_\varepsilon$. According to (4.5.23) and (4.5.26), this implies that

$$\lim_{\varepsilon \to 0} \sup_{(\eta, \xi) \in Z_\varepsilon} \|\alpha^*_\eta f(\xi, \eta) - (h_j^{-1})^* f(\eta)\| = 0 . \tag{4.5.27}$$

Since, by (4.5.19) and condition (iii) in Lemma 4.2.4, $\varphi(T^{-1}(\eta, 0)) = 1$, we can find $C < \infty$ such that

$$|\varphi(T^{-1}(\eta, \xi)) - 1| \leq C\varepsilon \quad \text{for} \quad 0 < \varepsilon \leq \varepsilon_1 \quad \text{and} \quad (\eta, \xi) \in Z_\varepsilon . \tag{4.5.28}$$

Now we are ready to prove (4.5.16). We denote by $(T^{-1})^* (f \wedge \theta \wedge v_j)$ the pull-back of $f \wedge \theta \wedge v_j$ with respect to T^{-1}. Then

$$\int_{(z,\zeta) \in (V_j \times U_j) \cap \partial U_\varepsilon} f(\zeta) \wedge \theta(z, \zeta) \wedge v_j(z) = \int_{(\eta, \xi) \in Z_\varepsilon} (T^{-1})^* (f \wedge \theta \wedge v_j) (\eta, \xi) . \tag{4.5.29}$$

Since

$$\theta(z, \zeta) \wedge v_j(z) = \frac{(n-1)!}{(2\pi\mathrm{i})^n} \frac{\varphi^\nu(z, \zeta) \, \overline{\omega}'_{z,\zeta}(w(z, \zeta)) \wedge \omega_{z,\zeta}(u(z, \zeta))}{\varepsilon^{2n}} \wedge v_j(z)$$

for $(z, \zeta) \in (U_j \times U_j) \cap \partial U_\varepsilon$, it follows from (4.5.17) and (4.5.18) that for $(\eta, \xi) \in Z_\varepsilon$

$$(T^{-1})^* (f \wedge \theta \wedge v_j) (\eta, \xi) = \frac{(n-1)!}{(2\pi\mathrm{i})^n} \alpha^* f(\eta, \xi)$$

$$\wedge \frac{\varphi^\nu(T^{-1}(\eta, \xi)) \, \overline{\omega}'_{\eta,\xi}(w(T^{-1}(\eta, \xi))) \wedge \omega_\xi(\xi)}{\varepsilon^{2n}} \wedge (h_j^{-1})^* v_j(\eta) . \tag{4.5.30}$$

Since the volume of Z_ε is of order ε^{2n-1}, it follows from (4.5.22), (4.5.28), (4.5.29) and (4.5.30) that the proof of (4.5.16) will be complete if we show that

$$\lim_{\varepsilon \to 0} \frac{(n-1)!}{(2\pi\mathrm{i})^n} \int_{(\eta, \xi) \in Z_\varepsilon} \alpha^* f(\eta, \xi) \wedge \frac{\overline{\omega}'_\xi(w(T^{-1}(\eta, \xi))) \wedge \omega_\xi(\xi)}{\varepsilon^{2n}} \wedge (h_j^{-1})^* v_j(\eta)$$

$$= (-1)^q \int_{\eta \in h_j(V_j)} (h_j^{-1})^* f(\eta) \wedge (h_j^{-1})^* v_j(\eta) . \tag{4.5.31}$$

Since, for every $\eta \in h_j(V_j)$, the set of all $\xi \in \mathbb{C}^n$ with $(\eta, \xi) \in Z_\varepsilon$ is of real dimension $2n - 1$, and since the form $\overline{\omega}'_\xi(w(T^{-1}(\eta, \xi))) \wedge \omega_\xi(\xi)$ is of degree $2n - 1$ in ξ, in (4.5.31) the form $\alpha^* f$ can be replaced by $\alpha^*_\eta f$. In view of (4.5.27) this implies that in (4.5.31) the form $\alpha^* f$ can be replaced by $(h_j^{-1})^* f$. Therefore, it is sufficient to prove that, for every fixed $\eta \in h_j(V_j)$,

$$\frac{(n-1)!}{(2\pi i)^n} \int\limits_{\{\xi \in \mathbb{C}^n : (\eta, \xi) \in Z_\varepsilon\}} \frac{\overline{\omega}'_\xi(w(T^{-1}(\eta, \xi))) \wedge \omega_\xi(\xi)}{\varepsilon^{2n}} = 1 .$$

This follows from the Leray formula (cp. Remark 1.10.3), because, for $(\eta, \xi) \in Z_\varepsilon$,

$$\overline{\omega}'_\xi(w(T^{-1}(\eta, \xi))) \wedge \omega_\xi(\xi) = \omega'_\xi(w(T^{-1}(\eta, \xi))) \wedge \omega_\xi(\xi)$$

and, by (4.5.18),

$$\langle w(T^{-1}(\eta, \xi)), \xi \rangle = \langle w(T^{-1}(\eta, \xi)), u(T^{-1}(\eta, \xi)) \rangle = \|s(T^{-1}(\eta, \xi))\|_\sigma^2 = \varepsilon^2 . \blacksquare$$

4.5.3. Theorem (Koppelman-Leray formula). *We use the notations from Subsection 4.3.1. Let $(s^*, \varkappa^*)$ be a Leray section for (D, s, φ)* (Subsection 4.3.2), *and let $\nu \geqq \max(2n\varkappa, n\varkappa^*)$ be an integer. Suppose, in addition, that all derivatives of $\varphi^\nu(z, \zeta)\, s^*(z, \zeta)/\langle s^*(z, \zeta), s(z, \zeta)\rangle$ which are of order $\leqq 2$ in z and of order $\leqq 1$ in ζ are continuous for all (z, ζ) in some neighbourhood $W \subseteq D \times X$ of $D \times \partial D$.*[1] *Then, for every continuous $(0, q)$-form f on $\overline{D}$ such that $\bar\partial f$ is also continuous on $\overline{D}$, $0 \leqq q \leqq n$, the forms $L_{\partial D} f$, $R_{\partial D}\, \bar\partial f$, $B_D\, \bar\partial f$, $R_{\partial D} f$, $\bar\partial f R_{\partial D} f$ and $\bar\partial B_D f$ are continuous in D, and we have*

$$(-1)^q f = L_{\partial D} f - (R_{\partial D} + B_D)\, \bar\partial f + \bar\partial (R_{\partial D} + B_D) f \quad \text{in} \quad D\,, \tag{4.5.32}$$

where $L_{\partial D} = L_{\partial D}(\varphi^\nu, s^*, s)$, $R_{\partial D} = R_{\partial D}(\varphi^\nu, s^*, \bar s, s)$ and $B_D = B_D(\varphi^\nu, \bar s, s)$ (cp. Subsection 4.5.1).

Remark. If $s^*(z, \zeta) = \bar s(z, \zeta)$, then $L_{\partial D} = B_{\partial D}$, $R_{\partial D} = 0$ and, therefore, (4.5.32) agrees with the Koppelman formula (4.5.9). If $q = 0$, then $B_D f = 0$ and $R_{\partial D} f = 0$ and, therefore, (4.5.32) agrees with the Leray formula (4.3.18).

To prove Theorem 4.5.3 we need the following lemma.

4.5.4. Lemma. *Let the hypotheses of Theorem 4.5.3 be fulfilled. Set*

$$\overline{\Omega}(z, \zeta, \lambda) := \varphi^\nu(z, \zeta)\, \overline{\omega}'_{z,\zeta,\lambda}(t^*(z, \zeta, \lambda)) \wedge \omega_\zeta(s(z, \zeta)) ,$$

where $t^* = t^*_{(s^*, \bar s, s)}$ (cp. Subsection 4.5.1). *Then for $(z, \zeta, \lambda) \in W \times [0, 1]$*

$$(\bar\partial_z + d_{\zeta,\lambda})\, (f(\zeta) \wedge \overline{\Omega}(z, \zeta, \lambda)) = \bar\partial f(\zeta) \wedge \overline{\Omega}(z, \zeta, \lambda) . \tag{4.5.33}$$

Proof. Since the form $\overline{\Omega}(z, \zeta, \lambda)$ contains the factor $d\zeta_1 \wedge \ldots \wedge d\zeta_n$ (when $\zeta_1, \ldots, \zeta_n$ are local holomorphic coordinates), we only have to prove that $(\bar\partial_{z,\zeta} + d_\lambda)\, \overline{\Omega}(z, \zeta, \lambda) = 0$. Let $\{(U_j, h_j)\}$ be the holomorphic atlas from Subsection 4.3.1, set $\mu := \max(\varkappa, \varkappa^*)$, let $v_j(z, \zeta, \lambda)$ be the expression of $\varphi^\mu(z, \zeta)\, t^*(z, \zeta, \lambda)$ with respect to (U_j, h_j) (cp. Subsection 4.3.1), and let $s_j(z, \zeta)$ be the expression of $s(z, \zeta)$ with respect to (U_j, h_j). Then, by (4.5.2) and (4.5.7),

$$\overline{\Omega}(z, \zeta, \lambda) = \varphi^{\nu - n\mu}(z, \zeta) \sum_{k=1}^{n} (-1)^{k+1}\, v_{jk}(z, \zeta, \lambda) \bigwedge_{l \neq k} (\bar\partial_{z,\zeta} + d_\lambda)\, v_{jl}(z, \zeta, \lambda)$$
$$\wedge\, \omega_\zeta(s_j(z, \zeta)) \quad \text{for} \quad z \in U_j,\ (z, \zeta, \lambda) \in W \times [0, 1] ,$$

where $v_{j1}, \ldots, v_{jn}$ are the components of v_j. Therefore,

$$(\bar\partial_{z,\zeta} + d_\lambda)\, \overline{\Omega}(z, \zeta, \lambda) = n\varphi^{\nu - n\mu}(z, \zeta) \bigwedge_{k=1}^{n} (\bar\partial_{z,\zeta} + d_\lambda)\, v_{jk}(z, \zeta, \lambda) \wedge \omega_\zeta(s_j(z, \zeta))$$
$$\text{for} \quad z \in U_j,\ (z, \zeta, \lambda) \in W \times [0, 1] . \tag{4.5.34}$$

[1] The theorem is valid also without this additional assumption, but then the proof becomes technically more complicated.

We have $\sum_{k=1}^{n} v_{jk}s_{jk} = \langle \varphi^\mu t^*, s\rangle = \varphi^\mu$, where $s_{j1}, \ldots, s_{jn}$ are the components of s_j. Since the functions s_{jk} and φ are holomorphic, this implies that

$$\sum_{k=1}^{n} s_{jk}(z,\zeta)\,(\bar\partial_{z,\zeta} + \mathrm{d}_\lambda)\, v_{jk}(z,\zeta,\lambda) = 0 \quad \text{for} \quad z \in U_j,\ (z,\zeta,\lambda) \in W \times [0,1]\,.$$

Therefore

$$\bigwedge_{k=1}^{n} (\bar\partial_{z,\zeta} + \mathrm{d}_\lambda)\, v_{jk}(z,\zeta,\lambda) = 0 \quad \text{for} \quad z \in U_j,\ (z,\zeta,\lambda) \in W \times [0,1]\,, \tag{4.5.35}$$

because the form on the left-hand side of (4.5.35) is continuous and the set $\{(z,\zeta,\lambda) \in W \times [0,1] : s(z,\zeta) \neq 0\}$ is dense in $W \times [0,1]$. It follows from (4.5.34) and (4.5.35) that $(\bar\partial_{z,\zeta} + \bar{\mathrm{d}}_\lambda)\, \bar\Omega(z,\zeta,\lambda) = 0$. ■

Proof of Theorem 4.5.3. By the remark following Theorem 4.5.3 we can assume that $1 \leqq q \leqq n$. It is clear that $L_{\partial D}f$, $R_{\partial D}\,\bar\partial f$ and $R_{\partial D}f$ are continuous in D. By Theorem 4.5.2, $B_D\,\bar\partial f$, B_Df and $\bar\partial B_Df$ are continuous in D. By differentiation under the sign of integration (this is admissible in view of the hypothesis of the theorem on smoothness of $\varphi^\nu s^*/\langle s^*, s\rangle$), we obtain that $\bar\partial R_{\partial D} f$ is continuous in D. It remains to prove (4.5.32). In view of the Koppelman formula (4.5.9), the proof will be complete if we show that

$$\bar\partial R_{\partial D} f = B_{\partial D} f - L_{\partial D} f + R_{\partial D}\,\bar\partial f \quad \text{in} \quad D\,. \tag{4.5.36}$$

By Stokes' formula we obtain from Lemma 4.5.4

$$\int\limits_{(\zeta,\lambda)\in\partial D\times[0,1]} \bar\partial f(\zeta) \wedge \bar\Omega(z,\zeta,\lambda) - \bar\partial_z \int\limits_{(\zeta,\lambda)\in\partial D\times[0,1]} f(\zeta) \wedge \bar\Omega(z,\zeta,\lambda)$$
$$= \int\limits_{\partial D} f(\zeta) \wedge \bar\Omega(z,\zeta,\lambda)|_{\lambda=0} - \int\limits_{\partial D} f(\zeta) \wedge \bar\Omega(z,\zeta,\lambda)|_{\lambda=1}\,.$$

This implies (4.5.36), because, by (4.5.6),

$$\bar\Omega(z,\zeta,\lambda)|_{\lambda=0} = \varphi^\nu(z,\zeta)\,\frac{\bar\omega'_{z,\zeta}(s^*(z,\zeta)) \wedge \omega_\zeta(s(z,\zeta))}{\langle s^*(z,\zeta), s(z,\zeta)\rangle^n}$$

and

$$\bar\Omega(z,\zeta,\lambda)|_{\lambda=1} = \varphi^\nu(z,\zeta)\,\frac{\bar\omega'_{z,\zeta}(\bar s(z,\zeta)) \wedge \omega_\zeta(s(z,\zeta))}{\|s(z,\zeta)\|_\sigma^{2n}}\,. \ ■$$

4.5.5. Corollary. *If, under the hypotheses of Theorem 4.5.3, $s^*(z,\zeta)$ depends holomorphically on $z \in D$ and if $1 \leqq q \leqq n$, then*

$$f = \bar\partial(-1)^q\,(R_{\partial D} + B_D) f + (-1)^{q+1}\,(R_{\partial D} + B_D)\,\bar\partial f \quad in \quad D\,. \tag{4.5.37}$$

In particular, for every continuous $(0,q)$-form f on $\bar D$ such that $\bar\partial f = 0$ in D,

$$u := (-1)^q\,(R_{\partial D} + B_D) f \tag{4.5.38}$$

is a continuous solution of $\bar\partial u = f$ in D.

Proof. If $s^*(z,\zeta)$ is holomorphic in z, then the form $\bar\omega'_{z,\zeta}(s^*(z,\zeta)) \wedge \omega_\zeta(s(z,\zeta))$ is of degree zero in z. Therefore, $L_{\partial D}f = 0$ if $q \geqq 1$. ■

4.6. The Koppelman-Leray-Norguet formula

The Koppelman-Leray-Norguet formula, which will be proved in the present section, is a generalization of the Koppelman-Leray formula (4.5.32) as well as of the Leray-Norguet formula (4.4.11).

4.6.1. The operators L_{S_K} and R_{S_K} for differential forms of arbitrary degree. We use the notations from Subsections 4.3.1 and 4.4.1. Further, let $(s_1^*, \dots, s_N^*, \varkappa^*)$ be a Leray-Norguet section for (D, s, φ) (Subsection 4.4.2), and let $K = (k_1, \dots, k_l)$ be a strictly increasing collection of integers $1 \leqq k_1 < \dots < k_l \leqq N$.

Then, by (4.5.2) and condition (ii)* in the definition of a Leray-Norguet section, for every integer $\nu \geqq n\varkappa^*$, the differential form (cp. (4.5.1))

$$\varphi^\nu(z, \zeta)\, \overline{\omega}'_{z,\zeta,\lambda}\left(\sum_{k\in K} \lambda_k \frac{s_k^*(z, \zeta)}{\langle s_k^*(z, \zeta), s(z, \zeta)\rangle}\right) \wedge \omega_\zeta(s(z, \zeta))$$

is continuous for all (z, ζ, λ) in some neighbourhood $\subseteqq D \times X \times \Delta_K$ of $D \times S_K \times \Delta_K$. For every bounded differential form f on S_K and every integer $\nu \geqq n\varkappa^*$,

$$\begin{aligned}&\left(L_{S_K}(\varphi^\nu, s_1^*, \dots, s_N^*, s) f\right)(z)\\ &:= \frac{(n-1)!}{(2\pi i)^n} \int\limits_{(\zeta,\lambda)\in S_K\times\Delta_K} f(\zeta)\, \varphi^\nu(z, \zeta)\, \overline{\omega}'_{z,\zeta,\lambda}\left(\sum_{k\in K} \lambda_k \frac{s_k^*(z, \zeta)}{\langle s_k^*(z, \zeta), s(z, \zeta)\rangle}\right) \wedge \omega_\zeta(s(z, \zeta))\,,\\ &\qquad z \in D\,. \qquad (4.6.1)\end{aligned}$$

Let $t^* = t^*_{(s_1^*, \dots, s_N^*, \bar{s}, s)}$ (see (4.4.8)). Then, for every integer $\nu \geqq \max(n\varkappa, n\varkappa^*)$, the differential form (cp. (4.5.1))

$$\varphi^\nu(z, \zeta)\, \overline{\omega}'_{z,\zeta,\lambda}(t^*(z, \zeta, \lambda)) \wedge \omega_\zeta(s(z, \zeta))$$

is continuous for all (z, ζ, λ) in some neighbourhood $\subseteqq D \times X \times \Delta_{0K}$ of $D \times S_K \times \Delta_{0K}$. For every bounded differential form f on S_K and every integer $\nu \geqq \max(n\varkappa, n\varkappa^*)$ we can therefore define

$$\begin{aligned}&\left(R_{S_K}(\varphi^\nu, s_1^*, \dots, s_N^*, \bar{s}, s) f\right)(z)\\ &:= (-1)^{|K|} \frac{(n-1)!}{(2\pi i)^n} \int\limits_{(\zeta,\lambda)\in S_K\times\Delta_{0K}} f(\zeta) \wedge \varphi^\nu(z, \zeta)\, \overline{\omega}'_{z,\zeta,\lambda}(t^*(z, \zeta, \lambda)) \wedge \omega_\zeta(s(z, \zeta))\,,\\ &\qquad z \in D\,. \qquad (4.6.2)\end{aligned}$$

Remark. If f is of degree q, then

$$L_{S_K} f = 0 \quad \text{and} \quad R_{S_K} f = 0 \quad \text{if} \quad |K| \geqq n + 1 - q\,. \qquad (4.6.3)$$

This follows from the fact that then the form under the sign of integration in (4.6.1) and (4.6.2) is of degree $\geqq n + q$ in ζ, whereas $\dim_R S_K = 2n - |K| \leqq n + q - 1$.

Remark. For $N = 1$, Definitions (4.6.1) and (4.6.2) agree with (4.5.5) and (4.5.8), respectively.

Remark. If f is a function, then Definition (4.6.1) agrees with (4.4.7) and $R_{S_K} f = 0$. If f is of degree 1, then Definition (4.6.2) agrees with (4.4.9).

4.6.2. Theorem (Koppelman-Leray-Norguet formula). *We use the notations from Subsections* 4.3.1 *and* 4.4.1. *Further, let* $(s_1^*, \dots, s_N^*, \varkappa^*)$ *be a Leray-Norguet section for* (D, s, φ) (cp. Subsection 4.4.2), *and let* $\nu \geqq \max(2n\varkappa, n\varkappa^*)$ *be an integer. Suppose, additionally, that, for* $k = 1, \dots, N$, *all derivatives of* $\varphi^\nu(z, \zeta)\, s^*(z, \zeta)/\langle s_k^*(z, \zeta), s(z, \zeta)\rangle$

which are of order $\leqq 2$ in z and of order $\leqq 1$ in ζ are continuous for all (z, ζ) in some neighbourhood $W_k \subseteq D \times X$ of $D \times S_k$.[1] *Let f be a continuous $(0, q)$-form on $\bar{D}$ such that $\bar{\partial} f$ is also continuous on $\bar{D}$, $0 \leqq q \leqq n$. Then, for every collection $K = (k_1, \dots, k_l)$ of strictly increasing integers $1 \leqq k_1 < \dots < k_l \leqq N$, the forms $L_{S_K} f$, $R_{S_K} \bar{\partial} f$, $B_D \bar{\partial} f$, $R_{S_K} f$, $B_D f$, $\bar{\partial} R_{S_K} f$ and $\bar{\partial} B_D f$ are continuous in D, and we have*

$$(-1)^q f = \sum_{|K| \leqq n-q} L_{S_K} f - (B_D + \sum_{|K| \leqq n-q-1} R_{S_K}) \bar{\partial} f + \bar{\partial}(B_D + \sum_{|K| \leqq n-q} R_{S_K}) f \quad \text{in } D, \tag{4.6.4}$$

where the summation is over all strictly increasing collections $K = (k_1, \dots, k_l)$ of integers $1 \leqq k_1 < \dots < k_l \leqq N$ with $l \leqq n - q - 1$ and $l \leqq n - q$, respectively, and $L_{S_K} = L_{S_K}(\varphi^\nu, s_1^*, \dots, s_N^*, s)$, $R_{S_K} = R_{S_K}(\varphi^\nu, s_1^*, \dots, s_N^*, \bar{s}, s)$ (cp. Subsection 4.6.1) *and $B_D = B_D(\varphi^\nu, \bar{s}, s)$* (cp. Subsection 4.5.1).

Remark. If $q = 0$, then $B_D f = 0$ and $R_{S_K} f = 0$ and, therefore, (4.6.4) agrees with the Leray-Norguet formula (4.4.11). If $(s^*, \varkappa^*)$ is a Leray section for (D, s, φ), then by setting $s_k^* = s^*$ we obtain a Leray-Norguet section $(s_1^*, \dots, s_N^*, \varkappa^*)$ for (D, s, φ). Then $L_{S_K} = 0$, $R_{S_K} = 0$ if $|K| \geqq 2$ and $L_{S_1} + \dots + L_{S_N} = L_{\partial D}$, $R_{S_1} + \dots + R_{S_N} = R_{\partial D}$, that is, in this case (4.6.4) agrees with the Koppelman-Leray formula (4.5.32).

To prove Theorem 4.6.2 we need the following lemma.

4.6.3. Lemma. *Let the hypotheses of Theorem 4.6.2 be fulfilled. For every collection $K = (k_1, \dots, k_l)$ of strictly increasing integers $1 \leqq k_1 < \dots < k_l \leqq N$, we write $W_K := W_{k_1} \cup \dots \cup W_{k_l}$.* Set

$$W := \bigcup_K W_K \times \Delta_{0K}, \tag{4.6.5}$$

where $\bigcup_K$ is over all collections $K = (k_1, \dots, k_l)$ of strictly increasing integers $1 \leqq k_1 < \dots < k_l \leqq N$, and define

$$\bar{\Omega}(z, \zeta, \lambda) := \varphi^\nu(z, \zeta)\, \bar{\omega}'_{z,\zeta,\lambda}(t^*(z, \zeta, \lambda)) \wedge \omega_\zeta(s(z, \zeta)) \quad \text{for} \quad (z, \zeta, \lambda) \in W, \tag{4.6.6}$$

where $t^* = t^*_{(s_1^*, \dots, s_N^*, \bar{s}, s)}$ (see (4.4.8) *and* (4.5.1)). *Then*

$$(\bar{\partial}_z + d_{\zeta,\lambda}) \left(f(\zeta) \wedge \bar{\Omega}(z, \zeta, \lambda)\right) = \bar{\partial} f(\zeta) \wedge \bar{\Omega}(z, \zeta, \lambda) \quad \text{for} \quad (z, \zeta, \lambda) \in W. \tag{4.6.7}$$

Proof. Repetition of that of Lemma 4.5.4. ■

Proof of Theorem 4.6.2. By the remark following Theorem 4.6.2, we can assume that $1 \leqq q \leqq n$. It is clear that $L_{S_K} f$, $R_{S_K} \bar{\partial} f$ and $R_{S_K} f$ are continuous in D. By Theorem 4.5.2, $B_D \bar{\partial} f$, $B_D f$ and $\bar{\partial} B_D f$ are continuous in D. By differentiation under the sign of integration we obtain that $\bar{\partial} R_{S_K} f$ is continuous in D. It remains to prove (4.6.4). In view of the Koppelman formula (4.5.9), the proof will be complete if we show that

$$\sum_{|K| \leqq n-q} \bar{\partial} R_{S_K} f = B_{\partial D} f - \sum_{|K| \leqq n-q} L_{S_K} f + \sum_{|K| \leqq n-q-1} R_{S_K} \bar{\partial} f \quad \text{in } D.$$

This follows by Stokes' formula from (4.4.4), (4.6.3) and Lemma 4.6.3. ■

[1] The theorem is valid also without this additional assumption, but then the proof becomes technically more complicated.

4.6.4. Corollary. *If, under the hypotheses of Theorem 4.6.2, the sections $s_1^*(z,\zeta), \dots, s_N^*(z,\zeta)$ are holomorphic in $z \in D$ and if $1 \leqq q \leqq n$, then*

$$f = \bar\partial(-1)^q \Big(B_D + \sum_{|K| \leqq n-q} R_{S_K}\Big) f + (-1)^{q+1} \Big(B_D + \sum_{|K| \leqq n-q-1} R_{S_K}\Big) \bar\partial f \quad \text{in} \quad D. \tag{4.6.8}$$

In particular, for every continuous $(0, q)$-form f on $\bar D$ such that $\bar\partial f = 0$ in D,

$$u := (-1)^q \Big(B_D + \sum_{|K| \leqq n-q} R_{S_K}\Big) f \tag{4.6.9}$$

is a continuous solution of $\bar\partial u = f$ in D.

Proof. If $s_1^*(z,\zeta), \dots, s_N^*(z,\zeta)$ are holomorphic in z, then, for every collection $K = (k_1, \dots, k_l)$ of strictly increasing integers $1 \leqq k_1 < \dots < k_l \leqq N$, the form

$$\varphi^\nu(z,\zeta)\, \bar\omega'_{z,\zeta,\lambda}\left(\sum_{k\in K} \lambda_k \frac{s_k^*(z,\zeta)}{\langle s_k^*(z,\zeta), s(z,\zeta)\rangle}\right) \wedge \omega_\zeta\big(s(z,\zeta)\big)$$

is of degree zero in z. Therefore, $L_{S_K} f = 0$ if $q \geqq 1$. ∎

4.7. The Weil formula

In this section we construct a Leray-Norguet section (cp. Subsection 4.4.2) for analytic polyhedra in Stein manifolds. This Leray-Norguet section is holomorphic in both variables z and ζ. Therefore, if such a polyhedron is real non-degenerate, then we obtain a formula for solving the $\bar\partial$-equation (cp. Corollary 4.6.4) as well as a version of formula (4.4.13), which is a generalization of the classical Weil formula (mentioned in the Preface to this book and in the Notes following Chapter 2) to Stein manifolds. Remark that in the construction given in the present section, results from the theory of coherent analytic sheaves are used. This is not necessary for the case of $\mathbb{C}^n$ (cp. Remark 4.7.5 below).

4.7.1. Definition. Let X be a Stein manifold. An open set $D \subset\subset X$ is called an *analytic polyhedron* if there are holomorphic functions $F_1, \dots, F_N$ in some neighbourhood $U_{\bar D}$ of $\bar D$ such that

$$D = \{z \in U_{\bar D} : |F_k(z)| < 1 \quad \text{for} \quad k = 1, \dots, N\}.$$

D is called *real non-degenerate* if, moreover, for every collection of integers $1 \leqq k_1 < \dots < k_l \leqq N$ and all $z \in U_{\bar D}$ with $|F_{k_1}(z)| = \dots = |F_{k_l}(z)| = 1$,

$$d\,|F_{k_1}(z)| \wedge \dots \wedge d\,|F_{k_l}(z)| \neq 0. \tag{4.7.1}$$

Remark. If D is a real non-degenerate analytic polyhedron in a Stein manifold, then the boundary of D is piecewise C^1 in the sense defined at the beginning of Section 4.3.1 (set $V_k := U_{\bar D}$ and $\varrho_k := |F_k| - 1$).

4.7.2. Lemma. *Let X be a Stein manifold, and let $s(z,\zeta)$, $\varphi(z,\zeta)$ be as in Lemma 4.2.4. Suppose that X is a relatively compact open subset of some larger Stein manifold. (However, cp. also the remark following Lemma 4.2.4.) Then, for every holomorphic function F on X there exists a holomorphic map $h^*: X \times X \to T^*(X)$* (cp. Subsection 4.2.3) *such that $h^*(z,\zeta) \in T_z^*(X)$ and*

$$\varphi(z,\zeta)\big(F(\zeta) - F(z)\big) = \langle h^*(z,\zeta), s(z,\zeta)\rangle \tag{4.7.2}$$

for all $z, \zeta \in X$ (for the definition of $\langle\cdot,\cdot\rangle$ see (4.3.2)).

Proof. It follows from conditions (ii)—(iv) in Lemma 4.2.4 that (4.7.2) can be solved locally. That is, we can find an open covering $\{W_j\}$ of $X \times X$ and holomorphic maps $h_j^*: W_j \to T^*(X)$ such that $h_j^*(z, \zeta) \in T_z^*(X)$ and

$$\varphi(z, \zeta) \left(F(\zeta) - F(z)\right) = \langle h_j^*(z, \zeta), s(z, \zeta)\rangle \tag{4.7.3}$$

for all $(z, \zeta) \in W_j$. If $U \subseteq X \times X$ is open, then we denote by $\mathscr{K}_s(U)$ the set of all holomorphic maps $s^*: U \to T^*(X)$ such that $s^*(z, \zeta) \in T_z^*(X)$ and $\langle s^*(z, \zeta), s(z, \zeta)\rangle = 0$ for all $(z, \zeta) \in U$. $\mathscr{K}_s$ is an analytic subsheaf of ${}_{X\times X}\mathcal{O}^B$, where B is the pull-back of $T^*(X)$ with respect to the map $X \times X \ni (z, \zeta) \to z \in X$. By the Oka Theorem 4.1.2, $\mathscr{K}_s$ is coherent. By (4.7.3), $h_j^* - h_i^* \in \mathscr{K}_s(W_j \cap W_i)$. Therefore, we obtain from Cartan's Theorem B (Theorem 4.1.3 (ii)) a collection $t_j^* \in \mathscr{K}_s(W_j)$ such that $h_j^* - h_i^* = t_j^* - t_i^*$ in $W_j \cap W_i$. Setting $h^* := h_j^* - t_j^*$ in W_j, we complete the proof. ■

4.7.3. A Koppelman-Leray-Norguet formula. Let X be a Stein manifold, and let $D \subset\subset X$ be a really non-degenerate special analytic polyhedron. Let $U_{\overline{D}}, F_1, \dots, F_N$ be as in Definition 4.7.1. Since $D \subset\subset U_{\overline{D}}$, it follows immediately from the definition of Stein manifolds that, for sufficiently small $\varepsilon > 0$, the set $\{z \in U_{\overline{D}}: |F_k(z)| < 1 + \varepsilon$ for $k = 1, \dots, N\}$ is a Stein manifold. Therefore, after shrinking $U_{\overline{D}}$, we can assume that $U_{\overline{D}}$ is a Stein manifold. Without loss of generality we can also assume that X is a relatively compact subset of some larger Stein manifold and we can apply Lemma 4.2.4 (cp. also the remark following Lemma 4.2.4). Let $s(z, \zeta)$, $\varphi(z, \zeta)$ and $\varkappa$ be as in Lemma 4.2.4. By Lemma 4.7.2, we can find holomorphic maps $h_1^*, \dots, h_N^*: U_{\overline{D}} \to T^*(X)$ such that $h_k^*(z, \zeta) \in T_z^*(X)$ and

$$\varphi(z, \zeta) \left(F_k(\zeta) - F_k(z)\right) = \langle h_k^*(z, \zeta), s(z, \zeta)\rangle \tag{4.7.4}$$

for all $z, \zeta \in U_{\overline{D}}$ and $k = 1, \dots, N$. Set $S_k := \{z \in \partial D: |F_k(z)| = 1\}$, $k = 1, \dots, N$, and define S_K and Δ_K as in Subsection 4.4.1. Since, by (4.7.4),

$$\frac{\varphi(z, \zeta)\, h_k^*(z, \zeta)}{\langle h_k^*(z, \zeta), s(z, \zeta)\rangle} = \frac{h_k^*(z, \zeta)}{F_k(\zeta) - F_k(z)} \quad \text{for} \quad z \in D \quad \text{and} \quad \zeta \in S_k,$$

$(h_1^*, \dots, h_N^*, 1)$ is a Leray-Norguet section for (D, s, φ) (cp. Subsection 4.4.2). Therefore, from Corollaries 4.4.5 and 4.6.4 we obtain the following

Theorem. *Let* $v \geqq \max(2n\varkappa, n)$ *be an integer,* $L_{S_K} = L_{S_K}(\varphi^v, h_1^*, \dots, h_N^*, s)$, $R_{S_K} = R_{S_K}(\varphi^v, h_1^*, \dots, h_N^*, \bar{s}, s)$ (cp. Subsections 4.4.2 and 4.6.1), *and let* $B_D = B_D(\varphi^v, \bar{s}, s)$ (cp. Subsections 4.3.2 and 4.5.1). *Then*

(i) *For every continuous function f on $\overline{D}$ such that $\bar\partial f$ is also continuous on $\overline{D}$, we have*

$$f = \sum_{|K|=n} L_{S_K} f - \sum_{|K|\leqq n-1} R_{S_K}\, \bar\partial f - B_D\, \bar\partial f \quad in \quad D. \tag{4.7.5}$$

In particular, for every continuous function f on $\overline{D}$ that is holomorphic in D, we have

$$f = \sum_{|K|=n} L_{S_K} f \quad in \quad D. \tag{4.7.6}$$

(ii) *Let $1 \leqq q \leqq n$. Then, for every continuous $(0, q)$-form f on $\overline{D}$ such that $\bar\partial f$ is also continuous on $\overline{D}$, we have*

$$f = \bar\partial(-1)^q \Big(B_D + \sum_{|K|\leqq n-q} R_{S_K}\Big) f + (-1)^{q+1} \Big(B_D + \sum_{|K|\leqq n-q-1} R_{S_K}\Big)\, \bar\partial f \quad in \quad D. \tag{4.7.7}$$

In particular, for every continuous $(0, q)$*-form* f *on* $\bar{D}$ *such that* $\bar{\partial} f = 0$ *in* D,

$$u := (-1)^q \left(B_D + \sum_{|K| \leq n-q} R_{S_K}\right) f \tag{4.7.8}$$

is a continuous solution of $\bar{\partial} u = f$ *in* D.

4.7.4. Theorem (Weil formula). *We use the notations from Subsection 4.7.3. Then, for every continuous function* f *on* $\bar{D}$ *that is holomorphic in* D *and for every integer* $\nu \geq \max(2n\varkappa - n, 0)$, *we have*

$$(2\pi i)^n f(z) = \sum_{|K|=n} \int_{\zeta \in S_K} f(\zeta)\, \varphi^\nu(z, \zeta) \frac{\det\left(h_{k_1}^*(z, \zeta), \dots, h_{k_n}^*(z, \zeta)\right) \omega_\zeta\left(s(z, \zeta)\right)}{\left(F_{k_1}(\zeta) - F_{k_1}(z)\right) \dots \left(F_{k_n}(\zeta) - F_{k_n}(z)\right)},$$

$$z \in D, \tag{4.7.9}$$

where the summation is over all collections $K = (k_1, \dots, k_n)$ *of strictly increasing integers* $1 \leq k_1 < \dots < k_n \leq N$, *and the form* $\det(h_{k_1}^*, \dots, h_{k_n}^*)\, \omega_\zeta(s)$ *is defined analogously as in* (4.3.4).

Proof. Consider a collection $K = (k_1, \dots, k_n)$ of strictly increasing integers $1 \leq k_1 < \dots < k_n \leq N$. Since the sections $h_k^*(z, \zeta)$ are holomorphic in ζ, we obtain (cp. (4.3.4), (1.7.5), Definition 1.7.6 and Proposition 1.7.5) that on $S_K \times \Delta_K$

$$(n-1)!\, \bar{\omega}'_{\zeta,\lambda}\left(\sum_{i=1}^n \lambda_{k_i} \frac{h_{k_i}^*}{\langle h_{k_i}^*, s\rangle}\right) \wedge \omega_\zeta(s)$$

$$= \det\nolimits_{1,n-1}\left(\sum_{i=1}^n \lambda_{k_i} \frac{h_{k_i}^*}{\langle h_{k_i}^*, s\rangle}, \sum_{i=1}^n \frac{h_{k_i}^*}{\langle h_{k_i}^*, s\rangle} d\lambda_{k_i}\right) \wedge \omega_\zeta(s)$$

$$= \frac{\det(h_{k_1}^*, \dots, h_{k_n}^*) \sum_\sigma \operatorname{sgn}(\sigma)\, \lambda_{k_{\sigma(1)}} d\lambda_{k_{\sigma(2)}} \wedge \dots \wedge d\lambda_{k_{\sigma(n)}} \wedge \omega_\zeta(s)}{\langle h_{k_1}^*, s\rangle \dots \langle h_{k_n}^*, s\rangle},$$

where the summation is over all permutations σ of $\{1, \dots, n\}$ and $\operatorname{sgn}(\sigma)$ is the signature of σ. By (4.7.4), this implies that on $S_K \times \Delta_K$

$$\varphi^{\nu+n} \bar{\omega}'_{\zeta,\lambda}\left(\sum_{i=1}^n \lambda_{k_i} \frac{h_{k_i}^*}{\langle h_{k_i}^*, s\rangle}\right) \wedge \omega_\zeta(s)$$

$$= \frac{\varphi^\nu \det(h_{k_1}^*, \dots, h_{k_n}^*) \sum_\sigma \operatorname{sgn}(\sigma)\, \lambda_{k_{\sigma(1)}} d\lambda_{k_{\sigma(2)}} \wedge \dots \wedge d\lambda_{k_{\sigma(n)}} \wedge \omega_\zeta(s)}{(n-1)!\, \left(F_{k_1}(\zeta) - F_{k_1}(z)\right) \dots \left(F_{k_n}(\zeta) - F_{k_n}(z)\right)}.$$

Hence, by (4.7.6) and (4.4.7),

$$(2\pi i)^n f(z) = \sum_{|K|=n} \int_{\zeta \in S_K} \left[\int_{\lambda \in \Delta_K} \sum_\sigma \operatorname{sgn}(\sigma)\, \lambda_{k_{\sigma(1)}} d\lambda_{k_{\sigma(2)}} \wedge \dots \wedge d\lambda_{k_{\sigma(n)}}\right]$$

$$\wedge f(\zeta)\, \varphi^\nu(z, \zeta) \frac{\det\left(h_{k_1}^*(z, \zeta), \dots, h_{k_n}^*(z, \zeta)\right) \wedge \omega_\zeta\left(s(z, \zeta)\right)}{\left(F_{k_1}(\zeta) - F_{k_1}(z)\right) \dots \left(F_{k_n}(\zeta) - F_{k_n}(z)\right)}, \quad z \in D.$$

Therefore, it remains to prove that

$$\int_{\lambda \in \Delta_K} \sum_\sigma \operatorname{sgn}(\sigma)\, \lambda_{k_{\sigma(1)}} d\lambda_{k_{\sigma(2)}} \wedge \dots \wedge d\lambda_{k_{\sigma(n)}} = 1. \tag{4.7.10}$$

Since $d\lambda_1 \wedge \dots \wedge d\lambda_N$ defines the orientation of Δ (cp. Subsection 4.4.1), it follows from (4.4.3) that the orientation of Δ_{0K} is defined by the form

$$d\lambda_{k_1} \wedge \dots \wedge d\lambda_{k_n} = \operatorname{sgn}(\sigma)\, d\lambda_{k_{\sigma(1)}} \wedge \dots \wedge d\lambda_{k_{\sigma(n)}}.$$

Since on Δ_{0K}

$$-\mathrm{d}\lambda_0 \wedge \mathrm{d}\lambda_{k_{\sigma(2)}} \wedge \dots \wedge \mathrm{d}\lambda_{k_{\sigma(n)}} = \mathrm{d}\lambda_{k_{\sigma(1)}} \wedge \mathrm{d}\lambda_{k_{\sigma(2)}} \wedge \dots \wedge \mathrm{d}\lambda_{k_{\sigma(n)}},$$

together with (4.4.3) this implies that the orientation of Δ_K is defined by

$$\operatorname{sgn}(\sigma)\, \mathrm{d}\lambda_{k_{\sigma(2)}} \wedge \dots \wedge \mathrm{d}\lambda_{k_{\sigma(n)}}.$$

Therefore, since

$$\lambda_{k_{\sigma(1)}} = 1 - \lambda_{k_{\sigma(2)}} - \dots - \lambda_{k_{\sigma(n)}} \quad \text{on} \quad \Delta_K,$$

(4.7.10) follows from the relation

$$\int\limits_{\substack{t_2, \dots, t_n \geqq 0 \\ t_2 + \dots + t_n \leqq 1}} (1 - t_2 - \dots - t_n)\, \mathrm{d}t_2 \dots \mathrm{d}t_n = \frac{1}{n!},$$

which can be easily verified. ■

4.7.5. Remark. For the case $X = \mathbb{C}^n$, we can set $s(z, \zeta) = \zeta - z$, $\varphi \equiv 1$, $\varkappa = 0$, and $h_1^*, \dots, h_N^*$ are $\mathbb{C}^n$-valued holomorphic functions. Then the Weil formula (4.7.9) can be written

$$(2\pi i)^n f(z) = \sum_{|K|=n} \int\limits_{\zeta \in S_K} f(\zeta) \frac{\det\left(h_{k_1}^*(z, \zeta), \dots, h_{k_n}^*(z, \zeta)\right) \mathrm{d}\zeta_1 \wedge \dots \wedge \mathrm{d}\zeta_n}{\left(F_{k_1}(\zeta) - F_{k_1}(z)\right) \dots \left(F_{k_n}(\zeta) - F_{k_n}(z)\right)}. \tag{4.7.11}$$

Remark that in this case the Weil formula can be proved without use of results from the theory of coherent analytic sheaves, because the vector functions $h_1^*, \dots, h_N^*$ then can be obtained from Theorem 2.5.4.

4.8. The support functions Φ and $\tilde{\Phi}$

In this section we construct the support functions $\Phi(z, \zeta)$ and $\tilde{\Phi}(z, \zeta)$ (cp. Theorem 2.4.3 and Construction 3.1.1) for strictly pseudoconvex open sets in Stein manifolds.

4.8.1. Notation and preliminaries. Let X be a Stein manifold of complex dimension n, let $D_0 \subset\subset X$ be a strictly pseudoconvex open set, let $\theta \subset\subset X$ be a neighbourhood of ∂D_0, and let ϱ_0 be a strictly plurisubharmonic C^2-function in a neighbourhood of $\bar{\theta}$ such that $D_0 \cap \bar{\theta} = \{z \in \bar{\theta} : \varrho_0(z) < 0\}$.

Remark. We do not assume that $\mathrm{d}\varrho_0(z) \neq 0$ for all $z \in \partial D_0$. Further, the case $D_0 = \emptyset$ is also admitted.

Set $N(\varrho_0) := \{z \in \bar{\theta} : \varrho_0(z) = 0\}$ and suppose that

$$N(\varrho_0) \subset\subset \theta. \tag{4.8.1}$$

We choose a metric dist $(\cdot, \cdot)$ on X as follows: Consider X as a C^∞-submanifold of some $\mathbb{R}^m$ and denote by dist $(\cdot, \cdot)$ the restriction of the Euclidean metric in $\mathbb{R}^m$ to X. If $Y, Z \subseteq X$, then we write $\operatorname{dist}(Y, Z) := \inf\{\operatorname{dist}(z, \zeta) : z \in Z, \zeta \in Y\}$.

Let $A_j \subset\subset B_j \subset\subset X$ $(j = 1, \dots, L)$ be finitely many open sets such that $\bar{\theta} \subseteq A_1 \cup \dots \cup A_L$ and, for every $1 \leqq j \leqq L$, there are holomorphic coordinates $b_j = (b_{j1}, \dots, b_{jn}) : B_j \to \mathbb{C}^n$. For $z \in B_j$ we write $z_j = (z_{j1}, \dots, z_{jn}) := b_j(z) = \left(b_{j1}(z), \dots, b_{jn}(z)\right)$, and we denote by $x_{jk} = x_{jk}(z)$, $k = 1, \dots, 2n$, the corresponding real coordinates such that $z_{jk} = x_{jk}(z) + i x_{jk+n}(z)$, $k = 1, \dots, n$. We can assume that the sets $b_1(B_1), \dots, b_L(B_L)$ are convex. For $z, \zeta \in B_j$ we set $[z, \zeta]_j := \{\xi \in B_j : \xi_j = \lambda z_j + (1-\lambda)\zeta_j, 0 \leqq \lambda \leqq 1\}$, $j = 1, \dots, L$. Since ϱ_0 is strictly plurisubharmonic on $\bar\theta$, since $A_j \subset\subset B_j$, and since $\bar\theta \subseteqq A_1 \cup \dots \cup A_L$, we can find numbers $\varepsilon, \beta > 0$ and C^1-functions a_{jkl} on B_j ($j = 1, \dots, L$; $k, l = 1, \dots, n$) such that the following conditions are fulfilled:

$$\operatorname{dist}(A_j, \partial B_j) > 2\varepsilon \quad \text{for} \quad j = 1, \dots, L. \tag{4.8.2}$$

$$\sum_{k,l=1}^{n} \frac{\partial^2 \varrho_0(\zeta)}{\partial \zeta_{jk}\, \partial \bar\zeta_{jl}} (z_{jk} - \zeta_{jk})(\bar z_{jl} - \bar\zeta_{jl}) > 3\beta[\operatorname{dist}(z, \zeta)]^2$$

$$\text{for} \quad \zeta \in \bar A_j \cap \bar\theta,\ z \in \bar\theta \quad \text{with} \quad 0 < \operatorname{dist}(z, \zeta) \leqq 2\varepsilon \quad \text{and} \quad j = 1, \dots, L. \tag{4.8.3}$$

$$\left| \sum_{k,l=1}^{n} \left(\frac{\partial^2 \varrho_0(\zeta)}{\partial \zeta_{jk}\, \partial \bar\zeta_{jl}} - a_{jkl}(\zeta) \right) (z_{jk} - \zeta_{jk})(z_{jl} - \zeta_{jl}) \right| < \beta[\operatorname{dist}(z, \zeta)]^2$$

$$\text{for} \quad \zeta \in \bar A_j \cap \bar\theta,\ z \in \bar\theta \quad \text{with} \quad \operatorname{dist}(z, \zeta) \leqq 2\varepsilon \quad \text{and} \quad j = 1, \dots, L. \tag{4.8.4}$$

$$\left| \frac{1}{2} \sum_{k,l=1}^{2n} \left(\frac{\partial^2 \varrho_0(\zeta)}{\partial x_{jk}\, \partial x_{jl}} - \frac{\partial^2 \varrho_0(\xi)}{\partial x_{jk}\, \partial x_{jl}} \right) (x_{jk}(z) - x_{jk}(\zeta))\, (x_{jl}(z) - x_{jl}(\zeta)) \right| < \beta[\operatorname{dist}(z, \zeta)]^2$$

$$\text{for} \quad \zeta \in \bar A_j \cap \bar\theta;\ \xi, z \in \bar\theta;\ \operatorname{dist}(z, \zeta) \leqq 2\varepsilon;\ \xi \in [z, \zeta]_j;\ j = 1, \dots, L. \tag{4.8.5}$$

Since $N(\varrho_0) \subset\subset \theta$, after shrinking ε, we can find open sets W, U such that $D_0 \cup U$ is strictly pseudoconvex and, moreover, the following conditions are fulfilled:

$$N(\varrho_0) \subset\subset U \subset\subset W \subset\subset \theta, \tag{4.8.6}$$

$$\max_{\zeta \in \bar W} |\varrho_0(\zeta)| < \frac{\varepsilon^2 \beta}{3}, \tag{4.8.7}$$

$$\operatorname{dist}(W, \partial\theta) > 2\varepsilon. \tag{4.8.8}$$

Since $\bar\theta \subseteqq A_1 \cup \dots \cup A_L$, we can find functions $\gamma_j \in C_0^\infty(A_j)$ such that $\gamma_1 + \dots + \gamma_L = 1$ on $\bar\theta$. If ψ is a C^2-function in some neighbourhood of a point $z \in \bar\theta$, then we define

$$||\psi(z)||_2 := |\psi(z)| + \sum_{j=1}^{L} \gamma_j(z) \left[\sum_{k=1}^{2n} \left| \frac{\partial \psi(z)}{\partial x_{jk}} \right| + \sum_{k,l=1}^{2n} \left| \frac{\partial^2 \psi(z)}{\partial x_{jk}\, \partial x_{jl}} \right| \right]. \tag{4.8.9}$$

If ψ is a C^2-function in some neighbourhood of a set $Y \subseteqq \bar\theta$, then we write

$$||\psi||_{2,Y} := \sup_{z \in Y} ||\psi(z)||_2. \tag{4.8.10}$$

We denote by $\mathfrak{M}_2(\bar\theta)$ the normed space of all C^2-functions in a neighbourhood of $\bar\theta$ endowed with the norm $||\cdot||_{2,\theta}$. Choose $\delta > 0$ so small that, for every $\varrho \in \mathfrak{M}_2(\bar\theta)$ with $||\varrho_0 - \varrho||_{2,\theta} \leqq \delta$,

$$\varrho_0(z)\, \varrho(z) > 0 \quad \text{for} \quad z \in \bar\theta \setminus U \tag{4.8.11}$$

and conditions (4.8.3)—(4.8.7) are fulfilled with ϱ instead of ϱ_0, where

$$N(\varrho) := \{z \in \bar\theta : \varrho(z) = 0\}. \tag{4.8.12}$$

We denote by $\mathfrak{U}_\delta(\varrho_0)$ the set of all $\varrho \in \mathfrak{M}_2(\bar\theta)$ with $||\varrho_0 - \varrho||_{2,\theta} \leqq \delta$. Then every $\varrho \in \mathfrak{U}_\delta(\varrho_0)$ is strictly plurisubharmonic on $\bar\theta$ (this follows from (4.8.3) with ϱ instead of ϱ_0) and it follows from (4.8.11) that, for every $\varrho \in \mathfrak{U}_\delta(\varrho_0)$,

$$D_\varrho := (D_0 \setminus U) \cup \{z \in U : \varrho(z) < 0\}$$

is a strictly pseudoconvex open set such that

$$D_\varrho \cup U = D_0 \cup U\,, \quad \partial D_\varrho \subseteq U \quad \text{and} \quad D_\varrho \cap U = \{z \in U : \varrho(z) < 0\}\,. \tag{4.8.13}$$

Let $C^1\big((D_0 \cup U) \times U\big)$ be the Fréchet space of all C^1-functions in $(D_0 \cup U) \times U$ endowed with the topology of uniform convergence together with the first-order derivatives on compact sets.

For $\varrho \in \mathfrak{U}_\delta(\varrho_0)$ we define

$$F_\varrho(z,\zeta) := -\sum_{j=1}^{L} \gamma_j(\zeta)\left[2\sum_{k=1}^{n} \frac{\partial\varrho(\zeta)}{\partial\zeta_{jk}}(z_{jk}-\zeta_{jk}) + \sum_{k,l=1}^{n} a_{jkl}(\zeta)\,(z_{jk}-\zeta_{jk})\,(z_{jl}-\zeta_{jl})\right]$$
$$\text{for} \quad z,\zeta \in \bar{\theta} \quad \text{with} \quad \operatorname{dist}(z,\zeta) \leqq 2\varepsilon\,. \tag{4.8.14}$$

This definition is correct, because, by (4.8.2), for all $z, \zeta \in \bar{\theta}$ with $\gamma_j(\zeta) \neq 0$ and dist $(z,\zeta) \leqq 2\varepsilon$, z belongs to B_j.

4.8.2. Lemma. *We use the notations from Subsection* 4.8.1. *Then, for every* $\varrho \in \mathfrak{U}_\delta(\varrho_0)$,

$$\operatorname{Re} F_\varrho(z,\zeta) \geqq \varrho(\zeta) - \varrho(z) + \beta\,[\operatorname{dist}(z,\zeta)]^2$$
$$\textit{for} \quad z,\zeta \in \bar{\theta} \quad \textit{with} \quad \operatorname{dist}(z,\zeta) \leqq 2\varepsilon\,. \tag{4.8.15}$$

Proof. Let $\zeta \in A_j$, $z \in \bar{\theta}$ and dist $(z,\zeta) \leqq 2\varepsilon$. Then it follows from Lemma 1.4.13 and Taylor's theorem that

$$-\operatorname{Re}\left[2\sum_{k=1}^{n} \frac{\partial\varrho(\zeta)}{\partial\zeta_{jk}}(z_{jk}-\zeta_{jk}) + \sum_{k,l=1}^{n} \frac{\partial^2\varrho(\zeta)}{\partial\zeta_{jk}\,\partial\zeta_{jl}}(z_{jk}-\zeta_{jk})\,(z_{jl}-\zeta_{jl})\right]$$
$$= \varrho(\zeta) - \varrho(z) + \sum_{k,l=1}^{n} \frac{\partial^2\varrho(\zeta)}{\partial\zeta_{jk}\,\partial\bar{\zeta}_{jl}}(z_{jk}-\zeta_{jk})\,(\bar{z}_{jl}-\bar{\zeta}_{jl}) + R(z,\zeta)\,,$$

where, by (4.8.5) (which holds, by definition of $\mathfrak{U}_\delta(\varrho_0)$, with ϱ instead of ϱ_0), $|R(z,\zeta)| \leqq \beta[\operatorname{dist}(z,\zeta)]^2$. By (4.8.3) and (4.8.4), this implies (4.8.15). ■

4.8.3. Theorem. *We use the notations of Subsection* 4.8.1. *Then there exist continuous maps* $a, \tilde{a}, b, \tilde{b}$ *from* $\mathfrak{U}_\delta(\varrho_0)$ *(endowed with the metric induced by the norm* $\|\cdot\|_{2,\theta}$*) into the Fréchet space* $C^1\big((D_0 \cup U) \times U\big)$ *with the following properties: If* $\varrho \in \mathfrak{U}_\delta(\varrho_0)$ *and* $\Phi_\varrho := a(\varrho)$, $\tilde{\Phi}_\varrho := \tilde{a}(\varrho)$, $M_\varrho := b(\varrho)$ *and* $\tilde{M}_\varrho := \tilde{b}(\varrho)$, *then the following conditions are fulfilled:*

(i) $\Phi_\varrho(z,\zeta)$ *and* $\tilde{\Phi}_\varrho(z,\zeta)$ *are holomorphic in* $z \in D_\varrho \cup U$ $\big(= D_0 \cup U$, cp. (4.8.13)$\big)$.

(ii) $\Phi_\varrho(z,\zeta) \neq 0$ *and* $\tilde{\Phi}_\varrho(z,\zeta) \neq 0$ *for* $\zeta \in U, z \in D_\varrho \cup U$ *with*

$$\operatorname{dist}(z,\zeta) \geqq \varepsilon\,. \tag{4.8.16}$$

$$M_\varrho(z,\zeta) \neq 0 \quad \textit{and} \quad \tilde{M}_\varrho(z,\zeta) \neq 0 \quad \textit{for} \quad \zeta \in U, z \in D_\varrho \cup U\,. \tag{4.8.17}$$

$$\Phi_\varrho(z,\zeta) = F_\varrho(z,\zeta)\,M_\varrho(z,\zeta) \quad \textit{and} \quad \tilde{\Phi}_\varrho(z,\zeta) = \big(F_\varrho(z,\zeta) - 2\varrho(\zeta)\big)\,\tilde{M}_\varrho(z,\zeta)$$
$$\textit{for} \quad \zeta \in U, z \in D_\varrho \cup U \quad \textit{with} \quad \operatorname{dist}(z,\zeta) \leqq \varepsilon\,. \tag{4.8.18}$$

(iii) $\tilde{\Phi}_\varrho(z,\zeta) = \Phi_\varrho(z,\zeta)$ *for* $\zeta \in N(\varrho), z \in D_\varrho \cup U\,.$ (4.8.19)

Proof. Conditions (4.8.3)–(4.8.7) with ϱ instead of ϱ_0 will be denoted by $(4.8.3)_\varrho$ to $(4.8.7)_\varrho$. First we prove that, for every $\varrho \in \mathfrak{U}_\delta(\varrho_0)$,

$$\operatorname{Re} F_\varrho(z,\zeta) > \frac{\varepsilon^2\beta}{3} \quad \text{and} \quad \operatorname{Re} F_\varrho(z,\zeta) - 2\varrho(\zeta) > \frac{\varepsilon^2\beta}{3}$$
$$\text{for} \quad \zeta \in \bar{W}, z \in \bar{W} \cup D_\varrho \quad \text{with} \quad \varepsilon \leqq \operatorname{dist}(z,\zeta) \leqq 2\varepsilon\,. \tag{4.8.20}$$

To do this we fix some points $\zeta \in \overline{W}$ and $z \in \overline{W} \cup D_\varrho$ with $\varepsilon \leqq \operatorname{dist}(z, \zeta) \leqq 2\varepsilon$. Then, by (4.8.8), $z \in \theta$ and it follows from (4.8.15) that

$$\operatorname{Re} F_\varrho(z, \zeta) \geqq \varrho(\zeta) - \varrho(z) + \beta\varepsilon^2 .$$

Together with $(4.8.7)_\varrho$ this implies

$$\operatorname{Re} F_\varrho(z, \zeta) > -\varrho(z) + \tfrac{2}{3}\beta\varepsilon^2 \tag{4.8.21}$$

and

$$\operatorname{Re} F_\varrho(z, \zeta) - 2\varrho(\zeta) > -\varrho(z) + \tfrac{2}{3}\beta\varepsilon^2 . \tag{4.8.22}$$

Further, since $z \in \theta$ and $z \in \overline{W} \cup D_\varrho$, at least one of the following relations holds: $z \in \overline{W}$ or $z \in \theta \cap D_\varrho$. If $z \in \overline{W}$, then it follows from $(4.8.7)_\varrho$ that $\varrho(z) < \varepsilon^2\beta/3$. If $z \in \theta \cap D_\varrho$, then it follows from (4.8.13) and (4.8.11) that $\varrho(z) < 0$. In both cases $-\varrho(z) > -\varepsilon^2\beta/3$. Together with (4.8.21) and (4.8.22) this implies (4.8.20).

In view of (4.8.20), for $\zeta \in \overline{W}$ and $z \in \overline{W} \cup D_\varrho$ with $\varepsilon \leqq \operatorname{dist}(z, \zeta) \leqq 2\varepsilon$, we can define $\ln F_\varrho(z, \zeta)$ and $\ln \big(F_\varrho(z, \zeta) - 2\varrho(\zeta)\big)$ for all $\varrho \in \mathfrak{U}_\delta(\varrho_0)$. Choose a C^∞-function χ on $X \times X$ such that $\chi(z, \zeta) = 1$ if $\operatorname{dist}(z, \zeta) \leqq \varepsilon + \varepsilon/4$ and $\chi(z, \zeta) = 0$ if $\operatorname{dist}(z, \zeta) \geqq 2\varepsilon - \varepsilon/4$. For $\zeta \in W$ and $z \in W \cup D_\varrho$ we define

$$f_\varrho(z, \zeta) := \begin{cases} \bar\partial_z[\chi(z, \zeta) \ln F_\varrho(z, \zeta)] & \text{if } \ \varepsilon \leqq \operatorname{dist}(z, \zeta) \leqq 2\varepsilon , \\ 0 & \text{otherwise} , \end{cases}$$

and

$$\tilde f_\varrho(z, \zeta) := \begin{cases} \bar\partial_z[\chi(z, \zeta) \ln \big(F_\varrho(z, \zeta) - 2\varrho(\zeta)\big)] & \text{if } \ \varepsilon \leqq \operatorname{dist}(z, \zeta) \leqq 2\varepsilon , \\ 0 & \text{otherwise} . \end{cases}$$

We denote by $Z^\infty_{(0,1)}(W \cup D_0)$ the Fréchet space of $\bar\partial$-closed $C^\infty_{(0,1)}$-forms on $W \cup D_0$ endowed with the topology of uniform convergence of all derivatives on compact sets. Since $W \cup D_0 = W \cup D_\varrho$ (cp. (4.8.13)), and since $F_\varrho(z, \zeta)$ is holomorphic in $z \in W \cup D_0$, for every $\zeta \in W$ and every $\varrho \in \mathfrak{U}_\delta(\varrho_0)$ the forms $f_\varrho(\cdot, \zeta)$ and $\tilde f_\varrho(\cdot, \zeta)$ belong to $Z^\infty_{(0,1)}(W \cup D_0)$. Since $U \subset\subset W$ and $U \cup D_0$ is strictly pseudoconvex, in the same way as in the case of Lemma 2.4.1 (using Theorem 2.12.3 (iv) instead of Lemma 2.3.4), we can prove that there exists a continuous linear operator $T\colon Z^\infty_{(0,1)}(W \cup D_0) \to C^\infty(U \cup D_0)$ such that $\bar\partial T g = g$ on $U \cup D_0$ for all $g \in Z^\infty_{(0,1)}(W \cup D_0)$. For all $\varrho \in \mathfrak{U}_\delta(\varrho_0)$ and all $\zeta \in U$, $z \in U \cup D$, we define

$$u_\varrho(z, \zeta) := \big(Tf_\varrho(\cdot, \zeta)\big)(z) , \qquad b(\varrho)(z, \zeta) := \exp[-u_\varrho(z, \zeta)] ,$$

$$\tilde u_\varrho(z, \zeta) := \big(T\tilde f_\varrho(\cdot, \zeta)\big)(z) , \qquad \tilde b(\varrho)(z, \zeta) := \exp[-\tilde u_\varrho(z, \zeta)] ,$$

$$a(\varrho)(z, \zeta) := \begin{cases} F_\varrho(z, \zeta)\, b(\varrho)(z, \zeta) & \text{if } \ \operatorname{dist}(z, \zeta) \leqq \varepsilon . \\ \exp[\chi(z, \zeta) \ln F_\varrho(z, \zeta) - u_\varrho(z, \zeta)] & \text{if } \ \operatorname{dist}(z, \zeta) \geqq \varepsilon , \end{cases}$$

$$\tilde a(\varrho)(z, \zeta) := \begin{cases} \big(F_\varrho(z, \zeta) - 2\varrho(\zeta)\big)\, \tilde b(\varrho)(z, \zeta) & \text{if } \ \operatorname{dist}(z, \zeta) \leqq \varepsilon . \\ \exp[\chi(z, \zeta) \ln \big(F_\varrho(z, \zeta) - 2\varrho(\zeta)\big) - \tilde u(z, \zeta)] & \text{if } \ \operatorname{dist}(z, \zeta) \geqq \varepsilon . \end{cases}$$

It is easy to show that the maps $a, \tilde a, b, \tilde b$ fulfil conditions (i)—(iii). ■

4.9. Formulas for real non-degenerate strictly pseudoconvex polyhedra

In this section we introduce the concept of a strictly pseudoconvex polyhedron in a Stein manifold. Special cases of such polyhedra are analytic polyhedra (cp. Section 4.7) as well as intersections of finitely many strictly pseudoconvex open sets.

By means of the support function Φ from the preceding section, for such polyhedra, we construct a Leray-Norguet section, which is holomorphic in z. By Corollary 4.6.4, this gives a formula for solving the $\bar{\partial}$-equation in real non-degenerate strictly pseudoconvex polyhedra.

4.9.1. Definition. Let X be a Stein manifold of complex dimension n. An open set $D \subset\subset X$ is called a *strictly pseudoconvex polyhedron* if there are a neighbourhood $U_{\bar{D}}$ of $\bar{D}$, finitely many Stein manifolds $X_1, \dots, X_N$ of complex dimension $\leqq n$, holomorphic maps $F_k: U_{\bar{D}} \to X_k$, $k = 1, \dots, N$, as well as strictly pseudoconvex open sets $D_k \subset\subset X_k$, $k = 1, \dots, N$, such that

$$D = F_1^{-1}(D_1) \cap \dots \cap F_N^{-1}(D_N) .$$

If $\varrho_1, \dots, \varrho_N$ are strictly plurisubharmonic C^2-functions in some neighbourhoods $\theta_1, \dots, \theta_N$ of $\partial D_1, \dots, \partial D_N$, respectively, such that

$$D_k \cap \theta_k = \{z \in \theta_k : \varrho_k(z) < 0\} , \qquad k = 1, \dots, N ,$$

then $\partial D \subseteqq F_1^{-1}(\theta_1) \cup \dots \cup F_N^{-1}(\theta_N)$ and a point $z \in F_1^{-1}(\theta_1) \cup \dots \cup F_N^{-1}(\theta_N)$ belongs to D if and only if, for every $1 \leqq k \leqq N$ with $z \in F_k^{-1}(\theta_k)$, $\varrho_k(F_k(z)) < 0$. D is called *real non-degenerate* if these functions F_k and ϱ_k can be chosen so that the following condition is fulfilled: For every collection of indices $1 \leqq k_1 < \dots < k_l \leqq N$, we have

$$d(\varrho_{k_1} \circ F_{k_1})(z) \wedge \dots \wedge d(\varrho_{k_l} \circ F_{k_l})(z) \neq 0$$

for all $z \in \partial D$ with $\varrho_{k_1}(F_{k_1}(z)) = \dots = \varrho_{k_l}(F_{k_l}(z)) = 0$.

Remark. The boundary of a real non-degenerate strictly pseudoconvex polyhedron is piecewise C^1 in the sense defined at the beginning of Subsection 4.3.1.

4.9.2. Proposition. *Let X be a Stein manifold. Then every strictly pseudoconvex polyhedron $D \subset\subset X$ is a Stein manifold.*

Proof. Let X_k, F_k, D_k be as in Definition 4.9.1. By Proposition 2.13.8, every D_k is a Stein manifold.

In the same way as the equivalence of conditions (iii) and (v) in Theorem 1.3.7 was proved, the following proposition can be proved: An open subset W of a Stein manifold is a Stein manifold if and only if, for every infinite set $Y \subseteqq W$ which is discrete in W, there exists $f \in \mathcal{O}(W)$ which is unbounded on Y. By means of this proposition, we can repeat the proof of Theorem 1.3.11 (ii) and obtain that the sets $F_1^{-1}(D_1), \dots, F_N^{-1}(D_N)$ are Stein manifolds. This implies that D is a Stein manifold. ■

Remark. Every analytic polyhedron in a Stein manifold is a strictly pseudoconvex polyhedron. Then, in Definition 4.9.1, we have to set $X_1 = \dots = X_N = \mathbb{C}^1$ and $\varrho_1(z) = \dots = \varrho_N(z) = |z|^2 - 1$. Further, if $D_1, \dots, D_N \subset\subset X$ are strictly pseudoconvex open sets in a Stein manifold X, then $D_1 \cap \dots \cap D_N$ is a strictly pseudoconvex polyhedron. Then in Definition 4.9.1 we have to set $X_1 = \dots = X_N = X$ and $F_1(z) = \dots = F_N(z) = z$, $z \in X$.

In the construction of a Leray-Norguet section for strictly pseudoconvex polyhedra, we use the following Lemma, which shows that in Lemma 4.7.2, after shrinking X, the section h^* can be given by a continuous linear operator (cp. Theorem 2.5.4, which is used in the proof of Theorem 2.5.5 for the corresponding construction in the case of strictly pseudoconvex open sets in $\mathbb{C}^n$).

4.9.3. Lemma. *Let X be a Stein manifold, and let $s(z, \zeta)$, $\varphi(z, \zeta)$ be as in Lemma 4.2.4. (Suppose that X is a relatively compact open subset of some larger Stein manifold —*

cp. also the remark following Lemma 4.2.4.) *Let $Y \subset\subset X$ be an open set. We denote by $\Gamma^*(X \times X)$ $[\Gamma^*(Y \times Y)]$ the Fréchet space of all holomorphic maps*

$$h^*: X \times X \to T^*(X) \quad [h^*: Y \times Y \to T^*(X)]$$

(cp. Subsection 4.2.3) *such that $h^*(z, \zeta) \in T_z^*(X)$ for all $z, \zeta \in X$ $[z, \zeta \in Y]$ endowed with the topology of uniform convergence on compact sets. Then there exists a continuous linear map $H: \mathcal{O}(X) \to \Gamma^*(Y \times Y)$ (here the topology of $\mathcal{O}(X)$ is also that of uniform convergence on compact sets), such that, for every $F \in \mathcal{O}(X)$,*

$$\varphi(z, \zeta) \left(F(\zeta) - F(z)\right) = \langle (HF)(z, \zeta), s(z, \zeta)\rangle \quad \text{for all} \quad z, \zeta \in Y .$$

Proof. Let $\mathcal{F}_s$ be the subsheaf of ${}_{X \times X}\mathcal{O}^1$ generated by s (cp. Example 4.1.7). Then by Theorem 4.1.4, $\mathcal{F}_s(X \times X)$ and $\mathcal{F}_s(Y \times Y)$ are Fréchet spaces endowed with the topology of uniform convergence on compact sets. Consider the commutative diagram

$$\begin{array}{ccccc} & & \Gamma^*(X \times X) & \xrightarrow{r_1} & \Gamma^*(Y \times Y) \\ & & \downarrow{\scriptstyle \varphi\langle\cdot, s\rangle} & & \downarrow{\scriptstyle \varphi\langle\cdot, s\rangle} \\ \mathcal{O}(X) & \xrightarrow{\vartheta} & \mathcal{F}_s(X \times X) & \xrightarrow{r_2} & \mathcal{F}_s(Y \times Y) \end{array}$$

where r_1, r_2 are the restriction maps and $\vartheta(F)(z, \zeta) := \varphi(z, \zeta)\left(F(\zeta) - F(z)\right)$, $z, \zeta \in X$. By Lemma 4.7.2, $\vartheta(\mathcal{O}(X)) \subseteq \varphi\langle \Gamma^*(X \times X), s\rangle$. Therefore, by Propositions 1 and 2, Appendix 2, the proof will be complete if we show that there are a Hilbert space $\mathfrak{H}$ and continuous linear maps $A': \Gamma^*(X \times X) \to \mathfrak{H}$, $A'': \mathfrak{H} \to \Gamma^*(Y \times Y)$ such that $r_1 = A''A'$.

To do this we introduce a Hermitean metric in the fibres of $T^*(X)$ as follows: Choose a locally finite open covering $\{W_j\}$ of X and a C^∞-partition of unity $\{\chi_j\}$ subordinate to $\{W_j\}$ such that, for every j, there is a C^∞-bundle isomorphism $G_j: T^*(W_j) \to W_j \times \mathbb{C}^n$ ($T^*(W_j)$ is the restriction of $T^*(X)$ to W_j). Let $g_j: T^*(W_j) \to \mathbb{C}^n$ be the corresponding C^∞-maps for which $G_j(a) = (z, g_j(a))$, $a \in T_z^*(W_j)$, $z \in W_j$. For all $z \in X$ and $a, b \in T_z^*(X)$, we define

$$(a, b) := \sum_j \chi_j(z) \left(g_j(a), g_j(b)\right)_{\mathbb{C}^n} ,$$

where $(v, w)_{\mathbb{C}^n} := v_1\overline{w}_1 + \ldots + v_n\overline{w}_n$ for $v, w \in \mathbb{C}^n$.

By means of this Hermitean metric, we define the required Hilbert space $\mathfrak{H}$ to be the space of all $h^* \in \Gamma^*(Y \times Y)$ such that

$$\int\limits_{(z,\zeta) \in Y \times Y} \left(h^*(z, \zeta), h^*(z, \zeta)\right) d\sigma_{4n} < \infty ,$$

where $d\sigma_{4n}$ is a volume form on $X \times X$, that is a real $4n$-form which defines the orientation of $X \times X$. Then the linear maps A' and A'' can be chosen to be the natural maps. ■

4.9.4. Corollary. *Let the hypotheses of Lemma 4.9.3 be fulfilled, and let $V \subseteq Y$ be an open set. We denote by $C^1(X \times V)$ the Fréchet space of all C^1-functions on $X \times V$ endowed with the topology of uniform convergence on compact sets. Let $\mathfrak{N}^1(X \times V)$ be the subspace of all $\psi(z, \zeta) \in C^1(X \times V)$ that are holomorphic in $z \in X$. Further, we denote by $\mathfrak{N}_*^1(Y \times V)$ the Fréchet space (endowed with the same topology) of all C^1-maps $h^*: Y \times V \to T^*(X)$ such that $h^*(z, \zeta) \in T_z^*(X)$ for all $(z, \zeta) \in Y \times V$ and such that, for every fixed $\zeta \in V$, $h^*(z, \zeta)$ is holomorphic in $z \in Y$. Then there exists a continuous linear map $\widetilde{H}: \mathfrak{N}^1(X \times V) \to \mathfrak{N}_*^1(Y \times V)$ such that, for every $\psi \in \mathfrak{N}^1(X \times V)$ with*

$\psi(z, z) = 0$ *for all* $z \in V$, *we have*

$$\varphi(z, \zeta)\, \psi(z, \zeta) = \langle (\tilde{H}\psi)(z, \zeta), s(z, \zeta) \rangle \quad \text{for all} \quad z \in Y,\ \zeta \in V.$$

Proof. Set $(\tilde{H}\psi)(z, \zeta) := -\big(H(\psi(\cdot, \zeta))\big)(z, \zeta)$, where H is the map from Lemma 4.9.3. ■

4.9.5. A Koppelman-Leray-Norguet formula. Let X be a Stein manifold, and let $D \subset\subset X$ be a real non-degenerate strictly pseudoconvex polyhedron. Let $U_{\bar{D}}$, X_k, F_k, D_k, θ_k, ϱ_k be as in Definition 4.9.1. By Proposition 4.9.2, for sufficiently small $\varepsilon > 0$, the set $D_\varepsilon := F_1^{-1}(D_1^\varepsilon) \cap \dots \cap F_N^{-1}(D_N^\varepsilon)$, where $D_k^\varepsilon := D_k \cup \{z \in \theta_k : \varrho_k(z) < \varepsilon\}$, is a Stein manifold. Therefore, after shrinking $U_{\bar{D}}$, we can assume that $U_{\bar{D}}$ is a Stein manifold. Moreover, without loss of generality, we can assume that X is a relatively compact open subset of some larger Stein manifold and apply Lemma 4.2.4 (cp. also the remark following Lemma 4.2.4). Let $s(z, \zeta)$, $\varphi(z, \zeta)$, $\varkappa$ be as in Lemma 4.2.4. By Theorem 4.8.3 and Lemma 4.8.2, after shrinking θ_k, we can find numbers $\varepsilon, \alpha > 0$ and C^1-functions $\Phi_k(z, \zeta)$ defined for $z \in D_k \cup \theta_k$ and $\zeta \in \theta_k$ such that the following conditions are fulfilled:

(i) $\Phi_k(z, \zeta)$ is holomorphic in $z \in D_k \cup \theta_k$.

(ii) $\Phi_k(z, \zeta) \neq 0$ for $z \in D_k \cup \theta_k, \zeta \in \theta_k$ with $\operatorname{dist}(z, \zeta) \geqq \varepsilon$. (4.9.1)

(iii) $|\Phi_k(z, \zeta)| \geqq \alpha\big(\varrho_k(\zeta) - \varrho_k(z) + [\operatorname{dist}(z, \zeta)]^2\big)$ for $z \in D_k \cup \theta_k, \zeta \in \theta_k$ with $\operatorname{dist}(z, \zeta) \leqq 2\varepsilon$. (4.9.2)

(iv) $\Phi_k(z, z) = 0$ for all $z \in \theta_k$. (4.9.3)

Set $\Psi_k(z, \zeta) := \Phi_k\big(F_k(z), F_k(\zeta)\big)$ for $z \in F_k^{-1}(D_k \cup \theta_k)$, $\zeta \in F_k^{-1}(\theta_k)$. Then, by Corollary 4.9.4, after shrinking θ_k, we can find $T^*(X)$-valued C^1-maps $h_k^*(z, \zeta)$ defined for $z \in F_k^{-1}(D_k \cup \theta_k)$ and $\zeta \in F_k^{-1}(\theta_k)$ such that the following conditions are fulfilled:

(1) $h_k^*(z, \zeta) \in T_z^*(X)$ for $z \in F_k^{-1}(D_k \cup \theta_k), \zeta \in F_k^{-1}(\theta_k)$.

(2) $h_k^*(z, \zeta)$ is holomorphic in $z \in F_k^{-1}(D_k \cup \theta_k)$.

(3) $\varphi(z, \zeta)\Psi_k(z, \zeta) = \langle h_k^*(z, \zeta), s(z, \zeta)\rangle$ for $z \in F_k^{-1}(D_k \cup \theta_k), \zeta \in F_k^{-1}(\theta_k)$. (4.9.4)

Set $S_k := \{z \in \partial D : \varrho_k(F_k(z)) = 0\}$, $k = 1, \dots, N$, and define S_K and Δ_K as in Subsection 4.4.1. Since, by (4.9.4),

$$\frac{\varphi(z, \zeta)\, h_k^*(z, \zeta)}{\langle h_k^*(z, \zeta), s(z, \zeta)\rangle} = \frac{h_k^*(z, \zeta)}{\Phi_k\big(F_k(z), F_k(\zeta)\big)} \quad \text{for} \quad z \in D \quad \text{and} \quad \zeta \in S_k,$$

where, by (4.9.1) and (4.9.2), $\Phi_k(F_k(z), F_k(\zeta)) \neq 0$, we see that $(h_1^*, \dots, h_N^*, 1)$ is a Leray-Norguet section for D. Therefore, from Theorem 4.4.3 and Corollary 4.6.4 we obtain the following

Theorem. *Let* $\nu \geqq 2n\varkappa$ *be an integer, let* $L_{S_K} = L_{S_K}(\varphi^\nu, h_1^*, \dots, h_N^*, s)$, $R_{S_K} = R_{S_K}(\varphi^\nu, h_1^*, \dots, h_N^*, \bar{s}, s)$ (cp. Subsections 4.4.2 and 4.6.1) *and* $B_D = B_D(\varphi^\nu, \bar{s}, s)$ (cp. Subsections 4.3.2 and 4.5.1). *Then*

(i) *For every continuous function f on* $\bar{D}$ *such that* $\bar{\partial}f$ *is also continuous on* $\bar{D}$, *we have*

$$f = \sum_{|K| \leqq n} L_{S_K} f - \sum_{|K| \leqq n-1} R_{S_K}\, \bar{\partial} f - B_D\, \bar{\partial} f \quad in \quad D. \tag{4.9.5}$$

(ii) *Let* $1 \leqq q \leqq n$. *Then for every continuous* $(0, q)$*-form f on* $\bar{D}$ *such that* $\bar{\partial}f$ *is also continuous on* $\bar{D}$, *we have*

$$f = \bar{\partial}(-1)^q \Big(B_D + \sum_{|K| \leqq n-q} R_{S_K}\Big) f + (-1)^{q+1} \Big(B_D + \sum_{|K| \leqq n-q-1} R_{S_K}\Big)\, \bar{\partial} f \quad in \quad D. \tag{4.9.6}$$

In particular, for every continuous $(0, q)$*-form* f *on* $\overline{D}$ *such that* $\bar\partial f = 0$ *in* D,

$$u := (-1)^q \left(B_D + \sum_{|K| \leqq n-q} R_{S_K}\right) f \tag{4.9.7}$$

is a continuous solution of $\bar\partial u = f$ *in* D.

Remark. It follows from Theorem 4.8.3 and Corollary 4.9.4 that the formulas (4.9.5)—(4.9.7) depend continuously on D in the following sense: If the functions ϱ_k and F_k from Definition 4.9.1 converge uniformly, together with the first and second-order derivatives, then, for every fixed $z \in D$ (or z in some fixed compact subset of D), the kernels of the corresponding integrals in (4.9.5)—(4.9.7) converge uniformly.

4.10. Formulas for strictly pseudoconvex open sets with non-smooth boundary

In this section we generalize the Koppelman-Leray formula constructed in Section 3.1 (Theorem 3.1.3) to strictly pseudoconvex open sets with not necessarily smooth boundary in Stein manifolds.

4.10.1. Notation and preliminaries. Let X be a Stein manifold, let $D \subset\subset X$ be a strictly pseudoconvex open set, and let ϱ be a strictly plurisubharmonic C^2-function in some neighbourhood θ of ∂D such that $D \cap \theta = \{z \in \theta : \varrho(z) < 0\}$. We set $N(\varrho) := \{z \in \overline{\theta} : \varrho(z) = 0\}$ and suppose that $N(\varrho) \subset\subset \theta$. (We do not assume that $d\varrho(z) \neq 0$ for $z \in \partial D$.) Since $D \subset\subset X$, without loss of generality, we can assume that X is a relatively compact open subset of some larger Stein manifold and apply Lemma 4.2.4 (cp. also the remark following Lemma 4.2.4). Let $s(z, \zeta)$, $\varphi(z, \zeta)$, $\varkappa$ be as in Lemma 4.2.4, and let $\bar{s}(z, \zeta) = \sigma s(z, \zeta)$ be as in Subsection 4.3.1. By Theorem 4.8.3 and Lemma 4.8.2, after shrinking θ, we can find numbers $\varepsilon, \alpha > 0$ as well as C^1-functions $\Phi(z, \zeta)$ and $\widetilde{\Phi}(z, \zeta)$ defined for $z \in D \cup \theta$, $\zeta \in \theta$ such that the following conditions are fulfilled:

(i) $\Phi(z, \zeta)$ and $\widetilde{\Phi}(z, \zeta)$ are holomorphic in $z \in D \cup \theta$.

(ii) $\Phi(z, \zeta) \neq 0$ and $\widetilde{\Phi}(z, \zeta) \neq 0$ for $z \in D \cup \theta, \zeta \in \theta$ with $\operatorname{dist}(z, \zeta) \geqq \varepsilon$; (4.10.1)

$$|\Phi(z, \zeta)| \geqq \alpha\big(\varrho(\zeta) - \varrho(z) + [\operatorname{dist}(z, \zeta)]^2\big)$$
for $z \in D \cup \theta, \zeta \in \theta$ with $\operatorname{dist}(z, \zeta) \leqq \varepsilon$; (4.10.2)

$$|\widetilde{\Phi}(z, \zeta)| \geqq \alpha(-\varrho(\zeta) - \varrho(z) + [\operatorname{dist}(z, \zeta)]^2)$$
for $z \in D \cup \theta, \zeta \in \theta$ with $\operatorname{dist}(z, \zeta) \leqq \varepsilon$; (4.10.3)

$$\Phi(z, z) = 0 \quad \text{for all} \quad z \in \theta\,. \tag{4.10.4}$$

(iii) $\widetilde{\Phi}(z, \zeta) = \Phi(z, \zeta)$ for $\zeta \in N(\varrho), z \in D \cup \theta$. (4.10.5)

By Corollary 4.9.4, after shrinking θ, we can find a $T^*(X)$-valued C^1-map $s^*(z, \zeta)$ defined for $z \in D \cup \theta, \zeta \in \theta$ such that the following conditions are fulfilled:

(1) $s^*(z, \zeta) \in T_z^*(X)$ for $z \in D \cup \theta, \zeta \in \theta$.

(2) $s^*(z, \zeta)$ is holomorphic in $z \in D \cup \theta$.

(3) $\varphi(z, \zeta)\, \Phi(z, \zeta) = \langle s^*(z, \zeta), s(z, \zeta)\rangle$ for $z \in D \cup \theta, \zeta \in \theta$. (4.10.6)

4.10.2. Remark. If $z \in D$ and $\zeta \in \partial D$, then, by (4.10.6),

$$\frac{\varphi(z,\zeta)\, s^*(z,\zeta)}{\langle s^*(z,\zeta), s(z,\zeta)\rangle} = \frac{s^*(z,\zeta)}{\Phi(z,\zeta)},$$

where, by (4.10.1) and (4.10.2), $\Phi(z,\zeta) \neq 0$. Therefore, $(s^*, 1)$ is a Leray section for (D, s, φ). In the case of smooth ∂D this gives the corresponding formulas stated in Corollary 4.5.5 (with $\nu \geqq \max(2n\varkappa, n)$, $L_{\partial D} = L_{\partial D}(\varphi^\nu, s^*, s)$, $R_{\partial D} = R_{\partial D}(\varphi^\nu, s^*, \bar{s}, s)$ and $B_D = B_D(\varphi^\nu, \bar{s}, s)$).

4.10.3. The operators L_D and R_D. Choose $\chi \in C_0^\infty(\theta)$ such that $\chi = 1$ in some neighbourhood of $N(\varrho)$. It follows from (4.10.1) and (4.10.3) that, for every $z \in D$, there is a neighbourhood $V_z \subseteq \theta$ of $N(\varrho)$ such that $\tilde{\Phi}(z,\zeta) \neq 0$ for $\zeta \in (D \cap \theta) \cup V_z$. Since $\operatorname{supp} \chi \subset\subset \theta$, this implies that, for every fixed $z \in D$, the map $\chi(\zeta)\, s^*(z,\zeta)/\tilde{\Phi}(z,\zeta)$ is C^1 for $\zeta \in D \cup V_z$.

Consequently, for every fixed $z \in D$, the differential form

$$\bar{\omega}_\zeta\left(\frac{\chi(\zeta)\, s^*(z,\zeta)}{\tilde{\Phi}(z,\zeta)}\right) \wedge \omega_\zeta(s(z,\zeta)) \tag{4.10.7}$$

is continuous for $\zeta \in \bar{D}$, where this form is defined as follows: Let $\{(U_j, h_j)\}$ be the holomorphic atlas from Subsection 4.3.1, let $v_j = (v_{j1}, \dots, v_{jn})$ be the expression of $\chi s^*/\tilde{\Phi}$ with respect to (U_j, h_j), and let $s_j = (s_{j1}, \dots, s_{jn})$ be the expression of s with respect to (U_j, h_j). Then we define

$$\bar{\omega}_\zeta\left(\frac{\chi(\zeta)\, s^*(z,\zeta)}{\tilde{\Phi}(z,\zeta)}\right) \wedge \omega_\zeta(s(z,\zeta)) := \bigwedge_{k=1}^{n} \bar{\partial}_\zeta v_{jk}(z,\zeta) \wedge \bigwedge_{k=1}^{n} d_\zeta s_{jk}(z,\zeta)$$

$$\text{for} \quad z \in U_j \cap D, \zeta \in \bar{D}. \tag{4.10.8}$$

By (4.3.1), this definition is independent of j. Therefore, for every integer $\nu \geqq 0$ and every measurable bounded function f on D, we can define

$$L_D f(z) := \frac{n!}{(2\pi i)^n} \int_{\zeta \in D} f(\zeta)\, \varphi^\nu(z,\zeta)\, \bar{\omega}_\zeta\left(\frac{\chi(\zeta)\, s^*(z,\zeta)}{\tilde{\Phi}(z,\zeta)}\right) \wedge \omega_\zeta(s(z,\zeta)), \quad z \in D. \tag{4.10.9}$$

Since $s^*(z,\zeta)$ and $\tilde{\Phi}(z,\zeta)$ are holomorphic in z, then $L_D f$ is a holomorphic function in D.

Set

$$\tilde{t}^*(z,\zeta,\lambda) := (1-\lambda)\frac{\chi(\zeta)\, s^*(z,\zeta)}{\tilde{\Phi}(z,\zeta)} + \lambda \frac{\bar{s}(z,\zeta)}{\|s(z,\zeta)\|_\sigma^2}. \tag{4.10.10}$$

Then, for $0 \leqq \lambda \leqq 1$, $z \in \bar{D}$ and ζ in some neighbourhood of $\bar{D} \setminus z$, the map $\varphi^\varkappa(z,\zeta) \times \tilde{t}^*(z,\zeta,\lambda)$ has continuous derivatives of order 1 in ζ and of order $\leqq 2$ in z (Lemma 4.2.4 (v)). Therefore, the differential form

$$\varphi^{n\varkappa}(z,\zeta)\, \bar{\omega}_{z,\zeta,\lambda}(\tilde{t}^*(z,\zeta,\lambda)) \wedge \omega_\zeta(s(z,\zeta)) \tag{4.10.11}$$

is continuous for $z \in \bar{D}, \zeta \in \bar{D} \setminus z, 0 \leqq \lambda \leqq 1$ (and C^1 with respect to z), where this form is defined as follows: If $\{(U_j, h_j)\}$ is the holomorphic atlas from Subsection 4.3.1, $\tilde{t}_j^* = (\tilde{t}_{j1}^*, \dots, \tilde{t}_{jn}^*)$ is the expression of $\tilde{t}^*$ and $s_j = (s_{j1}, \dots, s_{jn})$ is the expression of s with respect to (U_j, h_j), then, for $z \in U_j \cap \bar{D}, \zeta \in \bar{D} \setminus z$, $0 \leqq \lambda \leqq 1$, we define

$$\bar{\omega}_{z,\zeta,\lambda}(\tilde{t}^*(z,\zeta,\lambda)) \wedge \omega_\zeta(s(z,\zeta)) := \bigwedge_{k=1}^{n} (\bar{\partial}_{z,\zeta} + d_\lambda)\, \tilde{t}_{jk}^*(z,\zeta,\lambda) \wedge \bigwedge_{k=1}^{n} d_\zeta s_{jk}(z,\zeta). \tag{4.10.12}$$

By (4.3.1) this definition is independent of j. If $z \in D$, then the monomials in the form (4.10.11) which are of degree 1 in λ have a singularity of order $\leqq 2n - 1$ at $\zeta = z$. Therefore, for every integer $\nu \geqq n\varkappa$ and every bounded differential form f on D, we can define

$$R_D f(z) := \frac{n!}{(2\pi i)^n} \int\limits_{(\zeta,\lambda)\in D\times[0,1]} f(\zeta) \wedge \varphi^\nu(z,\zeta)\, \overline{\omega}_{z,\zeta,\lambda}(\tilde{t}^*(z,\zeta,\lambda)) \wedge \omega_\zeta(s(z,\zeta))\,, \quad z \in D\,. \tag{4.10.13}$$

Then it is easy to see that $R_D f$ is continuous in D and $R_D f$ is a $(0, q-1)$-form if f is a $(0, q)$ form ($R_D f = 0$ if $q = 0$).

4.10.4. Theorem (Koppelman-Leray formula). *We use the notations from Subsections* 4.10.1 *and* 4.10.3, *and we suppose that in Definitions* (4.10.9) *and* (4.10.13) *an integer* $\nu \geqq \max(2n\varkappa, n)$ *is used (the same* ν *for both definitions). Then*

(i) *For every continuous bounded function* f *on* D *such that* $\bar\partial f$ *is also continuous and bounded on* D, *we have*

$$f = L_D f + R_D \bar\partial f \quad \text{in} \quad D\,. \tag{4.10.14}$$

(ii) *Let* f *be a continuous bounded* $(0, q)$*-form on* D *such that* $\bar\partial f$ *is also continuous and bounded on* D, $1 \leqq q \leqq n$. *Then*

$$f = \bar\partial R_D f + R_D \bar\partial f \quad \text{in} \quad D\,. \tag{4.10.15}$$

In particular, if $\bar\partial f = 0$ *in* D, *then*

$$u := R_D f \tag{4.10.16}$$

is a continuous solution of $\bar\partial u = f$ *in* D.

Proof. Repetition of that of Theorem 3.1.3. ■

4.10.5. The Hörmander diameter. We use the notations of Subsection 4.10.1. For every $\xi \in \partial D$ and $\delta > 0$ we define

$$H_\xi^\Phi(\delta) := \{z \in \overline{D} : |\Phi(z,\xi)| < \delta^2\}. \tag{4.10.17}$$

$H_\xi^\Phi(\delta)$ is called the *Hörmander ball* of radius δ centered at ξ. If $W \subseteqq D$, then the number

$$\operatorname{diam}_H W := \inf\{\delta > 0 : \text{there exists } \xi \in \partial D \text{ such that } W \subseteqq H_\xi^\Phi(\delta)\}$$

will be called the *Hörmander diameter* of W.

Remark. For $X = \mathbb{C}^n$ this definition is equivalent to that of Subsection 3.2.1, that is, there are constants $c > 0$ and $C < \infty$ such that, for all $\xi \in \partial D$ and $\delta > 0$,

$$H_\xi^\Phi(c\delta) \subseteqq H_\xi(\delta) \subseteqq H_\xi^\Phi(C\delta)\,,$$

where $H_\xi(\delta)$ is as in Subsection 3.2.1. This follows from (3.2.2) and the fact that, for some $\alpha > 0$, $|\Phi(z,\zeta)| \geqq \alpha|\zeta - z|^2$ for all $z \in D$, $\zeta \in \partial D$.

4.10.6. Theorem. *We use the notations of Subsections* 4.10.1 *and* 4.10.3. *Let* $\nu \geqq n\varkappa$ *and* $1 \leqq q \leqq n$ *be integers. We denote by* $L^\infty_{(0,q)}(D)$ *the Banach space of bounded* $(0, q)$ *forms* f *on* D, *with the norm* $\|\cdot\|_{0,D} := \sup_{z\in D}\|f(z)\|$ (cp. Subsection 2.11.2).

(i) *For every bounded* $(0, q)$*-form* f *on* D *and each point* $z \in \overline{D}$, *the integral*

$$R_D f(z) := \frac{n!}{(2\pi i)^n} \int\limits_{(\zeta,\lambda)\in D\times[0,1]} f(\zeta) \wedge \varphi^\nu(z,\zeta)\, \overline{\omega}_{z,\zeta,\lambda}(\tilde{t}^*(z,\zeta,\lambda)) \wedge \omega_\zeta(s(z,\zeta)) \tag{4.10.18}$$

converges, and the so defined $(0, q-1)$*-form* $R_D f$ *is continuous on* $\overline{D}$.

(ii) *There is a constant $C < \infty$ such that, for every $f \in L^{\infty}_{(0,q)}(D)$,*

$$\|R_D f\|_{0,D} \leqq C \operatorname{diam}_H (\operatorname{supp} f) \|f\|_{0,D}, \tag{4.10.19}$$

and, moreover, if $\xi \in \partial D$, $\delta > 0$ such that $\operatorname{supp} f \subseteq H^{\Phi}_{\xi}(\delta)$, *then for all $z \in D$*

$$\|R_D f(z)\| \leqq C \left[\frac{\delta}{\operatorname{dist}(z, H^{\Phi}_{\xi}(\delta))}\right]^{2n-1} \delta \|f\|_{0,D}. \tag{4.10.20}$$

(iii) *By parts* (i) *and* (ii) *the integral R_D defines a bounded linear operator from $L^{\infty}_{(0,q)}(D)$ into $C^0_{(0,q-1)}(\overline{D})$* (cp. Subsection **2.11.2**). *This operator is compact.*

Proof. Repetition of that of Theorem **3.2.2.** ■

4.10.7. Corollary. *Let X be a Stein manifold and let $D \subset\subset X$ be a strictly pseudoconvex open set (with not necessarily smooth boundary). Further, let $U_1, \ldots, U_N \subseteq X$ be open sets such that $\overline{D} \subseteq U_1 \cup \ldots \cup U_N$. We denote by $H^{\infty}(D)$ the Banach space of bounded holomorphic functions in D endowed with the* sup-*norm. Then there exist bounded linear operators $L_j: H^{\infty}(D) \to H^{\infty}(D)$, $j = 1, \ldots, N$, with the following properties:*

(i) *For every $f \in H^{\infty}(D)$, $f = \sum_{j=1}^{N} L_j f$.*

(ii) *For every $f \in H^{\infty}(D)$ and $j = 1, \ldots, N$, the function $L_j f$ is bounded and holomorphic in some neighbourhood of $\overline{D} \setminus (\partial D \cap U_j)$.*

(iii) *If $f \in H^{\infty}(D)$ admits a continuous continuation to $\overline{D}$, then the functions $L_1 f, \ldots, L_N f$ admit continuous continuations to $\overline{D}$ also.*

Proof. Choose $\chi_j \in C^{\infty}_0(U_j)$ such that $\sum \chi_j = 1$ on $\overline{D}$, and define $L_j f := L_D \chi_j f$, where L_D is as in Theorem **4.10.4** (see the proof of Proposition **3.4.1** for further details). ■

Remark. Corollary 4.10.7 can be proved also without use of global integral formulas on Stein manifolds by means of a construction given in the proof of Theorem 3 in Henkin [5].

4.11. Bounded extension of holomorphic functions from complex submanifolds

In this section we prove

4.11.1. Theorem. *Let X be a Stein manifold, let $D \subset\subset X$ be a strictly pseudoconvex open set (with not necessarily smooth boundary), and let Y be a closed complex submanifold of some neighbourhood of $\overline{D}$. Then*

(i) *For every bounded holomorphic function f on $Y \cap D$, there exists a bounded holomorphic function F on D such that $F = f$ on $Y \cap D$.*

(ii) *For every continuous function f on $Y \cap \overline{D}$ that is holomorphic in $Y \cap D$, there exists a continuous function F on $\overline{D}$ that is holomorphic in D such that $F = f$ on $Y \cap \overline{D}$.*

Proof. We prove parts (i) and (ii) simultaneously. Let f be a

$$\begin{cases} \text{bounded holomorphic function on } Y \cap D\,. \\ \text{continuous function on } Y \cap \overline{D} \text{ that is holomorphic in } Y \cap D\,. \end{cases}$$

For $\xi \in \partial D$ and $\delta > 0$, we set $E_{\xi}(\delta) := \{z \in X : \operatorname{dist}(z, \xi) < \delta\}$, where $\operatorname{dist}(\cdot, \cdot)$ is the metric introduced at the beginning of Subsection 4.8.1. Since $Y \cap \overline{D}$ is compact, we can choose $\delta > 0$ so small that for every $\xi \in Y \cap \bar{D}$ there is a biholomorphic map $h_\xi: E_\xi(\delta) \to \mathbb{C}^n$ such that $h_\xi(Y \cap E_\xi(\delta))$ is the intersection of $h_\xi(E_\xi(\delta))$ with a complex plane in $\mathbb{C}^n$. In view of Corollary 4.10.7, it is sufficient to prove Theorem 4.11.1 for the case when f has the following property: There are a point $\xi \in Y \cap \partial D$ and an open set $D_0 \subseteqq X$ such that

$$\bar{D} \setminus (\partial D \cap E_\xi(\delta/3)) \subseteq D_0 \tag{4.11.1}$$

and

$$\begin{cases} f \text{ is bounded and holomorphic in } Y \cap D_0 . \\ f \text{ is continuous on } Y \cap \bar{D}_0 \text{ and holomorphic } Y \cap D_0 . \end{cases}$$

By Corollary 1.5.20 (which can be easily proved also for strictly pseudoconvex open sets in Stein manifolds), we can assume that, moreover, D_0 is strictly pseudoconvex and Y is a closed complex submanifold in some neighbourhood of $\bar{D}_0$. Further, by Lemma 1.5.23 we can choose a strictly pseudoconvex open set $D_\xi \subseteqq X$ such that

$$E_\xi(\delta/3) \cap D_0 \subseteqq D_\xi \subseteqq E_\xi(\delta/2) . \tag{4.11.2}$$

It follows from (4.11.1) that $\bar{D} \subseteqq D_0 \cup E_\xi(\delta/3)$. Therefore, we can choose a strictly pseudoconvex open set D_1 such that $D \subset\subset D_1 \subset\subset D_0 \cup E_\xi(\delta/3)$. Then the sets $U_\xi := E_\xi(\delta/3) \cap D_1$ and $U_0 := D_0 \cap D_1$ form an open covering of D_1. Since δ is chosen so small that with respect to appropriate local holomorphic coordinates on $E_\xi(\delta)$ the manifold $Y \cap E_\xi(\delta)$ is a complex plane, and since $D_\xi \subset\subset E_\xi(\delta)$, it follows from Theorem 3.6.8 that there exists a

$$\begin{cases} \text{bounded holomorphic function } f_\xi \text{ on } D_\xi \\ \text{continuous function } f_\xi \text{ on } \bar{D}_\xi \text{ that is holomorphic in } D_\xi \end{cases}$$

such that $f_\xi = f$ on $Y \cap D_\xi$. Further, since D_0 is a Stein manifold (Proposition 2.13.8), we obtain from Corollary 4.1.8 a holomorphic function f_0 on D_0 such that $f_0 = f$ in $Y \cap D_0$. Then $f_0 - f$ is holomorphic in $D_\xi \cap D_0$ and $f_0 - f = 0$ on $Y \cap D_\xi \cap D_0$, that is, $f_0 - f \in \mathcal{F}_Y(D_\xi \cap D_0)$ (cp. Example 4.1.5). Since, by (4.11.2), $U_\xi \cap U_0 \subseteqq D_\xi \cap D_0 \cap D_1$, we can therefore apply Theorem 4.1.3 (ii) to the open covering $\{U_\xi, U_0\}$ of D_1 and the function $f_0 - f$ restricted to $U_\xi \cap U_0$. Thus we obtain $\tilde{f}_\xi \in \mathcal{F}_Y(U_\xi), \tilde{f}_0 \in \mathcal{F}_Y(U_0)$ with $f_0 - f = \tilde{f}_0 - \tilde{f}_\xi$ on $U_\xi \cap U_0$. Setting $F := f_0 - \tilde{f}_0$ in $U_0 \cap D_0$ and $F := f_\xi - \tilde{f}_\xi$ in $U_\xi \cap D_\xi$, we complete the proof. ■

Remark. Recall that by Lemmas 3.6.2 and 3.6.7 the extension of bounded holomorphic functions stated in Theorem 3.6.8 (i) can be given by a bounded linear operator. Moreover, the decomposition of singularities stated in Corollary 4.10.7 is obtained by bounded linear operators. Combining this with similar arguments as in the proof of Lemma 4.9.3 (connected with Propositions 1 and 2 in Appendix 2), the proof of Theorem 4.11.1 given above can be easily modified to a proof of the following theorem:

Under the hypotheses of Theorem 4.11.1 *there exists a bounded linear operator* E: $H^\infty(Y \cap D) \to H^\infty(D)$ *such that* $Ef = f$ *on* $Y \cap D$ *for all* $f \in H^\infty(Y \cap D)$. *Here* $H^\infty(D)$ *and* $H^\infty(Y \cap D)$ *are the spaces of bounded holomorphic functions on* D *and* $Y \cap D$, *respectively.*

It is not clear whether such a bounded linear operator exists for part (ii) of Theorem 4.11.1 (cp. also point 3 in the Exercises, remarks and problems at the end of Chapter 3). If ∂D is smooth and intersects Y transversally, then the answer is affirmative (Henkin [5]).

4.12. Formulas in holomorphic vector bundles

In this section we show that the integral formulas presented in this book can be generalized to $(0, q)$-forms with values in holomorphic vector bundles. The idea is as follows: 1. This is trivial for product bundles. 2. It can be proved that, for every holomorphic vector bundle B over a Stein manifold X, there exists an injective holomorphic homomorphism of vector bundles $\alpha: B \to X \times \mathbb{C}^N$ (for certain N) as well as a surjective holomorphic homomorphism of vector bundles $\beta: X \times \mathbb{C}^N \to B$ such that $\beta \circ \alpha = \mathrm{id}$ (id: identity map). 3. Since the $\bar\partial$-operator commutes with α and β, the formulas in B follow from the corresponding formulas in $X \times \mathbb{C}^N$. Let us explain this in detail.

4.12.1. Lemma. *Let B be a holomorphic vector bundle over a Stein manifold X. Suppose, additionally, that X is a relatively compact open subset of some larger Stein manifold. Then there exist an integer N and a surjective holomorphic homomorphism of vector bundles $\beta: X \times \mathbb{C}^N \to B$.*

Proof. In view of the hypothesis that X is a relatively compact open subset of some larger Stein manifold, this follows immediately from Cartan's Theorem A (Theorem 4.1.3 (i)). ∎

Remark. For our purpose (formulas on compact sets) it can be assumed without loss of generality that X is a relatively compact open subset of some larger Stein manifold. However, we remark that Lemma 4.12.1 is valid also without this assumption. Forster/Ramspott [1] proved that this is possible for $N \geqq \dim_{\mathbb{C}} B + [(\dim_{\mathbb{C}} X)/2]$.

4.12.2. Lemma. *Let B, X, β, N be as in Lemma 4.12.1. Then there exists a holomorphic homomorphism of vector bundles $\alpha: B \to X \times \mathbb{C}^N$ such that $\beta\alpha = \mathrm{id}$.*

Proof. The proof is similar to the proof of Lemma 4.2.1 and is left to the reader. ∎

Now it is easy to generalize the integral formulas of the preceding section to $(0, q)$-forms with values in holomorphic vector bundles. Consider, for example, the Koppelman-Leray formula stated in Theorem 4.10.4. Let X, B, α, β, N be as in Lemmas 4.12.1 and 4.12.2. Let $D \subset\subset X$ be a strictly pseudoconvex open set (with not necessarily smooth boundary) and let L_D, R_D be as in Theorem 4.10.4. We denote by L_D^N and R_D^N the corresponding operators for forms with values in the product bundle $X \times \mathbb{C}^N$. Set

$$L_D^B := \beta \circ L_D^N \circ \alpha \quad \text{and} \quad R_D^B := \beta \circ R_D^N \circ \alpha . \tag{4.12.1}$$

Since α, β are holomorphic and, therefore, $\bar\partial \circ \alpha = \alpha \circ \bar\partial$ and $\bar\partial \circ \beta = \beta \circ \bar\partial$, and since $\beta \circ \alpha = \mathrm{id}$, then we obtain from Theorem 4.10.4 the following

4.12.3. Corollary. (i) *For every continuous bounded section $f: D \to B$ such that $\bar\partial f$ is also continuous and bounded in D, we have*

$$f = L_D^B f + R_D^B \bar\partial f \quad \text{in} \quad D . \tag{4.12.2}$$

(ii) *Let $1 \leqq q \leqq n$, and let f be a B-valued continuous and bounded $(0, q)$-form on D such that $\bar\partial f$ is also continuous and bounded on D. Then*

$$f = \bar\partial R_D^B f + R_D^B \bar\partial f \quad \text{in} \quad D . \tag{4.12.3}$$

In particular, if $\bar\partial f = 0$ in D, then

$$u := R_D^B f \tag{4.12.4}$$

is a continuous solution of $\bar\partial u = f$ in D.

Finally, we remark that, since $D \subset\subset X$ and, therefore, α, β are bounded on D, Theorem 4.10.6 is valid also for the operator R_D^B.

Notes

Integral formulas for solving the $\bar\partial$-equation on Stein manifolds X were first obtained in 1975/76 by Palm [1] and Stout [1], however under the additional condition that X is a complete intersection in $\mathbb{C}^n$. The formulas on general Stein manifolds presented in this chapter (without those of Sections 4.10–4.12) were obtained in 1979 by Henkin/Leiterer [1] by means of ideas of Dynin (personal communication, 1974) and Bishop [1] (1961). Hortmann [1] (1979) used a construction of Toledo/Tong [1] and obtained a formula for solving the $\bar\partial$-equation on strictly pseudoconvex C^∞-domains in Stein manifolds.

The Weil formula in $\mathbb{C}^n$ was obtained in Bergman [1] (1934–36) and A. Weil [1] (1935) (see also Sommer [1], Gleason [1], Harvey [2]). The Koppelman-Leray-Norguet formula for analytic polyhedra in $\mathbb{C}^n$ was obtained in 1971 by Henkin [6] for (0, 1)-forms and then by Poljakov [1] for $(0, q)$-forms by means of ideas of Henkin [1], Leray [1], Lieb [2], Koppelman [1], and Norguet [2]. Then Range/Siu [1] and Poljakov [3] proved this formula for domains with piecewise smooth strictly pseudoconvex boundary in $\mathbb{C}^n$. For the general case of strictly pseudoconvex polyhedra in $\mathbb{C}^n$ this formula was obtained by Henkin (see Čirka/Henkin [1]), Sergeev [1] and Henkin/Leiterer [1], see also Øvrelid [3].

Under certain additional "complex" non-degeneration conditions the solution of the $\bar\partial$-equation on real non-degenerate strictly pseudoconvex polyhedra D in a Stein manifold X which is given by the Koppelman-Leray-Norguet formula (4.9.7) admits uniform estimates. The first such estimates were obtained in 1971 by Henkin [6] for the case of analytic polyhedra in $\mathbb{C}^n$ and (0, 1)-forms. Poljakov generalized this result to $(0, q)$-forms (see Fuks [1]). For domains with piecewise smooth strictly pseudoconvex boundaries in $\mathbb{C}^n$ such estimates were obtained in 1973 by Range/Siu [1] and Poljakov [3]. Henkin/Sergeev [1] proved uniform estimates for the $\bar\partial$-equation in strictly pseudoconvex polyhedra in $\mathbb{C}^n$ of a more general class containing both classes mentioned above. The class of strictly pseudoconvex polyhedra considered in Henkin/Sergeev [1] is defined by two conditions (C) and (CR) which can be formulated as follows: Let X, D, $U_{\bar D}$, X_k, F_k, θ_k, ϱ_k, N be as in Definition 4.9.1. For every collection $K = (k_1, \dots, k_l)$ of integers $1 \leqq k_1 < \dots < k_l \leqq N$ we set

$$S_K := \{z \in \partial D : \varrho_{k_1}(F_{k_1}(z)) = \dots = \varrho_{k_l}(F_{k_l}(z)) = 0\} .$$

Condition (C). *For every collection $K = (k_1, \dots, k_l)$ of integers $1 \leqq k_1 < \dots < k_l \leqq N$ the complex Jacobi matrix (with respect to local holomorphic coordinates) of the map $(F_{k_1}, \dots, F_{k_l}) : U_{\bar D} \to X_{k_1} \times \dots \times X_{k_l}$ has constant rank on S_K, which is $\geqq \min\{\dim_{\mathbb{C}} X_{k_1}, \dots, \dim_{\mathbb{C}} X_{k_l}\}$. We denote this rank by r_K.*

Condition (CR). *Let $K = (k_1, \dots, k_l)$ and $M = (m_1, \dots, m_t)$ be two collections of integers $1 \leqq k_1 < \dots < k_l \leqq N$ and $1 \leqq m_1 < \dots < m_t \leqq N$ such that $r_{K \cup \{m_s\}} > r_K$ for every $s = 1, \dots, t$. Then the rank of the real Jacobi matrix (with respect to local real coordinates) of the map*

$$(\varrho_{m_1} \circ F_{m_1}, \dots, \varrho_{m_t} \circ F_{m_t}, F_{k_1}, \dots, F_{k_l}) : U_{\bar D} \to \mathbb{R}^t \times X_{k_1} \times \dots \times X_{k_l}$$

is equal to $t + 2r_K$ on $S_{K \cup M}$.

Observe that condition (CR) implies that D is real non-degenerate.

Henkin/Petrosjan [1] (1978) obtained uniform estimates for the $\bar\partial$-equation in real non-degenerate analytic polyhedra $D \subset\subset \mathbb{C}^2$ satisfying the condition that $\bigcup_{|K|=2} S_K$ is the Šilov boundary of D.

Frumin [1] (1981) obtained an appropriate generalization of this result to the case of $\mathbb{C}^n$.

By means of the integral formulas on Stein manifolds presented in this chapter, all these estimates for the $\bar\partial$-equation can be generalized to Stein manifolds.

Observe also the following recent result of Heunemann [1] which gives another (very simple) possibility to generalize the results of Henkin/Sergeev [1] to Stein manifolds: *Let X be a closed complex submanifold of $\mathbb{C}^m$, and let $D \subset\subset X$ be a strictly pseudoconvex* *polyhedron satisfying conditions* (C) *and* (CR). *Then there exists a strictly pseudoconvex polyhedron* $D_1 \subset\subset \mathbb{C}^m$ *satisfying conditions* (C) *and* (CR) *as well as a neighbourhood* $U_1 \subseteq \mathbb{C}^m$ *of* $\overline{D}_1$ *and a holomorphic map h from* U_1 *onto* $U_1 \cap X$ *such that* $h(z) = z$ *for* $z \in U_1 \cap X$ *and* $h(D_1) = D$.

Remark that, for the case of an analytic polyhedron D, the corresponding strictly pseudoconvex polyhedron D_1 need not be also an analytic polyhedron. Thus, in this way, estimates for the $\bar{\partial}$-equation in analytic polyhedra in Stein manifolds are obtained by means of estimates for a more general class of polyhedra in $\mathbb{C}^m$.

The Extension Theorem 4.11.1 (i) was obtained in 1980 by Henkin/Leiterer [2]. Part (i) of this theorem was proved independently by Amar [1] (Amar assumed that ∂D is of class C^∞ whereas Y is allowed to be a C^∞-submanifold of some neighbourhood of $\overline{D}$ such that $Y \cap D$ is complex). For the case that ∂D is C^2 and the intersection $\partial D \cap Y$ is transversal, Theorem 3.11.1 was proved in 1972 by Henkin [5]. Observe also that Cumenge [1] (for the case of transversal intersection $Y \cap \partial D$) and Amar [1] (for the general case) obtained a version of Theorem 4.11.1 for functions in Hardy classes. "Partially", Theorem 4.11.1 (ii) is a generalization of some of the known results on interpolation sets (cp. points 6 and 7 in the Exercises, remarks and problems at the end of Chapter 3 — a set $N \subseteq \partial D$ is called an interpolation set if every continuous function on N extends to a continuous function on $\overline{D}$ that is holomorphic in D).

Exercises, remarks and problems

1. Let X be a Stein manifold of complex dimension n. Suppose that there is a holomorphic vector bundle B of complex dimension n over X as well as a holomorphic map $h: X \times X \to B$ such that the following conditions are fulfilled:

(1) $h(z, \zeta) \in B_z$ for all $z, \zeta \in X$ (B_z is the fibre of B over z).

(2) $h(z, \zeta) \neq 0$ for $z \neq \zeta$.

(3) For every fixed $z \in X$, the map $h(z, \zeta)$ (considered as a B_z-valued map) is biholomorphic for ζ in some neighbourhood of z, where $h(z, z) = 0$.

Then it is easily seen that in the construction of the integral formulas presented in this chapter, B and h can be used instead of $T(X)$ and s, where the factor φ^ν can be omitted.

Open problem. Let X be an arbitrary Stein manifold of complex dimension n. Do there exist a holomorphic vector bundle B of complex dimension n over X and a holomorphic map $h: X \times X \to B$ such that conditions (1)—(3) are fulfilled?

2. Prove the Extension Theorem 4.11.1 for the case that Y is a closed analytic subset of some neighbourhood of $\overline{D}$ whose singularities are isolated and contained in D (for the definition of analytic sets see, for example, Grauert/Fritzsche [1] and Gunning/Rossi [1]).

3. (A counterexample to the extension Theorem 4.11.1 for the case that Y has a singularity on ∂D.) Let $D := \{z \in \mathbb{C}^2 : |z_1|^2 + |z_2 - 1|^2 < 1\}$, $Y_1 := \{z \in D : z_1 = 0\}$, $Y_2 := \{z \in D : z_1 = z_2^2\}$. Then $Y_1 \cup Y_2$ is a closed complex submanifold of D (which cannot be continued smoothly into a neighbourhood of $\overline{D}$ at the point $(0, 0)$). Then there does not exist a bounded holomorphic function F on D such that $F(z) = 1$ for $z \in Y_1$ and $F(z) = 0$ for $z \in Y_2$.

Proof. If F is a bounded holomorphic function in D, then it follows from the Cauchy formula applied to the discs $G_\varepsilon := \{z \in D : z_2 = \varepsilon\}$, $\varepsilon > 0$, that there is a constant $C < \infty$ such that

$$\left| \frac{\partial F(\zeta, \varepsilon)}{\partial z_1} \right| \leqq C\varepsilon \quad \text{for all} \quad \varepsilon > 0 .$$

This is impossible if $F(0, \varepsilon) = 1$ and $F(\varepsilon^2, \varepsilon) = 0$ for all $\varepsilon > 0$.

4. Open problem. Does there exist a (not necessarily real non-degenerate) analytic polyhedron $D \subset\subset \mathbb{C}^2$ without uniform estimates for the $\bar{\partial}$-equation?

5. a) (Henkin [7], Lieb [2], Øvrelid [2]). Let X be a Stein manifold, and let $D \subset\subset X$ be a strictly pseudoconvex polyhedron satisfying conditions (C) and (CR) mentioned in the Notes above. Let $A(\overline{D})$ be the algebra of all continuous functions on $\overline{D}$ that are holomorphic in D. Prove that for every point $z \in D$ the ideal $\{f \in A(\overline{D}) : f(z) = 0\}$ is finitely generated.

b) **Open problem.** Does there exist a domain of holomorphy $D \subset\subset \mathbb{C}^2$ such that, for some $z \in D$, the ideal $\{f \in A(\overline{D}), f(z) = 0\}$ is not finitely generated?

6 (Gromov/Eliašberg [1]). Prove that every Stein manifold of complex dimension n is biholomorphically equivalent to a closed complex submanifold of $\mathbb{C}^{[3n/2]+2}$. Remark. A proof of the Gromov-Eliašberg theorem is not yet published. It was proved by Remmert, Bishop [1] and Narasimhan [3] (see Hörmander [1]) that every Stein manifold of complex dimension n is biholomorphically equivalent to some closed complex submanifold of $\mathbb{C}^{2n+1}$.

7. Let $D_n = \{Z \in \mathbb{C}^{n^2} : ZZ^* < I\}$ (cp. point 25 in the Exercises, remarks and problems at the end of Chapter 1), and let $SD_n = \{Z : ZZ^* = I\}$ be the Šilov boundary of D_n.

a) Prove the Bochner formula (Bochner [2]):

$$f(Z) = c_n \int_{SD_n} \frac{f(G)\, dG}{[\det(I - G^*Z)]^n}, \qquad Z \in D_n,$$

where f is a continuous function on $\overline{D}_n$ that is holomorphic in D_n.

b) Find a formula for solving $\bar\partial g = f$, where f is a continuous (0, 1)-form on $\overline{D}_n$ with $\bar\partial f = 0$, such that

$$Sf := \int_{SD_n} \frac{g(G)\, dG}{[\det\,(I - G^*Z)]^n} = 0 \quad \text{for all} \quad Z \in D_n.$$

c) (Dautov/Henkin (unpublished)). Prove that, for some $C < \infty$, the solution from b) satisfies the estimate

$$\sup_{Z \in D_n} |g(Z)|\, [\operatorname{dist}(Z, SD_n)]^{n^2-2} \leq C \sup_{Z \in D_n} |f(Z)|.$$

d) Prove that the solution from 10b) does not admit uniform estimates if $n > 1$.

e) **Open problem.** Do there exist uniform estimates for the $\bar\partial$-equation in D_n if $n > 1$?

f) **Open problem.** Let $z \in D_n$. Is the ideal $\{f \in A(\overline{D}_n) : f(z) = 0\}$ finitely generated? Here $A(\overline{D}_n)$ denotes the algebra of continuous functions on $\overline{D}_n$ that are holomorphic in D_n.

8. Let X be a Stein manifold, let $D \subset\subset X$ be a strictly pseudoconvex polyhedron satisfying conditions (C) and (CR) mentioned in the Notes above. Let SD be the Bergman-Šilov boundary of D, and let K be the Cauchy-Leray-Norguet projection defined by the first sum in (4.9.5) for every continuous function on SD. For $0 < \alpha < 1$ we denote by $H^\alpha(SD)$ the space of α-Hölder continuous functions on SD and by $A^\alpha(\overline{D})$ the space of α-Hölder continuous functions on $\overline{D}$ that are holomorphic in D.

a) (Ajrapetjan [1], Jöricke [1]). Prove that, for every $0 < \alpha < 1$ and $f \in H^\alpha(SD)$, $Kf \in A^\beta(\overline{D})$ for all $0 < \beta < \alpha$.

b) (Jöricke [1]). Let D be a polydisc and $0 < \alpha < 1$. Find $f \in H^\alpha(B)$ such that $Kf \notin A^\alpha(\overline{D})$.

c) (Ahern/Schneider [1], Phong/Stein [1]). Let D be a smooth strictly pseudoconvex domain and $0 < \alpha < 1$. Prove that, for every $f \in H^\alpha(\partial D)$, $Kf \in A^\alpha(\overline{D})$.

d) Let D_n, SD_n be as in point 10, and let S be the Cauchy-Szegö projection for D_n, that is, the orthogonal projection from $L_2(D_n)$ onto the subspace spanned by the holomorphic functions in a neighbourhood of $\overline{D}_n$. Theorem (Jöricke [2]). Then, for each $0 < \alpha < 1$ and $f \in H^\alpha(SD_n)$, $Sf(z) = O\,(\delta^{\alpha-[(n/2)^2]}(\ln \delta)^\sigma)$ where $\delta = \operatorname{dist}(z, \partial D_n)$ and $\sigma = -1$ for n odd, $\sigma = 0$ for n even. This estimate cannot be improved.

9. **Open problem** (Bedford/Fornaess [1]). Let X be a Stein manifold, and let $D \subset\subset X$ be a domain of holomorphy with smooth real-analytic boundary.

a) Let f be an arbitrary continuous function on $\overline{D}$ that is holomorphic in D. Can f be approximated uniformly on $\overline{D}$ by holomorphic functions in a neighbourhood of $\overline{D}$? The case $D \subset\subset \mathbb{C}^2$ is considered in Fornaess/Nagel [1] and Fornaess/Øvrelid [1].

b) Do there exist uniform estimates for the $\bar{\partial}$-equation in D?

10 (Bishop [1]). Let X be a connected Stein manifold and $P \subset\subset X$ an open set of the form $P = \{z \in X : |f_j(z)| < 1;\ j = 1, \ldots, n\}$, where $f_1, \ldots, f_n$ are holomorphic functions on X and $n = \dim_{\mathbb{C}} X$. (Such sets are called *special* analytic polyhedra.) Set $f := (f_1, \ldots, f_n)$. Let X_f be the set of points $z \in X$ such that f is biholomorphic in a neighbourhood of z. Observe that by Sard's theorem the Lebesgue measure of $f(X \setminus X_f)$ vanishes.

a) Prove that there is an integer $k \geqq 1$ (the multiplicity of f) such that for each $w \in f(X \setminus X_f)$ the set $f^{-1}(w)$ consists of k points.

b) For each continuous function φ on X and each $w \in f(X \setminus X_f)$, we define $\tilde{\varphi}(w) := \varphi(z^1) + \ldots + \varphi(z^k)$, where $\{z^1, \ldots, z^k\} = f^{-1}(w)$. Prove that $\tilde{\varphi}$ admits a continuous extension to $f(X)$, which will be denoted also by $\tilde{\varphi}$. If φ is holomorphic on X, then $\tilde{\varphi}$ is holomorphic on $f(X)$.

c) Let $S := \{z \in X : |f_1(z)| = \ldots = |f_n(z)| = 1\}$. Denote by $df_1 \wedge \ldots \wedge df_n$ the measure on S which is defined by the differential form $df_1 \wedge \ldots \wedge df_n$ in the smooth points of S and by zero in the other points. Prove that this measure is finite.

d) Prove that for each holomorphic function φ on X and for each $z \in P$

$$\tilde{\varphi}(f(z)) = \frac{1}{(2\pi i)^n} \int\limits_S \frac{df_1 \wedge \ldots \wedge df_n}{(f_1 - f_1(z)) \ldots (f_n - f_n(z))}.$$

Remark. If $X = \mathbb{C}^2$ and $S \subseteqq X_1$ this formula was obtained by Martinelli [3] (see also Ajsenberg/Južakov [1]).

e) Let ψ be a holomorphic function on X. For $z, v \in X_f$, we put $\tilde{\psi}(z, v) =: (\psi(z^2) - \psi(v)) \ldots (\psi(z^k) - \psi(v))$, where $\{z, z^2, \ldots, z^k\} = f^{-1}(f(z))$. Prove that $\tilde{\psi}$ admits a holomorphic extension to $f(X) \times f(X)$. Prove that for each holomorphic function φ on X and each $z \in P$

$$\varphi(z)\,\tilde{\psi}(z, z) = \frac{1}{(2\pi i)^n} \int\limits_S \varphi\tilde{\psi}(\cdot, z) \frac{df_1 \wedge \ldots \wedge df_n}{(f_1 - f_1(z)) \ldots (f_n - f_n(z))}.$$

Hint. Apply the formula in d) to the function $\varphi\tilde{\psi}(\cdot, z)$.

11. For some further results in the theory of functions on complex manifolds obtained by the method of integral formulas let us refer to Anderson/Berndtsson [1], Ajrapetjan/Henkin [1], Ajsenberg/Južakov [1], Berndtsson [2], Bishop [2], Dautov/Henkin [1], Gindikin/Henkin [1], Griffiths/Harris [1], Harvey/Polking [1], Henkin [3], Krantz [2], Rudin [3], Skoda [2], Vladimirov [1].

Appendix 1. Estimation of some integrals

In this appendix we give estimates for some integrals in $\mathbb{R}^n$, which are used in the book.

Notation. By n we denote a positive integer. If $x \in \mathbb{R}^n$, then we denote the canonical coordinates of x by $x_1, \dots, x_n$, and we assume that the orientation of $\mathbb{R}^n$ is defined by the form $dx_1 \wedge \dots \wedge dx_n$, $x \in \mathbb{R}^n$. We will also write $dx_1 \dots dx_n$ and $d\sigma_n$ instead of $dx_1 \wedge \dots \wedge dx_n$. If $S \subseteq \mathbb{R}^n$ is a smooth surface of real dimension $k \leqq n$, then we denote the Euclidean volume form on S by $d\sigma_k$. For $x \in \mathbb{R}^n$ we set

$$x' := (x_2, \dots, x_n)\,, \qquad x'' := (x_3, \dots, x_n) \quad \text{and} \quad |x| := \Big(\sum_{j=1}^{n} |x_j|^2\Big)^{1/2}.$$

In the proofs we denote all "large" constants by $C, C', \dots$, and all "small" constants will be denoted by $\alpha, \alpha', \dots$. An expression of the form $a(x) \lesssim C b(x)$ $(a(x) \gtrsim \alpha b(x))$ must be read as follows: There exists a constant $C < \infty$ $(\alpha > 0)$ such that $a(x) \leqq C b(x)$ $(a(x) \geqq \alpha b(x))$ for all x considered in the corresponding situation.

Proposition 1. *Let $n \geqq 1$ and $0 < R < \infty$. Then there is a constant $C < \infty$ such that for all $t, s \in \mathbb{R}^n$ with $|t|, |s| \leqq R$*

$$\int\limits_{x \in \mathbb{R}^n, |x| < R} \left| \frac{x_1 - t_1}{|x - t|^n} - \frac{x_1 - s_1}{|x - s|^n} \right| d\sigma_n \leqq C\,|t - s|\,\big|\ln|t - s|\big|\,.$$

Proof

$$\int\limits_{|x-t| \leqq |t-s|/2} \left| \frac{x_1 - t_1}{|x - t|^n} - \frac{x_1 - s_1}{|x - s|^n} \right| d\sigma_n$$

$$\leqq \int\limits_{|x-t| \leqq |t-s|/2} |x - t|^{1-n}\, d\sigma_n + C\,|t - s|^{1-n} \int\limits_{|x-t| \leqq |t-s|/2} d\sigma_n \leqq C'\,|t - s| \tag{1}$$

and

$$\int\limits_{|x-s| \leqq |t-s|/2} \left| \frac{x_1 - t_1}{|x - t|^n} - \frac{x_1 - s_1}{|x - s|^n} \right| d\sigma_n \leqq C\,|t - s|\,. \tag{2}$$

Further,

$$\left|\frac{x_1 - t_1}{|x-t|^n} - \frac{x_1 - s_1}{|x-s|^n}\right|$$

$$\leqq \left|\frac{(x_1 - t_1)\,(|x-s| - |x-t|) \sum\limits_{\nu=0}^{n-1} |x-s|^\nu\,|x-t|^{n-\nu-1} - (t_1 - s_1)\,|x-t|^n}{|x-s|^n\,|x-t|^n}\right|$$

$$\leqq \big||x-s| - |x-t|\big| \sum_{\nu=0}^{n-1} \frac{|x-s|^\nu\,|x-t|^{n-\nu}}{|x-s|^n\,|x-t|^n} + \frac{|t-s|}{|x-s|^n}$$

$$\leqq n\,|t-s| \left(\max\left\{\frac{1}{|x-s|^n}, \frac{1}{|x-t|^n}\right\} + \frac{1}{|x-s|^n}\right).$$

Consequently, since $|t|, |s| \leqq R$,

$$\int\limits_{|x-t|,\,|x-s| \geqq |t-s|/2,\,|x|<R} \left|\frac{x_1 - t_1}{|x-t|^n} - \frac{x_1 - s_1}{|x-s|^n}\right| \mathrm{d}\sigma_n$$

$$\leqq C\,|t-s| \int\limits_{|t-s|/2 \leqq |y| \leqq 2R} \frac{\mathrm{d}\sigma_n}{|y|^n} \leqq C'\,|t-s| \int\limits_{|t-s|/2}^{2R} \frac{\mathrm{d}r}{r} \leqq C''\,|t-s|\,\big|\ln|t-s|\big|\,. \;\blacksquare$$

Proposition 2. *Let $n \geqq 1$ and let D be a bounded domain in $\mathbb{R}^n$ with piecewise smooth boundary* (for the definition of a piecewise smooth boundary see Section 1.8). *Then there is a constant $C < \infty$ with the following property: If f is a C^1-function in D such that for some $K < \infty$ and $0 < \alpha < 1$, $\|\mathrm{d}f(x)\| \leqq K[\operatorname{dist}(x, \partial D)]^{-\alpha}$ for $x \in D$,*[1]) *then*

$$|f(x) - f(y)| \leqq CK\,|x-y|^{1-\alpha} \quad \textit{for} \quad x, y \in D\,.$$

Proof. Since ∂D is piecewise smooth, we can find a number $\varepsilon > 0$ so small that, for all $x, y \in D$ with $\operatorname{dist}(x, \partial D) \leqq \varepsilon$ and $\operatorname{dist}(y, \partial D) \leqq \varepsilon$, there exists a C^1-function $\gamma_{xy}\colon [0, 3\,|x-y|] \to D$ such that

$$\begin{aligned}
&\gamma_{xy}(0) = x\,, \qquad \gamma_{xy}(3\,|x-y|) = y\,,\\
&\operatorname{dist}(\gamma_{xy}(\lambda), \partial D) \geqq \lambda \quad \text{for} \quad \lambda \in [0, |x-y|]\,,\\
&\operatorname{dist}(\gamma_{xy}(\lambda), \partial D) \geqq |x-y| \quad \text{for} \quad \lambda \in [|x-y|, 2\,|x-y|]\,,\\
&\operatorname{dist}(\gamma_{xy}(\lambda), \partial D) \geqq 3\,|x-y| - \lambda \quad \text{for} \quad \lambda \in [2\,|x-y|, 3\,|x-y|]\,,\\
&\left|\frac{\mathrm{d}}{\mathrm{d}\lambda}\gamma_{xy}(\lambda)\right| \leqq C \quad \text{for} \quad \lambda \in [0, 3\,|x-y|]\,.
\end{aligned}$$

Clearly, we only have to prove the proposition for points $x, y \in D$ with $\operatorname{dist}(x, \partial D) \leqq \varepsilon$ and $\operatorname{dist}(y, \partial D) \leqq \varepsilon$. For such points we obtain

$$|f(x) - f(y)| = \left|\int\limits_0^{3|x-y|} \frac{\mathrm{d}}{\mathrm{d}\lambda} f(\gamma_{xy}(\lambda))\,\mathrm{d}\lambda\right| \leqq C \int\limits_0^{3|x-y|} \|\mathrm{d}f(\gamma_{xy}(\lambda))\| \left|\frac{\mathrm{d}}{\mathrm{d}\lambda}\gamma_{xy}(\lambda)\right| \mathrm{d}\lambda$$

$$\leqq C'K \left(\int\limits_0^{|x-y|} \lambda^{-\alpha}\,\mathrm{d}\lambda + \int\limits_{|x-y|}^{2|x-y|} |x-y|^{-\alpha}\,\mathrm{d}\lambda + \int\limits_{2|x-y|}^{3|x-y|} (3\,|x-y| - \lambda)^{-\alpha}\,\mathrm{d}\lambda\right)$$

$$\leqq 3C'K\,|x-y|^{1-\alpha}\,. \;\blacksquare$$

[1]) $\|\mathrm{d}f(x)\| := \left(\sum\limits_{j=1}^{n} |\partial f(x)/\partial x_j|^2\right)^{1/2}$, $\quad \operatorname{dist}(x, \partial D) := \inf\{|x-y| : y \in \partial D\}$.

Proposition 3. *Let $n \geqq 1$ and $0 < R < \infty$. Then there exists a constant $C < \infty$ such that for all $\varepsilon > 0$*

(i) $$\int\limits_{x \in \mathbb{R}^n, |x| \leqq R} \frac{d\sigma_n}{(\varepsilon + |x|^2)\,|x|^{n-1}} \leqq C/\sqrt{\varepsilon}\,,$$

(ii) $$\int\limits_{x \in \mathbb{R}^n, |x| \leqq R} \frac{d\sigma_n}{(\varepsilon + |x_1| + |x|^2)^2\,|x|^{n-2}} \leqq C/\sqrt{\varepsilon}\,.$$

Proof. (i)

$$\int\limits_{|x| \leqq R} \frac{d\sigma_n}{(\varepsilon + |x|^2)\,|x|^{n-1}} \leqq \int\limits_0^R dr \int\limits_{|x|=r} \frac{d\sigma_{n-1}}{(\varepsilon + r^2)\,r^{n-1}} \leqq C \int\limits_0^R \frac{dr}{\varepsilon + r^2} \leqq C'/\sqrt{\varepsilon}\,.$$

(ii). For $n = 1$ we have

$$\int\limits_{|x| \leqq R} \frac{dx}{(\varepsilon + |x| + |x|^2)^2\,|x|^{-1}} \leqq \int\limits_0^R \frac{2x\,dx}{(\varepsilon + x)^2} \leqq C \int\limits_0^R \frac{2x\,dx}{\varepsilon^2 + x^2} \leqq C' \ln \varepsilon^2 \leqq C''/\sqrt{\varepsilon}\,.$$

For $n \geqq 2$ we obtain

$$\int\limits_{|x| \leqq R} \frac{d\sigma_n}{(\varepsilon + |x_1| + |x|^2)^2\,|x|^{n-2}} \leqq \int\limits_{|x'| \leqq R} d\sigma_{n-1} \int\limits_0^R \frac{dx_1}{(\varepsilon + |x_1| + |x'|^2)\,|x'|^{n-2}}$$

$$\leqq C \int\limits_{|x'| \leqq R} \frac{d\sigma_{n-1}}{(\varepsilon + |x'|^2)\,|x'|^{n-2}} \leqq C \int\limits_0^R dr \int\limits_{|x'|=r} \frac{d\sigma_{n-2}}{(\varepsilon + r^2)\,r^{n-2}}$$

$$\leqq C' \int\limits_0^R \frac{dr}{\varepsilon + r^2} \leqq C''/\sqrt{\varepsilon}\,. \blacksquare$$

Proposition 4. *Let $n, d \geqq 1$ be integers. Then there exists a constant $C = C(n, d) < \infty$ such that the following statement is true: Let $p_1, \ldots, p_s$ $(1 \leqq s \leqq n)$ be real-valued polynomials of degree $\leqq d$ defined on $\mathbb{R}^n$, let $\varepsilon, \mu, \nu, \delta, \gamma \geqq 0$, and let $D \subseteqq \mathbb{R}^n$. Then* (for the definition of $\|\cdot\|$ see Subsection 1.6.3)

$$\int\limits_{x \in D} \frac{\|dp_1(x) \wedge \ldots \wedge dp_s(x)\|\, dx_1 \wedge \ldots \wedge dx_n}{[\varepsilon + \sum_{j=1}^{s} |p_j(x)| + \gamma(|x| + \sum_{j=1}^{s} |p_j(x)|)^\mu]^\delta\,[|x| + \sum_{j=1}^{s} |p_j(x)|]^\nu}$$

$$\leqq C \int\limits_{t \in T(D)} \frac{dt_1 \wedge \ldots \wedge dt_n}{[\varepsilon + \sum_{j=1}^{s} |t_j| + \gamma\,|t|^\mu]^\delta\,|t|^\nu}\,, \tag{3}$$

where $T(D) := \{(|p_1(x)|, \ldots, |p_s(x)|) : x \in D\}$ *if* $s = n$ *and*

$$T(D) := \bigcup_{1 \leqq i_1 < \ldots < i_{n-s} \leqq n} \{(|p_1(x)|, \ldots, |p_s(x)|, |x_{i_1}|, \ldots, |x_{i_{n-s}}|) : x \in D\} \quad \textit{if} \quad s < n\,.$$

Proof. From the definition of $\|dp_1 \wedge \ldots \wedge dp_s\|$ it follows that

$$\|dp_1(x) \wedge \ldots \wedge dp_s(x)\| \leqq C \max_{1 \leqq j_1 < \ldots < j_s \leqq n} |\det[(\partial p_l(x)/\partial x_{j_k})_{k,l=1}^s]|\,.$$

On the other hand, if $1 \leqq j_1 < \dots < j_s \leqq n$ and $1 \leqq i_1 < \dots < i_{n-s} \leqq n$ such that $\{j_1, \dots, j_s\} \cup \{i_1, \dots, i_{n-s}\} = \{1, \dots, n\}$, then

$$|\mathrm{d}p_1(x) \wedge \dots \wedge \mathrm{d}p_s(x) \wedge \mathrm{d}x_{i_1} \wedge \dots \wedge \mathrm{d}x_{i_{n-s}}| = |\det [(\partial p_l(x)/\partial x_{j_k})_{k,\,l=1}^s]\, \mathrm{d}x_1 \wedge \dots \wedge \mathrm{d}x_n| .$$

This implies that

$$||\mathrm{d}p_1(x) \wedge \dots \wedge \mathrm{d}p_s(x)||\, \mathrm{d}x_1 \wedge \dots \wedge \mathrm{d}x_n$$
$$\leqq C \max_{1 \leqq i_1 < \dots < i_{n-s} \leqq n} |\mathrm{d}p_1(x) \wedge \dots \wedge \mathrm{d}p_s(x) \wedge \mathrm{d}x_{i_1} \wedge \dots \wedge \mathrm{d}x_{i_{n-s}}| .$$

Consequently, the integral on the left-hand side of (3) can be estimated by the maximum over all collections $1 \leqq i_1 < \dots < i_{n-s} \leqq n$ of

$$\int_D \frac{|\mathrm{d}p_1(x) \wedge \dots \wedge \mathrm{d}p_s(x) \wedge \mathrm{d}x_{i_1} \wedge \dots \wedge \mathrm{d}x_{i_{n-s}}|}{[\varepsilon + \sum_{j=1}^{s} |p_j(x)| + \gamma(\sum_{j=1}^{s} |p_j(x)| + \sum_{k=1}^{n-s} |x_{i_k}|)^\mu]^\delta [\sum_{j=1}^{s} |p_j(x)| + \sum_{k=1}^{n-s} |x_{i_k}|]^\nu} .$$

For every collection $1 \leqq i_1 < \dots < i_{n-s} \leqq n$, the integrand of the last integral is the pull-back of the integrand of the integral on the right-hand side of (3) with respect to the map $\alpha_{i_1,\dots,i_{n-s}}\colon D \to T(D)$ defined by

$$\alpha_{i_1,\dots,i_{n-s}}(x) := (|p_1(x)|, \dots, |p_s(x)|, |x_{i_1}|, \dots, |x_{i_{n-s}}|) .$$

It is a corollary of Bezout's theorem that for every $y \in \mathbb{R}^n$ there are no more than d^s points $x \in \mathbb{R}^n$ such that $(p_1(x), \dots, p_s(x), x_{i_1}, \dots, x_{i_{n-s}}) = y$. Consequently, for every $y \in \mathbb{R}^n$ there are no more than $2^n d^s$ points $x \in \mathbb{R}^n$ such that $\alpha_{i_1,\dots,i_{n-s}}(x) = y$. This completes the proof of the proposition. ■

Proposition 5. *Let $n \geqq 1$. Then there exists a constant $C < \infty$ such that for all $\delta > 0$ and $-\infty < \alpha < \infty$ the following estimates hold:*

(i) $$\int_{x \in \mathbb{R}^n, |x| < \delta} \frac{\mathrm{d}\sigma_n}{|x|^{n-\varkappa}} \leqq C\delta^\varkappa \quad \text{for} \quad \varkappa = 1, 2 ,$$

(ii) $$\int_{\substack{x \in \mathbb{R}^n, |x| < \delta, \\ \alpha < x_1 < \alpha + \delta^2}} \frac{\mathrm{d}\sigma_n}{(|x_1| + |x|^2)\, |x|^{n-1-\varkappa}} \leqq C\delta^\varkappa \quad \text{for} \quad \varkappa = 1, 2 ,$$

(iii) $$\int_{\substack{x \in \mathbb{R}^n, |x| < \delta, \\ \alpha < x_1 < \alpha + \delta^2}} \frac{\mathrm{d}\sigma_n}{(|x_1| + |x_2| + |x|^2)^2\, |x|^{n-2-\varkappa}} \leqq C\delta^\varkappa \quad \text{for} \quad \varkappa = 1, 2 .$$

Proof. (i)

$$\int_{|x| < \delta} \frac{\mathrm{d}\sigma_n}{|x|^{n-\varkappa}} = \int_0^\delta \mathrm{d}r \int_{|x| = r} \frac{\mathrm{d}\sigma_{n-1}}{r^{n-\varkappa}} \leqq C \int_0^\delta r^{\varkappa-1}\, \mathrm{d}r \leqq C'\delta^\varkappa .$$

(ii)

$$\int_{\substack{|x| < \delta, \\ \alpha < x_1 < \alpha + \delta^2}} \frac{\mathrm{d}\sigma_n}{(|x_1| + |x|^2)\, |x|^{n-1-\varkappa}} \leqq \int_\alpha^{\alpha+\delta^2} \mathrm{d}x_1 \int_0^\delta \mathrm{d}r \int_{|x'| = r} \frac{\mathrm{d}\sigma_{n-2}}{(|x_1| + r^2)\, r^{n-1-\varkappa}}$$

$$\leqq C\delta^{\varkappa-1} \int_\alpha^{\alpha+\delta^2} \mathrm{d}x_1 \int_0^\delta \frac{\mathrm{d}r}{|x_1| + r^2} \leqq C'\delta^{\varkappa-1} \int_\alpha^{\alpha+\delta^2} \frac{\mathrm{d}x_1}{\sqrt{|x_1|}} \leqq C''\delta^\varkappa .$$

(iii)

$$\int\limits_{\substack{|x|<\delta,\\ \alpha<x_1<\alpha+\delta^2}} \frac{d\sigma_n}{(|x_1| + |x_2| + |x|^2)^2\, |x|^{n-2-\varkappa}}$$

$$\leqq C \int\limits_{\alpha}^{\alpha+\delta^2} dx_1 \int\limits_0^{\delta} dx_2 \int\limits_0^{\delta} dr \int\limits_{|x''|=r} \frac{d\sigma_{n-3}}{(|x_1| + x_2 + r^2)^2\, r^{n-2-\varkappa}}$$

$$\leqq C'\delta^{\varkappa-1} \int\limits_{\alpha}^{\alpha+\delta^2} dx_1 \int\limits_0^{\delta} dr \int\limits_0^{\delta} \frac{dx_2}{(|x_1| + x_2 + r^2)^2}$$

$$\leqq C''\delta^{\varkappa-1} \int\limits_{\alpha}^{\alpha+\delta^2} dx_1 \int\limits_0^{\delta} \frac{dr}{|x_1| + r^2} \leqq C'''\delta^{\varkappa}. \blacksquare$$

Proposition 6. *Let $n \geqq 2$. Then there exists a constant $C < \infty$ such that for every $\varepsilon > 0$ the following estimates hold:*

(i) $$\int\limits_{|x|=\varepsilon} \frac{d\sigma_{n-1}}{(|x_1| + \varepsilon^2)^2} \leqq C\varepsilon^{n-4},$$

(ii) $$\int\limits_{|x|=\varepsilon} \frac{d\sigma_{n-1}}{(|x_1| + |x_2| + \varepsilon^2)^3} \leqq C\varepsilon^{n-5},$$

(iii) $$\int\limits_{|x|=\varepsilon} \frac{d\sigma_{n-1}}{|x_1| + \varepsilon^2} \leqq C\varepsilon^{n-2} |\ln \varepsilon| .$$

Proof. (i)

$$\int\limits_{|x|=\varepsilon} \frac{d\sigma_{n-1}}{(|x_1| + \varepsilon^2)^2} \leqq C \int\limits_0^{\varepsilon} dx_1 \int\limits_{|x'|=\sqrt{\varepsilon^2-x_1^2}} \frac{d\sigma_{n-2}}{(x_1 + \varepsilon^2)^2} \leqq C'\varepsilon^{n-2} \int\limits_0^{\varepsilon} \frac{dx_1}{(x_1 + \varepsilon^2)^2} \leqq C''\varepsilon^{n-4}.$$

(ii)

$$\int\limits_{|x|=\varepsilon} \frac{d\sigma_{n-1}}{(|x_1| + |x_2| + \varepsilon^2)^3} = \iint\limits_{x_1^2+x_2^2\leqq\varepsilon^2} dx_1\, dx_2 \int\limits_{|x''|=\sqrt{\varepsilon^2-x_1^2-x_2^2}} \frac{d\sigma_{n-3}}{(|x_1| + |x_2| + \varepsilon^2)^3}$$

$$\leqq C\varepsilon^{n-3} \int\limits_0^{\varepsilon}\int\limits_0^{\varepsilon} \frac{dx_1\, dx_2}{(x_1 + x_2 + \varepsilon^2)^3} \leqq C'\varepsilon^{n-5}.$$

(iii)

$$\int\limits_{|x|=\varepsilon} \frac{d\sigma_{n-1}}{|x_1| + \varepsilon^2} \leqq C \int\limits_0^{\varepsilon} dx_1 \int\limits_{|x'|=\sqrt{\varepsilon^2-x_1^2}} \frac{d\sigma_{n-2}}{x_1 + \varepsilon^2}$$

$$\leqq C'\varepsilon^{n-2} \int\limits_0^{\varepsilon} \frac{dx_1}{x_1 + \varepsilon^2} \leqq C''\varepsilon^{n-2} |\ln \varepsilon| . \blacksquare$$

Proposition 7. *Let $n \geqq 2$ and $0 < R < \infty$. Then there exists a constant $C < \infty$ such that for all $\varepsilon > 0$ the following estimates hold:*

(i) $$\int\limits_{x \in R^n, |x| < R} \frac{d\sigma_n}{(\varepsilon^2 + |x|^2)\,|x|^{n-1}} \leqq C/\varepsilon\,,$$

(ii) $$\int\limits_{x \in R^n, |x| < R} \frac{d\sigma_n}{(\varepsilon^2 + |x_1| + |x|^2)^2\,|x|^{n-2}} \leqq C/\varepsilon\,,$$

(iii) $$\int\limits_{x \in R^n, |x| < R} \frac{d\sigma_n}{(\varepsilon^2 + |x_1| + |x_2| + |x|^2)^3\,|x|^{n-3}} \leqq C/\varepsilon\,,$$

(iv) $$\int\limits_{x \in R^n, |x| < R} \frac{d\sigma_n}{(\varepsilon + |x|^2)^2\,|x|^{n-2}} \leqq C/\varepsilon\,,$$

(v) $$\int\limits_{x \in R^n, |x| < R} \frac{d\sigma_n}{(\varepsilon + |x_1| + |x|^2)^3\,|x|^{n-3}} \leqq C/\varepsilon\,.$$

Proof. Parts (i) and (ii) follow from Proposition 3 if we replace ε by ε^2.

Proof of (iii) for $n = 2$.

$$\int\limits_{|x| < R} \frac{|x|\, d\sigma_2}{(\varepsilon^2 + |x_1| + |x_2| + |x|^2)^3} \leqq C \int\limits_0^R\!\!\int\limits_0^R \frac{dx_1\, dx_2}{(\varepsilon^2 + |x_1| + |x_2|)^{5/2}} \leqq C'/\varepsilon\,.$$

Proof of (iii) for $n \geqq 3$.

$$\int\limits_{|x| < R} \frac{d\sigma_n}{(\varepsilon^2 + |x_1| + |x_2| + |x|^2)^3\,|x|^{n-3}}$$

$$\leqq C \int\limits_0^R dx_1 \int\limits_0^R dx_2 \int\limits_0^R dr \int\limits_{|x''| = r} \frac{d\sigma_{n-3}}{(\varepsilon^2 + x_1 + x_2 + r^2)^3\, r^{n-3}}$$

$$\leqq C' \int\limits_0^R dr \int\limits_0^R\!\!\int\limits_0^R \frac{dx_1\, dx_2}{(\varepsilon^2 + x_1 + x_2 + r^2)^3} \leqq C'' \int\limits_0^R \frac{dr}{\varepsilon^2 + r^2} \leqq C'''/\varepsilon\,.$$

(iv)

$$\int\limits_{|x| < R} \frac{d\sigma_n}{(\varepsilon + |x|^2)^2\,|x|^{n-2}} = \int\limits_0^R dr \int\limits_{|x| = r} \frac{d\sigma_{n-1}}{(\varepsilon + r^2)^2\, r^{n-2}} \leqq C \int\limits_0^R \frac{r\, dr}{(\varepsilon + r^2)^2} = C'/\varepsilon\,.$$

(v)

$$\int\limits_{|x| < R} \frac{d\sigma_n}{(\varepsilon + |x_1| + |x|^2)^3\,|x|^{n-3}} \leqq C \int\limits_0^R dx_1 \int\limits_0^R dr \int\limits_{|x'| = r} \frac{d\sigma_{n-2}}{(\varepsilon + x_1 + r^2)^3\, r^{n-3}}$$

$$\leqq C' \int\limits_0^R dx_1 \int\limits_0^R \frac{r\, dr}{(\varepsilon + x_1 + r^2)^3} \leqq C'' \int\limits_0^R \frac{dx_1}{(\varepsilon + x_1)^2} \leqq C'''/\varepsilon\,. \blacksquare$$

Proposition 8. *There exists a constant $C < \infty$ such that for all $\varepsilon > 0$ the following estimates hold:*

(i)
$$\int\limits_{x \in R^2,\ |x| < \varepsilon} \frac{d\sigma_2}{(|x_1| + |x_2|)^{3/2}} \leqq C\sqrt{\varepsilon}\,,$$

(ii) *If $n \geqq 3$, then*
$$\int\limits_{\substack{x \in R^n,\ |x| < \sqrt{\varepsilon},\\ |x_1|,\, |x_2| < \varepsilon}} \frac{d\sigma_n}{(|x_1| + |x_2|)^{3/2}\, |x|^{n-3}} \leqq C\varepsilon\,.$$

Proof. (i)
$$\int\limits_{|x| < \varepsilon} \frac{d\sigma_2}{(|x_1| + |x_2|)^{3/2}} \leqq C \int_0^\varepsilon\!\!\int_0^\varepsilon \frac{dx_1\, dx_2}{(x_1 + x_2)^{3/2}} \leqq C' \int_0^\varepsilon \frac{dx_1}{\sqrt{x_1}} \leqq C''\sqrt{\varepsilon}\,.$$

(ii)
$$\int\limits_{\substack{x \in R^n,\ |x| < \sqrt{\varepsilon},\\ |x_1|,\, |x_2| < \varepsilon}} \frac{d\sigma_n}{(|x_1| + |x_2|)^{3/2}\, |x|^{n-3}} \leqq C \int_0^\varepsilon\!\!\int_0^\varepsilon dx_1\, dx_2 \int_0^{\sqrt{\varepsilon}} dr \int\limits_{|x''| = r} \frac{d\sigma_{n-3}}{(x_1 + x_2)^{3/2}\, r^{n-3}}$$

$$\leqq C' \int_0^\varepsilon\!\!\int_0^\varepsilon \frac{dx_1\, dx_2}{(x_1 + x_2)^{3/2}} \int_0^{\sqrt{\varepsilon}} dr \leqq C''\sqrt{\varepsilon} \int_0^\varepsilon \frac{dx_1}{\sqrt{x_1}} \leqq C'''\varepsilon\,. \blacksquare$$

Appendix 2. On Banach's open mapping theorem

If $A: E \to F$ is a continuous linear map between Fréchet spaces E and F, then we set $\operatorname{Ker} A := \{x \in E : Ax = 0\}$ and $\operatorname{Im} A := A(E)$.

Proposition 1. *Let*

$$\begin{array}{ccccc} & & E_1 & \xrightarrow{A_{21}} & E_2 \\ & & \downarrow{\scriptstyle A_{31}} & & \downarrow{\scriptstyle A_{42}} \\ E_5 & \xrightarrow{A_{35}} & E_3 & \xrightarrow{A_{43}} & E_4 \end{array} \tag{1}$$

be a commutative diagram of Fréchet spaces E_j and continuous linear maps A_{jk} such that the following conditions are fulfilled:

(i) $\operatorname{Im} A_{35} \subseteqq \operatorname{Im} A_{31}$. (2)

(ii) *There exists a continuous linear operator $P_{21}: E_1 \to E_2$ such that*

$$\operatorname{Im} P_{21} \subseteqq \operatorname{Ker} A_{42}, \tag{3}$$

$$P_{21}x = A_{21}x \quad \textit{for all} \quad x \in \operatorname{Ker} A_{42}A_{21}. \tag{4}$$

Then there exists a continuous linear operator A_{25} such that the diagram

$$\begin{array}{ccc} E_5 & \xrightarrow{A_{25}} & E_2 \\ {\scriptstyle A_{35}}\downarrow & & \downarrow{\scriptstyle A_{42}} \\ E_3 & \xrightarrow{A_{43}} & E_4 \end{array} \tag{5}$$

is commutative.

Proof. Set $A_{21}^0 := A_{21} - P_{21}$. Then, by (3), the commutative diagram (1) implies the commutative diagram

$$\begin{array}{ccccc} & & E_1 & \xrightarrow{A_{21}^0} & E_2 \\ & & \downarrow{\scriptstyle A_{31}} & & \downarrow{\scriptstyle A_{42}} \\ E_5 & \xrightarrow{A_{35}} & E_3 & \xrightarrow{A_{43}} & E_4 \end{array}. \tag{6}$$

It follows from (4) that $\operatorname{Ker} A_{21}^0 \subseteqq \operatorname{Ker} A_{42}A_{21}$. Since (1) is commutative, this implies that

$$\operatorname{Ker} A_{21}^0 \supseteqq \operatorname{Ker} A_{31}. \tag{7}$$

We denote by K the canonical map from E_1 onto the factor space $E_1/\mathrm{Ker}\, A_{31}$; let $\hat{A}_{31}$ be the operator which is defined by the condition that the diagram

$$\begin{array}{ccc} E_1 & \xrightarrow{A_{31}} & E_3 \\ {\scriptstyle K}\downarrow & & \uparrow{\scriptstyle \hat{A}_{31}} \\ & E_1/\mathrm{Ker}\, A_{31} & \end{array} \tag{8}$$

is commutative. Further, by (7), there is a unique operator $\hat{A}^0_{21}$ such that the diagram

$$\begin{array}{ccc} E_1 & \xrightarrow{A^0_{21}} & E_2 \\ {\scriptstyle K}\downarrow & & \uparrow{\scriptstyle \hat{A}^0_{21}} \\ & E_1/\mathrm{Ker}\, A_{31} & \end{array} \tag{9}$$

is commutative. From the commutative diagrams (6), (8) and (9) we obtain the following commutative diagram

$$\begin{array}{ccccc} E_1/\mathrm{Ker}\, A_{31} & \xleftarrow{K} & E_1 & \xrightarrow{A^0_{21}} & E_2 \\ & & {\scriptstyle A_{31}}\downarrow & & \downarrow{\scriptstyle A_{42}} \\ & & E_3 & \xrightarrow{A_{43}} & E_4 \\ & & \uparrow{\scriptstyle A_{35}} & & \\ & & E_5 & & \end{array} \tag{10}$$

(with $\hat{A}^0_{21}\colon E_1/\mathrm{Ker}\, A_{31} \to E_2$ and $\hat{A}_{31}\colon E_1/\mathrm{Ker}\, A_{31} \to E_3$).

Since, by (2), $\mathrm{Im}\, A_{35} \subseteqq \mathrm{Im}\, A_{31} = \mathrm{Im}\, \hat{A}_{31}$, there is a unique operator $\hat{A}_{15}\colon E_5 \to E_1/\mathrm{Ker}\, A_{31}$ such that the diagram

$$\begin{array}{ccc} E_5 & \xrightarrow{A_{35}} & E_3 \\ {\scriptstyle \hat{A}_{15}}\downarrow & & \uparrow{\scriptstyle \hat{A}_{31}} \\ & E_1/\mathrm{Ker}\, A_{31} & \end{array} \tag{11}$$

is commutative. Since (11) is commutative, $\hat{A}_{15}$ is closed and therefore, by Banach's open mapping theorem, continuous. Define $A_{25} := \hat{A}^0_{21}\hat{A}_{15}$. Then the commutative diagrams (10) and (11) imply that (5) is commutative. ■

Proposition 2. *Condition* (ii) *in Proposition* 1 *is fulfilled if there exist a Hilbert space H and continuous linear operators $A''\colon H \to E_2$, $A'\colon E_1 \to H$ such that $A_{21} = A''A'$.*

Proof. Set $P_{21} := A''PA'$, where P is the orthogonal projection from H onto $\mathrm{Ker}\,(A_{42}A'')$. ■

A bounded linear operator A in a Banach space E is called a *Fredholm operator*, if $\dim \mathrm{Ker}\, A < \infty$ and $\dim (E/\mathrm{Im}\, A) < \infty$ (then $\mathrm{Im}\, A$ is a topologically closed subspace of E — cp. Proposition 4 below). A linear operator V in E is called *compact* if, for every bounded subset U of E, $V(U)$ is a relatively compact subset of E. Recall that for every compact operator V in E, the operator $\mathrm{id} + V$ is a Fredholm operator, where $\dim \mathrm{Ker}\,(\mathrm{id} + V) = \dim \big(E/\mathrm{Im}\,(\mathrm{id} + V)\big)$ ($\mathrm{id} :=$ identity operator).

Proposition 3. *Let E, F be Banach spaces and let A be a closed linear operator between E and F with the domain of definition $D(A) \subseteqq E$. Suppose that the following conditions are fulfilled:*

(i) $A(D(A)) = F$.

(ii) *There exists a bounded linear operator* $B: F \to E$ *such that* $B(F) \subseteqq D(A)$ *and* $\mathrm{id} - AB$ *is compact.*

Then there exists a boundet linear operator $A^{(-1)}: F \to E$ *sucht that* $A^{(-1)}F \subseteqq D(A)$ *and* $AA^{(-1)} = \mathrm{id}$.

Proof. Since $\mathrm{id} - AB$ is compact, we can find closed linear subspaces M and N of F such that $M \cap \operatorname{Ker} AB = 0$, $AB(M) \cap N = 0$, $M + \operatorname{Ker} AB = AB(M) + N = F$. Since $A(D(A)) = F$, and since $\dim N = \dim \operatorname{Ker} AB$, we can find a linear operator $S: \operatorname{Ker} AB \to D(A)$ such that $AS(\operatorname{Ker} AB) = N$ and $ASx \neq 0$ for all $0 \neq x \in \operatorname{Ker} AB$. Let $B': F \to E$ be the bounded linear operator defined by $B'x = Bx$ for $x \in M$ and $B'x = Sx$ for $x \in \operatorname{Ker} AB$. Then AB' is invertible. Set $A^{(-1)} := B'(AB')^{-1}$. ■

Proposition 4. *Let A be a closed operator between Banach spaces E and F with the domain of definition $D(A) \subseteqq E$. If $A(D(A))$ is finitely codimensional in F, then $A(D(A))$ is a topologically closed subspace of F.*

Proof. Let n be the codimension of $A(D(A))$ in F. Choose a linear operator $B: \mathbb{C}^n \to F$ such that $\operatorname{Im} B + A(D(A)) = F$. Let A' be the operator with the domain of definition $D(A') := D(A) \oplus \mathbb{C}^n$ which is defined by $A'x = Ax$ for $x \in D(A) \oplus 0$ and $A'x := Bx$ for $x \in 0 \oplus \mathbb{C}^n$. Then A' is a closed linear operator between $E \oplus \mathbb{C}^n$ and F such that $A'(D(A')) = F$. Set $\operatorname{Ker} A' := \{x \in D(A') : A'x = 0\}$. Then $\operatorname{Ker} A'$ is a topologically closed subspace of $E \oplus \mathbb{C}^n$. Let $\hat{A}'$ be the closed linear operator between $E \oplus \mathbb{C}^n/\operatorname{Ker} A'$ and F with the domain of definition $D(A')/\operatorname{Ker} A'$ which is induced by A'. This operator is bijective from $D(A')/\operatorname{Ker} A'$ onto F. By Banach's open mapping theorem its inverse $(\hat{A}')^{-1}$ is a bounded linear operator from F into $E \oplus \mathbb{C}^n/\operatorname{Ker} A'$. Since $A(D(A)) = [(\hat{A}')^{-1}]^{-1} (E \oplus 0/\operatorname{Ker} A')$, this implies that $A(D(A))$ is topologically closed. ■

Bibliography

Agranovskij, M. L., and R. E. Val'skij
[1] Maximality of invariant algebras of functions (Russian). Sibirsk. Mat. Ž. **12** (1971), 3—12; Engl. transl. in Siberian Math. J. **12** (1971), no. 1, 1—7.

Ahern, P., and R. Schneider
[1] Holomorphic Lipschitz functions in pseudoconvex domains. Amer. J. Math. **101**, (1979), no. 3 543—565.

Ajrapetjan, R. A.
[1] Boundary properties of an integral of Cauchy-Leray type in piecewise smooth domains in $\mathbb{C}^n$ (Russian). Dokl. Akad. Nauk SSSR **246** (1979); no. 4, 779—782; Engl. transl. in Soviet Math. Dokl. **20** (1979), no. 3, 519—522.

Ajrapetjan, R. A., and G. M. Henkin
[1] Integral representations of differential forms on CR-manifolds and the theory of CR-functions (Russian). Uspehi Mat. Nauk. In press.

Ajsenberg, L. A., and A. P. Južakov
[1] Integral representations and residues in multidimensional complex analysis (Russian). Nauka, Sibirsk. Otdel., Novosibirsk 1979.

Aleksandrov, A. B.
[1] The existence of inner functions in the ball (Russian). Mat. Sb. **118** (1982), 147—163.

Amar, E.
[1] Extension de fonctions holomorphes et courants. Preprint, Université de Paris-Sud, Département de Mathématique, December 1980; Bull. Sci. Math. (2) **107** (1983), 25—48.
[2] Generalisation d'un théorème de Wolff à la boule de $\mathbb{C}^n$. Preprint, Université de Paris-Sud, Département de Mathématique, December 1980.

Anderson, M., and B. Berndtsson
[1] Henkin-Ramirez formulas with weight factors. Ann. Inst. Fourier. To appear.

Andreian Cazacu, C.
[1] Theorie der Funktionen mehrerer komplexer Veränderlicher. VEB Deutscher Verlag der Wissenschaften, Berlin 1975.

Arenson, E. A.
[1] Gleason valleys and the Choquet boundary of the algebra of functions on a convex compactum. Zap. Naučn. Sem. Leningrad. Otdel. Mat. Inst. Steklov. (LOMI) **183** (1981), 204—207.

Basener, R.
[1] Peak points, barriers and pseudoconvex boundary points. Proc. Amer. Math. Soc. **65** (1977), 89—92.

Bedford, K., and J. E. Fornaess
[1] Approximation on pseudoconvex domains. Preprint, Princeton University 1979.

Bergman, S.
[1] Über eine Integraldarstellung von Funktionen zweier komplexer Veränderlicher. Mat. Sb. **1** (1936), 851—861.

[2] Über eine in gewissen Bereichen mit Maximumfläche gültige Integraldarstellung der Funktionen zweier komplexer Variabler, I. Math. Z. **39** (1934), 76–94; II. Math. Z. **39** (1934), 605–608.

BERNDTSSON, B.
[1] Integral formulas for the $\partial\bar{\partial}$-equation and zeros of bounded holomorphic functions in the unit ball. Math. Ann. **249** (1980), 163–176.
[2] A formula for interpolation and division in $\mathbb{C}^n$. Preprint 1981.

BISHOP, E.
[1] Mappings of partially analytic spaces. Amer. J. Math. **83** (1961), 209–242.
[2] Some global problems in the theory of functions of several complex variables. Amer. J. Math. **83** (1961), 479–498.
[3] Partially analytic spaces. Amer. J. Math. **83** (1961), 669–692.

BOCHNER. S.
[1] Analytic and meromorphic continuation by means of Green's formula. Ann. Math. **44** (1943), 652–673.
[2] Group invariance of Cauchy's formula in several variables. Ann. Math. **45** (1944), 686–707.

BREMERMANN, H.
[1] Über die Äquivalenz der pseudokonvexen Gebiete und der Holomorphiegebiete im Raum von n komplexen Veränderlichen. Math. Ann. **128** (1954), 63–91.
[2] On the conjecture of the equivalence of the plurisubharmonic functions and the Hartogs functions. Math. Ann. **131** (1956), 76–86.

BURNS, D., and E. L. STOUT
[1] Extending functions from submanifolds of the boundary. Duke Math. J. **43** (1976), 391–404.

BYČKOV, S. N.
[1] Geometric properties of the boundary of a domain of holomorphy (Russian). Izv. Akad. Nauk SSSR Ser. Mat. **44** (1980), no. 1, 46–62; Engl. transl. in Math. USSR Izv. **16** (1981), no. 1, 41–54.

CARLESON, L.
[1] Selected problems on exeptional sets. Van Nostrand, Princeton—Toronto—London—Melbourne 1967.
[2] Interpolations by bounded analytic functions and the corona problem. Ann. Math. **76** (1962), 547–559.
[3] The corona theorem. In: Lecture Notes in Math., vol. 118. Springer-Verlag, Berlin—Heidelberg—New York 1970, 121–132.

CARTAN, H.
[1] Sur les matrices holomorphes de n variables complexes. J. Math. Pures Appl. **19** (1940), 1–26.
[2] Espaces fibrés analytiques. Symp. Internat. Topol. Algebraica, Agosto 1956, Mexico, 1958, 97–121.
[3] Séminaires E. N. S., 1951/52.

CARTAN, H., and P. THULLEN
[1] Zur Theorie der Singularitäten der Funktionen mehrerer komplexer Veränderlicher. Regularitäts- und Konvergenzbereiche. Math. Ann. **106** (1932), 617–647.

CHAUMAT, J., and A. M. CHOLLET
[1] Ensembles pics pour $A^\infty(D)$. Ann. Inst. Fourier **29** (1979), no. 3, 171–200.

ČIRKA, E. M., and G. M. HENKIN
[1] Boundary properties of holomorphic functions of several complex variables (Russian). Current problems in mathematics, vol. 4, Moscow 1975, 12–142.

CORNALBA, M., and P. GRIFFITHS
[1] Analytic cycles and vector bundles on non-compact algebraic varieties. Invent. math. **28** (1975), 1–106.

Cumenge, A.
[1] Extension dans les classes de Hardy de fonctions holomorphes définies sur des sous-variétés en position générique. Ann. Inst. Fourier. To appear.

Dautov, Š. A., and A. M. Kytmanov
[1] Boundary properties of an integral of Martinelli-Bochner type (Russian). Properties of holomorphic functions of several complex variables, Inst. Fiz. Sibirsk. Otdel. Akad. Nauk SSSR, Krasnoyarsk 1973, 49—54.

Dautov, Š. A., and G. M. Henkin
[1] The zeroes of holomorphic functions of finite order and weight estimates for solutions of the $\bar{\partial}$-equation (Russian). Mat. Sb. **107** (1978), 163—174.

Debiard, A., and B. Gaveau
[1] Frontière de Shilov des domaines faiblement pseudoconvexes de $\mathbb{C}^n$. Bull. Sci. Math. **100** (1976), 17—31.

Diederich, K., and J. E. Fornaess
[1] A strange bounded smooth domain of holomorphy. Bull. Amer. Math. Soc. **82** (1976), 74—76.
[2] Pseudoconvex domains with real analytic boundary. Ann. Math. **107** (1978), 371—384.

Docquier, F., and H. Grauert
[1] Levisches Problem und Rungescher Satz für Teilgebiete Steinscher Mannigfaltigkeiten. Math. Ann. **140** (1960), 94—123.

Dolbeault, P.
[1] Formes différentielles et cohomologie sur une variété analytique complexe, I. Ann. Math. **64** (1956), 83—130; II. Ann. Math. **65** (1957), 282—330.
[2] Sur la cohomologie des variétés analytiques complexes. C. R. Acad. Sci. Paris **236** (1953), 175—177.

Dyn'kin, E. M.
[1] Smoothness of Cauchy-type integrals (Russian, English summary). Investigations on linear operators and the theory of functions, IX. Zap. Naučn. Sem. Leningrad. Otdel. Mat. Inst. Steklov. (LOMI) **92** (1979), 115—133, 320—321.

Ehrenpreis, L.
[1] A new proof and an extension of Hartog's theorem. Bull. Amer. Math. Soc. **67** (1961), 507—509.

Fantappiè, L.
[1] L'indicatrice proiettiva dei funzionali lineari e i prodotti funzionali proiettivi. Annali di Mat. **22**, Serie 4 (1943), 181—289.

Fefferman, C.
[1] The Bergman kernel and biholomorphic mappings of pseudoconvex domains. Invent. Math. **26** (1974), 1—65.

Fefferman, C., and E. M. Stein
[1] H^p spaces of several variables. Acta Math. **129** (1972), 137—193.

Folland, G. B., and E. M. Stein
[1] Estimates for the $\bar{\partial}_b$-complex and analysis on the Heisenberg group. Comm. Pure Appl. Math. **27** (1974), 429—522.

Fornaess, J. E.
[1] Embedding strictly pseudoconvex domains in convex domains. Amer. J. Math. **98** (1976), 529—569.

Fornaess, J., and A. Nagel
[1] The Mergelyan property for weakly pseudoconvex domains. Manuscripta Math. **22** (1977), 199—208.

Fornaess, J. E., and N. Øvrelid
[1] Finitely generated ideals in $A(\Omega)$. Preprint 1982.

FORSTER, O.
[1] Plongements des variétés de Stein. Comment. Math. Helv. **45** (1970), no. 2, 170—184.

FORSTER, O., and K. J. RAMSPOTT
[1] Über die Anzahl der Erzeugenden von projektiven Steinschen Moduln. Arch. Math. **19** (1968), 417—422.

FUKS, B. A.
[1] The theory of analytic functions of several complex variables. To appear.

FRUMIN, I. L.
[1] Estimates for solutions of the $\bar{\partial}$-equation in analytic polyhedra in general position (Russian). Uspehi Mat. Nauk **37** (1982), no. 4, 189—190.

GAMELIN, T. W.
[1] Wolff's proof of the corona theorem. Israel J. Math. **37** (1980), no. 1—2, 113—119.

GINDIKIN, S. G., and G. M. HENKIN
[1] Integral geometry for the $\bar{\partial}$-cohomology in q-linearly concav domains in CP^n (Russian). Funkcional. Anal. i Priložen. **12** (1978), no. 4, 6—23.

GLEASON, A.
[1] The Cauchy-Weil theorem. J. Math. and Mech. **12** (1963), 429 —444.
[2] Finitely generated ideals in Banach algebras. J. Math. and Mech. **13** (1964), 125—132.

GOHBERG, I., and J. LEITERER
[1] Holomorphic vector-valued functions of one variable. I. Functions on a compact set (Russian). Mat. Issled. **7** (1972), no. 4, 60—84.
[2] Holomorphic vector-valued functions of one variable. II. Functions in a domain (Russian). Mat. Issled. **8** (1973), no. 1, 37—58.
[3] Cocycles, operator-valued functions and families of subspaces (Russian). Mat. Issled. **8** (1973), no. 2, 23—56.

GORIN, E. A., S. V. HRUŠČEV, and S. A. VINOGRADOV
[1] Free interpolation according to Jones (Russian). Zap. Naučn. Sem. Leningrad. Otdel. Mat. Inst. Steklov. (LOMI) **113** (1981).

GRAUERT, H.
[1] Approximationssätze für holomorphe Funktionen mit Werten in komplexen Räumen. Math. Ann. **133** (1957), 139—159.
[2] Holomorphe Funktionen mit Werten in komplexen Lieschen Gruppen. Math. Ann. **133** (1957), 450—472.
[3] Analytische Faserungen über holomorph-vollständigen Räumen. Math. Ann. **135** (1958), 263—273.
[4] On Levi's problem and the imbedding of real-analytic manifolds. Ann. Math. **68** (1958), 460—472.

GRAUERT, H., und K. FRITZSCHE
[1] Einführung in die Funktionentheorie mehrerer Veränderlicher. Springer-Verlag, Berlin—Heidelberg—New York 1974.

GRAUERT, H., und I. LIEB
[1] Das Ramirezsche Integral und die Lösung der Gleichung $\bar{\partial} f = \alpha$ im Bereich der beschränkten Formen. Rice Univ. Studies **56** (1970), no. 2, 29—50.

GRAUERT, H., und R. REMMERT
[1] Zur Spaltung lokal-freier Garben über Riemannschen Flächen. Math. Z. **144** (1975), 35—43.
[2] Kohärente analytische Garben. To appear.

GREINER, P. C., and E. STEIN
[1] Estimates for the $\bar{\partial}$-Neumann problem. Princeton University Press, Princeton 1977.

GRIFFITHS, P., and J. HARRIS.
[1] Principles of algebraic geometry. John Wiley, New York 1978.

GROMOV, M. L., and JA. M. ELIAŠBERG
[1] Nonsingular mappings of Stein manifolds (Russian). Funkcional. Anal. i Priložen. **5** (1971), no. 2, 82—83; Engl. transl. in Func. Anal. Appl. **5** (1971), 156—157.

GUNNING, R. C., and H. ROSSI
[1] Analytic functions of several complex variables. Prentice-Hall, Englewood Cliffs (N.J.) 1965.

HAKIM, M., and N. SIBONY
[1] Frontière de Shilov et spectra de $A(\bar{D})$ pour des domaines faiblement pseudoconvexes. C.R. Acad. Sci. Paris **281** (1975), A 959—A 962.
[2] Spectre de $A(\bar{\Omega})$ pur des domaines bornés faiblement pseudoconvexes réguliers, J. Funct. Anal. **37** (1980), no. 2, 127—135.

HARTOGS, F.
[1] Einige Folgerungen aus der Cauchyschen Integralformel bei Funktionen mehrerer Veränderlicher. Münchener Sitz. Ber. **36** (1906), 223—242.

HARVEY, R.
[1] Holomorphic chains and their boundaries. Proc. Symp. Pure Math. **30** (1977), no. 1, 309—382.
[2] Integral formulas connected by Dolbeault's isomorphism. Rice University Studies **56** (1970), 77—97.

HARVEY, R., and H. B. LAWSON
[1] On boundaries of complex analytic varieties, I. Ann. Math. **102** (1975), 233—290.

HARVEY, R., and J. POLKING
[1] Fundamental solutions in complex analysis. Duke Math. J. **46** (1979), no. 2, 253—340.

HARVEY, F. R., and R. O. WELLS, Jr.
[1] Zero sets of non-negative strictly plurisubharmonic functions. Math. Ann. **201** (1973), 165—170.
[2] Holomorphic approximation and hyperfunction theory on a C^1 totally real submanifold of a complex manifold. Math. Ann. **197** (1972), 287—318.

HAVIN, V. P., and S. A. VINOGRADOV
[1] Free interpolation in H^∞ and in certain other classes of functions (Russian). Investigations on linear operators and the theory of functions. Zap. Naučn. Sem. Leningrad. Otdel. Mat. Inst. Steklov. (LOMI) **47** (1974), 15—54, **56** (1976), 12—58, and **92** (1979), 317.

HEFER, H.
[1] Zur Funktionentheorie mehrerer Veränderlicher. Math. Ann. **122** (1950), 276—278.

HENKIN, G. M.
[1] Integral representation of functions in strongly pseudoconvex domains and applications to the $\bar{\partial}$-problem (Russian). Mat. Sb. **82** (1970), 300—308; Engl. transl. in Math. USSR-Sb. **11** (1970), 273—281.
[2] Solutions with bounds for the equations of H. Lewy and Poincaré-Lelong. Construction of functions of Nevanlinna class with given zeroes in a strongly pseudoconvex domain (Russian). Dokl. Akad. Nauk SSSR **224** (1975), no. 4, 771—774; Engl. transl. in Soviet. Math. Dokl. **16** (1975), no. 5, 1310—1314.
[3] H. Lewy's equation and analysis on pseudoconvex manifolds, I (Russian). Uspehi, Mat. Nauk. **32** (1977), no. 3, 57—118; Engl. transl. in Russian Math. Surveys **32** (1977), no. 3, 59—130; and II, Mat. Sb. **102** (1977), no. 1, 71—108; Engl. transl. in Math. USSR-Sb. **102** (1977)., no. 1, 63—64.
[4] Integral representation of functions holomorphic in strictly pseudoconvex domains and some applications (Russian). Mat. Sb. **78** (1969), 611—632; Engl. transl. in Math. USSR-Sb. **7** (1969), 597—616.
[5] Continuation of bounded holomorphic functions from submanifolds in general position in a strictly pseudoconvex domain (Russian). Izv. Akad. Nauk SSSR Ser. Mat. **36** (1972), 540—567; Engl. transl. in Math. USSR Izv. **6** (1972), no. 3, 536—563.

[6] A uniform estimate for the solution of the $\bar{\partial}$-problem in a Weil domain (Russian). Uspehi Mat. Nauk **26** (1971), no. 3, 211—212.
[7] Approximation of functions in pseudoconvex domains and a theorem of Z. L. Leĭbenzon (Russian). Bull. Acad. Polon. Sci. Sér. Sci. Math. Astronom. Phys. **19** (1971), 37—42.

Henkin, G. M., and J. Leiterer
[1] Global integral formulas for solving the $\bar{\partial}$-equation on Stein manifolds. Preprint, Berlin and Warszawa 1979; Ann. Polonici Math. **39** (1981), 93—116.
[2] Theory of functions on strictly pseudoconvex sets with non-smooth boundary. Report R-MATH-02/81 IMath der AdW der DDR, Berlin 1981.

Henkin, G. M., and A. I. Petrosjan
[1] Solution with a uniform estimate of the $\bar{\partial}$-equation in a real non-degenerate Weil polyhedron (Russian). Izv. Akad. Nauk Armjan. SSSR Ser. Mat. **13** (1978), no. 5—6, 428—441.

Henkin, G. M., and A. V. Romanov
[1] Exact Hölder estimates of solutions of the $\bar{\partial}$-equation (Russian). Izv. Akad. Nauk SSSR Ser. Mat. **35** (1971), 1171—1183; Engl. transl. in Math. USSR Izv. **5** (1971), 1180—1192.

Henkin, G. M., and A. G. Sergeev
[1] Uniform estimates for solutions of the $\bar{\partial}$-equation in pseudoconvex polyhdera (Russian). Mat. Sb. **112** (1980), no. 4, 522—567; Engl. transl. in Math. USSR-Sb. **40** (1981), 469—507.

Henkin, G. M., and A. E. Tumanov
[1] Interpolation submanifolds of pseudoconvex manifolds (Russian). Mathematical programming and related questions (Proc. Seventh Winter School, Drogobych, 1974). Theory of functions and functional analysis, Central. Ekonom.-Mat. Inst. Akad. Nauk SSSR, Moscow 1976, 74—86.

Henriksen, S.
[1] A peak set of Hausdorff dimension 1 for the algebra $A(D)$ in the boundary of a domain D with C^∞-boundary in $\mathbb{C}^n$. Preprint, Univ. of Oslo 1981.

Heunemann, D.
[1] Estimates for the $\bar{\partial}$-equation on Stein manifolds. Math. Nachr. To appear.

Hoffman, K.
[1] Banach spaces of analytic functions. Prentice-Hall, Englewood Cliffs (N.J.) 1962.

Hörmander, L.
[1] An introduction to complex analysis in several variables. D. Van Nostrand, Princeton—Toronto—London 1966.
[2] L^2 estimates and existence theorems for the $\bar{\partial}$-operator. Acta Math. **113** (1965), 89—152.
[3] L^p estimates for (pluri)-subharmonic functions. Math. Scand. **20** (1967), no. 1, 65—78.

Hörmander, L., and J. Wermer
[1] Uniform approximation on compact sets in $\mathbb{C}^n$. Math. Scand. **23** (1968), no. 1, 5—21.

Hortmann, M.
[1] Globale holomorphe Kerne zur Lösung der Cauchy-Riemannschen Differentialgleichungen. Preprint 1979; In: Fornaess, J. (ed.): Recent developments in several complex variables. Princeton University Press, Princeton 1981.

Jöricke, B.
[1] The multidimensional analogue of Privalov's theorem (Russian). Math. Nachr. **107** (1982), 221—233.
[2] The analogue of Tamrasov's theorem and continuity of the Cauchy projection in Hölder norms for certain multidimensional domains (Russian). Math. Nachr. **112** (1983).

Kerzman, N.
[1] Hölder and L^p estimates for solutions of $\bar{\partial}u = f$ in strongly pseudoconvex domains. Bull. Amer. Math. Soc. **76** (1970), 860—864.

[2] Hölder and L^p estimates for solutions of $\bar{\partial} u = f$ in strongly pseudoconvex domains. Comm. Pure Appl. Math. **24** (1971), 301–379.

KERZMAN, N., and E. M. STEIN
[1] The Szegö kernel in terms of Cauchy-Fantappiè kernels. Duke Math. J. **45** (1978), 197–224.

KNESER, H.
[1] Die singulären Kanten bei analytischen Funktionen mehrerer komplexer Veränderlicher. Math. Ann. **106** (1932), 656–660.

KOHN, J. J.
[1] Harmonic integrals on strongly pseudo-convex manifolds I. Ann. Math. **78** (1963), 112–148; II. Ann. Math. **79** (1964), 450–472.
[2] Boundaries of complex manifolds. In: Proc. Conf. Complex Analysis, Minneapolis 1964, Springer-Verlag, Berlin—Heidelberg—New York 1965, 81–94.
[3] Global regularity for $\bar{\partial}$ on weakly pseudoconvex manifolds. Trans. Amer. Math. Soc. **181** (1973), 273–292.
[4] Subellipticity of the $\bar{\partial}$-Neumann problem on pseudo-convex domains: sufficient conditions. Acta Math. **142** (1979), 79–122.

KOHN, J. J., and L. NIRENBERG
[1] A pseudoconvex domain not admitting a holomorphic support function. Math. Ann. **201** (1973), 265–268.

KOOSIS, P.
[1] Introduction to H^p-spaces. London Math. Soc. Lecture Note series, vol. **40**, Cambridge University Press, Cambridge 1980.

KOPPELMAN, W.
[1] The Cauchy integral for differential forms. Bull. Amer. Math. Soc. **73** (1967), 554–556.

KRANTZ, S. G.
[1] Characterizations of various domains of holomorphy via $\bar{\partial}$-estimates and applications to a problem of Kohn. Illinois J. Math. **23** (1979), 267–285.
[2] Function theory of several complex variables. John Wiley, New York 1982.

LEITERER, J.
[1] Über analytische Operatorfunktionen einer komplexen Veränderlichen mit stetigen Randwerten. Math. Nachr. **72** (1976), 247–274.
[2] Theorem B für analytische Funktionen mit stetigen Randwerten. In: Beiträge zur Analysis, Band 8. VEB Deutscher Verlag der Wissenschaften, Berlin 1976, 95–102.
[3] Equivalence of Steinness and validity of Oka's principle for Subdomains of Stein manifolds. Math. Nachr. **89** (1979), 181–183.
[4] Über das Okasche Prinzip für Schnitte in holomorphen Vektorraumbündeln. Preprint, ZIMM der AdW der DDR, Berlin 1977.

LELONG, P.
[1] Les fonctions plurisousharmoniques. Ann. Ecole Norm. Sup. **62** (1945), 301–338.
[2] Sur quelques problèmés de la thèorie des fonctions de deux variables complexes. Ann. Ecole Norm. Sup. **58** (1941), 83–177.
[3] Propriétés métriques des variétés analytiques complexes définies par une équation. Ann. Ecole Norm. Sup. **67** (1950), 393–419.

LERAY, J.
[1] Fonction de variables complexes: sa représentation comme somme de puissances négatives de fonctions linéaires. Rend. Accad. Naz. Lincei. serie 8, **20** (1956), 589–590.
[2] Le calcul différentiel et intégral sur une variété analytique complexe. Bull. Soc. Math. France 87 (1959), 81–180.

LEVI, E.
[1] Studii sui punti singolari essenziali delle funzioni analitiche di due o più variabili complesse. Ann. mat. pura ed appl. **17** (1910), no. 3, 61–87.

LEWY, H.
[1] On the local character of the solutions of an atypical linear differential equation in three variables and a related theorem for regular functions of two complex variables. Ann. Math. **64** (1956), 514–522.
[2] An example of a smooth linear partial differential equation without solution. Ann. Math. **66** (1957), 155–158.

LIEB, I.
[1] Ein Approximationssatz auf streng pseudokonvexen Gebieten. Math. Ann. **184** (1969), 56–60.
[2] Die Cauchy-Riemannschen Differentialgleichungen auf streng pseudokonvexen Gebieten. Math. Ann. **190** (1970/71), 6–44; **199** (1972), 241–256.

MALGRANGE, B.
[1] Lectures on the theory of functions of several complex variables. Bombay 1958.
[2] Ideals of differentiable functions. Oxford University Press, Oxford 1966.

MARTINELLI, E.
[1] Alcuni teoremi integrali per le funzioni analitiche die più variabili complesse. Mem. r. Acad. Ital. **9** (1938), 269–283.
[2] Sopra una dimostrazione di R. Fueter per un teorema di Hartogs. Comment. Math. Helv. **15** (1943), 340–349.
[3] Contributi alla teoria dei residui per le funsioni di due variabili complesse. Ann. mat. pura ed appl. **39** (1955), 4, 335–343.

MILNOR, J.
[1] Lectures on the h-cobordism theorem. Princeton University Press, Princeton 1965.

MORREY, C. B.
[1] The analytic embedding of abstract real analytic manifolds. Ann. Math. **68** (1958), 159–201.

MUSKHELIŠVILI, N. I.
[1] Singular integral equations (Russian). Nauka, Moscow 1968. Engl. transl. ed. J. R. M. RADOK, Noordhoff, Leyden 1977.

NAGEL, A.
[1] Smooth zero sets and interpolation sets for some algebras of holomorphic functions on strictly pseudoconvex domains. Duke Math. J. **43** (1976), 323–348.

NAGEL, A., and W. RUDIN
[1] Local boundary behavior of bounded holomorphic functions. Canad. J. Math. **30** (1978), 583–592.

NARASIMHAN, R.
[1] Compact analytic variéties. L'Enseignement Math. **14**, (1968), 75–98.
[2] Analysis on real and complex manifolds. North Holland, Amsterdam 1968.
[3] Imbedding of holomorphically complete complex spaces. Amer. J. Math. **82** (1960), 917–934.

NIRENBERG, R., and R. O. WELLS, Jr.
[1] Approximation theorems on differentiable submanifolds of a complex manifold. Trans. Amer. Math. Soc. **142** (1969), 15–35.

NORGUET, F.
[1] Sur les domaines d'holomorphie des fonctions uniformes de plusieurs variables complexes. Bull. Soc. Math. France **82** (1954), 137–159.
[2] Problèmes sur les formes différentielles et les courants. Ann. Inst. Fourier **11** (1961), 1–82.

OKA, K.
[1] Sur les fonctions analytiques de plusieurs variables complexes. IV. Domaines d'holomorphie et domaines rationellement convexes. Jap. J. Math. **17** (1941), 517–521.
[2] Sur les fonctions analytiques de plusieurs variables complexes. II. Domaines d'holomorphie. J. Sci. Hiroshima Univ. **7** (1937), 115–130.

[3] Sur les fonctions analytiques de plusieurs variables complexes. III. Deuxième problème de Cousin. J. Sci. Hiroshima Univ. **9** (1939), 7—19.
[4] Sur les fonctions analytiques de plusieurs variables complexes. VI. Domaines pseudoconvexes. Tohoku Math. J. **49** (1942), 15—22.
[5] Sur les fonctions analytiques de plusieurs variables complexes. VII. Sur quelques notions arithmétiques. Bull. Soc. Math. France **78** (1950), 1—27.
[6] Sur les fonctions analytiques de plusieurs variables complexes. IX. Domaines finis sans point critique intérieur. Jap. J. Math. **23** (1953), 97—155.
[7] Sur les fonctions analytiques de plusieurs variables complexes. I. Domaines convexes par rapport aux fonctions rationelles. J. Sci. Hiroshima Univ. **6** (1936), 245—255.
[8] Sur les fonctions analytiques de plusieurs variables complexes. V. L'intégrale de Cauchy. Jap. J. Math. **17** (1941), 525—531.

ØVRELID, N.
[1] Integral representation formulas and L^p-estimates for the $\bar\partial$-equation. Math. Scand. **29** (1971), no. 1, 137—160.
[2] Generators of the maximal ideals of $A(\bar{D})$. Pac. J. Math. **39** (1971) 1, 219—223.
[3] Integral representation formulas for differential forms and solutions of the $\bar\partial$-equation. Colloque International du C.N.R.S., N 208, Agora Mathematica, Gauthier-Villars, Paris 1974, 180—198.

PALM, A.
[1] Integral representation formulas on strictly pseudoconvex domains in Stein manifolds. Duke Math. J. **43** (1976), 41—62.

PETROSJAN, A. I.
[1] Uniform approximation of functions by polynomials on Weil polyhedra (Russian). Izv. Akad. Nauk SSSR Ser. Mat. **34** (1970), 1241—1261; Engl. transl. in Math. USSR Izv. **34** (1970), no. 6, 1250—1271.

PHONG, D. H., and E. M. STEIN
[1] Estimates for the Bergman and Szegö projections on strongly pseudoconvex domains. Duke Math. J. **44** (1977), 695—704.

PINČUK, S. I.
[1] A boundary uniqueness theorem for holomorphic functions of several complex variables (Russian). Mat. Zametki **15** (1974), 205—212; Engl. transl. in Math. Notes **15** (1974), no. 2, 116—120.

POLJAKOV, P. L.
[1] The Cauchy-Weil formula for differential forms (Russian). Mat. Sb. **85** (1971), 388—402; Engl. transl. in Math. USSR-Sb. **14** (1971), no. 3, 383—398.
[2] Extension of bounded holomorphic functions from an analytic curve in general position in a polydisc. Funkcional. Anal. i Priložen. In press.
[3] Banach cohomology on piecewise strictly pseudoconvex domains (Russian). Mat. Sb. **88** (1972), 238—255; Engl. transl. in Math. USSR-Sb. **17** (1972), no. 2, 237—256.

POMPEIJU, D.
[1] Sur les singularités des fonctions analytiques uniformes. C.R. Acad. Sci. Paris **139** (1904), 914—915.

RANGE, R. M.
[1] Hölder estimates for $\bar\partial$ on convex domains in $\mathbb{C}^2$ with real analytic boundary. Proc. Symp. Pure Math. **30** (1977), no. 2, 31—33.

RANGE, R. M., and Y. T. SIU
[1] Uniform estimates for the $\bar\partial$-equation on domains with piecewise smooth strictly pseudoconvex boundaries. Math. Ann. **206** (1973), 325—354.

RAMIREZ DE ARELLANO, E.
[1] Ein Divisionsproblem und Randintegraldarstellungen in der komplexen Analysis. Math. Ann. **184** (1970), 172—187.

RÖHRL, H.
[1] Das Riemann-Hilbertsche Problem der Theorie der linearen Differentialgleichungen. Math. Ann. **133** (1957), 1—25.

RONKIN, L. I.
[1] Introduction to the theory of entire functions of several variables (Russian). Moscow 1971. Engl. transl., Amer. Math. Soc., Providence 1974.

ROSSI, H.
[1] Holomorphically convex sets in several complex variables. Ann. Math. **74** (1961), 470—493.

RUDIN, W.
[1] Peak interpolation sets of class C^1. Pacific J. Math. **75** (1978), no. 1, 267—279.
[2] Projections on invariant subspaces. Proc. Amer. Math. Soc. **13** (1962), 429—432.
[3] Function theory in the unit ball of $\mathbb{C}^n$. Springer-Verlag, New York 1980.

ŠABAT, B. V.
[1] Introduction to complex analysis (Russian). Nauka, Moscow 1976.

SAKAI, A.
[1] Uniform approximation in several complex variables. Osaka J. Math. **15** (1978), 589—611.

SALAEV, V. V.
[1] Direct and inverse estimates for a singular Cauchy integral along a closed curve (Russian). Mat. Zametki **19** (1976), no. 3, 365—380; Engl. transl. in Math. Notes **19** (1976), no. 3, 221—231.

SCHNEIDER, M.
[1] Vollständige Durchschnitte in Steinschen Mannigfaltigkeiten. Math. Ann. **186** (1970), 191—200.

SERGEEV, A. G.
[1] Uniform estimates for the $\bar{\partial}$-problem in a Levi-Weil domain (Russian). Dokl. Akad. Nauk SSSR **236** (1977), 1080—1082; Engl. transl. in Soviet Math. Dokl. **18** (1977), no. 5, 1335—1338.

SIBONY, N.
[1] Un exemple de domaine pseudoconvexe régulier ou l'équation $\bar{\partial}u = f$ n'admet pas de solution bornée pour f bornée. Invent. math. **62** (1980), 235—242.
[2] Prolongement des fonctions holomorphes bornées et métrique de Carathéodory. Invent. math. **29** (1975), 205—230.
[3] Valeurs au bord de fonctions holomorphes et ensembles polynomialement convexes. In: Lecture Notes in Math., vol. 578, Springer-Verlag Berlin—Heidelberg—New York 1977, 300—313.

SKODA, H.
[1] Valeurs au bord pour les solutions de l'opérateur d'' dans les ouverts strictement pseudoconvexes. C.R. Acad. Sci. Paris **280** (1975), 633—636.
[2] Valeurs au bord pour les solution de l'operateur d'' et charactérisation des zéros des fonctions de la classe de Nevanlinna. Bull. Soc. Math. France **104** (1976). 225—299.

SOMMER, F.
[1] Über die Integralformeln in der Funktionentheorie mehrerer komplexer Veränderlicher. Math. Ann. **125** (1952), 172—182.

STOUT, E. L.
[1] An integral formula for holomorphic functions on strictly pseudoconvex hypersurfaces. Duke Math. J. **42** (1975), 347—356.
[2] The boundary values of holomorphic functions of several complex variables. Duke Math. J. **44** (1977), 105—108.

TOLEDO, D., and Y. L. TONG
[1] A parametrix for $\bar{\partial}$ and Riemann-Roch in Čech-theory. Topology **15** (1976), 273—301.

TUMANOV, A. E.
[1] On the boundary values of holomorphic functions of several complex variables (Russian). Uspehi Mat. Nauk **29** (1974), no. 4, 158—159.
[2] Peak sets of metric dimension 2.5 for algebras of holomorphic functions on the 3-dimensional sphere in C^2 (Russian). Izv. Akad. Nauk SSSR Ser. Mat. **41** (1977), 115—133.

Val'skij, R. Ė.
[1] On measures orthogonal to analytic functions in $\mathbb{C}^n$ (Russian). Dokl. Akad. Nauk SSSR **198** (1971), 502—505; Engl. transl. in Soviet Math. Dokl. **12** (1971), no. 3, 808—812.

Varopoulos, N. Th.
[1] B.M.O. functions and the $\bar{\partial}$-equation. Pacific J. Math. **71** (1977), no. 1, 221—273.

Vituškin, A. G.
[1] Analytic capacity of sets in problems of approximation theory (Russian). Uspehi Mat. Nauk **22** (1967), no. 6, 141—199; Engl. transl. in Russian Math. Surveys **22** (1967), no. 6, 139—1200.
[2] On a problem of W. Rudin (Russian). Dokl. Akad. Nauk SSSR **213** (1973), no. 1, 14—15; Engl. transl. in Soviet Math. Dokl. **14** (1973), no. 6, 1618—1619.

Vladimirov, V. S.
[1] Methods of the theory of many complex variables (Russian). Moscow 1964. Engl. transl., MIT Press, Cambridge 1966.

Weil, A.
[1] L'intégrale de Cauchy et les fonctions de plusieurs variables. Math. Ann. **111** (1935), 178—182.

Weinstock, B. M.
[1] Continuous boundary values of analytic functions of several complex variables. Proc. Amer. Math. Soc. **21** (1969), 463—466.

Wells, Jr., R.O.
[1] Differential analysis on complex manifolds. Prentice Hall, Englewood Cliffs (N.J.) 1973.

List of symbols

$\mathbb{R}$, $\mathbb{C}$	real, complex numbers
$\mathbb{R}^n$, $\mathbb{C}^n$	real, complex Euclidean space of dimension n
$\overline{X}$	topological closure of the set X
$\bar{z}$	$:= (\bar{z}_1, \dots, \bar{z}_n)$ for $z = (z_1, \dots, z_n) \in \mathbb{C}^n$
$\lvert z \rvert$	$:= (\lvert z_1\rvert^2 + \dots + \lvert z_n\rvert^2)^{1/2}$ for $z = (z_1, \dots, z_n) \in \mathbb{C}^n$
$Y \subset\subset X$	means that Y is contained in a compact subset of X
$\dim_{\mathbb{R}} X$, $\dim_{\mathbb{C}} X$	real, complex dimension of X
$\deg f$	degree of the differential form f
$\lvert f \rvert$	where f is a differential form; 42
$\lVert f(x) \rVert$	where f is a differential form; 42,88
$\lvert I \rvert$	length of the multi-index I; 43
supp f	the smallest closed set outside of which f vanishes
$\sum'$	summation over strictly increasing multi-indices; 43
∂D	boundary of D
dist (X, Y)	Euclidean distance if $X, Y \subseteq \mathbb{R}^m$; see p. 185 if $X, Y \subseteq$ manifold
$\hat{K}_D^P$	$P^0(D)$-hull of K; 31
$\hat{K}_D^{\mathcal{O}}$	$\mathcal{O}(D)$-hull of K; 17, 97
$\hat{F}_\varrho(z, \zeta)$	Levi polynomial of ϱ; 25
$F(z, \zeta)$	76, 106, 187
$\Phi(z, \zeta)$	75–77, 106, 107, 187
$\tilde{\Phi}(z, \zeta)$	106, 107, 187
$\tilde{H}_\xi(\delta)$	116
$H_\xi(\delta)$, $E_\xi(\delta)$	Hörmander, Euclidean ball; 116
$H_\xi^\Phi(\delta)$	Hörmander ball; 194
diam_H	Hörmander diameter; 116
$d(\xi_1, \dots, \xi_N; \delta)$	order of the system $H_{\xi_j}(\delta)$; 132
$d\sigma_k$	Lebesgue measure in $\mathbb{R}^k$ or Euclidean volume form on k-dimensional surfaces of $\mathbb{R}^m$
$\sigma_k(U)$	$:= \int_U d\sigma_k$
det, $\det_{s_1, \dots, s_m}$	for matrices of differential forms; 45, 46
ω, ω'	44
$\lVert\cdot\rVert_{0,Y}$	sup-norm; 48, 88
$\lVert\cdot\rVert_{\alpha,Y}$	α-Hölder norm; 49, 88
$T(X)$, $T^*(X)$	complex tangent, cotangent bundle of X; 162
B_z	fiber of the vector bundle B over z
$\langle b, a \rangle$	$:= b_1 a_1 + \dots + b_n a_n$ if $b, a \in \mathbb{C}^n$; see p. 164 if $b \in T_z^*(X)$, $a \in T_z(X)$

${}_X\mathcal{O}^B$	sheaf of germs of holomorphic sections in the holomorphic vector bundle B; 160
${}_X\mathcal{O}^N$	$:= {}_X\mathcal{O}^{X\times\mathbb{C}^N}$; 260
S_K	smooth pieces of a piecewise C^1-boundary; 169

Differential operators

d	exterior differential operator
∂	holomorphic component of d; 9, 43, 44
$\bar\partial$	antiholomorphic component of d; 9, 43, 44, 86
$\partial_j = \partial/\partial z_j$	9
$\bar\partial_j = \partial/\partial \bar z_j$	
∂^k	$:= \partial_1^{k_1} \dots \partial_n^{k_n}$ if $k = (k_1, \dots, k_n)$ is a multi-order; 9

Integral operators

B_D, $B_{\partial D}$	47, 165, 173	L_D, R_D	111, 112, 193
$L^w_{\partial D}$, $R^w_{\partial D}$	50, 51	R^*_D	125
$L_{\partial D}$	135, 137, 165	L_{S_K}, R_{S_K}	171, 180
$R_{\partial D}$	166, 174		

Spaces of functions or differential forms

$C^0(Y)$	continuous complex-valued functions on Y; 9, 48
$BC^0(Y)$	bounded functions in $C^0(Y)$; 48
$C^k(Y)$	k times continuously differentiable functions in $C^0(Y)$; 9, 75
$C^k_0(Y)$	functions $f \in C^k(Y)$ with $\operatorname{supp} f \subset\subset D$; 9
$\mathcal{O}(Y)$	holomorphic functions on Y; 11, 14, 78
$\mathcal{O}^m(Y)$	$:= \mathcal{O}(Y) \oplus \dots \oplus \mathcal{O}(Y)$ (m times); 12, 14, 78
$P^0(Y)$	continuous plurisubharmonic functions on D; 23
$L^\infty(Y)$	bounded measurable complex-valued functions on Y; 48
$H^\alpha(Y)$	functions $f \in C^0(Y)$ with $\|f\|_{\alpha,Y} < \infty$; 49
$C^\alpha(Y)$	functions $f \in C^0(Y)$ with $\|f\|_{\alpha,K} < \infty$ for each $K \subset\subset Y$; 49
$C^{k+\alpha}(Y)$	functions $f \in C^k(Y)$ whose derivatives of order k belong to $C^\alpha(Y)$; 49
$S_{(p,q)}(Y)$	differential forms of bidegree (p, q) with coefficients in $S(Y)$, where S is one of the symbols L^∞, C^0, BC^0, H^α, C^k, $C^{k+\alpha}$; 43, 44, 49
$Z^\infty_{(0,1)}(D)$	forms $f \in C^\infty_{(0,1)}(D)$ with $\bar\partial f = 0$; 75
$Z^0_{(0,q)}(\bar D)$	forms $f \in C^0_{(0,q)}(\bar D)$ with $\bar\partial f = 0$; 72
$F^{1/2}_{(0,q)}(\bar D)$	forms $f \in H^{1/2}_{(0,q)}(\bar D)$ with $\bar\partial f \in C^0_{(0,q+1)}(\bar D)$; 72
$E^{1/2}_{(0,q)}(\bar D)$	$:= \bar\partial F^{1/2}_{(0,q-1)}(\bar D)$; 72

Spaces defined for holomorphic vector bundles *B*

$\mathcal{O}(D, B)$	holomorphic sections of B over D
$C^0(\bar D, B)$	continuous sections of B over $\bar D$; 88
$H^\alpha(\bar D, B)$	sections $f \in C^0(\bar D, B)$ with $\|f\|_{\alpha,\bar D} < \infty$; 88
$C^0_{(0,q)}(\bar D, B)$	continuous B-valued $(0, q)$-forms on $\bar D$; 88
$Z^0_{(0,q)}(\bar D, B)$	forms $f \in C^0_{(0,q)}(\bar D, B)$ with $\bar\partial f = 0$; 88
$F^{1/2}_{(0,q)}(\bar D, B)$	forms $f \in H^{1/2}_{(0,q)}(\bar D, B)$ with $\bar\partial f \in C^0_{(0,q+1)}(\bar D, B)$; 88
$E^{1/2}_{(0,q)}(\bar D, B)$	$:= \bar\partial F^{1/2}_{(0,q-1)}(\bar D, B)$; 88

Subject index

Bei Fragen zur Produktsicherheit wenden Sie sich bitte an:
If you have any questions regarding product safety,
please contact:

Walter de Gruyter GmbH
Genthiner Straße 13
10785 Berlin
productsafety@degruyterbrill.com